Discrete Mathematical Structures
for Computer Science

RONALD E. PRATHER
UNIVERSITY OF DENVER

Discrete
Mathematical Structures
for Computer Science

HOUGHTON MIFFLIN COMPANY   BOSTON
Atlanta   Dallas   Geneva, Illinois   Hopewell, New Jersey   Palo Alto   London

To John and Marianne

Printed in the United States of America
Library of Congress Catalog Card Number: 75-25014
ISBN: 0-395-20622-7

Contents

Preface (to the Instructor)

The traditional mathematics sequence for computer science and engineering students tends to emphasize continuum analysis and the infinitesimal calculus. Not to minimize the importance of these topics, but to call attention to a resulting gap in the student's preparation, a group of foremost educators has recommended a course "Introduction to Discrete Structures" for inclusion in a modern computer science program. In their report "Curriculum '68" [*Communications of the Association for Computing Machinery*, **11** (March 1968)] this course—labeled B3—occupies an early and central position in the overall scheme. The relative sophistication of the recommended material and the early placement of the course—sophomore or junior level—present certain difficulties, both for the instructor and for authors of prospective texts. I have found this to be quite a challenge on both accounts. If this text helps to make the job easier and more satisfying for some of us, particularly the students, then I shall have been amply rewarded for my efforts.

I have included a long and leisurely preliminary chapter as a device for achieving flexibility. The less experienced students, say a class of sophomores, would profit from a detailed study of these introductory ideas concerning sets, relations, functions, and the like. When teaching the course to a somewhat more advanced group, I find that I can almost get by in treating this opening chapter (Chapter 0, Preliminaries) as a reading assignment. Even then it is often necessary to go back and review certain portions from time to time.

The real Chapter 1, Algebras and Algorithms, represents something of a departure in its approach to the study of discrete algebraic structures. I have found that few computer science students are motivated toward a study of abstract algebra unless it can be related to their perception of the computing field. To the beginning student particularly, computer science is often equated with programming. Indeed, we spend a great deal of our time trying to correct this mistaken impression. It occurred to me that we might try to capitalize on this state of affairs by integrating the algebra with an informal introduction to algorithms. This idea led finally to the presentation of an "algebraic flowchart language" suitable for the design and communication of algorithms. This is exactly what is done in the succeeding chapters. I have used this language or its sequential equivalent in order to describe the typical and most familiar algorithmic processes one might encounter, whether in graphs, lattices, groups, or wherever.

Except for this one innovation, the selection of topics and their order of presentation is quite standard. Although the development of the text proceeds through chapters mathematically oriented toward graphs, monoids, lattices, groups, and logic, such chapter headings would be misleading to the mathematically inclined. I have tried to relate each of these formal subject areas to the computer science applications whenever the opportunity arises. Thus the chapter on monoids provides an introduction to finite state machines, and the chapter on lattices leads finally to the applications of Boolean algebra to switching theory. Even Chapter 1 has its applications—an introduction to structured programming and the analysis of algorithms. In some cases, the applications are more advanced than others; such sections have been starred (*). Particularly with a beginning group of students, one might want to skip some of these sections, and this can easily be done without loss of continuity. Again this relates to the overall flexibility of the text. I think that the instructor will find that (except for Chapters 1 and 2) even the individual chapters can be juggled without serious disruption. Beyond Chapters 1 and 2, I have envisioned the test as splitting into two separate tracks, as follows:

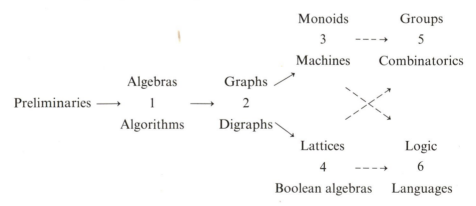

At that stage, there are only minor dependencies among the various chapters (as indicated by the dotted arrows), and these are easily discernible. With these remarks as a guide, one can design various arrangements of the chapters to suit particular requirements. For this reason, I have not thought a great deal about the arrangement of the chapters into semester courses. Certainly there is more than enough material for a single course. It seems more appropriate, however, that the individual instructor decide which chapters to stress, which starred sections to omit, etc. I have only tried to build in the flexibility which will help carry out this process.

There are several unifying threads which weave their way through the text. Certainly the spirit of Chapter 1 would be one of these. There is also the special role of combinatorics. There is a continuing debate in academic computer science circles as to whether an algebraic or a combinatorial approach is best suited for the course in discrete structures. Not wishing to take one side or the other on this issue, I have tried to steer a course down the middle of the road. I have included the more elementary enumerative techniques as an integral part of the preliminary chapter. Continuing in the chapters on graphs and lattices, right on through the introduction to the Pólya theory of enumeration, a combinatorial flavor is nearly always in evidence. But all in all, I have tried to provide a balanced treatment, favoring neither algebra nor combinatorics.

I have included well over 500 exercises. Actually there are several times this number because many of the exercises have separate parts. This allows the instructor to assign different parts from one time to the next, some having answers provided for the student and others not. Most of the exercises are designed as routine reinforcements of the text material or are of only moderate difficulty. These are easily distinguished by the instructor. Those exercises which are somewhat more difficult are marked with an asterisk. In this way the instructor can adjust the assignments to fit the level of a particular class.

In most respects, the writing of a text is a solitary endeavor, and so every author appreciates each spark of interest on the part of his colleagues. In my own case, I would like to acknowledge the cooperation and encouragement of my two department chairmen during the period of my writing, Herbert J. Greenberg and William S. Dorn. At the same time, James W. Burgmeier is especially to be thanked along with Professor Dorn for a careful reading of early versions of the manuscript and for the numerous suggested improvements. The final reviews by C. L. Liu, University of Illinois; E. J. Neuendorf, University of Dayton; and M. Edelberg, Sperry Research Center, were most constructive. I would also like to thank Brent Hailpern for checking all of the examples and exercises for accuracy; any mistakes that remain are my own responsibility. In one way or another, all of these people have contributed to the improvement of the text, but none in greater measure than my students at the University of Denver. I thank them all.

Ronald E. Prather

Introduction (to the Student)

The first time that I used the notes for this text in teaching a course in discrete mathematics to a group of sophomore computer science students, I gave the first few sections as a reading assignment, saying that I would answer any questions at the next class meeting. Well, the next meeting came, and as I asked whether or not there were any questions, a voice from the first row said, "Yeah, an innocent question, what does the title of the book mean?" After delving a little further, I found out that he was bothered by the word "discrete." We should not be overly critical of this student, since he later went on to earn an A in the course. In fact, I now see that his question was entirely reasonable and seems to strike at the heart of the matter. I thought that you might benefit from this story, or at least from the explanation that I am about to give. Of course you must bear in mind that we will necessarily be speaking about ideas that are to be developed much more fully in the preliminary chapter. However, if you are willing to overlook some of the details, I am quite confident that with a little patience you will be able to appreciate the main thrust of this discussion in advance.

How do we measure the size of a set? How do we classify sets according to their size, and, then, which of the resulting categories are appropriate to our study? If for some n, we can label all the elements of a set with the integers $1, 2, \ldots, n$, then the set is said to be *finite*, of size or cardinality n. Otherwise the set is *infinite*. But that is not the only size differentiation we employ. It may be that the set is infinite but a labeling can yet be achieved if we use all of the positive integers as labels. Thus we say that a set is *countable* (or better, *countably infinite*) if a one-to-one correspondence can be arranged between all its elements and all of the positive integers. In the standard usage, the term *countable* usually means finite or countably infinite, and the word *denumerable* is sometimes used in referring only to the infinite case. There are any number of examples of denumerable or countably infinite sets: all the even integers, the prime numbers, all the "words" on a finite alphabet, and so on. With the existence of one-to-one correspondences taken as the criterion for size equality, we would have to say that all of these examples are sets "of the same size," the size or cardinality of the set of all positive integers.

You might think that this is about as big as a set could be, but nothing is further from the truth. Consider the collection of real numbers in the range from zero to one. Think of each of these as being expressed as an infinite decimal, that is, a countably infinite sequence of digits

$$0 . d_1 d_2 d_3 \cdots$$

as in $\pi/10 = 0.314 \cdots$ with $d_1 = 3$, $d_2 = 1$, $d_3 = 4$, and so on. Interestingly enough, the collection of all such sequences (real numbers) is *uncountable*, as we will show by a now famous argument known as "Cantor's diagonalization method." You will perhaps find this argument to be a bit puzzling on first sight. That should not discourage you in the slightest, however, especially after it is pointed out that this same reaction was expressed by many if not most of Cantor's nineteenth-century mathematical peers. For reasons which need not concern us here, it was some time before the mathematical community recognized and fully appreciated his line of reasoning.

Suppose, in direct contrast with what we intend to prove, that there is a labeling or listing of *all* these real numbers in one-to-one correspondence with the positive integers:

$$\text{real number } \#1 \to 0 . \mathbf{d}_{11} d_{12} d_{13} \cdots$$
$$\text{real number } \#2 \to 0 . d_{21} \mathbf{d}_{22} d_{23} \cdots$$
$$\text{real number } \#3 \to 0 . d_{31} d_{32} \mathbf{d}_{33} \cdots$$
$$\vdots \qquad\qquad \vdots$$
$$\text{real number } \#i \to 0 . d_{i1} d_{i2} d_{i3} \cdots \mathbf{d}_{ii} \cdots$$
$$\vdots \qquad\qquad \vdots$$

Note that the first subscript here carries the index or label for the various real numbers while the second identifies the digit position within the respective numbers of the listing. We will show that however one claims to have accomplished this listing, *not all* of the real numbers in the range zero to one are present. To see this, we construct an infinite decimal $0 . c_1 c_2 c_3 \cdots$ in which c_i ($i \geq 1$) is always chosen to be different from d_{ii}. Thus we take $c_1 \neq d_{11}$, $c_2 \neq d_{22}$, $c_3 \neq d_{33}$, and so on. Obviously we have considerable latitude here, but for definiteness we might say, if $d_{ii} = 3$ choose $c_i = 4$, whereas if $d_{ii} \neq 3$ then choose $c_i = 3$. The point is, the infinite decimal so constructed cannot be in the list because it differs in at least one digit position (the boldface diagonal one) from each and every member! It follows that a complete listing of the real numbers cannot exist, that is, the set of real numbers is not countable. Yet this set provides the whole basis for calculus and the classical mathematical analysis, so it is not as though we could wish it away.

We do the next best thing. In *discrete* mathematics, in direct contrast with calculus, we include the countably infinite as well as the finite, but we specifically exclude the uncountable sets from our consideration. This is not simply a matter of convenience, for it is easy to see why we would primarily be concerned with finite sets in computer science. After all, the computer is inherently finite in all its important aspects. Each of its computations is limited by the finite word length and finite memory capacity. Yet the reason we enlarge our sphere of interest to include the countably infinite sets is perhaps not so apparent. Consider the addition and multiplication of integers as represented in a large-scale computer. The process is one of "simulation" of the ordinary arithmetic. Unless we exceed certain very large but finite bounds on the magnitude of the integers involved, the simulation is perfect, machine errors notwithstanding. Moreover, in considering larger and larger machines, there is no way we can establish a fixed bound on the range of simulation that might be possible. For these reasons primarily, we are willing to include the countable sets as within our area of interest.

True, in numerical analysis one tries to operate within the uncountable continuum of the real number system. Certainly this is a legitimate and important part of computer science. But the inexact computer simulation introduces a host of new and quite difficult problems, mainly having to do with "round-off" error analysis. It is convenient to separate this area as a course of study in and of itself.

If this is done, we are left again with the discrete—as opposed to the continuum—mathematics, an area which is quite tidy by comparison. Here the approach is less analytical, more "algebraic," more of a "combinatorial" nature. If the meaning of these terms is not altogether clear at this stage, you have two alternatives. You can read the book or you can ask your instructor an "innocent" question. Of course, I hope you will read the book in either case.

Discrete Mathematical Structures
for Computer Science

0

Preliminaries

In our everyday world, we tend to associate objects with one another when they have common attributes because experience shows that such organization is a definite aid to our thinking. Whether we speak of the string quartets of Beethoven, all brown-eyed American children, or the members of the New York Stock Exchange, the same technique of classification is being applied. The advantages of having a precise notation and a meaningful theory for dealing with such classification should be no less apparent to the student than to a philosopher. In the mathematical sciences especially such precision is of the utmost importance and it is provided by axiomatic set theory.

The formal aspects of set theory need not concern us here. In fact, we take the notion of a set as essentially undefined. This is the usual approach when elementary set theory is to be employed primarily as a notational convenience. Though we have somewhat more in mind in this book, our purposes are still best served by an informal approach. Thus we use the words *set, collection*, and *class* synonymously, as in "set of elements," "collection of members," and "class of objects." Just as *line, point*, and *incidence* are undefined or primitive terms in geometry, so, too, in our use of set theory the notions of set, element, and membership will remain undefined. Suffice it to say that a *set* is a collection of distinguishable objects or elements, ordinarily sharing some common feature which qualifies them for membership in the set.

Though our development of set theory is rather informal, we will nevertheless provide a degree of precision that will more than suffice for a meaningful introduction to the main ideas of interest: relations, functions, partitions, etc. Many of the difficult aspects of a formal approach to set theory have to do with the need for treating infinite, even uncountable, sets. Because the sets we consider are at most countable, and more often finite, we are on relatively safe ground in choosing a more informal approach. Nevertheless, problems of a different sort appear almost immediately when concentrating on finite sets. They have to do with enumeration or counting, if you wish. In this preliminary chapter we shall review a few elementary counting techniques for particular configurations or arrangements of elements into sets. These *combinatorial* techniques, however, are only the first manifestations of a thread, one of several, that weaves its way through the text.

0.1   SETS AND SUBSETS

Though we leave the concept of set virtually undefined, we will pay very close attention to the notational aspects of set theory. The set-theoretic notations have a precise meaning throughout the mathematical sciences and have become, with the rules of logic, the essence of the language of mathematics. Students will do well to thoroughly familiarize themselves with these conventions if they have not done so already.

Whenever possible, we will use capital letters as *names* for the sets we introduce and small letters for elements that may or may not be members of a particular set. If S is a set, we indicate the fact that a is an element or member of S by the notation $a \in S$ (read "a is in S" or "a in S," whichever is grammatically correct in the context). On the other hand, we write $a \notin S$ if a is not in S.

If S has only finitely many elements, it is possible though not always practical to enumerate or list its members by enclosing these elements in braces. Thus

$$S = \{a, b, c\}$$

indicates that a, b, c, and only they, are members of the set S. The order of appearance in the braces is of no consequence, neither is duplication. Thus $\{a, c, b\}$ and $\{a, b, c, b\}$ are but two other representations of the same set. If we use representations without duplication, then two finite sets are said to be of the same size if the elements they each contain can be paired in a *one-to-one correspondence*. In this way, the concept of number itself can be defined in set-theoretic terms. Notationally, we let $|S|$ represent the size or *cardinality* of S, that is, the number of elements in S. For instance, $|S| = 3$ for the set S just introduced.

Whenever a listing is either not possible or not convenient, we will usually resort to the use of a generic symbol such as x followed by a colon and a mathematical and/or English language statement or proposition $P(x)$ that serves as a rule for deciding just which x's belong to S. Again the braces are used in writing

$$S = \{x: P(x)\}$$

In a precise axiomatic approach to set theory, there are certain restrictions on the type of propositions $P(x)$ that can be employed for this purpose, but we will not address ourselves to such technicalities here. We merely point out that this style of presentation coincides with the idea that the elements of a set share some common property P. In the case of $S = \{a, b, c\}$ the property is simply of belonging to S. Membership in a set is itself a property! Therefore, the propositional notation is more general than simple listing.

EXAMPLE 0.1   The set

$$S = \{x: x \text{ is an integer, } x > 5\}$$

consists of all integers that are greater than 5. Thus we may write $7 \in S$ and $2\pi \notin S$. If the reader is not accustomed to the propositional notation, it may be advisable that he or she first render a linguistic interpretation, for example: read

$=$	as	"is"
$\{x$	as	"the set of all elements x"
$:$	as	"such that" or "having the property that"
$,$	as	"and"
$\}$	as	(silent)

Then the original sentence becomes "S is the set of all elements x having the property that x is an integer and x is greater than 5." From this expression we can easily obtain our first description of the set. Incidentally, it would be almost as clear to write

$$S = \{6, 7, 8, \dots\}$$

with the enumeration fairly evident even though S is not finite.   △

EXAMPLE 0.2   The set of all real numbers, usually denoted by

$$R = \{x: x \text{ is a real number}\}$$

is the basis for "continuous" mathematics. Since this book is concerned mainly with discrete mathematical structures, the set R will only rarely appear. (See the Introduction for the distinction between discrete and continuous mathematics.) We note particularly that it is not possible to list the members of R, and so the form $\{x: P(x)\}$ is the only alternative. Even without a listing, however, we can claim that $2\pi \in R$, for example, because 2π satisfies the property P—that is, $P(2\pi)$ is true for the given proposition

$$P(x): x \text{ is a real number.} \qquad △$$

EXAMPLE 0.3   A set that is of only passing interest (as that of Example 0.1) might be denoted by any letter, S, say; and in a different context where confusion would not arise, S might represent a different set altogether. Thus we could introduce the set

$$S = \left\{x: \begin{matrix} x \text{ is a word in the English language and} \\ \text{there are at least 3 occurrences of } x \text{ in this book} \end{matrix}\right\}$$

In this case it would be possible to list the members of S, but surely this would not be convenient. In any case,

$$set \in S \qquad kiss \notin S$$
$$algorithm \in S \qquad mimsy \notin S$$
$$machine \in S \qquad hobbit \notin S$$

etc.   △

EXAMPLE 0.4   The set

$$S = \{x: x \text{ is a primary color}\}$$

could just as well be enumerated or listed as

$$S = \{red, yellow, blue\}$$

even though we seem to lose some information in the process; the common property of the elements is no longer so apparent. Incidentally, we are not bothered by the fact that these elements are only words (or symbols, if you like) which represent the primary colors. For that much is true even in representing numbers, as in the following example.   △

EXAMPLE 0.5   The set of *natural numbers* is denoted by

$$N = \{0, 1, 2, 3, \dots\}$$

and the positive integers are designated as

$$Z^+ = \{1, 2, 3, \dots\}$$

The latter notation stems from the rather common usage of the symbol Z for the set of all integers, positive, negative, and zero; that is,

$$Z = \{0, -1, 1, -2, 2, \dots\}$$

Similarly, the set of the first n positive integers is written

$$n^+ = \{1, 2, \dots, n\}$$

Here is one instance where we depart from the use of capital letters as names for sets. The set n^+ is most illustrative of the idea of an *n-set*, that is a set S with $|S| = n$. We may speak of 3-sets, 2-sets, and even 1-sets (which are often called *singleton* sets). But does there exist a 0-set?   △

EXAMPLE 0.6   A particular symbol $\varnothing$ is used universally to denote the set with no members, and for obvious reasons, it is called the *empty* or *null* set. Evidently $x \in \varnothing$ is false for each and every x. It is sometimes convenient to think of a set as a kind of "box" with its elements inside, in which case, because there is nothing mysterious about an empty box, the idea of an empty set should not be particularly puzzling. In any event, the real necessity for the admittance of $\varnothing$ as a set in good standing will become increasingly apparent.   △

EXAMPLE 0.7   In the set of decimal digits

$$D = \{0, 1, 2, 3, 4, 5, 6, 7, 8, 9\}$$

each element $d \in D$ is also a FORTRAN character. In the sense of the terminology we are about to introduce, we say that *D is a subset of*

$$F = \{x: x \text{ is a FORTRAN character}\}$$ △

The set A is said to be a *subset* of the set S if

$$a \in A \Rightarrow a \in S$$

that is, if every element of A is also an element of S. The visual image conveyed by Figure 0.1 may be helpful in understanding this concept.

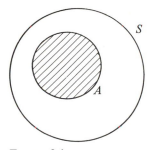

FIGURE 0.1

Note   The symbolism $P \Rightarrow Q$ for statements P and Q is read "P implies Q" or "whenever P is true, Q is also true," whereas $P \Leftrightarrow Q$ means "P if and only if Q" or "whenever P is true, Q is also true, and conversely." In case the readers are not very familiar with such logical notations, they are referred to the leisurely discussion of these ideas in Section 0.10. The discussion is more or less self-contained, so the students may jump ahead to read parts of that section whenever they feel the need.

As a kind of shorthand, we most often write $A \subseteq S$ when A is a subset of S or, as we shall also say, A is *contained* in S. Again the picture of Figure 0.1 should spring quickly to mind whenever one sees the notation $A \subseteq S$ (or equivalently, $S \supseteq A$). The notation $A = B$ for sets should be taken to mean that A and B have precisely the same members. That is, the *equality of sets* is defined as follows:

$$A = B \Leftrightarrow A \subseteq B \quad \text{and} \quad B \subseteq A$$

If $A \subseteq B$ but $A \neq B$, we say that A is a *proper* subset of B and write $A \subset B$.

EXAMPLE 0.8   In Examples 0.1 through 0.7,

$$D \subseteq F \quad \text{and} \quad S \subseteq Z^+ \subseteq Z$$

where S is the set of Example 0.1. Let us pay particular attention to the claim that $S \subseteq Z^+$. We see that

$$x \in S \quad \Rightarrow \quad \begin{array}{c} x \text{ an integer} \\ x > 5 \end{array} \quad \Rightarrow \quad \begin{array}{c} x \text{ an integer} \\ x > 0 \end{array} \quad \Rightarrow \quad x \in Z^+$$

Note that we cannot similarly show that $x \in Z^+ \Rightarrow x \in S$. Thus, if we let $x = 4$, then $x \in Z^+$, but $x \notin S$. Therefore, we do not have $S = Z^+$, only $S \subseteq Z^+$, so that S is a proper subset of Z^+.   △

EXAMPLE 0.9   For any set S we have

$$\varnothing \subseteq S \quad \text{and} \quad S \subseteq S$$

The verification of the first of these containments presents certain logical difficulties. We have to argue that for every $x \in \varnothing$ it is also true that $x \in S$. Since there are no such x (where $x \in \varnothing$), we say that the implication $x \in \varnothing \Rightarrow x \in S$ holds *vacuously*. Thus it is true that every element of the empty set is purple. The reader who is disturbed by this line of reasoning may want to defer his final judgment until after he has read parts of Section 0.10. (See the "note" preceding Example 0.8.) Alternatively, he may temporarily find it easier to accept the equivalent *contrapositive* statement: $x \notin S \Rightarrow x \notin \varnothing$.   △

Note   Given statements P and Q, $P \Rightarrow Q$ is the same as (not Q) $\Rightarrow$ (not P). Again the remark in the earlier note is appropriate.

EXERCISES

1 Use the standard set notation to represent the following sets $A, B, \ldots, H$, respectively.
   (a) All postwar (World War II) United States presidents
   (b) All English words beginning with the letter a
   (c) All real numbers between (and including) 0 and 1
   (d) All even integers
   (e) All perfect squares (of integers)
   (f) All FORTRAN variable names
   (g) All rational numbers
   (h) All pairs (n, m) of positive integers in which n is smaller than m

2 Describe each of the following sets in words; that is, do the reverse of Exercise 1.
   (a) $A' = \{m : m = 4n, n \in Z\}$
   (b) $B' = \{x : x \in R, x \neq x\}$
   (c) $C' = \{x : x \in R, 0 < x \leq 1\}$

(d) $D' = \{x: x \in R, x^2 = 1\}$

(e) $E' = \{(x, n): x$ is an English word whose length is $n\}$

(f) $F' = \{m: m$ is a perfect square, $m \le 100\}$

(g) $G' = \{\text{GOGO, ALPHA2, DUMBO}\}$

(h) $H' = \{x: x \in R, ax^2 + bx + c = 0\}$

3 Given the sets described in Exercises 1 and 2, show that:

(a) $B' \subseteq B$ (b) $G' \subseteq F$ (c) $F' \subseteq E$ (d) $D' \subseteq G$

(e) $A' \subseteq D$ (f) $D' \nsubseteq C'$ (g) $E \nsubseteq F'$ (h) $C' \nsubseteq D'$

(i) $D' \nsubseteq B'$ (j) $D \nsubseteq E$

[*Hint*: To show that $S \nsubseteq T$, it suffices to produce an element of S that is not in T.]

4 Give examples of sets, in standard set notation, whose elements are

(a) living persons   (b) birds   (c) sets   (d) nations

(e) coins   (f) students at your college or university

(g) professional football players   (h) mountains   (i) French words

(j) colors

5 How many subsets does an n-set have? Why?

6 Which of the sets in Exercises 1 and 2 are finite? Which are countable? [*Hint*: See the Introduction.]

7 How should we define the size or cardinality of the empty set?

8 Which of the following statements is true?

(a) $\varnothing = 0$ (b) $\varnothing = \{\varnothing\}$ (c) $\varnothing = \{0\}$ (d) $\varnothing \subseteq \{0\}$

0.2 ALGEBRA OF SETS

We are going to discuss several ways of composing new sets from old ones, building finally toward an *algebra of sets*. Of the many ways in which it is possible to build new sets out of several given sets, our algebra is concerned mainly with the union and intersection. Each of these constructions has its roots in our everyday thinking about classifications of objects, and so the concepts are quite intuitive from the start. If a meeting is called for all mathematics and/or computer science faculty, the meeting involves a set union. If a club restricts its membership to residents of Colorado over the age of 21, the club membership involves a set intersection.

With the definitions that follow we can begin to appreciate more fully the advantages of the propositional form for denoting sets. Thus if A and B are sets, the collection

$$A \cup B = \{x: x \in A \quad \text{or} \quad x \in B\}$$

is called the *union* of A and B, whereas

$$A \cap B = \{x: x \in A \quad \text{and} \quad x \in B\}$$

is their *intersection*. Note particularly that we use the word "or" in the inclusive sense; that is, we include in the union those elements that happen to be both in A and in B. See Figure 0.2.

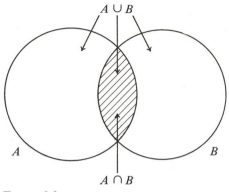

FIGURE 0.2

EXAMPLE 0.10   If $A = \{1, 2, 3, 7\}$ and $B = \{2, 7, 8, 9\}$, then

$$A \cup B = \{1, 2, 3, 7, 8, 9\}$$
$$A \cap B = \{2, 7\}$$

Note that the properties observed here,

(a) $A \subseteq A \cup B, B \subseteq A \cup B,$ (a') $A \cap B \subseteq A, A \cap B \subseteq B,$
(b) $A \subseteq C, B \subseteq C \;\Rightarrow\; A \cup B \subseteq C,$ (b') $C \subseteq A, C \subseteq B \;\Rightarrow\; C \subseteq A \cap B,$

are quite generally true of unions and intersections. For example, consider (b'). Suppose $C \subseteq A$ and $C \subseteq B$. Then if we take any element $c \in C$, we will also have $c \in A$ and $c \in B$, so that $c \in A \cap B$ by the definition of intersection. Thus $C \subseteq A \cap B$ is a logical consequence of $C \subseteq A$ and $C \subseteq B$. The reader should infer that $A \cap B$ is the largest set that is contained both in A and in B. What similar inference about the union can be drawn? $\triangle$

We can make several observations about the four properties listed in the preceding. First of all, they exhibit a certain *duality* that is characteristic of set algebra, as we shall soon learn. Secondly, suppose that A and B are subsets of a given set S. Then properties (b) and (a') imply that the same is true of their union and intersection, that is,

$$A \subseteq S, B \subseteq S \;\Rightarrow\; A \cup B \subseteq S \quad \text{by (b)}$$
$$A \cap B \subseteq A \subseteq S \quad \text{by (a')}$$

It follows that we will never leave the *system of subsets* of S if we continue to perform unions and intersections of the subsets of S. In arithmetic or algebra, this phenomenon is usually described by saying that the system is *closed*. Let us pursue this idea a bit further by describing this system, called the *power set* of S:

$$\mathscr{P}(S) = \{A : A \subseteq S\}$$

We observe that $\mathscr{P}(S)$ is a set whose elements are themselves sets (all the subsets of S). Within the framework of our discussion, the union and intersection behave very much like the arithmetic operations of addition (+) and multiplication (·) in ordinary algebra, but with some notable exceptions.

EXAMPLE 0.11   If $S = \{a, b, c\}$, then

$$\mathscr{P}(S) = \{\varnothing, \{a\}, \{b\}, \{c\}, \{a, b\}, \{a, c\}, \{b, c\}, S\}$$

Note that $\varnothing \in \mathscr{P}(S)$ and $S \in \mathscr{P}(S)$, in accordance with Example 0.9. Also observe that in computing the intersection of $\{a, c\}$ and $\{b\}$ we need the idea of an empty set. Following the definition of the intersection to the letter, we look for all elements x such that $x \in \{a, c\}$ and $x \in \{b\}$, and as there are no such elements to be found, we must write $\{a, c\} \cap \{b\} = \varnothing$.   △

EXAMPLE 0.12   Again, if $S = \{a, b, c\}$, we can indicate all of the unions and intersections in $\mathscr{P}(S)$ by a tabulation, much as in setting up ordinary addition and multiplication tables. We have

$$\{a, b\} \cup \{a, c\} = \{a, b, c\} = S$$
$$\{a, b\} \cap \{a, c\} = \{a\}, \text{etc.}$$

All together we have the following tables:

$\cup$	$\varnothing$	$\{a\}$	$\{b\}$	$\{c\}$	$\{a, b\}$	$\{a, c\}$	$\{b, c\}$	S
$\varnothing$	$\varnothing$	$\{a\}$	$\{b\}$	$\{c\}$	$\{a, b\}$	$\{a, c\}$	$\{b, c\}$	S
$\{a\}$	$\{a\}$	$\{a\}$	$\{a, b\}$	$\{a, c\}$	$\{a, b\}$	$\{a, c\}$	S	S
$\{b\}$	$\{b\}$	$\{a, b\}$	$\{b\}$	$\{b, c\}$	$\{a, b\}$	S	$\{b, c\}$	S
$\{c\}$	$\{c\}$	$\{a, c\}$	$\{b, c\}$	$\{c\}$	S	$\{a, c\}$	$\{b, c\}$	S
$\{a, b\}$	$\{a, b\}$	$\{a, b\}$	$\{a, b\}$	S	$\{a, b\}$	S	S	S
$\{a, c\}$	$\{a, c\}$	$\{a, c\}$	S	$\{a, c\}$	S	$\{a, c\}$	S	S
$\{b, c\}$	$\{b, c\}$	S	$\{b, c\}$	$\{b, c\}$	S	S	$\{b, c\}$	S
S	S	S	S	S	S	S	S	S

$\cap$	$\varnothing$	$\{a\}$	$\{b\}$	$\{c\}$	$\{a, b\}$	$\{a, c\}$	$\{b, c\}$	S
$\varnothing$	$\varnothing$	$\varnothing$	$\varnothing$	$\varnothing$	$\varnothing$	$\varnothing$	$\varnothing$	$\varnothing$
$\{a\}$	$\varnothing$	$\{a\}$	$\varnothing$	$\varnothing$	$\{a\}$	$\{a\}$	$\varnothing$	$\{a\}$
$\{b\}$	$\varnothing$	$\varnothing$	$\{b\}$	$\varnothing$	$\{b\}$	$\varnothing$	$\{b\}$	$\{b\}$
$\{c\}$	$\varnothing$	$\varnothing$	$\varnothing$	$\{c\}$	$\varnothing$	$\{c\}$	$\{c\}$	$\{c\}$
$\{a, b\}$	$\varnothing$	$\{a\}$	$\{b\}$	$\varnothing$	$\{a, b\}$	$\{a\}$	$\{b\}$	$\{a, b\}$
$\{a, c\}$	$\varnothing$	$\{a\}$	$\varnothing$	$\{c\}$	$\{a\}$	$\{a, c\}$	$\{c\}$	$\{a, c\}$
$\{b, c\}$	$\varnothing$	$\varnothing$	$\{b\}$	$\{c\}$	$\{b\}$	$\{c\}$	$\{b, c\}$	$\{b, c\}$
S	$\varnothing$	$\{a\}$	$\{b\}$	$\{c\}$	$\{a, b\}$	$\{a, c\}$	$\{b, c\}$	S

$\triangle$

Upon closer examination, we find that our set algebra has some rather unusual features.   The identities or properties

(P1a) $A \cup A = A$ (P1b) $A \cap A = A$
(P2a) $A \cup (B \cup C) = (A \cup B) \cup C$ (P2b) $A \cap (B \cap C) = (A \cap B) \cap C$
(P3a) $A \cup B = B \cup A$ (P3b) $A \cap B = B \cap A$
(P4a) $A \cup (A \cap B) = A$ (P4b) $A \cap (A \cup B) = A$
(P5a) $A \cap (B \cup C)$ (P5b) $A \cup (B \cap C)$
$\quad = (A \cap B) \cup (A \cap C)$ $\quad = (A \cup B) \cap (A \cup C)$

can be verified for all sets A, B, C. Note the complete duality between the properties in the left-hand column and those in the right-hand column upon replacing $\cup$ by $\cap$ or vice versa. Of particular interest are properties (P1), (P4), and (P5b), since they do not hold in ordinary algebra, such as when we replace $\cup, \cap$ respectively by $+, \cdot$. In fact, this whole collection of properties, when taken together as an axiomatic system, leads to a distinctive mathematical structure known as a distributive lattice. We shall study these properties in some detail in Chapter 4 and learn that set algebra is by no means the only example of such a mathematical system.

EXAMPLE 0.13   Suppose that we want to verify the set identity (P4a). It is important to remember that here the equal sign denotes a set equality, so that we must appeal to the definition of set equality (each side must be shown to be contained in the other).   The containment

$$A \cup (A \cap B) \supseteq A$$

is easy enough to establish, for it follows from the property (a) in Example 0.10. To establish the reverse inclusion, we let $x \in A \cup (A \cap B)$. Then $x \in A$ or $x \in A \cap B$. If the first is true, then we are finished already; if the second is true, we have $x \in A$ again, according to property (a′) of Example 0.10 $\triangle$

EXAMPLE 0.14   In (P5b) we must show that for arbitrary sets A, B, C we have

$$A \cup (B \cap C) \subseteq (A \cup B) \cap (A \cup C)$$

and vice versa. If $x \in A \cup (B \cap C)$, then $x \in A$ or $x \in B \cap C$. In the first instance, we have both $x \in A \cup B$ and $x \in A \cup C$ by property (a) of Example 0.10. But then $x \in (A \cup B) \cap (A \cup C)$ by the definition of the intersection. In the second instance, we have $x \in B$ and $x \in C$ so that again by property (a), $x \in A \cup B$ and $x \in A \cup C$. Then x is in the intersection as before. We leave the verification of the reverse inclusion as an exercise. △

We must not forget that the empty set is among the members of $\mathscr{P}(S)$. Both $\varnothing$ and S enjoy important special properties in connection with the union and intersection operations, in particular in connection with the set complementation operator. Unlike the union and intersection, this operator is usually applied to a single argument $A \subseteq S$, where the *complementary set* $\sim A$ simply consists of those elements (of S) which are not in A; that is,

$$\sim A = \{x : x \in S, x \notin A\}$$

The *relative complement* $S \sim A$ (or *set difference*) has the same definition whether A is a subset of S or not. A convenient picture of the relationship between A and $\sim A$ is shown in Figure 0.3. This figure clearly suggests some of the properties mentioned earlier: for any $A \subseteq S$,

(P6a) $A \cup \varnothing = A$ (P6b) $A \cap \varnothing = \varnothing$
(P7a) $A \cup S = S$ (P7b) $A \cap S = A$
(P8a) $A \cup (\sim A) = S$ (P8b) $A \cap (\sim A) = \varnothing$

Of course, we cannot prove these identities with a picture; we must argue from the definition of set equality as before.

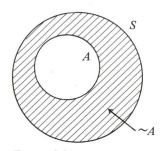

FIGURE 0.3

EXAMPLE 0.15   In property (P8a) we have $A \subseteq S$ and $\sim A \subseteq S$ so that

$$A \cup (\sim A) \subseteq S$$

because of property (b) in Example 0.10. Conversely, if $x \in S$, we surely have $x \in A$ or $x \notin A$. But $x \in \sim A$ in case of the latter, so that $x \in A \cup (\sim A)$ by virtue of the definition of the union. △

In view of the *associative* properties (P2), we can generalize the operations of union and intersection to n arguments without regard to the order of composition. In complete analogy with the use of the summation symbol in arithmetic, we may therefore define

$$\bigcup_{i=1}^{n} A_i = A_1 \cup A_2 \cup \cdots \cup A_n$$

$$= \{a: a \in A_i \text{ for some } i = 1, 2, \ldots, n\}$$

$$\bigcap_{i=1}^{n} A_i = A_1 \cap A_2 \cap \cdots \cap A_n$$

$$= \{a: a \in A_i \text{ for all } i = 1, 2, \ldots, n\}$$

As an application, let us show how the generalized union and intersection can be used to define the intuitive notion of the partitioning of a set.

Let us begin by agreeing to say that the sets A and B are *disjoint* if $A \cap B = \varnothing$ (intuitively, they do not overlap). More generally, any finite number of sets $A_1, A_2, \ldots, A_n$ are said to be disjoint (or, for the sake of clarity, *mutually disjoint*) if

$$A_i \cap A_j = \varnothing \qquad (i \neq j)$$

Using our generalized union, we may say that a collection

$$\pi = \{A_1, A_2, \ldots, A_n\}$$

of nonempty subsets $A_i \subseteq S$ is a *partition* of S if

(I) $\displaystyle\bigcup_{i=1}^{n} A_i = S$ (the A_i are a *covering* of S),

(II) $A_i \cap A_j = \varnothing$ for all $i \neq j$ (they are mutually disjoint).

The subsets A_i are called the *blocks* of the partition. A partition of S into k nonempty blocks is called a *k-partition* of S. Readers should consult Figure 0.4 to see that the definition of a partition agrees with their intuition.

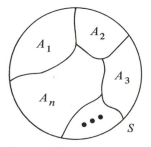

FIGURE 0.4

EXAMPLE 0.16   Suppose that $S = \{1, 2, 3, 4, 5, 6, 7, 8, 9\}$ and

$$A_1 = \{1, 2, 5, 6\}, \qquad A_2 = \{3, 7, 9\}, \qquad A_3 = \{4, 8, 9\}$$

Then $\pi = \{A_1, A_2, A_3\}$ is a covering of S, but not a partition since $A_2 \cap A_3 = \{9\} \neq \emptyset$. If, instead, $A_2 = \{3, 7\}$, then π is a 3-partition of S.   △

EXERCISES

1 If $S = \{a, b, c, d\}$, what are the elements of the power set $\mathscr{P}(S)$?

2 Given the following subsets of $S = \{a, b, c, d, e, f, g\}$,

$$A = \{a, b, d, f\}, \qquad B = \{a, c, f, g\}, \qquad C = \{b, e, g\}$$

determine the sets:
(a) $A \cap B$ (b) $A \cup C$ (c) $(\sim A) \cup B$
(d) $(A \cup B) \cap C$ (e) $(A \cup (\sim B)) \cap C$ (f) $A \cap B \cap C$
(g) $A \cap C$ (h) $B \cup (\sim C)$

3 List all the partitions of the set $S = \{1, 2, 3, 4\}$.

4 For arbitrary sets A, B, C prove
(a) Property (a) in Example 0.10
(b) Property (b) in Example 0.10
(c) Property (a′) in Example 0.10

5 For arbitrary subsets A, B, C of the set S prove:
(a) Property (P1a)   (b) Property (P2b)
(c) Property (P3a)   (d) Property (P4b)
(e) Property (P5a)   (f) Property (P6a)
(g) Property (P7b)   (h) Property (P8b)

6 Show that

$$A \subseteq C \text{ and } B \subseteq D \Rightarrow A \cap B \subseteq C \cap D$$

7 Let

$$\pi = \{A_1, A_2, \ldots, A_n\} \qquad \text{and} \qquad \gamma = \{B_1, B_2, \ldots, B_m\}$$

be partitions of the set S. Show that

$$\{A_i \cap B_j : A_i \in \pi, B_j \in \gamma, A_i \cap B_j \neq \emptyset\}$$

is again a partition of S.

8 Illustrate the construction of Exercise 7 for
(a) $\pi = \{\{1, 2, 4\}, \{3, 5\}\}, \gamma = \{\{1, 3, 4\}, \{2, 5\}\}$
(b) $\pi = \{\{a, b, d, e\}, \{c, f\}, \{g\}\}, \gamma = \{\{a, d, g\}, \{c, f\}, \{b, e\}\}$
(c) $\pi = \{\{1, 5\}, \{2, 6\}, \{3, 7\}, \{4, 8\}\}, \gamma = \{\{1, 2, 3, 4\}, \{5, 6, 7, 8\}\}$

9 A collection

$$\mu = \{A_1, A_2, \ldots, A_n\}$$

of subsets $A_i \subseteq S$ is called a *set system* of S if

(1) $\displaystyle\bigcup_{i=1}^{n} A_i = S,$ (2) $A_i \nsubseteq A_j$ for all $i \neq j$

(a) Show that every partition is a set system.
(b) Show that the converse is not true.

10 List all set systems (see Exercise 9) of the set
(a) $S = \{a, b, c\}$ (b) $S = \{1, 2, 3, 4\}$

*11 Modify Exercise 7 to obtain an analogous statement that holds for set systems (see Exercise 9).

0.3   RELATIONS AND FUNCTIONS

The simplest mathematical structure of all is the unstructured set. Therefore, our opening sections, being devoted to just such unstructured sets, provide a point of departure for the study of discrete structures. We can superimpose more and more structure on our sets as we proceed.

The image of our large-scale computers as simple "number crunchers" is by now pretty much a thing of the past. Computers are now being used to process highly structured information of ever-increasing complexity. Even the most elementary applications in business deal with associated items of information in a file. The information may consist of an employee name, social security number, and job classification, or it may consist of descriptions, part number, etc., of items in an inventory. Such associations can be treated more formally with the aid of the set product construction, which in turn will lead quite naturally to the important definitions of relations and functions. Without the latter, a meaningful analysis of our subsequent mathematical structures would hardly be possible.

The products are again set compositions; but unlike the union and intersection, the product $A \times B$ falls outside the system of subsets of S, even though $A, B \subseteq S$. Given any sets A and B, the *product* $A \times B$ is defined to be the collection of all ordered pairs (a, b) such that $a \in A$ and $b \in B$. That is,

$$A \times B = \{(a, b): a \in A, b \in B\}$$

By an *ordered* pair, we mean that (a, b) and (a', b') are regarded as equal only when $a = a'$ and $b = b'$. Thus $(a, b) \neq (b, a)$, in general. This should remind the reader of the definition of equality for complex numbers. In fact, when $A = B = R$, the set of real numbers, then $A \times B$ is recognized as the set of points in the Cartesian plane (or as the set of complex numbers). For this reason, the product $A \times B$ is quite often referred to as the *Cartesian product* of the sets A and B.

EXAMPLE 0.17   If $S = \{a, b, c\}$ and $T = \{1, 2\}$, then

$$S \times T = \{(a, 1), (a, 2), (b, 1), (b, 2), (c, 1), (c, 2)\} \qquad \triangle$$

As in the case of unions and intersections, we disregard parentheses in extending our definition of products to arbitrarily many factors. We set, for $n \geq 2$,

$$\underset{i=1}{\overset{n}{\times}} A_i = A_1 \times A_2 \times \cdots \times A_n$$

$$= \{(a_1, a_2, \ldots, a_n) : a_i \in A_i \text{ for each } i = 1, 2, \ldots, n\}$$

In the special case where all the A_i are the same, say $A_1 = A_2 = \cdots = A_n = A$, we denote this product set by A^n. Thus for any set A we have

$$A^n = \{(a_1, a_2, \ldots, a_n) : a_i \in A \text{ for all } i\}$$

and the elements of such a product set are usually called the *n-tuples* or *n-vectors* from the set A. The term n-tuple is just the obvious generalization of the terminology: pair, triple, quadruple, $\ldots$, n-tuple. How would you define A^1? or A^0?

EXAMPLE 0.18   If $B = \{0, 1\}$, then

$$B^n = \{(b_1, b_2, \ldots, b_n) : b_i = 0 \text{ or } 1 \ (1 \leq i \leq n)\}$$

In words, B^n consists of all the n-tuples or n-vectors whose coordinates are each either 0 or 1. Thus in the special case of $n = 3$, we have

$$B^3 = \{(0, 0, 0), (0, 0, 1), (0, 1, 0), (0, 1, 1), (1, 0, 0), (1, 0, 1), (1, 1, 0), (1, 1, 1)\} \qquad \triangle$$

As the reader can easily appreciate in reviewing our discussion of employee files and inventories, we are most often interested only in a subset of a product set. But let us give a more vivid illustration in case there is still confusion on this point. Suppose $T = \{t_1, t_2, \ldots, t_m\}$ is a collection of unfilled jobs and $S = \{s_1, s_2, \ldots, s_n\}$ is a set of applicants. We might then let the ordered pair (s_i, t_j) signify that the person s_i is qualified for the job t_j. We may consider this ordered pair as an element of the product set $S \times T$. But the personnel office will rarely have a file that indicates the complete set of correspondences signified by $S \times T$. Unless every applicant is qualified for every job, it is a proper subset of $S \times T$ that underlies the personnel "matching" problem.

In more abstract terms, a *relation* R from a set S to a set T is simply a subset $R \subseteq S \times T$. We say that s *is related to* t (by the relation R) if $(s, t) \in R$. The notation $s \, R \, t$ is used in the same connection. If S and T are finite, say

$$S = \{s_1, s_2, \ldots, s_n\} \qquad T = \{t_1, t_2, \ldots, t_m\}$$

as before, then we may represent each relation $R \subseteq S \times T$ as a rectangular $n \times m$ array or *relation matrix* $R = (r_{ij})$ such that

$$r_{ij} = \begin{cases} 1 & \text{if } s_i \ R \ t_j \qquad ((s_i, t_j) \in R) \\ 0 & \text{otherwise} \end{cases}$$

The actual entries zero and one have no particular numerical significance. Any other pair of symbols would serve equally well. But conversely, any $n \times m$ rectangular array of zeros and ones can be used to define a relation from an n-set to an m-set, as shown by the following example.

EXAMPLE 0.19   The 4×5 array

R	t_1	t_2	t_3	t_4	t_5
s_1	0	1	0	0	0
s_2	1	1	0	0	1
s_3	0	0	0	0	0
s_4	1	0	0	1	1

can serve to define a relation R from

$$S = \{s_1, s_2, s_3, s_4\} \quad \text{to} \quad T = \{t_1, t_2, t_3, t_4, t_5\}$$

by indicating

$$s_i \ R \ t_j \Leftrightarrow r_{ij} = 1$$

Then we have

$$(s_1, t_2) \in R \qquad (s_1, t_4) \notin R$$
$$(s_4, t_1) \in R \qquad (s_1, t_3) \notin R$$
$$\text{etc.}$$

and in fact

$$R = \{(s_1, t_2), (s_2, t_1), (s_2, t_2), (s_2, t_5), (s_4, t_1), (s_4, t_4), (s_4, t_5)\}$$

It should be clear that the relation matrix for this relation is the matrix we had at the beginning of the example.   △

A *graphical representation* can also be given for relations from one finite set to another. We can use a dot or node for each element of S, say at the left, and similarly for each element of T at the right. Then a *directed line* can be drawn from the point s_i to the point t_j just to indicate $s_i \ R \ t_j$. In the case of Example 0.19 we obtain the graphical representation shown in Figure 0.5.   △

We are all familiar with the arithmetic of the integers in which the elements of Z are ordered. Thus we say that

$$n < m \Leftrightarrow m - n \in Z^+$$

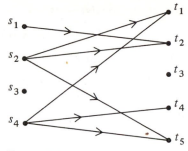

FIGURE 0.5

that is, if n precedes m in the usual listing

$$Z = \{\ldots, -2, -1, 0, 1, 2, \ldots\}$$

Even this simple comparison, however, can be viewed as a relation from Z to Z. Thus as a subset of $Z \times Z$, the relation contains $(2, 5)$, $(-3, 7)$, and $(0, 3)$, but not $(4, 1)$ or $(4, 4)$. Similarly, "less than or equal to" is a relation $\leq$ from Z to Z. Such relations from a set S to itself are quite common, and we speak of them in each case as a *relation on a set S*. We recall that the last relation referred to satisfies the familiar properties

(i) $n \leq n$

(ii') $n \leq m, m \leq n \Rightarrow n = m$

(iii) $n \leq m, m \leq p \Rightarrow n \leq p$

for all $n, m, p \in Z$, which are usually called the *reflexive, antisymmetric,* and *transitive* properties, respectively. They belong to a class of properties that are useful for comparing or classifying relations on a set S. In the case of finite sets S, we are able to characterize the properties according to their effects on the corresponding relation matrices or the graphical representations.

Throughout the following three groups of characterizations, R is an arbitrary relation on a finite set $S = \{x_1, x_2, \ldots, x_n\}$. Nevertheless, all of the definitions are applicable to infinite sets. In either case, we use the symbols x, y, z to denote elements of S. Again, in view of the fact that S is finite, as just described, we need a few preliminary observations regarding the special nature of the relation matrices for a relation R on the set S. For one thing, the matrices are square arrays because the same elements (those of the set S) serve both as row and as column headings. When we speak of the *diagonal* of such matrices, we mean the one going from the upper left-hand corner to the lower right-hand corner. As for graphical representations, a major simplification comes about by our not duplicating the dots or nodes corresponding to the elements of S. They are only drawn once, and consequently, it is entirely possible that a directed line or arc may begin and end at the same point, for example, when we have $x R x$ for some $x \in S$. Such an arc is called a *loop*.

Characterization of reflexive and irreflexive relations

A relation R on a set S is said to be *reflexive* (*irreflexive*) provided that

(i) $x R x$ for all x $((i')$ $x R x$ for no $x)$

In words, every element is related to itself when the reflexive relation holds, whereas no element is related to itself when the relation is irreflexive. If S is a set of (well-adjusted) people, then "likes" is a reflexive relation, while "is a grandfather of" is irreflexive.

The relation matrix for a reflexive relation should have all ones along the diagonal. It should have all zeros on the diagonal when the relation is irreflexive. The graphical representation of a reflexive relation is characterized by a loop at every node x. (See Figure 0.6.) But there should not be any of these loops in the case of an irreflexive relation.

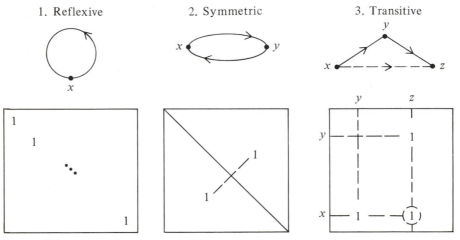

FIGURE 0.6

Characterization of symmetric and antisymmetric relations

A relation R on a set S is *symmetric* (*antisymmetric*) whenever

(ii) $x R y \Rightarrow y R x$ $((ii')$ $x R y, y R x \Rightarrow x = y)$

Again, if S is a set of persons, then "is a sibling of" is a symmetric relation. We have already observed that "less than or equal to" is an antisymmetric relation on the set of integers. In fact, it turns out that there are a great many relations which satisfy properties (i), (ii'), and (iii). An abstract treatment of such relations will lead to the notion of *partially ordered sets*, which are discussed in Chapter 4.

In the matrix of a symmetric relation the diagonal acts like a mirror in that $r_{ij} = r_{ji}$. If we have a one in a particular location of the array, then a one must also be found in the symmetric position across the diagonal, as shown in Figure

0.6. The same is true of zeros. In contrast, we can never have symmetrically located ones in the matrix of an antisymmetric relation, for the *contrapositive* of condition (ii') reads:

$$x \neq y \Rightarrow x \not{R} y \quad \text{or} \quad y \not{R} x \quad ((x, y) \notin R \quad \text{or} \quad (y, x) \notin R)$$

in matrix terminology we have

$$i \neq j \Rightarrow r_{ij} = 0 \quad \text{or} \quad r_{ji} = 0$$

As for the graphical representations, we must have a directed line or arrow going both ways (from x to y and from y to x) or none at all in the case of a symmetric relation, but never such two-way arrows for an antisymmetric relation. In view of these remarks, can a relation be at once symmetric and antisymmetric?

Characterization of transitive relations

A relation R on a set S is said to be *transitive* provided that

(iii) $x R y, y R z \Rightarrow x R z$

If S is a set of persons, then "is a sibling of" is a transitive relation, and it would probably be a better world if "likes" or "is a friend of" were transitive. ·

All that we can say about the relation matrix of a transitive relation is that two ones often necessitate a third (the one with dotted circle in Figure 0.6), for in matrix terminology, condition (iii) translates to

$$r_{ij} = r_{jk} = 1 \Rightarrow r_{ik} = 1$$

This is a reasonable rule for use in showing that a particular relation is not transitive. It could even be used to exhaustively verify transitivity, but perhaps the graphical characterization is more vivid. Again we find that two arrows necessitate a third (as indicated by the dashed arrow in Figure 0.6).

EXAMPLE 0.20   In order to illustrate the use of these characterizations, consider the following relation matrix as describing a relation R on the set $S = \{x_1, x_2, x_3, x_4\}$:

R	x_1	x_2	x_3	x_4
x_1	1	0	1	1
x_2	0	1	0	1
x_3	1	0	1	0
x_4	1	1	0	1

Here R is reflexive because the diagonal consists entirely of ones. Also R is symmetric because all the ones in the array are located symmetrically with respect to the diagonal. On the other hand, the relation is not transitive because

$r_{41} = r_{13} = 1$, whereas $r_{43} = 0$. Obviously the relation is not irreflexive, nor is it antisymmetric.   △

EXAMPLE 0.21   In the same way, the following 3×3 array represents a relation R on the set $S = \{x_1, x_2, x_3\}$:

R	x_1	x_2	x_3
x_1	1	0	0
x_2	1	1	0
x_3	1	1	0

This relation is neither reflexive nor irreflexive (thus showing that these two properties are not opposites). It is an antisymmetric relation because there are no off-diagonal symmetric ones. It is also transitive, as can be verified by exhaustive listing; thus $r_{32} = r_{21} = 1$ and $r_{31} = 1$ as required.

As a useful exercise, the student might draw the graphical representations for the last two examples and then try to obtain the same results by looking at the graphical characterization of the various properties.   △

EXAMPLE 0.22   In the algebra of sets the question as to whether $A \subseteq B$ (yes or no) can be viewed as a relation on $\mathscr{P}(S)$. If $S = \{a, b, c\}$, then this relation corresponds to the matrix,

$\subseteq$	$\varnothing$	$\{a\}$	$\{b\}$	$\{c\}$	$\{a, b\}$	$\{a, c\}$	$\{b, c\}$	S
$\varnothing$	1	1	1	1	1	1	1	1
$\{a\}$	0	1	0	0	1	1	0	1
$\{b\}$	0	0	1	0	1	0	1	1
$\{c\}$	0	0	0	1	0	1	1	1
$\{a, b\}$	0	0	0	0	1	0	0	1
$\{a, c\}$	0	0	0	0	0	1	0	1
$\{b, c\}$	0	0	0	0	0	0	1	1
S	0	0	0	0	0	0	0	1

Interestingly enough, for any set S, this relation, "is contained in," satisfies the same three properties as the "less than or equal to" relation on the integers. Property (i) follows from Example 0.9, (ii′) is a consequence of the definition of set equality, and (iii) is obvious.   △

The reader does not need to be reminded of the importance of the idea of functions in any mathematical investigation. In computer science the idea of relations plays an equally important role. It is fortunate, therefore, that one (the concept of function) is but a special case of the other. A relation from S to T

is called a *function* if every element $s \in S$ appears once and only once as the left-hand entry of an ordered pair $(s, t) \in R$. Now, this definition may not coincide exactly with the reader's previous understanding, but a few words of clarification will set everything straight. Consider the graphical representation of such a relation R (function f). There will be exactly one arrow emanating from each node $s \in S$, leading ultimately to a unique element $t \in T$. If we now change our notation slightly, writing

$$f(s) = t \quad \text{in place of} \quad s\,R\,t$$

and

$$f\colon S \to T \text{ or } S \overset{f}{\to} T \quad \text{rather than} \quad R \subseteq S \times T$$

then the pieces will begin to fall into place. (See Figure 0.7.) Here S is called the *domain* and T the *range* of the function $f\colon S \to T$.

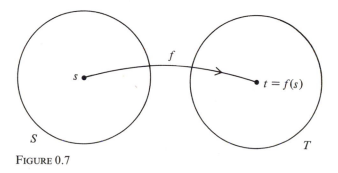

FIGURE 0.7

EXAMPLE 0.23 The relation given in Example 0.19 is not a function. For one thing, the element s_3 does not appear as a left-hand entry. For another, s_2 appears more than once. A function should give a *value* for each argument in its domain S. And this value should be a unique element of the range T if we are to write $f\colon S \to T$. $\triangle$

EXAMPLE 0.24 The relation graphically represented in Figure 0.8 does define a function $f\colon S \to T$. As a relation, it is a subset

$$\{(s_1, t_5), (s_2, t_3), (s_3, t_2), (s_4, t_3)\} \subseteq S \times T$$

But as a function, we prefer the notation

$$f(s_1) = t_5 \qquad f(s_2) = t_3 \qquad f(s_3) = t_2 \qquad f(s_4) = t_3 \qquad \triangle$$

For the time being, we defer any further presentation of examples. We also postpone our discussion of the special properties that serve to distinguish one function from another. At this stage, it is enough to be sure that we have grasped

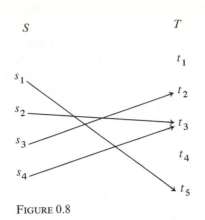

FIGURE 0.8

the basic definition of function as a special case of a relation, so that both of these notions can be used in our subsequent discussion.

EXERCISES

1 With A, B, C as given in Exercise 2 of Section 0.2, determine the sets:
  (a) $A \times C$ (b) $B \times C$ (c) $C \times C = C^2$.
  (d) C^3 (e) $A \times \emptyset$ (f) $A \times \{\emptyset\}$

2 If $B = \{0, 1\}$, what are the elements of B^4?

3 If S is an n-set and T is an m-set, how many relations are there from S to T? Why?

4 Given the sets A, B, C, D, show that
  (a) $(A \cap B) \times (C \cap D) = (A \times C) \cap (B \times D)$
  (b) $A \times (B \cup C) = (A \times B) \cup (A \times C)$

5 Draw the graphical representations corresponding to the following relation matrices:

(a)

R	t_1	t_2	t_3	t_4
s_1	1	1	0	1
s_2	0	1	1	0
s_3	0	1	0	0

(b)

R	t_1	t_2	t_3
s_1	1	0	0
s_2	0	1	1
s_3	1	1	1
s_4	1	0	1

Then express each relation as a subset $R \subseteq S \times T$.

6 Decide whether each of the following relations is

  (i) reflexive, irreflexive
  (ii) symmetric, antisymmetric

(iii) transitive

(a)

	x_1	x_2	x_3	x_4
x_1	1	0	0	1
x_2	0	1	1	0
x_3	0	1	1	0
x_4	1	0	0	1

(b)

	x_1	x_2	x_3	x_4
x_1	1	1	1	0
x_2	1	1	1	1
x_3	1	1	1	1
x_4	0	1	1	1

(c)

	x_1	x_2	x_3	x_4
x_1	1	1	0	1
x_2	1	1	1	0
x_3	0	1	1	0
x_4	1	0	0	1

(d)

	x_1	x_2	x_3	x_4
x_1	1	0	0	1
x_2	1	1	0	0
x_3	1	1	1	1
x_4	0	1	0	1

7 Draw the graphical representation of the relations in Exercise 6. Then express each relation as a subset $R \subseteq S \times S$.

8 For any set S, show that the relation $A \subseteq B\,(?)$ is a reflexive, antisymmetric, and transitive relation on $\mathcal{P}(S)$.

9 If S is an n-set and T is an m-set, how many functions are there from S to T? Why?

0.4 SAMPLES AND SELECTIONS

Up to now there is little to distinguish our presentation from that which would be considered prerequisite for almost any intermediate course in the mathematical sciences, whether it be group theory, probability theory, information theory, or whatever. Therein lies the beauty of the set-theoretic notions, however. These notions are so widely applicable as almost to constitute a universal language. Having introduced the ingredients of the language, we can now begin to specialize and to exploit the features of this language in the areas of more direct concern to us. By way of illustration, let us take a brief detour to introduce a few elementary ideas from combinatorial mathematics, important for their application to computer science.

Undoubtedly the student has learned something about permutations and combinations in high school or early college mathematics. As usually presented, these ideas are used to answer questions of the sort: "In how many ways can we choose such and such from this and that . . . ?" The decision as to whether permutations or combinations are involved will usually depend on whether or not the order of the choices is significant. If we now allow repetitions in our choices, then we are led to the more general idea of "samples and selections."

We are aware of the fact that the numerical answer to such questions as "In how many ways can we choose ..." is not dependent on the nature of the objects themselves. They may be playing cards, friends, colors, coins, or whatever. So it is just as well to remain uncommitted on this point and suppose simply that we are choosing from a nonempty but finite set of *symbols* σ_i, the set being denoted by

$$\Sigma = \{\sigma_1, \sigma_2, \ldots, \sigma_n\}$$

and called, appropriately, an *alphabet*. The results of our reasoning will be the same for an arbitrary *n*-set, whether its elements are interpreted as symbols or not. All that really matters is that the elements (symbols) be distinct. This convention, however, is already inherent in the idea of a set.

EXAMPLE 0.25   If

$$\Sigma = \{a, b, c, \ldots, z\}$$

the set consists of the ordinary lowercase English alphabet. It is a 26-set whose elements are interpreted as symbols.   △

EXAMPLE 0.26   The 52-set

$$\Sigma = \{1H, 2H, \ldots, 9H, 0H, JH, QH, KH, 1S, 2S, \ldots, 9S, 0S, JS, QS, KS,$$
$$1D, 2D, \ldots, 9D, 0D, JD, QD, KD, 1C, 2C, \ldots, 9C, 0C, JC, QC, KC\}$$

is an alphabet that could serve as a deck of playing cards. On the other hand, the 6-set

$$\Sigma = 6^+ = \{1, 2, 3, 4, 5, 6\}$$

would be appropriate for representing the symbols of a die. Note the "multi-character" symbols of the playing card alphabet. Also note that it is essentially a product set.   △

EXAMPLE 0.27   Formally the FORTRAN character set (a 47-set) is the alphabet

$$\Sigma = \{A, B, \ldots, Z, 0, 1, \ldots, 9, +, -, *, /, =, ,, ., \$, (,), \square\}$$

where the last "symbol" is a space or blank character.   △

EXAMPLE 0.28   The alphabet for ALGOL is considerably larger than that for FORTRAN. The reference version of ALGOL uses 113 symbols as follows:

letters: *a   b   ...   z   A   B   ...   Z*
digits: 0   1   ...   9
logical values: **true   false**
arithmetic operators: +   −   ×   /   ÷   ↑

relational operators: $<$ $\leq$ $=$ $\geq$ $>$ $\neq$
logical operators: $\equiv$ $\supset$ $\vee$ $\wedge$ $\neg$
sequential operators: **go to** **if** **then** **else** **for** **do**
brackets: **( )** **[ ]** **' '** **begin** **end**
declarators: **own** **Boolean** **integer** **real** **array** **switch** **procedure**
separators: **,** **.** $_{10}$ **:** **;** **:=** **step** **until** **while** **comment** $\square$

Note the use once more of multicharacter symbols. $\triangle$

EXAMPLE 0.29 At any given moment, the English dictionary is itself an alphabet

$$E = \{a, aardvark, aardwolf, \ldots, zymurgy\}$$

Of course, we should include a blank symbol, punctuation symbols, etc. But the real problem for lexicographers is the rate of expansion of the dictionary. For example, sportscasters are adding to it everyday with "words" such as no show (football), red dog (football), blueliner (hockey), knuckler (baseball), airball (basketball), etc. $\triangle$

A finite sequence of symbols from the alphabet Σ written one after another without any intervening punctuation is called a *word* x on the alphabet Σ. Thus

$$x = \sigma_{i_1}\sigma_{i_2} \cdots \sigma_{i_k}$$

Note that we must use two-level subscripts here to allow each symbol to be any of the members of the alphabet. Suppose that we denote the *length* of x by $|x| = k$. Then we admit the *null word* ϵ (a sequence of length zero) in defining

$$\Sigma^* = \{x: x \text{ is a word on the alphabet } \Sigma\}$$

This notation is common in the literature of computer science, and is derived primarily from the use of Σ^k to denote the set of all words of length k. Note that there is only a punctuational distinction between this set and the product set Σ^k as defined in Section 0.3. In any case, the set equality

$$\Sigma^* = \{\epsilon\} \cup \Sigma \cup \Sigma^2 \cup \cdots$$

indicates the suitability of the notation Σ^*.

EXAMPLE 0.30 Given the alphabet $\Sigma = \{a, b, c, \ldots, z\}$ of Example 0.25, we have

$$\cdots \quad \begin{matrix} at \in \Sigma^2 & lbz \in \Sigma^3 \\ by \in \Sigma^2 & cat \in \Sigma^3 \\ gz \in \Sigma^2 & dog \in \Sigma^3 \\ \vdots & \vdots \end{matrix} \quad \cdots$$

Altogether Σ^* is the infinite union

$$\Sigma^* = \{\epsilon\} \cup \Sigma \cup \Sigma^2 \cup \cdots$$
$$= \{\epsilon, a, b, \ldots, z, at, by, gz, \ldots, lbz, cat, dog, \ldots\} \qquad \triangle$$

EXAMPLE 0.31   If, instead, Σ is the FORTRAN alphabet (Example 0.27), then we obtain

$$\Sigma^* = \{\text{GO TO 75, IF (X .LE. Y) Z} = Z + 1,$$
$$\text{FI ((X .LE) Z}+1 = \text{CRUD}, \ldots\}$$

The reader who is familiar with FORTRAN programming will realize how very few of these character strings (words) are recognized as legitimate statements in the FORTRAN language. $\qquad \triangle$

The k-letter words x (elements of Σ^k) are called *k-samples* from the alphabet Σ. Note in particular that we allow repetitions, which fact alone distinguishes the k-samples from the *k-permutations*. Ordinarily we think of our alphabet as an *n*-set. It follows that the idea of a k-permutation will make sense only if $k \leq n$, for we cannot choose more than n elements from an *n*-set without eventual duplication. In effect, this line of reasoning makes use of the so-called *pigeonhole principle* (see Section 0.7).

EXAMPLE 0.32   If $\Sigma = \{a, b, c, d\}$ $(n = 4)$, then we have the following 3-samples:

aaa	*aca*	*baa*	*bca*	*caa*	*cca*	*daa*	*dca*
aab	*acb*	*bab*	*bcb*	*cab*	*ccb*	*dab*	*dcb*
aac	*acc*	*bac*	*bcc*	*cac*	*ccc*	*dac*	*dcc*
aad	*acd*	*bad*	*bcd*	*cad*	*ccd*	*dad*	*dcd*
aba	*ada*	*bba*	*bda*	*cba*	*cda*	*dba*	*dda*
abb	*adb*	*bbb*	*bdb*	*cbb*	*cdb*	*dbb*	*ddb*
abc	*adc*	*bbc*	*bdc*	*cbc*	*cdc*	*dbc*	*ddc*
abd	*add*	*bbd*	*bdd*	*cbd*	*cdd*	*dbd*	*ddd*

$64 = 4^3$ samples in all. Of these, only the following are 3-permutations:

abc	*acd*	*bac*	*bcd*	*cab*	*cbd*	*dab*	*dbc*
abd	*adb*	*bad*	*bda*	*cad*	*cda*	*dac*	*dca*
acb	*adc*	*bca*	*bdc*	*cba*	*cdb*	*dba*	*cdb*

$\triangle$

EXAMPLE 0.33   Taking the deck of cards of Example 0.26 as our alphabet, we can obtain such 5-samples as

$$2H\ 9C\ 2H\ JS\ QS \qquad 2H\ 2H\ 2H\ JS\ JD$$

which cannot be interpreted as poker hands because of the repetitions. If we are counting distinct poker hands, the concept of a 5-permutation will not be appropriate either. We will need the idea of a 5-combination (5-subset) where the order of occurrence of the symbols is immaterial.

On the other hand if we were considering "runs" on a die, then repetitions in the occurrence would be meaningful. The word 21236342 as an 8-sample from the alphabet $6^+ = \{1, 2, 3, 4, 5, 6\}$ might represent a person's performance in a certain dice game. Each symbol of the word could correspond to a throw of the die, but it might be that the run 26332214 is considered as having the same score as the previously mentioned run; that is, the order of occurrence may have no bearing whatsoever. Both runs should then be treated as representations of the same *selection*, 12223346 (one 1, three 2's, two 3's, one 4, one 6). △

We have seen the need for unordered counterparts to the notions of sample and permutation. The samples (words) being sequences, the order of appearance of the symbols is significant; thus *abba* ≠ *baba*. If we disregard the order, we will arrive at the idea of a selection. Thus a *k-selection* from the alphabet (*n*-set) Σ is a list of (not necessarily distinct) symbols chosen from Σ without regard to order. If we do not allow replications, then the resulting selections are called *k-combinations*. It should be clear that except for notations, a *k*-combination is the same thing as a *k*-element subset of Σ.

EXAMPLE 0.34   Given the alphabet of Example 0.32, we have the following 3-selections:

$$
\begin{array}{ccccc}
aaa & bbb & ccc & ddd & abc \\
aab & abb & acc & add & abd \\
aac & bbc & bcc & bdd & acd \\
aad & bbd & ccd & cdd & bcd
\end{array}
$$

twenty in all, which should be compared with the sixty-four 3-samples of Example 0.32. When we disregard the order, we are in effect "identifying" those samples which agree with one another except for the order of appearance of the symbols; thus *aab*, *aba*, *baa* are treated as one.

In the same connection, there are only four 3-combinations, namely *abc*, *abd*, *acd*, *bcd*, corresponding to the 3-element subsets

$$\{a, b, c\} \qquad \{a, b, d\} \qquad \{a, c, d\} \qquad \{b, c, d\}$$

of the alphabet. △

EXERCISES

1 Give a definition for the *countable union*

$$A_1 \cup A_2 \cup \cdots \quad \left(\text{usually abbreviated } \bigcup_{i=1}^{\infty} A_i\right)$$

of sets $A_1, A_2, \ldots$, and then prove that

$$\Sigma^* = \{\epsilon\} \cup \Sigma \cup \Sigma^2 \cup \cdots$$

2 Given the alphabet $\Sigma = \{a, b, c, d\}$, list
   (a) all its 2-samples   (b) all its 2-permutations
   (c) all its 2-selections   (d) all its 2-combinations
   (e) all its 4-samples   (f) all its 4-permutations
   (g) all its 4-selections   (h) all its 4-combinations

3 From the alphabet $\Sigma = \{1, 2, 3, 4, 5\}$, list
   (a) all the 2-samples   (b) all the 2-permutations
   (c) all the 2-selections   (d) all the 2-combinations
   (e) all the 3-samples   (f) all the 3-permutations
   (g) all the 3-selections   (h) all the 3-combinations

4 For a definite k as suggested, express each of the following as a k-sample, k-permutation, k-selection, or k-combination from an appropriate alphabet.
   (a) A four-digit decimal number
   (b) A five-bit binary number
   (c) A bridge hand
   (d) A baseball lineup of a 15-member team
   (e) A series of 100 coin tosses, measured only according to the number of heads and the number of tails
   (f) An arrangement of 20 books on a bookshelf
   (g) A punched IBM card
   (h) A 10-city itinerary
   (i) A coloring of the fingernails with 16 bottles (shades) of nail polish
   (j) The solution to a 20-question multiple-choice exam

0.5 EQUIVALENCE RELATIONS

We recall that a relation on a set S is simply a subset $R \subseteq S \times S$. Thus R establishes a dichotomy on $S \times S$, so that any ordered pair $(a, b) \in S \times S$ satisfies the condition of either

$$(a, b) \in R \quad \text{or} \quad (a, b) \notin R$$

In the first case, we say that a is related to b (by the relation R) and we write $a \, R \, b$. In the latter case, we write $a \, \not{R} \, b$.

EXAMPLE 0.35   Let $S = Z = \{\text{integers}\}$ and for a fixed $n \geq 1$ define

$$a \, R \, b \Leftrightarrow a - b \text{ is a multiple of } n$$

For instance, if $n = 4$, then we can cite

$$12 \, R \, 4 \qquad 13 \, \not{R} \, 4$$
$$6 \, R \, 30 \qquad 5 \, \not{R} \, 7$$

as illustrations of related and unrelated pairs. We are referring to integer multiples, so that

$$12 - 4 = 8 = 2 \cdot 4$$

whereas

$$13 - 4 = 9 = p \cdot 4$$

does not have an integer solution for p.

Note that for any $a \in Z$, there are infinitely many integers related to a, namely

$$a, a \pm n, a \pm 2n, \ldots$$

Again, if $n = 4$ and $a = 1$, then the integers related to a will comprise the set

$$A_1 = \{\ldots, -3, 1, 5, 9, \ldots\}$$

On the other hand, if $a = 2$, then we obtain a collection

$$A_2 = \{\ldots, -2, 2, 6, 10, \ldots\}$$

which is entirely disjoint from the set A_1. The reader is invited to conclude that in general

$$\pi = \{A_0, A_1, A_2, \ldots, A_{n-1}\}$$

is a partition of Z. △

Now let S be any set and let R be a relation on S. Then the relation R is called an *equivalence relation* (in which case, the generic symbol $\sim$ often replaces R in written notations) if it satisfies the following three conditions:

(i) $a \sim a$ (reflexivity)
(ii) $a \sim b \Rightarrow b \sim a$ (symmetry)
(iii) $a \sim b, b \sim c \Rightarrow a \sim c$ (transitivity)

for all $a, b, c \in S$. For any $a \in S$ we define the *equivalence class* $[a]$ determined by a to be that subset of S consisting of all the elements that are related (equivalent) to a. That is,

$$[a] = \{x : x \sim a\}$$

Note that every equivalence class is nonempty because the reflexive property assures that $a \in [a]$.

In high school algebra we learn that equality is an equivalence relation; we often hear of transitivity being expressed as "Quantities equal to the same thing are equal to each other." Actually the three properties of reflexivity, symmetry, and transitivity are found in a variety of situations: similarity for triangles, equivalence among all persons having the same last name or among all those who are alumni of the same university, etc.; the list of examples is inexhaustible.

EXAMPLE 0.36 We mentioned just prior to Example 0.17 that if R is the set of real numbers, then $S = R \times R$ is the Cartesian coordinate plane. For points $a = (r_1, r_2)$ and $b = (r_1', r_2')$ we write

$$a \sim b \Leftrightarrow r_2 = r_2'$$

This is just a formal way of saying that two points are related by the relation $\sim$ if they are at the same elevation (ordinate) in the plane. The reader can easily verify that this relation is reflexive, symmetric, and transitive, and so is an equivalence relation on the set $S = R \times R$.

We next determine that if $a = (r_1, r_2)$, then

$$[a] = \{x : x \sim a\} = \{x : x = (r, r_2)\}$$

so that in effect $[a]$ is the horizontal line through a. The collection of all these horizontal lines is a partition of the plane into nonoverlapping subsets. That this phenomenon is a uniform consequence of equivalence relations we will soon establish.   △

EXAMPLE 0.37   As in Example 0.35, we let

$$a \sim b \Leftrightarrow a - b \text{ is a multiple of } n$$

for integers a, b. Then we have

(i) $a \sim a$ since $a - a = 0 = 0 \cdot n$
(ii) $a \sim b \Rightarrow a - b = p \cdot n \qquad (p \in Z)$
$\qquad \Rightarrow b - a = (-p) \cdot n \Rightarrow b \sim a$
(iii) $a \sim b, b \sim c \Rightarrow a - b = p \cdot n, \quad b - c = q \cdot n \qquad (p, q \in Z)$
$\qquad \Rightarrow a - c = (a - b) + (b - c) = (p + q) \cdot n \Rightarrow a \sim c$

So this relation, which is called *congruence* (*modulo n*), is an equivalence relation on the integers. In telling time, we use a system of congruence (modulo 12).

Suppose that we again let $n = 4$. Then we have equivalence (congruence) classes:

$$[0] = \{\dots, -4, 0, 4, 8, \dots\}$$
$$[1] = \{\dots, -3, 1, 5, 9, \dots\}$$
$$[2] = \{\dots, -2, 2, 6, 10, \dots\}$$
$$[3] = \{\dots, -1, 3, 7, 11, \dots\}$$

which partition the integers. We should call attention to the fact that $[4] = [0]$, $[5] = [1], [6] = [2]$, etc., which, however, would follow immediately from the lemma we are about to state—true of any equivalence relation.   △

Lemma 1

$a \sim b$ iff $[a] = [b]$.

Proof The notation "iff" is a shorthand for "if and only if." We therefore must show that each assertion implies the other. So suppose $a \sim b$. We shall see whether we can establish the set equality $[a] = [b]$. If $x \in [a]$, then $x \sim a$. We are assuming $a \sim b$, so we have $x \sim b$ by transitivity. However, this means that $x \in [b]$, so we have shown $(a \sim b \Rightarrow) [a] \subseteq [b]$. Completely analogously, we can show that $[b] \subseteq [a]$ as well. Altogether, the definition of equality for sets shows that we have established

$$a \sim b \Rightarrow [a] = [b]$$

Conversely, if $[a] = [b]$, we have $a \in [a] = [b] = \{x : x \sim b\}$ so that $a \sim b$. □

Lemma 2

$[a] \cap [b] \neq \emptyset \Rightarrow [a] = [b]$. □

We let the reader try his hand at proving this second result; it should be an easy consequence of Lemma 1. Its meaning, however, is best understood in terms of the contrapositive statement: two different equivalence classes are disjoint. In any case, that is enough to imply the following main result, which summarizes all the preceding discussion.

Characterization of equivalence relations

An equivalence relation $\sim$ on a set S partitions that set into equivalence classes, that is,

$$\pi = \{[a] : a \in S\}$$

is a partition of S. Conversely, each partition π of a set S defines a unique equivalence relation on the set:

$$a \sim b \Leftrightarrow a \text{ and } b \text{ are in the same block of } \pi$$

It follows that there is a one-to-one correspondence between the equivalence relations and the partitions of any set S. □

EXERCISES

1 Describe the partition of Z induced by the congruence relation:
   (a) $a \sim b$ (modulo 5)   (b) $a \sim b$ (modulo 2)
   (c) $a \sim b$ (modulo 10)   (d) $a \sim b$ (modulo 12)

2 On the set $Q = \{\text{rational numbers}\}$ define

$$a \, R \, b \Leftrightarrow a - b \in Z$$

   Show that R is an equivalence relation and describe the resulting equivalence classes.

*3 Consider the function (differentiation)

$$S \xrightarrow{D} T$$

from the set of "continuously differentiable" functions to the set of continuous functions.

Suppose that we define in the set S

$$f \sim g \Leftrightarrow D(f - g) = 0$$

Show that this is an equivalence relation on S and describe the resulting equivalence classes.

4 Prove Lemma 2. [*Hint*: Use Lemma 1.]

5 Show how the characterization of equivalence relations follows from the two lemmas.

6 A relation R is *circular* if

$$a \, R \, b \quad \text{and} \quad b \, R \, c \Rightarrow c \, R \, a$$

Show that a relation is reflexive and circular iff it is an equivalence relation.

*7 Let R be any relation on a set S. Its *transitive closure* $\bar{R}$ is defined by

$$x \, \bar{R} \, y \Leftrightarrow \text{there is a sequence } x_1, x_2, \ldots, x_k \text{ in } S \text{ such that}$$
$$x = x_1 \, R \, x_2 \, R \cdots R \, x_k = y$$

(See Section 0.6.) Show that $\bar{R}$ is transitive. Show that if R is reflexive and symmetric, then $\bar{R}$ is an equivalence relation.

*8 Given that $Q = \{\text{rational numbers}\}$ and for each fixed $q \in Q$ we define

$$A_q = \{q + n : n \in Z\}$$

show that

$$\pi = \{A_q : q \in Q \cap [0, 1)\}$$

is a partition of Q. Use the characterization of equivalence relations to associate an equivalence relation with this partition and show that the resulting relation coincides with that given in Exercise 2.

9 If

$$S = \{x : x \text{ is a student at this college/university}\}$$

which of the following are equivalence relations on S?
(a) $x \, R \, y$ means that x and y have the same major
(b) $x \, R \, y$ means that x and y are siblings
(c) $x \, R \, y$ means that x and y are enrolled in a course together
(d) $x \, R \, y$ means that x and y have the same initials
(e) $x \, R \, y$ means that x and y are friends

10 Show that if $\pi = \{A_\alpha\}$ is a partition of a set S, then

$$a \, R \, b \Leftrightarrow a \text{ and } b \text{ are in the same block of } \pi$$

is an equivalence relation on S. (This is part of the characterization of equivalence relations.)

0.6 ALGEBRA OF RELATIONS

We intend that this section be more or less parallel to the earlier section on set algebra, where we discussed the three operations $\cup$, $\cap$, and $\sim$, all closely connected to our everyday thinking about collections of objects. The inclusion relation $\subseteq$ among subsets is similarly connected to our ordinary thinking. The algebra of relations is a useful companion to these ideas, although it is somewhat closer to formalized logical discourse. Our choice of symbols—$\vee$, $\wedge$, $\neg$, and $\leq$—makes for an easy comparison with the symbols used in set algebra (we need only to straighten out the curves). Taken together, these preliminary sections provide concrete examples for the general discussion of algebras beginning in Chapter 1.

As in the preceding section, we restrict our attention here to relations R on a (fixed) set S. We also favor for the time being the notation $x R y$ over the equivalent $(x, y) \in R$. As in set algebra, we seek to form new relations from old ones and find it convenient to build them around the familiar terms "or," "and," and "not" from ordinary logical thought. Thus if R_1 and R_2 are relations on S, we define the *join* ($\vee$) and *meet* ($\wedge$) operations, respectively, as follows:

$$x(R_1 \vee R_2)y \Leftrightarrow x R_1 y \quad \text{or} \quad x R_2 y$$

$$x(R_1 \wedge R_2)y \Leftrightarrow x R_1 y \quad \text{and} \quad x R_2 y$$

Finally, the *negation* of R is the relation

$$x(\neg R)y \Leftrightarrow x \overset{\backprime}{R} y, \text{ that is, } (x, y) \notin R$$

The results of these operations are easily obtained if the relation matrices for R_1, R_2, R are given. They are also easily interpreted. For example, we have the relation $x(R_1 \vee R_2)y$ between the elements $x, y \in S$ if either (or both) of the assertions $(x, y) \in R_1$ and $(x, y) \in R_2$ is true. Not coincidentally, these interpretations are consistent with those required for a proper understanding of the logical (Boolean) expressions in the computer programming languages. (See the ALGOL logical operators in Example 0.27.) This is reason enough for our trying to gain an appreciation for relational algebra.

EXAMPLE 0.38 Consider the relations R_1, R_2, R on $S = \{x_1, x_2, x_3, x_4\}$ as given by the three relation matrices shown below.

R_1	x_1	x_2	x_3	x_4	R_2	x_1	x_2	x_3	x_4	R	x_1	x_2	x_3	x_4
x_1	1	0	1	0	x_1	1	1	1	0	x_1	1	1	1	1
x_2	0	0	1	1	x_2	1	0	1	0	x_2	0	1	0	0
x_3	1	1	0	0	x_3	1	0	1	0	x_3	1	0	0	0
x_4	1	0	0	1	x_4	0	0	1	1	x_4	0	0	1	0

The matrices for $R_1 \vee R_2$, $R_1 \wedge R_2$, and $\neg R$ are as follows:

$R_1 \vee R_2$	x_1	x_2	x_3	x_4	$R_1 \wedge R_2$	x_1	x_2	x_3	x_4	$\neg R$	x_1	x_2	x_3	x_4
x_1	1	1	1	0	x_1	1	0	1	0	x_1	0	0	0	0
x_2	1	0	1	1	x_2	0	0	1	0	x_2	1	0	1	1
x_3	1	1	1	0	x_3	1	0	0	0	x_3	0	1	1	1
x_4	1	0	1	1	x_4	0	0	0	1	x_4	1	1	0	1

$\triangle$

The perceptive student may ask, why are we talking as though this were a new algebra when it is merely the algebra of subsets all over again? He will be absolutely correct. If the relations R are viewed as subsets $R \subseteq S \times S$, then the join, meet, and negation operations are exactly the same as the union, intersection, and complementation (in $\mathscr{P}(S \times S)$). If we let $\mathscr{R}(S)$ denote all the relations on S, that is,

$$\mathscr{R}(S) = \{R: R \text{ is a relation on } S\}$$

then $\mathscr{R}(S)$ and $\mathscr{P}(S \times S)$ are "abstractly equivalent" (you might even say identical). There is a one-to-one correspondence between their elements that "preserves" all the algebraic operations. We need not go into more detail here, for it is more appropriate to treat such matters at a later point. But how do we answer the original question if all of this is true? For one thing, two different points of view can often be a distinct advantage. Thus imagine the confusion if we had to discuss equivalence relations on S in the context of $\mathscr{P}(S \times S)$! Also the new perspective shows the importance of the connection between relations and (directed) graphs, a connection which is hardly visible from the other point of view. Graphs are sufficiently important for computer science applications that we shall devote more than a chapter to their study. What is a graph? Well, we have been using them all along, whenever we refer to the "graphical representation" of a relation. Though they are only a descriptive aid at this point, their range of application is exceedingly broad.

EXAMPLE 0.39   The graphical representation of the relation R of Example 0.38 is shown in Figure 0.9. Note that where the relations are on a single set S, we do not duplicate the vertices. $\triangle$

How are we to compare one relation against another? Is there some notion about relations that is analogous to the containment of one set within another? The easiest way to approach this question is by considering the equivalence of $\mathscr{R}(S)$ and $\mathscr{P}(S \times S)$. Consider the relations R_1, R_2 as sets [elements of $\mathscr{P}(S \times S)$]. Then it may be that one of these is a subset of the other, say $(x, y) \in R_1 \Rightarrow (x, y) \in R_2$. We are thus led to the definition R_1 is *covered* by R_2; symbolically

$$R_1 \leq R_2 \Leftrightarrow (x \, R_1 \, y \Rightarrow x \, R_2 \, y)$$

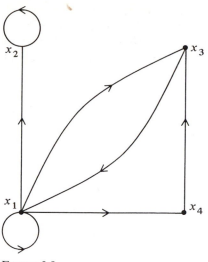

FIGURE 0.9

In the relational matrices, R_2 will have ones at each position in which R_1 has ones (and perhaps others as well). Automatically, $\leq$ inherits the properties (i), (ii'), (iii) because (Example 0.22) they are satisfied by $\subseteq$ in the algebra of sets. As a matter of fact, $\mathcal{R}(S)$ should satisfy all of the properties (P1) through (P8) listed in Section 0.2 if we make the replacements

$$\cup \leftrightarrow \vee$$
$$\cap \leftrightarrow \wedge$$
$$\sim \leftrightarrow \neg$$
$$S \leftrightarrow 1$$
$$\varnothing \leftrightarrow 0$$

Here 1 and 0 are the trivial relations in which everything is related or nothing is related; that is, in the relation matrices all the entries are ones or zeros, respectively.

To see how the covering relation is used, we return to some of the ideas considered in the preceding section on equivalence. It happens very often that a relation R is reflexive and symmetric, but not transitive. In computer sciences, reflexive and symmetric relations are called *compatibility relations*. For any relation R, however, there is an associated transitive relation $\bar{R}$ called the *transitive closure of R* such that

$$x \, \bar{R} \, y \Leftrightarrow \text{there is a sequence } x = x_1, x_2, \ldots, x_k \text{ in } S \text{ so that}$$
$$x = x_1 \, R \, x_2 \, R \cdots R \, x_k = y$$

Note   This is a kind of short-hand notation for

$$x_1 \, R \, x_2 \quad \text{and} \quad x_2 \, R \, x_3 \quad \text{and} \quad \ldots \quad \text{and} \quad x_{k-1} \, R \, x_k$$

where $x_1 = x$ and $x_k = y$. Then the dashed arrow of Figure 0.10 becomes a part of the graphical representation of $\bar{R}$. Readers should easily see that $\bar{R}$ is transitive and they should try to show that if R is a compatibility relation, then $\bar{R}$ is an equivalence relation. Why is it still reflexive and symmetric?

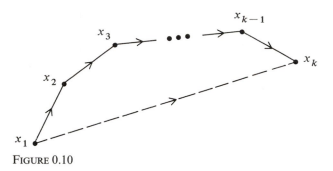

FIGURE 0.10

The transitive closure also satisfies the properties

(a) $R \le \bar{R}$
(b) $R \le Q \Rightarrow \bar{R} \le \bar{Q}$
(c) $\bar{\bar{R}} = \bar{R}$
(d) $R \le Q, Q$ transitive $\Rightarrow \bar{R} \le Q$

typical of closure operators in topology (in particular, in connection with the concept of a "closed set" of real numbers). The last property shows that $\bar{R}$ is the "smallest" transitive relation that covers R; it is a typical illustration of the usefulness of the covering relation in comparisons. It is a relation on relations!

EXAMPLE 0.40   In Example 0.38 we have $R_1 \le R_1 \vee R_2$; wherever there is a one in the relation matrix for R_1, we find a one in the same position in the matrix for $R_1 \vee R_2$. This is indication that the algebra $\mathscr{R}(S)$ will satisfy the properties listed in Example 0.10 if we make the usual replacements.   △

EXAMPLE 0.41   Consider the relation R of Example 0.38. The graphical representation of R and its transitive closure is shown in Figure 0.11. As examples of the necessity for the inclusion of the various (dashed) arrows for $\bar{R}$, we cite the following:

$$x_3 \; \bar{R} \; x_2 \quad \text{because} \quad x_3 \; R \; x_1 \; R \; x_2$$
$$x_3 \; \bar{R} \; x_3 \quad \text{because} \quad x_3 \; R \; x_1 \; R \; x_3$$
$$x_4 \; \bar{R} \; x_2 \quad \text{because} \quad x_4 \; R \; x_3 \; R \; x_1 \; R \; x_2$$

(See the note following the definition of transitive closure.)   △

EXAMPLE 0.42   Of two persons x, y, suppose we agree to say that $x \; R \; y$ if x is an acquaintance of y. Assuming that we all know ourselves, R is a reflexive

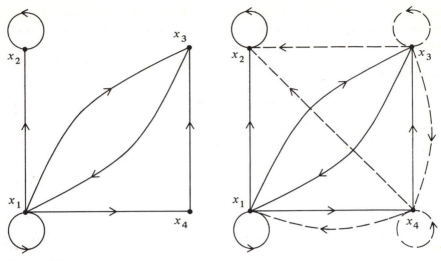

FIGURE 0.11

relation among the persons of the world, and certainly R is symmetric. On the other hand, it is surely not a transitive relation. Therefore, R is only a compatibility relation. It would be interesting to know whether its transitive closure $\bar{R} = 1$, which means that there is a chain of acquaintances between any two persons. Perhaps the existence of hermits prevents this from becoming a fact.   △

EXAMPLE 0.43   Let us see whether we can verify property (d),

$$R \le Q, Q \text{ transitive} \Rightarrow \bar{R} \le Q$$

for an arbitrary relation R. We assume that R is covered by the transitive relation Q. To see that $\bar{R}$ is also covered by Q, we only have to consider the following implications:

$$x \bar{R} y \Rightarrow x = x_1 R x_2 R \cdots R x_k = y$$
$$\Rightarrow x = x_1 Q x_2 Q \cdots Q x_k = y \Rightarrow x \bar{Q} y \Rightarrow x Q y$$

where the last implication is due to the transitivity of Q; that is, $\bar{Q} = Q$ for any transitive relation Q.   △

While we are still on the subject of relations, one final application comes to mind. In view of our characterization of equivalence relations (Section 0.5), it might be thought that the algebra of relations induces an algebra on the partitions as well. That is, if we are given the partitions π, γ on the set S, we might think of constructing the corresponding equivalence relations $r(\pi), r(\gamma)$, with which we could then perform joins and meets to obtain $r(\pi) \vee r(\gamma)$ and $r(\pi) \wedge r(\gamma)$, then transform back to partitions. However, there is a problem. The relation $r(\pi) \vee r(\gamma)$ may not be an equivalence relation, because the join of transitive relations is not transitive in general. (The meet of two transitive relations is transitive, however.) But a solution is close at hand: we can use the transitive closure!

As a result of our characterization of equivalence relations we have a pair of functions:

$$\{\text{partitions}\} \overset{r}{\underset{p}{\rightleftarrows}} \{\text{equivalence relations}\}$$

which establish a one-to-one correspondence between partitions and equivalence relations on a set S. Following the lead of the foregoing discussion, we define the *sum* and *product* of the partitions π and γ as follows:

$$\pi + \gamma = p(\overline{r(\pi) \vee r(\gamma)})$$

$$\pi \cdot \gamma = p(r(\pi) \wedge r(\gamma))$$

These operations are important in certain areas of computer science, most notably in the theory of automata.

Having more or less transformed the join and meet operations of the equivalence relations in $\mathscr{R}(S)$ into sums and products for

$$\Pi(S) = \{\pi \colon \pi \text{ is a partition on } S\}$$

might we not do the same with the comparison $\leq$ among relations? Again we can make use of our characterization result in defining

$$\pi_1 \leq \pi_2 \Leftrightarrow r(\pi_1) \leq r(\pi_2)$$

In such cases we say that π_1 is a *refinement* of π_2. If we study the meaning of all this,

$$\pi_1 \leq \pi_2 \Leftrightarrow r(\pi_1) \leq r(\pi_2)$$
$$\Leftrightarrow (xr(\pi_1)y \Rightarrow xr(\pi_2)y)$$

we find that the latter is equivalent to

$$x, y \text{ in same block of } \pi_1 \Rightarrow x, y \text{ in same block of } \pi_2$$

which means that every block of π_1 is entirely contained in some block of π_2. More than anything else, this meaning seems to justify the term "refinement." In this sense the "smallest" partition would be that in which every block is a singleton subset of S. If we denote this partition by θ, then we evidently have $\theta \leq \pi$ for every partition π.

EXAMPLE 0.44   Let

$$\pi = \{\overline{1, 2}; \overline{3}; \overline{4, 5, 6}; \overline{7}\} \qquad \pi_1 = \{\overline{1, 2}; \overline{3, 4}; \overline{5}; \overline{6, 7}\}$$
$$\gamma = \{\overline{1}; \overline{2, 3}; \overline{4}; \overline{5, 6, 7}\} \qquad \pi_2 = \{\overline{1, 2, 3, 4}; \overline{5, 6, 7}\}$$

be partitions of $S = 7^+$. Here we have introduced a convenient notation for partitions in which the inner braces are replaced by overlines and semicolons. This notation generally makes for easier reading. Now note that every block of π_1 is entirely contained in some block of π_2, for example,

$$\{1, 2\} \subseteq \{1, 2, 3, 4\} \qquad \{3, 4\} \subseteq \{1, 2, 3, 4\}$$
$$\{5\} \subseteq \{5, 6, 7\} \qquad \{6, 7\} \subseteq \{5, 6, 7\}$$

and so, $\pi_1 \leq \pi_2$, or as we say, π_1 is a refinement of π_2. Note also that π and γ are *incomparable* in that we have neither $\pi \leq \gamma$ nor $\gamma \leq \pi$. On the other hand, the partition

$$\theta = \{\overline{0}; \overline{1}; \overline{2}; \overline{3}; \overline{4}; \overline{5}; \overline{6}; \overline{7}\}$$

is a refinement of all the partitions. Among all the partitions on S, which is the "largest"?

The relational matrices for $r(\pi)$ and $r(\gamma)$ are shown in Figure 0.12. Note the *block diagonal* form, which always occurs in partitions, at least upon permuting the elements of S.

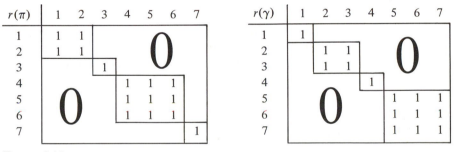

FIGURE 0.12

Now, in relational algebra joins and meets will yield the matrices shown in Figure 0.13. The transitive closure of the matrix on the left-hand side will have

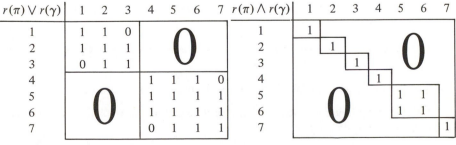

FIGURE 0.13

ones in place of the four zeros in the diagonal blocks. It follows that

$$\pi + \gamma = \{\overline{1, 2, 3}; \overline{4, 5, 6, 7}\}$$
$$\pi \cdot \gamma = \{\overline{1}; \overline{2}; \overline{3}; \overline{4}; \overline{5, 6}; \overline{7}\} \qquad \triangle$$

EXERCISES

1 Given the relations R_1, R_2, R_3 on $S = \{x_1, x_2, x_3, x_4\}$ described as follows:

R_1	x_1	x_2	x_3	x_4		R_2	x_1	x_2	x_3	x_4		R_3	x_1	x_2	x_3	x_4
x_1	0	1	0	1		x_1	1	1	0	1		x_1	1	1	1	1
x_2	1	0	1	1		x_2	1	1	0	0		x_2	1	0	1	1
x_3	1	0	1	0		x_3	0	0	1	0		x_3	1	1	0	0
x_4	0	0	1	0		x_4	0	1	0	1		x_4	1	0	0	1

determine the relation matrices corresponding to
(a) $R_1 \vee R_2$
(b) $R_1 \wedge R_2$
(c) $\neg R_3$
(d) $R_1 \vee (\neg R_3)$
(e) $(R_1 \vee R_2) \wedge R_3$
(f) $R_1 \vee (\neg R_2)$
(g) $\neg(R_1 \vee (R_2 \wedge (\neg R_3)))$
(h) $(R_2 \vee R_3) \wedge (R_1 \vee (\neg R_3))$

2 Show how the relations R_1, R_2, R_3 of Exercise 1 correspond to subsets of $S \times S$. Given that this correspondence is accomplished by a function

$$f: \mathcal{R}(S) \to \mathcal{P}(S \times S)$$

show that
(a) $f(R_1 \vee R_2) = f(R_1) \cup f(R_2)$
(b) $f(R_1 \wedge R_2) = f(R_1) \cap f(R_2)$
(c) $f(\neg R_3) = \sim f(R_3)$.

3 Show that the covering relation satisfies the Properties (i), (ii'), and (iii) mentioned in connection with the relation "less than or equal to" on the integers (Section 0.3).

4 Show that Properties (P1) through (P8) of Section 0.2 are satisfied in the algebra $\mathcal{R}(S)$ if we make the proper replacements ($\vee$ for $\cup$, $\wedge$ for $\cap$, etc.).

5 Show that the properties of Example 0.10 hold in $\mathcal{R}(S)$ after suitable replacements have been made ($\leq$ for $\subseteq$, $\vee$ for $\cup$, etc.).

6 Verify the Properties (a), (b), (c), (d) claimed for the transitive closure.

7 Determine the (relation matrix of the) transitive closure of (a) relation R_1,   (b) relation R_2,   (c) relation R_3 of Exercise 1.

8 Determine the (relation matrix of the) relation $r(\pi)$ where π is the partition
   (a) $\{\overline{1, 3}; \overline{2, 4, 7}; \overline{5, 6}\}$ (b) $\{\overline{1, 2, 5, 6}; \overline{3}; \overline{4}\}$ (c) $\{\overline{1, 3}; \overline{2, 5}; \overline{4}\}$

9 Determine the partition $p(R)$ where R is the (equivalence) relation given by the following matrix:

(a)

	x_1	x_2	x_3	x_4	x_5
x_1	1	0	1	0	0
x_2	0	1	0	0	1
x_3	1	0	1	0	0
x_4	0	0	0	1	0
x_5	0	1	0	0	1

(b)

	x_1	x_2	x_3	x_4	x_5	x_6
x_1	1	0	1	1	0	0
x_2	0	1	0	0	0	1
x_3	1	0	1	1	0	0
x_4	1	0	1	1	0	0
x_5	0	0	0	0	1	0
x_6	0	1	0	0	0	1

*10 Given a set S, show that

$$p(r(\pi)) = \pi \qquad r(p(R)) = R$$

for any partition π (respectively, any equivalence relation R). [*Hint:* Use the fact that the refinement (respectively, covering) relation satisfies condition (ii′).]

11 Compute $\pi + \gamma$ and $\pi \cdot \gamma$, given

(a) $\pi = \{\overline{1, 3}; \overline{2, 4}; \overline{5, 6}\}$ (b) $\pi = \{\overline{1, 2, 4}; \overline{3, 5, 6}\}$ (c) $\pi = \{\overline{1, 4, 5, 6}; \overline{2; 3}\}$
   $\gamma = \{\overline{1, 2}; \overline{3, 4}; \overline{5, 6}\}$ $\gamma = \{\overline{1, 3}; \overline{2; 4}; \overline{5; 6}\}$ $\gamma = \{\overline{1, 2}; \overline{3}; \overline{4, 5; 6}\}$

0.7   PROPERTIES OF FUNCTIONS

As we have seen, functions (or "mappings," as they are also called) simply transform elements of one set into elements of another. The functions r and p transform partitions into equivalence relations, and vice versa. Examples abound in the computer sciences. A compiler transforms programs written in some high-level language (for example, ALGOL or FORTRAN) into "machine language." Codes or encodings transform numbers or alphanumeric data into binary sequences. In the last case, one would probably want the encoding process to be reversible, which raises the question of the existence or nonexistence of an *inverse* for the function at hand. This concept of the inverse of a function is one that we would like to investigate in our survey of functional properties.

   Given a function $f: S \to T$ with domain S and range T, we recall that f associates each $x \in S$ with one and only one element $f(x)$, called the *image* of x, in the range T. Thus functions are by nature single valued in that

$$f(x) \neq f(y) \Rightarrow x \neq y$$

Thus we cannot have a picture such as Figure 0.14 anywhere in the representation of a function. Conversely, if the situation shown in Figure 0.15 never occurs, that is, if

$$f(x) = f(y) \Rightarrow x = y$$

then f is said to be *injective* (or in the older terminology, *one-to-one*).

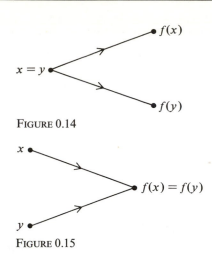

FIGURE 0.14

FIGURE 0.15

In what seems to be a different matter altogether, the function $f: S \to T$ is called *surjective* (older terminology, *onto*) if each $t \in T$ can be written as $t = f(s)$ for some $s \in S$, that is, if every element of the range is an image. We will later see that for finite sets S and T of the same cardinality, the notions of injective and surjective coincide. In general, however, a function that is both injective and surjective is called *bijective*. All along we have been using the expression *one-to-one correspondence* to describe this situation. It is probably a good idea for the student to familiarize himself with the newer terminology (injective, surjective, and bijective), since it is used quite extensively in the more theoretical areas of computer science.

EXAMPLE 0.45   It is instructive to see that even for finite sets all four possibilities in Figure 0.16 can arise.   △

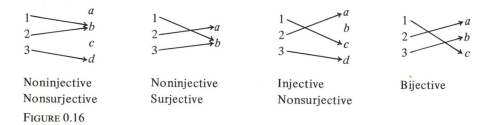

Noninjective   Noninjective   Injective   Bijective
Nonsurjective   Surjective   Nonsurjective

FIGURE 0.16

EXAMPLE 0.46   Let $A \subseteq S$. Then the *inclusion* mapping $i: A \to S$ given by

$$i(a) = a$$

for each $a \in A$ is always injective.   △

If $A = S$ in the last example, we write instead

$$1_S: S \to S$$

This particular inclusion mapping is called the *identity* function for the set S. At the same time, it is an instance of what we call a *transformation* of the set S, that is, a mapping $f: S \to S$. We use the same terminology for transformations (injective, surjective, and bijective) as for functions in general.

EXAMPLE 0.47   Consider the set

$$n^+ = \{1, 2, \ldots, n\}$$

The bijections of n^+ are called *permutations* (of the symbols $1, 2, \ldots, n$). When $n = 3$ we have $3! = 6$ permutations (Figure 0.17). As 3-permutations of the alphabet 3^+ (in the sense of Section 0.4) these permutations are denoted by 123, 213, 321, 132, 231, and 312, respectively. That is, we simply list the images of 1, 2, 3 in order.   △

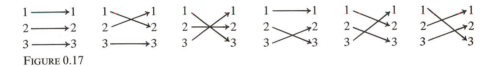
FIGURE 0.17

In order to describe the concept of invertibility, let us return briefly to the idea that a function $f: S \to T$ is really a special kind of relation $f \subseteq S \times T$: special in that

(i) for each $x \in S$ there is a pair $(x, y) \in f$
(ii) $(x, y) \in f$ and $(x, z) \in f \Rightarrow y = z$

In this context, we define the *inverse*, sometimes called the *converse*, $f^{-1} \subseteq T \times S$ (note the reversal) to be the relation

$$f^{-1} = \{(y, x): (x, y) \in f\}.$$

It is clear that we could just as well define the inverse for relations in general. In any event, we say that the original function f is *invertible* if f^{-1} is again a function. Further, as the definition of the inverse relation suggests, we have only to reverse the directions of the arrows in the graphical representation in order to describe the *inverse function* (also called f^{-1}).

EXAMPLE 0.48   Consider the function $f: S \to T$ described in Figure 0.18. As a relation $f \subseteq S \times T$ we have

$$f = \{(a, 3), (b, 1), (c, 2), (d, 3)\}$$
$$f^{-1} = \{(3, a), (1, b), (2, c), (3, d)\}$$

$$S \qquad T$$

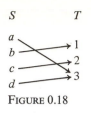

FIGURE 0.18

Now f^{-1} is not a function because condition (ii) fails; for example, when $a \neq d$ we have $(3, a) \in f^{-1}$ and $(3, d) \in f^{-1}$. It certainly appears that this failure is caused by the noninjectivity of f.   △

EXAMPLE 0.49   This time suppose that $S = \{1, 2, 3\}$, $T = \{a, b, c, d\}$, and let $f: S \to T$ have the graphical representation (from Example 0.45) shown in Figure 0.19. Then we have

$$f = \{(1, c), (2, a), (3, d)\}$$
$$f^{-1} = \{(c, 1), (a, 2), (d, 3)\}$$

Again $f^{-1} \subseteq T \times S$ is not a function, but in this case it is condition (i) that fails to hold. There exists an element, namely $b \in T$, for which there is not a pair $(b, s) \in f^{-1}$. We could have predicted this fact on the basis of the non-surjectivity of f.   △

$$S \qquad T$$

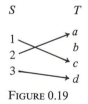

FIGURE 0.19

Reasoning as in the last two examples, we see that if $f: S \to T$ is not bijective, it is not invertible. Conversely, if f is injective and surjective and we consider the inverse relation $f^{-1} \subseteq T \times S$, then

(i) $y \in T \Rightarrow$ there is a pair $(x, y) \in f$ (f is surjective)
         $\Rightarrow$ there is a pair $(y, x) \in f^{-1}$ (definition of f^{-1})

(ii) $(z, x) \in f^{-1}, (z, y) \in f^{-1} \Rightarrow (x, z) \in f, (y, z) \in f$ (definition of f^{-1})
         $\Rightarrow f(x) = f(y)$ $(= z)$
         $\Rightarrow x = y$ (f is injective)

which show that f^{-1} is a function, that is, f is invertible. Altogether we have obtained the following important fact regarding functions $f: S \to T$.

Characterization of invertible functions

The function f is invertible iff it is bijective (injective and surjective)   □

We will see that the full meaning of invertibility comes through only in the context of functional composition. But first, it is important, as with sets and relations, to decide on a reasonable definition for equality of functions. To get some feeling for what is involved here, ask yourself why we should write $f = g$ when

$$f(x) = x^2 \sin^2 x + (2x^2 - x^2) \cos^2 x$$
$$g(x) = x^2$$

Notwithstanding the applicable trigonometric and algebraic identities, there is equality because, regardless of the real number one chooses to substitute for x, the two expressions will yield the same value.

In dealing with the more general set functions, we first require that the domains and ranges of the two functions coincide. Then if both $f, g \colon S \to T$ we regard f and g as *equal* ($f = g$) if

$$f(x) = g(x)$$

for all $x \in S$. The point is that, just as before, f and g may be defined quite differently. But if we consider that they are relations $f, g \subseteq S \times T$, then we have done nothing more than use our definition of set equality (on the subsets f and g) to decide when to write $f = g$. So all is consistent.

Now if $f \colon S \to T$ and $g \colon T \to U$, then we can define a function (call it $f \circ g$) from S to U, namely

$$(f \circ g)(x) = g(f(x)).$$

This new mapping $f \circ g \colon S \to U$ is called the *composition* of f and g. The illustration in Figure 0.20 may help to convey the idea.

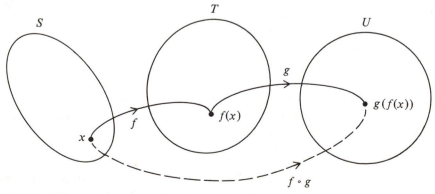

FIGURE 0.20

EXAMPLE 0.50 The notion of *composite function* in calculus fits the definition given here. Thus in the case of the function

$$h(x) = e^{\sin x}$$

where

$$f(x) = \sin x \qquad g(u) = e^u$$

we may write

$$h(x) = e^{\sin x} = e^{f(x)} = g(f(x)) = (f \circ g)(x)$$

It follows that $h = f \circ g$ according to our definition of equality of functions. △

EXAMPLE 0.51 Suppose that $f \colon S \to T$ and $g \colon T \to S$ are the functions represented in Figure 0.21. Then $f \circ g \colon S \to S$ is a transformation of S as shown in Figure 0.22. In detail, we have

$$(f \circ g)(a) = g(f(a)) = g(3) = d$$
$$(f \circ g)(b) = g(f(b)) = g(1) = c$$
$$(f \circ g)(c) = g(f(c)) = g(2) = a$$
$$(f \circ g)(d) = g(f(d)) = g(3) = d \qquad\qquad △$$

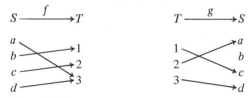

FIGURE 0.21

FIGURE 0.22

Using the idea of functional composition, we can now summarize those properties of the identity and inverse functions which tend to justify their names. If $f \colon S \to T$ is any function, then

$$f \circ 1_T = 1_S \circ f = f$$

so that the identity functions act like the number 1 in multiplication. Furthermore if f is bijective so that the inverse function $f^{-1} \colon T \to S$ exists, then

$$f \circ f^{-1} = 1_S \quad \text{and} \quad f^{-1} \circ f = 1_T$$

showing that the inverse function acts like a reciprocal with respect to composition.

EXAMPLE 0.52   If f is the permutation

of 3^+, then f^{-1} exists and is described by arrow reversal, that is,

In either order, the compositions $f \circ f^{-1}$ and $f^{-1} \circ f$ are identity functions, as shown in Figure 0.23.   △

FIGURE 0.23

EXAMPLE 0.53   Suppose we verify the property

$$f \circ f^{-1} = 1_S.$$

Then according to the definition of equality for functions, we must show that

$$(f \circ f^{-1})(x) = 1_S(x)$$

for all $x \in S$. But if $f(x) = y$, then $f^{-1}(y) = x$, and so

$$(f \circ f^{-1})(x) = f^{-1}(f(x)) = f^{-1}(y) = x = 1_S(x)$$

as required.   △

   Finally, let us consider invertibility in the special case of the function $f: S \to T$ where S and T are finite sets and $|S| = |T|$. We will find that either one of the conditions—injective or surjective—alone will suffice. Thus, if f is not injective, then there exist elements x, y such that $x \ne y$ in S and $f(x) = f(y)$. The remaining $|S| - 2$ elements have images that cannot include all the remaining $|T| - 1 = |S| - 1$ elements of T, so that f is also not surjective. Conversely, if f is not surjective, then there is some element $t \in T$ which is not an image. It follows that the images of the $|S|$ elements of S are at most $|T| - 1 = |S| -$ in number, so that two elements of S must have the same image. Hence f not injective. We have established the following principle, called *pigeonho*

principle because the last line in our argument could have been stated in the form, "One cannot put n pigeons into $n - 1$ pigeonholes without two pigeons occupying the same hole."

Pigeonhole principle

If $f: S \to T$ is a function on finite sets of the same cardinality, then f is injective iff it is surjective. □

Let us make one final observation that connects functions with the notion of an equivalence relation. Let $f: S \to T$ be any function whatsoever. If for any two elements x, y in the domain S, we write

$$x \, R \, y \Leftrightarrow f(x) = f(y)$$

then R is an obvious equivalence relation on S. It is called the *kernel* of f, abbreviated ker (f).

EXAMPLE 0.54   Let $S = \{1, 2, 3, 4, 5, 6\}$ and $T = \{a, b, c\}$. Then if $f: S \to T$ is the function represented in Figure 0.24, we have $1 \, R \, 4$, $1 \, R \, 3$, $2 \, R \, 6$, etc., for the relation $R = \ker(f)$. In the notation introduced in Example 0.44, the associated partition is $\{\overline{1, 3, 4}; \overline{2, 6}; \overline{5}\}$. △

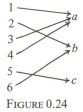

FIGURE 0.24

EXERCISES

1 Determine whether the functions shown in Figure 0.25 are injective, surjective, or bijective, and give reasons for your answers.

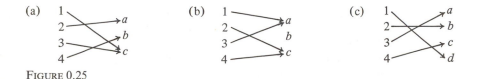

FIGURE 0.25

2 Write each function in Exercise 1 as a relation $f \subseteq S \times T$ and then determine the inverse relations $f^{-1} \subseteq T \times S$. Which of the latter are functions?

*3 How many functions are there from an n-set to an m-set? If $n \le m$, how many of these are injective? If $n \ge m$, how many are surjective? [*Hint*: See Section 0.8.]

4 Which of the following functions are injective? surjective? Justify your assertions.

(a) $f: Z \to Z$ where $f(n) = n + 1$
(b) $g: Z \to Z$ where $g(n) = 2n$
(c) $h: Z \to Z$ where $h(n) = n^2$
(d) $i: Z \to Z$ where $i(n) = an + b$ $(a, b \in Z)$

Describe the inverse function in case the given function is bijective.

5 Which of the following functions are injective? surjective? Justify your assertions.

(a) $f: Z \to B = \{0, 1\}$ where

$$f(n) = \begin{cases} 0 & \text{if } n \text{ is odd} \\ 1 & \text{if } n \text{ is even} \end{cases}$$

(b) $f: Z \times Z^+ \to Q = \{\text{rational numbers}\}$ where

$$g(n, m) = n/m$$

(c) $h: B^n \to \mathscr{P}(n^+)$ where

$$j \in h(b_1, b_2, \dots, b_n) \Leftrightarrow b_j = 1$$

(d) $i: \Pi(S) \to \mathscr{R}(S)$ where

$$x i(\pi) y \Leftrightarrow x, y \text{ together in a block of } \pi$$

[*Hint*: See Section 0.6.]   Describe the inverse functions whenever they exist.

6 Show that $f \circ g \ne g \circ f$ for the functions f and g of Exercise 4.

7 Show that the composition of two injective functions is injective. Show the same thing for surjective functions. Conclude that the composition of two permutations is again a permutation.

8 Describe the composite function of calculus

$$f(x) = \log x^2 \qquad (x \ne 0)$$

as a composition of functions. Give the domain and range of each function.

9 Determine the (associated partition of the) kernel for each of the functions shown in Figure 0.26.

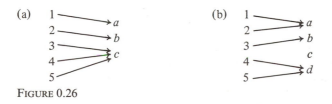

FIGURE 0.26

(c) Exercise 5(a)   (d) Exercise 5(b)
(e) Exercise 5(c)   (f) Exercise 4(b)
(g) Exercise 4(c)

*10 Consider the differentiation mapping of Exercise 3 of Section 0.5.
  (a) Show that D is not injective.
  (b) Show that D is surjective.
  (c) Use the language of partitions, kernels, equivalence classes, etc., to describe the concept of the "arbitrary constant of integration."

11 Prove that two people in Denver, Colorado, have the same initials.

12 Prove that two adults in San Francisco have the same date of birth.

0.8 ELEMENTARY COUNTING TECHNIQUES

For the design and analysis of algorithms and for other purposes as well, it is useful to have at one's disposal a bag of techniques for counting finite sets. In many instances, these tools will be necessary for determining the running time of a computer program or useful in estimating the storage requirements. So these enumeration techniques, though properly belonging to the subject called combinatorial mathematics, are becoming increasingly important in applications to computer science.

We shall illustrate in this section how the "combinatorial structure" of a set may facilitate its enumeration in the more elementary situations. Generally speaking, this structure consists in the arrangement or configuration of the elements, their common features, etc., and is ordinarily determined by the manner in which the set is formed. If one is careful to observe the symmetries inherent in the set formation, he will be in a better position to carry out the enumeration. This is particularly true with the advanced counting techniques to be presented in Chapter 5; here it is true only to a lesser degree.

We shall state our counting techniques as if they were "rules" of calculation, though we will provide brief justifications for these rules as we go along. Our first two rules have to do with set composition as discussed in Sections 0.2 and 0.3. The next four give the necessary formulas for counting samples and selections, permutations and combinations. Finally, there is a recursion formula for counting partitions.

Rule 1   (Rule of the product)

$$|A \times B| = |A| \cdot |B|$$

Justification There are $|A|$ ways to choose the first coordinate of each $(a, b) \in A \times B$; the choice of b may be made independently, there being $|B|$ choices for each a.   □

Corollary   For any finite set A,

$$|A^n| = |A|^n$$ □

EXAMPLE 0.55   In Example 0.17, where $S = \{a, b, c\}$ and $T = \{1, 2\}$, we should expect $2 \cdot 3 = 6$ elements in the product set $S \times T$. Of course, that is exactly the way it turned out.   △

EXAMPLE 0.56   In Example 0.18 we introduced the product set

$$B^n = \{(b_1, b_2, \ldots, b_n): b_i \in B\}$$

where $B = \{0, 1\}$. According to the corollary, we should have

$$|B^n| = |B|^n = 2^n$$

Thus when $n = 3$, the set B^n should have $2^3 = 8$ elements. Indeed, we were able to list just 8 elements for B^3 in Example 0.18.   △

Rule 2   (Rule of the sum)

$$|A \cup B| = |A| + |B| - |A \cap B|$$

Justification   When the number of elements in A and the number of elements in B are added together, the elements of the intersection are counted twice. So we must compensate by subtracting $|A \cap B|$.   □

A similar argument will establish the following more general result. The reader may want to refer to Figure 0.27 to see the situation for $n = 3$. This figure indicates the number of tallies, in particular for the overlapping areas, when adding the number of elements in each of the A_i.

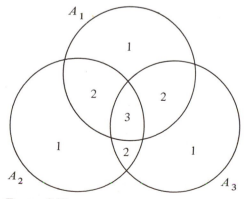

FIGURE 0.27

Corollary   (Principle of inclusion-exclusion)

$$\left| \bigcup_{i=1}^{n} A_i \right| = \sum_{i=1}^{n} |A_i| - \sum_{i<j} |A_i \cap A_j| + \sum_{i<j<k} |A_i \cap A_j \cap A_k|$$

$$+ \cdots + (-1)^{n-1} \left| \bigcap_{i=1}^{n} A_i \right| \qquad \square$$

EXAMPLE 0.57   There are ten listed faculty members in the mathematics department and seven in computer science at a certain university. On closer inspection, however, we find that three persons have joint appointments and are listed by both departments. Using the rule of the sum, we conclude that the number of persons attending a joint departmental faculty meeting should be $10 + 7 - 3 = 14$. $\triangle$

EXAMPLE 0.58   How many integers from 1 to 100 are evenly divisible by 3 and also by 5, but not by 6? Suppose we let

$$A_1 = \{n: 1 \le n \le 100, n \text{ is divisible by } 3\}$$

$$A_2 = \{n: 1 \le n \le 100, n \text{ is divisible by } 5\}$$

$$A_3 = \{n: 1 \le n \le 100, n \text{ is not divisible by } 6\}$$

The inclusion-exclusion formula for $n = 3$ becomes

$$|A_1 \cup A_2 \cup A_3| = (|A_1| + |A_2| + |A_3|)$$
$$- (|A_1 \cap A_2| + |A_1 \cap A_3| + |A_2 \cap A_3|) + |A_1 \cap A_2 \cap A_3|$$

If an integer is not divisible by 3 (not in A_1), then it is not divisible by 6 either (hence in A_3). It follows that the expression on the left-hand side of the equality is 100. On the other hand, the expression $|A_1 \cap A_2 \cap A_3|$ is our unknown. Simple calculations then show that we obtain

$$100 = (33 + 20 + 84) - (6 + 17 + 17) + |A_1 \cap A_2 \cap A_3|$$

and finally, $|A_1 \cap A_2 \cap A_3| = 100 - 137 + 40 = 3$. $\triangle$

The next four rules will provide formulas for counting the number of samples and selections, permutations and combinations, as introduced in Section 0.4. Most of these are best expressed in terms of the *factorial notation*

$$n! = n(n - 1) \cdots (2)(1)$$

with which the reader is no doubt already familiar.

Rule 3

The number of k-samples of an n-set is n^k.

Justification   Recall that the k-samples from the alphabet Σ are simply the elements of the set

$$\Sigma^k = \{x: x \text{ is a word of length } k \text{ on } \Sigma\}$$

and except for difference in notation, this is the same as the product Σ^k. Our rule now follows from the corollary to the rule of the product.   □

Rule 4

The number of k-permutations of an n-set is

$$P(n, k) = \frac{n!}{(n - k)!}$$

Justification   Observe that for each of the n choices for the first symbol σ_{i_1} in the k-letter word

$$x = \sigma_{i_1}\sigma_{i_2} \cdots \sigma_{i_k}$$

we have only $n - 1$ choices for the second symbol, etc., because we are not permitted to have repetitions. It follows that

$$P(n, k) = n(n - 1) \cdots (n - k + 1) = \frac{n!}{(n - k)!}$$

as claimed.   □

Corollary

There are $n!$ permutations of an n-set.   □

EXAMPLE 0.59   If $n = 4$ and $k = 3$, as in Example 0.32, then

$$n^k = 4^3 = 64$$

$$P(n, k) = \frac{n!}{(n - k)!} = \frac{4!}{(4 - 3)!} = 24$$

in agreement with the number of 3-samples and 3-permutations listed for the alphabet $\Sigma = \{a, b, c, d\}$ in Example 0.32.   △

EXAMPLE 0.60   The number of four-letter "words" on the English alphabet (Example 0.25) is $26^4 = 456,976$.   △

EXAMPLE 0.61   The number of possible batting orders for a twelve-member baseball team is $P(12, 9) = 12!/3! = 79,833,600$.   △

EXAMPLE 0.62   A FORTRAN identifier (for variables, arrays, functions, etc.) is a string of from one to six characters (Example 0.27). The first character must be alphabetic, but the others may be alphanumeric. Using Rule 1 and Rule 3, we find that the number of FORTRAN identifiers is

$$26 + 26 \cdot 36 + 26 \cdot 36^2 + 26 \cdot 36^3 + 26 \cdot 36^4 + 26 \cdot 36^5 = 1,617,038,306.$$ △

Rule 5

The number of k-combinations of an n-set is

$$C(n, k) = \frac{n!}{k!(n - k)!}$$

Justification   According to the Corollary to Rule 4, there are $k!$ permutations of each k-combination. Since we obtain all of the k-permutations exactly once in this way, we may simply divide $P(n, k)$ by $k!$ to obtain

$$C(n, k) = \frac{P(n, k)}{k!} = \frac{n!}{k!(n - k)!}. \qquad \Box$$

These numbers $C(n, k)$ are also to be denoted by $\binom{n}{k}$ and are often called *binomial coefficients*. This terminology derives from the binomial theorem of elementary algebra for multiplying $(a + b)^n$. Since $(a + b)^n$ means

$$(a + b)(a + b) \cdots (a + b)$$

with n factors, the coefficient of a particular term $a^{n-k}b^k$ in the product is easily determined. It is the number of ways of choosing k of the n parentheses from which to take b's. But since this is the number $C(n, k)$ or $\binom{n}{k}$, we have

$$(a + b)^n = a^n + \binom{n}{1} a^{n-1} b + \binom{n}{2} a^{n-2} b^2 + \cdots + b^n$$

We have yet to count the number of k-selections of an n-set. At first glance, it might be thought that we have only to divide n^k (the number of k-samples) by an appropriate factor. Thus a k-selection could be considered an equivalence class of k-samples, relative to an equivalence relation that simply identifies those words differing only in the order of occurrence of the symbols. These equivalence classes, however, have different sizes. So the problem is not so easy as one might think.

EXAMPLE 0.63   Consider the sixty-four 3-samples of $\Sigma = \{a, b, c, d\}$ listed in Example 0.32. According to the equivalence relation just described, aab is equivalent to three samples (including itself)

$$\text{aab} \sim \text{aba} \sim \text{baa}$$

whereas abc is equivalent to six samples and aaa is equivalent only to itself.   △

Rule 6

The number of k-selections of an n-set is $\binom{n+k-1}{k}$.

Justification   We shall establish a one-to-one correspondence (bijection) between the k-selections of n^+ and the k-element subsets of $(n + k - 1)^+$. If this can be done, the reader will see that the result follows easily from Rule 5.

Suppose $i_1 i_2 \cdots i_k$ is a k-selection from n^+. We may assume that $1 \leq i_1 \leq i_2 \leq \cdots \leq i_k \leq n$ and we set

$$f(i_1 i_2 \cdots i_k) = \{i_1 + 0, i_2 + 1, \ldots, i_k + k - 1\}$$

with the goal of showing that this mapping is bijective. The following string of implications shows that f is injective:

$$f(i_1 i_2 \cdots i_k) = f(j_1 j_2 \cdots j_k) \Rightarrow \begin{aligned} &\{i_1 + 0, i_2 + 1, \ldots, i_k + k - 1\} \\ &= \{j_1 + 0, j_2 + 1, \ldots, j_k + k - 1\} \end{aligned}$$

$$\Rightarrow i_1 = j_1, \quad i_2 = j_2, \ldots, i_k = j_k$$

$$\Rightarrow i_1 i_2 \cdots i_k = j_1 j_2 \cdots j_k.$$

Now if $\{n_1, n_2, \ldots, n_k\}$ is a k-subset of $(n - k + 1)^+$ with its elements listed in ascending order, then the k-selection $n_1 \, n_2 - 1 \cdots n_k - k + 1$ has the property

$$f(n_1 \, n_2 - 1 \cdots n_k - k + 1) = \{n_1, n_2, \ldots, n_k\}$$

showing that f is surjective as well. The student should show that $n_1, n_2 - 1, \ldots, n_k - k + 1$ are all in n^+, since this fact is crucial to the argument.   □

EXAMPLE 0.64   The number of distinct poker hands is

$$C(52, 5) = \frac{52!}{5! \, 47!} = 2{,}598{,}960. \qquad \triangle$$

EXAMPLE 0.65   In Example 0.34 we listed twenty 3-selections for the alphabet $\Sigma = \{a, b, c, d\}$ $(n = 4)$. This result agrees with the number given by Rule 6,

$$\binom{n + k - 1}{k} = \binom{4 + 3 - 1}{3} = \binom{6}{3} = \frac{6!}{3! \, 3!} = 20. \qquad \triangle$$

EXAMPLE 0.66   A throw of a set of k dice may be viewed as a k-selection from a 6-set. It follows from Rule 6 that the number of distinct throws is $\binom{5+k}{k}$.   △

The reader is perhaps familiar with the "Pascal triangle" and its use in computing the binomial coefficients $C(n, k)$. Suppose we think of $C(n, k)$ as the number of k-element subsets A of an n-set, say n^+. Then consider the two mutually exclusive cases: $n \in A$, $n \notin A$. In the first case, $A = B \cup \{n\}$, where B is a $(k - 1)$-subset of $(n - 1)^+$. In the second case, A is a k-subset of $(n - 1)^+$. If we account for both possibilities in a summation, then we arrive at the *recurrence equation* for the binomial coefficients:

$$C(n, k) = C(n - 1, k - 1) + C(n - 1, k)$$

A tabulation of these coefficients, beginning with the boundary conditions

$$C(n, 0) = C(n, n) = 1$$

gives rise to the *Pascal triangle* (Figure 0.28).

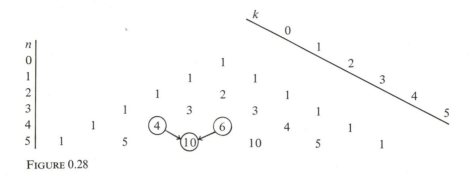

FIGURE 0.28

EXAMPLE 0.67   Assuming that the Pascal triangle has already been completed through $n = 4$, we may compute

$$C(5, 2) = C(4, 1) + C(4, 2) = 4 + 6 = 10$$

(see Figure 0.28). Of course, the same result could be obtained directly from the formula of Rule 5. △

Recurrence equations occupy an important position in combinatorial investigations. They are especially valuable when a closed formula is not available for the enumeration, such as in the case of counting the k-partitions of an n-set. When we can give a recurrence equation, analogous to the equation for the binomial coefficients, we can make successive determinations of as many of the terms $S(n, k)$ as desired.

Rule 7

The number of k-partitions of an n-set satisfies the recurrence equation:

$$S(n, k) = S(n - 1, k - 1) + kS(n - 1, k)$$

Justification   The argument is much like that given in obtaining the recurrence equation for the binomial coefficients. A k-partition of n^+ can arise in two exclusive ways. First of all, we may add a singleton block $\{n\}$ to a $(k - 1)$-partition of $(n - 1)^+$. Secondly, we may augment any existing block of a k-partition of $(n - 1)^+$ by adjoining n. But since this can be done in k ways, the above equation results. □

For any n, there is only one partition with one block and only one partition with n blocks. We are thus led to write

$$S(n, 1) = S(n, n) = 1$$

Together with the recurrence equation, these equalities allow for the construction of the *Stirling triangle* (Figure 0.29). The entries $S(n, k)$ are called *Stirling numbers* (of the second kind).

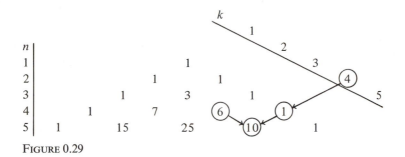

FIGURE 0.29

EXAMPLE 0.68   Supposing that the Stirling triangle has been determined through $n = 4$, we can compute

$$S(5, 4) = S(4, 3) + 4S(4, 4) = 6 + 4 \cdot 1 = 10$$

(see Figure 0.29). This result agrees with the following list:

$$\{\overline{1, 5}; \overline{2}; \overline{3}; \overline{4}\} \qquad \{\overline{1, 3}; \overline{2}; \overline{4}; \overline{5}\}$$
$$\{\overline{1}; \overline{2, 5}; \overline{3}; \overline{4}\} \qquad \{\overline{1, 4}; \overline{2}; \overline{3}; \overline{5}\}$$
$$\{\overline{1}; \overline{2}; \overline{3, 5}; \overline{4}\} \qquad \{\overline{1}; \overline{2, 3}; \overline{4}; \overline{5}\}$$
$$\{\overline{1}; \overline{2}; \overline{3}; \overline{4, 5}\} \qquad \{\overline{1}; \overline{2, 4}; \overline{3}; \overline{5}\}$$
$$\{\overline{1}; \overline{2}; \overline{3}; \overline{4}; \overline{5}\} \qquad \{\overline{1}; \overline{2}; \overline{3, 4}; \overline{5}\} \qquad \triangle$$

EXERCISES

1 A survey of 100 students in a certain dormitory reveals that:

47 students subscribe to *Playboy*
44 students subscribe to *Newsweek*
32 students subscribe to *Time*
11 students subscribe to *Playboy* and *Newsweek*
12 students subscribe to *Playboy* and *Time*
12 students subscribe to *Newsweek* and *Time*
 3 students subscribe to all three magazines

How many of the students subscribe to none of these magazines?

2 Two classes, one in mathematics and one in computer science, meet at successive hours. There are 24 students in the mathematics course and 27 in the computer science course. How many students are represented here if 8 students are enrolled in both courses?

3 Compute the number of possible initials (first name, last name) of a person's name.

4 Compute the number of possible social security numbers.

5 At a hamburger stand, it is observed that:

7% of the patrons use no condiments whatsoever
58% use mustard
58% use ketchup
47% use onions
29% use both ketchup and mustard
30% use mustard and onions
25% use ketchup and onions

What percentage of the patrons order hamburgers "with everything"?

6 Compute the number of
   (a) Four-digit decimal numbers
   (b) Five-bit binary numbers
   (c) Bridge hands
   (d) Baseball lineups for a 15 member team
   (e) Coin toss sequences (with 100 tosses) measured only as to the number of heads and the number of tails
   (f) Arrangements of 20 books on a bookshelf
   (g) Punched IBM cards
   (h) Ten-city itineraries (from among the 50 largest cities in the United States)
   (i) Fingernail colorings from 16 shades of nail polish
   (j) Solutions to a 20-question multiple choice (A, B, C, D, or E) examination

7 A quarterback has a repertoire of 12 plays and runs 50 plays in the course of a game. In the postgame analysis, the coach may be interested in the frequency distribution of the play-calling showing how many times each of the various plays were called. How many such "distributions" are there?

8 A computer program makes exactly seven procedure calls chosen from three different procedures. How many ways are there of calling the procedures without regard to the order?

9 Compute the number of possible outcomes for three successive throws of three dice when the dice are indistinguishable but the throws are not.

10 Compute the number of Colorado license plates (two letters followed by four digits) that could be issued.

11 Compute one additional row of the
   (a) Pascal triangle (Figure 0.28)
   (b) Stirling triangle (Figure 0.29)

12 Compute the number of partitions of a 6-set.

13 Prove the following identities involving the binomial coefficients:

(a) $\sum_{k=0}^{n} \binom{n}{k} = 2^n$ (b) $\sum_{k=0}^{n} (-1)^k \binom{n}{k} = 0$

[*Hint*: Choose particular values for a and b in the binomial theorem.]

14 Obtain a formula for the number of relations on an n-set. Of these, how many are
(a) reflexive? (b) symmetric? (c) irreflexive? (d) antisymmetric?

15 Compute the following:
   (a) $C(7, 4)$ (b) $C(8, 5)$
   (c) $C(9, 7)$ (d) $S(6, 3)$
   (e) $S(6, 5)$ (f) $S(7, 5)$

16 Prove that there are $\binom{n+k-1}{k}$ terms in the multinomial expansion $(x_1 + \cdots + x_n)^k$.

17 Give an alternative proof for

$$\binom{n}{k} = \binom{n-1}{k-1} + \binom{n-1}{k}$$

[*Hint*: Express the summands over a common denominator.]

0.9 COUNTABLE SETS

In Section 0.8 we discussed the most elementary techniques for counting finite sets. We recall that two finite sets are said to be of the same size or cardinality if their elements can be paired in a one-to-one correspondence. It is customary to extend this definition to infinite sets, particularly to the countably infinite sets. According to the discussion in the Introduction, which the student may want to reread, the set Z^+ is taken as the "prototype" for what is meant by a countable (that is, countably infinite) set. Let C be any infinite set. Then C is a countable set if and only if a one-to-one correspondence exists between C and the set Z^+.

EXAMPLE 0.69   As our set C suppose we consider in turn

$$C = Z \qquad C = Q(0, 1) \qquad C = B^*$$

where $Q(0, 1)$ is the set of all rational numbers between but excluding zero and one, and B^* is the set of all "words" on the alphabet $B = \{0, 1\}$. Since we are able to "list" the members in each case, as indicated by the following mappings:

$$
\begin{array}{lll}
1 \xrightarrow{f} 0 & 1 \xrightarrow{g} \frac{1}{2} & 1 \xrightarrow{h} \epsilon \\
2 \xrightarrow{f} 1 & 2 \xrightarrow{g} \frac{1}{3} & 2 \xrightarrow{h} 0 \\
3 \xrightarrow{f} -1 & 3 \xrightarrow{g} \frac{2}{3} & 3 \xrightarrow{h} 1 \\
4 \xrightarrow{f} 2 & 4 \xrightarrow{g} \frac{1}{4} & 4 \xrightarrow{h} 00 \\
5 \xrightarrow{f} -2 & 5 \xrightarrow{g} \frac{3}{4} & 5 \xrightarrow{h} 01
\end{array}
$$

$$6 \xrightarrow{f} 3 \qquad 6 \xrightarrow{g} \tfrac{1}{5} \qquad 6 \xrightarrow{h} 10$$
$$7 \xrightarrow{f} -3 \qquad 7 \xrightarrow{g} \tfrac{2}{5} \qquad 7 \xrightarrow{h} 11$$
$$8 \xrightarrow{f} 4 \qquad 8 \xrightarrow{g} \tfrac{3}{5} \qquad 8 \xrightarrow{h} 000$$
$$\vdots \qquad\qquad \vdots \qquad\qquad \vdots$$

the reader must certainly feel that Z, $Q(0, 1)$, and B^* are all countable. Without a bit of "hand-waving," however, it may yet be difficult to definitely establish that such mappings indeed represent one-to-one correspondences (that is, bijective functions), because at this stage we may be at somewhat of a loss to provide an explicit definition of the respective functions. Although in one case, the mapping $f: Z^+ \to Z$, we can at least write

$$f(n) = \begin{cases} \tfrac{1}{2}n & (n \text{ even}) \\ \tfrac{1}{2}(1 - n) & (n \text{ odd}) \end{cases}$$

even this description may prove to be somewhat awkward for establishing a bijection. Compare this with our next example! △

EXAMPLE 0.70   In the case of the function $F(n) = 2n$ from Z^+ to the set of all positive even integers, we have an easier time. Using the explicit definition of the function, we are able to deduce that

$$F(n) = F(m) \Rightarrow 2n = 2m \Rightarrow n = m$$

showing that F is injective. Then, since we can express each positive even integer p in the form $p = 2n$, where $n \in Z^+$, we have

$$p = 2n = F(n)$$

showing that F is surjective as well. This definitely establishes the countability of the set of positive even integers. △

It seems that we are not able to proceed in such a straightforward manner in the case of the sets Z, $Q(0, 1)$, and B^* as discussed in Example 0.69. Fortunately, however, there are certain general principles that apply to these situations and others as well. The first of these principles is intuitively clear, so we leave the justification as an exercise for the student.

First principle of countability

A subset of a countable set is countable. □

Second principle of countability

The set product of two countable sets is countable.

Justification   Using ideas to be developed in Section 1.1, the student will be able to extend this second principle to the case of an arbitrary finite number of factors. For the time being, however, we shall consider only two factors.

With only a slight loss of generality, whose ramifications should be examined by the reader, we will suppose that each factor is Z^+ itself. That is, we will show that $Z^+ \times Z^+$ is countable. For this purpose we define a function

$$f: Z^+ \times Z^+ \to Z^+$$

by writing

$$f(n, m) = 2^n 3^m$$

Surely this establishes a one-to-one correspondence between $Z^+ \times Z^+$ and a subset of Z^+ (why?) and we may therefore use the first principle of countability to conclude that $Z^+ \times Z^+$ is countable. □

Finally and perhaps most importantly, we are able to state the following:

Third principle of countability

A countable union of countable sets is countable.

Justification   A simple diagram is useful in support of this assertion. Supposing that we are given countably many sets $S_1, S_2, \ldots, S_n, \ldots$ each of which is countable. We may arrange their elements in an infinite array:

$$
\begin{array}{llll}
S_1: & x_{11} & x_{12} & \cdots & x_{1n} & \cdots \\
S_2: & x_{21} & x_{22} & \cdots & x_{2n} & \cdots \\
\vdots & \vdots & \vdots & & \vdots \\
S_n: & x_{n1} & x_{n2} & \cdots & x_{nn} & \cdots \\
& \vdots & \vdots & & \vdots
\end{array}
$$

(Some of these rows may terminate after finitely many terms.) Allowing for such terminations (in columns as well as rows) and the possible duplications, we see from the subscripts alone that $S = \bigcup_n S_n$ is in one-to-one correspondence with a subset of $Z^+ \times Z^+$. Then from both of the preceding principles, the third principle follows. □

EXAMPLE 0.71   Now the way is clear for us to return to the problem of establishing the countability of Z, $Q(0, 1)$, and B^* (Example 0.69). We may write

$$Z = N \cup Z^-$$

where Z^- is the set of negative integers. Then two obvious bijective functions can be given:

$$f_1(n) = n + 1 \qquad f_2(n) = -n$$

from N and Z^- to Z^+, respectively. Thus N and Z^- are countable, and the countability of Z follows from our third principle. In the case of $Q(0, 1)$, we have only to observe that (except for notation)

$$Q(0, 1) \subseteq Z^+ \times Z^+$$

and the first two principles apply. Finally, we recall from Section 0.4 that

$$B^* = \{\epsilon\} \cup B \cup B^2 \cup \cdots$$

where each B^k is finite, hence countable. Again by the third principle, B^* is countable. $\triangle$

EXERCISES

1 Show that the set of all rational numbers is countable.

*2 Justify or prove the first principle of countability. [*Hint*: In the case of an infinite subset S of a countably infinite set

$$C = \{c_1, c_2, \dots\}$$

let $i(1)$ be the smallest positive integer i such that $c_i \in S$. Then assuming that $i(1)$, $i(2), \dots, i(n-1)$ have been defined, let $i(n)$ be the smallest positive integer $i > i(n-1)$ such that $c_i \in S$. Set

$$f(n) = c_{i(n)}$$

and show that $f: Z^+ \to S$ is bijective.]

3 Show that the intersection of two countable sets is countable.

4 Show that with each fixed integer $n \geq 1$, the set of all n-tuples with integer coordinates is countable.

5 Show that the set of all words on the alphabet $B = \{0, 1\}$ that end in a zero, is countable.

*6 Show that the set of all infinite sequences whose terms are zero or one, is uncountable. [*Hint*: See the Introduction.]

7 Show that if A is countable, then A^n is countable for each integer $n \geq 1$.

8 Supply the missing arguments in the justification of the second principle of countability.

9 Show that the set of all intervals of the real line having rational endpoints is countable.

10 Show that the set of all finite sequences with integer terms is countable.

11 Show that the set of all prime numbers is countable.

0.10   THEOREMS AND PROOFS

We have intended that this preliminary chapter be kept as informal as possible. The same objective applies to the text as a whole. Nevertheless, it is unavoidable, as we introduce one mathematical system after another, that we gradually begin to account for our statements according to accepted forms of logical reasoning. In order that we should usually know what we are talking about, we try to use only a very few undefined terms. In geometry the terms "line" and "point" go undefined. In this chapter the terms "set" and "element" remain undefined. We make no apologies. In fact, we cannot avoid the use of such undefined terms altogether without falling into a trap of circularity in reasoning.

In this text new terms being formally (or even informally) defined are always italicized. A *definition*, in the mathematical sense, is simply an agreement to regard one word or phrase as having the same meaning as another, usually more complex phrase. Ultimately, these definitions depend on the sense of the undefined terms. It is no wonder that the philosopher and mathematician Bertrand Russell once described mathematics as "a subject in which we never know what we are talking about, nor whether what we are saying is true." Fortunately, we do not need to dwell on this aspect of the subject, because the theories we develop are rather more concerned with the logical relationships between statements than with the abstract philosophical meaning of an isolated statement.

A similar developmental scheme can be described for the theories of the various mathematical systems. It is necessary that certain statements not be subject to proof. We may prove Theorem N by showing that it follows logically from Theorem $N - 1$, but we cannot continue such reasoning *ad infinitum*. Sooner or later some statement must be taken without proof. Otherwise, we are again in danger of circularity in our reasoning. Thus the whole structure of any theory must rest on certain unproved statements, called *axioms* or *postulates*.

EXAMPLE 0.72   Considering the positive integers as an algebra with operations of addition and multiplication, one will often see the following properties taken as axioms:

identity $\qquad\qquad n \cdot 1 = n$

associativity $\quad n + (m + p) = (n + m) + p \qquad n \cdot (m \cdot p) = (n \cdot m) \cdot p$

commutativity $\qquad n + m = m + n \qquad\qquad n \cdot m = m \cdot n$

distributivity $\quad n \cdot (m + p) = n \cdot m + n \cdot p$

for all $n, m, p \in Z^+$. Actually, these axioms may be proved by means of the so-called principle of mathematical induction (Section 1.1), given suitable inductive definitions for the operations of addition and multiplication, but we have to stop somewhere. $\triangle$

Just what is the nature of the theorems that one encounters in this text, or elsewhere for that matter? Can they be classified somehow? As to their content, the answer is no, at least not easily. Of course, some of our theorems will be called *theorems*. Others will be called *lemmas*. Still others will be called *corollaries*. What is the difference? None, really, except that lemmas are usually rather technical assertions, often not important in themselves but useful as a stepping stone to proving a "theorem." And the name "corollary" is generally reserved for statements that are immediate logical consequences of some "theorem," ordinarily one that has just been proven. Sometimes, as in the principle of inclusion and exclusion (Section 0.8), we extend the usage of the term "corol-lary" to include results that are not direct logical consequences of a preceding result, but that can be derived along similar lines of reasoning.

As to the logical framework of theorems, there is a classification that can be described which may be helpful to the student. Many theorems have the *implication* structure: "if P is true, then Q is true" (or $P \Rightarrow Q$, as it is usually abbreviated) for statements P and Q.

EXAMPLE 0.73   A typical statement of this type in Z^+ is the following:

If n is prime and $n \neq 2$, then n is odd.

The student, however, must be able to detect such a logical structure in disguise, as for example, in "Every prime number except 2 is odd." In any case, P and Q are the propositions

$$P(n) = \langle (n \text{ is prime}) \wedge (n \neq 2) \rangle$$
$$Q(n) = \langle n \text{ is odd} \rangle$$

Note that the "theorem" does not assert that P (or Q) is true, nor that it is false; rather, it says something about the relationship between the simultaneous truth values of P and Q. △

In a *direct* proof of a theorem $P \Rightarrow Q$, one shows that Q is true whenever P is true. Accordingly, the method or technique is to assume that P is true, then to see whether one can thereby deduce Q as a consequence. Nothing need to be established about Q in case P is false. Direct proof, however, is only one possibility. Sometimes it is more convenient to prove the equivalent *contra-positive* statement: $\neg Q \Rightarrow \neg P$ (not $Q \Rightarrow$ not P).

EXAMPLE 0.74   Suppose that we want to give a contrapositive proof of the "theorem" of Example 0.73, that is, show that

n even $\Rightarrow n$ is composite or $n = 2$

As explained before, we suppose that n is even, say

$$n = 2m$$

for some integer $m \in Z^+$. Then there are two cases: either $m = 1$ or $m \neq 1$. We examine these separately:

$$m = 1 \Rightarrow n = 2 \cdot 1 = 2$$
$$m \neq 1 \Rightarrow n = 2m \quad \text{is composite (not prime)}$$

and arrive at the desired conclusion. △

EXAMPLE 0.75   Lemma 0.2 of Section 0.5 had the form $P \Rightarrow Q$. In a direct proof of this implication, one would assume that $[a] \cap [b] \neq \varnothing$. What, however, does it mean to say that a set is not empty? It means that there is some element in the set. So we may write $x \in [a] \cap [b]$. This in turn means that $x \in [a]$ and $x \in [b]$. Hereon the reader should be able to use the definition of equivalence classes and the preceding result (Lemma 0.1) to reach the desired conclusion $[a] = [b]$.   △

Understandably, the student will ask the question: How do I choose the next step in my attempted proof of a theorem, supposing that there are several legitimate steps that could be taken? Lewis Carroll gave a delightful if somewhat negative answer in *Alice in Wonderland.*

> Alice: "Would you tell me, please, which way I ought to go from here?"
> Cheshire Cat: "That depends a good deal on where you want to get to."
> Alice: "I don't much care where."
> Cheshire Cat: "Then it doesn't matter which way you go."

One should heed the advice of the Cheshire Cat in proving an implication $P \Rightarrow Q$. Though we cannot assume Q, we look at Q in order to see which way we would like to go in the midst of our proof.

Occasionally we resort to a proof of $P \Rightarrow Q$ *by contradiction*. We assume P, as in a direct proof, but suppose as well that Q is false. If this should lead to a contradiction (for example, $1 = 2$) then the implication $P \Rightarrow Q$ is considered proved. The reason is that we have shown that it is not possible, without invalidating some known truth, to have P true and Q false. Thus whenever P is true, Q must also be true, as we sought to establish.

EXAMPLE 0.76   Consider once more the implication $P \Rightarrow Q$ of Example 0.73:

$$P(n) = \langle (n \text{ is prime}) \wedge (n \neq 2) \rangle$$
$$Q(n) = \langle n \text{ is odd} \rangle$$

Assuming that P is true and Q is false, so n is even, we may write

$$n = 2m$$

Since n is prime, we must have $m = 1$ and $n = 2 \cdot 1 = 2$. Thus

$$n = 2 \quad \text{and} \quad n \neq 2$$

which is a contradiction ($2 \neq 2$!)   △

Other theorems we may encounter have the logical *equivalence* structure: "P is true if and only if Q is true" (or $P \Leftrightarrow Q$, or P iff Q, however we may choose to abbreviate). These statements simply combine the two implications $P \Rightarrow Q$ and $Q \Rightarrow P$. So there is really nothing new here except to point out the several alternative available methods of proof. Thus we may choose to prove that $P \Rightarrow Q$ directly, but to give a contrapositive proof $\neg P \Rightarrow \neg Q$ for the other implication.

EXAMPLE 0.77   In our proof of the pigeonhole principle (Section 0.7), both of the implications

$$f \text{ is injective} \quad \text{iff} \quad f \text{ is surjective}$$

were demonstrated in the contrapositive form.   △

Sometimes in a theorem expressing a logical equivalence $P \Leftrightarrow Q$, the statement Q is a "condition" or a set of conditions that we say are *necessary and sufficient* in order that P hold. In speaking of the separate implications, we will say that $P \Rightarrow Q$ establishes the *necessity* of the condition(s), whereas $Q \Rightarrow P$ establishes the *sufficiency* of the condition(s). Of course, this is more a matter of terminology than anything else.

Occasionally, we see theorems in the form of a *multiple equivalence*: "The following n statements are logically equivalent: $P_1, P_2, \ldots, P_n$." Such a statement is only an abbreviation for the statement that if any one of the statements is true, so are the $n - 1$ others. According to Rule 5 of Section 0.8, the proof of such a theorem would then seem to require a verification of $2C(n, 2) = n(n - 1)$ implications. Actually, what one tries to do instead is to prove some cycle of n implications, say

$$P_1 \Rightarrow P_2 \Rightarrow \cdots \Rightarrow P_n \Rightarrow P_1$$

or some permutation of this cycle. A little reflection will convince the student that such verifications will indeed suffice.

In this brief summary we have tried to indicate how the proof that one attempts for a given theorem is generally dictated by the logical form of the statement of the theorem. Nowhere is this more important to realize than in the case of theorems of the form:

"For all positive integers n, $P(n)$ is true."

We devote the first section of Chapter 1 to the method of mathematical induction, so often found suitable for the proof of such statements. It is a most powerful proof technique whose importance could not be overemphasized.

The reader must bear in mind that our entire discussion here has been carried on on an informal level so far as the logical notions are concerned. Nevertheless, even for those whose background in logic is not substantial, it should provide sufficient orientation for our subsequent development. Any second thoughts that a student may have will almost certainly be resolved when we return to the questions of logic in Chapter 6.

Throughout all of this discussion one question is likely to have remained unanswered in the student's mind: How does one arrive at the statement of a theorem? That is, what causes the mathematician to try to prove a certain statement in the first place? Why does a person expect a given assertion to be true? This is not an easy question to answer, but some attempts at clarification might be helpful. First of all, the experienced mathematician does not begin willy nilly from some arbitrary axiomatic structure and try to establish an avalanche of theorems. As the nineteenth-century French mathematician Henri Poincaré said, "A collection of facts is no more a science than a heap of stones is a house." For one thing, the important axiomatic schemes are not set down in a vacuum, so to speak. Though they may· be the logical beginning of the study of a theory, they are not the beginning in any chronological sense. What happens, instead, is that the axioms evolve as a means for explaining our observations of some limited aspect of our surroundings—reality if you wish. These axioms, then, constitute what is called a *mathematical model* for that particular subject area in which the observations came to light. As for which statements one would make or hope to make in a developing theory, intuition plays a role, to be sure, but mathematicians are not to be classified as wizards in this pursuit. Their intuitions are guided by observations and feelings about that portion of the real world that the model is meant to represent. This is how a meaningful theory comes about. This is what prevents it from becoming simply a "heap of stones."

EXERCISES

*1 Derive the properties of identity, associativity, commutativity, and distributivity in Z^+ by means of the principle of mathematical induction. [*Hint*: See Section 1.1.]

2 Give a contrapositive proof of Lemma 2, Section 0.5.

3 Give a proof by contradiction for Lemma 2, Section 0.5.

4 Alter the proof of the pigeonhole principle (Section 0.7) so that one of the two implications is demonstrated positively (that is, directly).

5 Why does a proof of the form

$$P_1 \Rightarrow P_2 \Rightarrow \cdots \Rightarrow P_n \Rightarrow P_1$$

establish that $P_1, P_2, \ldots, P_n$ are equivalent? That is, how does it follow that $P_i \Rightarrow P_j$ for all i, j?

6 Why is the contrapositive statement: not $P \Rightarrow$ not Q "logically equivalent" to $P \Rightarrow Q$?

SUGGESTED REFERENCES

0.1 Stoll, R. R. *Sets, Logic and Axiomatic Theories*. Freeman, San Francisco, 1961.

0.2 Lipschultz, S. *Theory and Problems of Set Theory and Related Topics*. Schaum, New York, 1961.

0.3 Halmos, P. R. *Naive Set Theory*. Van Nostrand, Princeton, N.J., 1960.

0.4 Liu, C. L. *Introduction to Combinatorial Mathematics*. McGraw-Hill, New York, 1968.

0.5 Ryser, H. J. *Combinatorial Mathematics.* M.A.A. Carus Monograph Series, distributed
by Wiley, New York, 1963.

In our list of suggested readings, it is inevitable that a fair number of the selections be
rather more advanced than are the treatments in the text. After all, we are trying to present
this material at a level appropriate for computer science undergraduates. In light of this
fact, it is perhaps better to defer the detailed study of some of these selections until a later
time. Others, however, can be viewed as collateral reading. Our remarks should help the
reader decide into which category a particular reference might fall.

The first reference gives a compact and readable introduction to the foundations of
mathematics; it is designed especially to prepare undergraduates for the study of abstract
theories. The second selection includes a number of illustrative examples from set theory,
chosen with the beginning student in mind. The book by Halmos is a beautiful and concise
introduction to the axioms of set theory but may be a bit too sophisticated for the beginning
student.

The two references in combinatorics offer entirely different points of view. The first
has quite a broad scope and is particularly appealing from the standpoint of the applications. On the other hand, the classical monograph by Ryser gives an elegant account of
certain areas of combinatorial theory but requires a definite mathematical maturity on the
part of the reader.

1

Algebras and Algorithms

The words *algebra* and *algorithm* may both be traced to an ancient Arabic textbook, *Kitab al jabr w'al-muqabala* by Mohammed al-Khwarizmi, but they have rarely been so closely linked in the intervening years. In fact, the word "algorithm" might have been lost altogether, but for its importance to computer science. It is interesting to note that its first appearance in the standard dictionaries (in the 1950s) coincides with the period of rapid growth of the computer industry. The reasons for reuniting the concepts of algebra and algorithms here will soon be explained, but they will be fully appreciated only as the material unfolds.

In a sense, what we shall present initially is more a matter of terminology and concept than anything else. The more detailed examples and applications will be given in the following chapters as we try to use the algebraic and algorithmic points of view as unifying themes. Of course, it is important for the student to realize that the more sophisticated applications must await his further development as a computer scientist, but if we are successful here, we will have provided a useful foundation for this development in future computer science courses.

The study of algorithmic processes (their design, analysis, implementation, etc.) is practically the essence of computer science. Loosely speaking, an *algorithm* is a specification of a mechanical or systematic procedure for carrying out a computation. There are certain "ground rules", to be sure, and for performing the computation, one generally has a definite mechanism in mind, which in turn determines the "language" used for specifying the algorithm. At this stage, however, we should not become entangled in such technicalities. The concept itself is more important. We will be introducing a simple and explicit, yet very general "flowchart language" for expressing algorithms in a variety of discrete algebraic structures. For this reason, we cannot easily separate algorithms from algebra in our presentation, nor would it be desirable.

We use the term *algebra* here in a very general sense. By comparison, high school algebra is essentially the study of but a few quite specific algebraic systems: the arithmetic of the real numbers, for instance. In computer science, quite diverse algebraic structures are encountered, in particular those which are discrete or even finite. Therefore, we will define various operations different from the ordinary additions and multiplications (though we may use similar symbols $+$, $\cdot$, etc., to describe them) primarily because the sets on which these operations are defined will rarely be the familiar numbers used in high school algebra. The elements of these sets may be relations, subsets, "strings" of symbols, "states" of a machine, or whatever. Still, the familiar concept of an algebraic expression will persist even in these diverse situations. We use them as ingredients in our algorithmic processes and as aids to our understanding of the particular algebraic structure at hand.

1.1 INTEGERS AND INDUCTION

Among the most important of all the algebraic systems, in spite of their apparent simplicity, are those involving the integers. By themselves, the integers are indispensable for any mathematical discussion. Indeed, we would have had difficulty in formulating the various counting techniques of Section 0.8 if we could not appeal to the reader's intuitive understanding of the set of natural numbers,

$$N = \{0, 1, 2, 3, \ldots\}$$

"intuitive" because, as the nineteenth-century mathematician Kronecker said, "God made the integers; all the rest is the work of man."

Again we let Z^+ denote the set of positive integers, and we remark once more that this notation derives from the use of Z to represent the set of all integers, positive, negative, and zero. In elementary school one is taught the rules for addition and multiplication in N. The extension of the number system to Z is designed to permit subtraction as an operation. Strictly speaking, there is no division in this system, but later in this section, we shall introduce a kind of integer division, important for its computer applications.

Since it is fairly certain that the reader has encountered the so-called *principle of mathematical induction* in earlier courses, we shall merely state the principle here and content ourselves with a few elementary examples. It is possible to base a proof of this important principle on Peano's postulates for the positive integers (as outlined in Exercise 8 of this section), but instead, the reader is invited to adopt the view that we are taking the principle itself as a postulate. As it is ordinarily applied, the principle is concerned with the set Z^+, and especially with proving that various mathematical statements are true of *all* of its members. Once one realizes that an exhaustive case-by-case verification is not possible, the power of the *inductive proof* becomes clear. In a somewhat different guise, the principle has now come to be used extensively for the analysis of algorithms, and consequently, for demonstrating the correctness of computer programs. Later in the chapter we will discuss this most interesting application.

Principle of mathematical induction

Let there be given for each integer $n \geq k$ an associated statement or proposition $P(n)$. Suppose further that

(i) $P(k)$ is true
(ii) For all $n \geq k$, $P(n) \Rightarrow P(n + 1)$

Then the principle asserts that

$$P(n) \text{ is true for all } n \geq k. \qquad \square$$

In other words, we have stated that in order to prove that a certain statement or proposition is true for *all* integers greater than or equal to some fixed integer k, it suffices to establish two things:

(i) the proposition is true for the smallest integer under consideration (in this case, k);
(ii) the assumption (often called the *inductive hypothesis*) that the proposition is true for *some* integer $n \geq k$ implies that it is true for the next largest integer $n + 1$.

EXAMPLE 1.1   As an analogy, try to imagine an endless single file of dominoes standing on end. We provide Figure 1.1 as an aid to your imagination. Now suppose that we want to be sure that all of these dominoes will fall. It is not

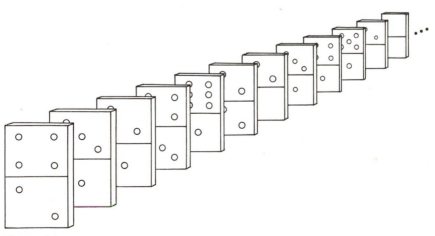

FIGURE 1.1

conclusive to know that the first thousand or even the first million will fall or have fallen, but we could be sure that they would all fall if we were certain of two things:

(i) the first domino is knocked over (in the direction of the others, of course);
(ii) the dominoes are so spaced that whenever any one domino falls, it will automatically knock over its neighbor.

Some politicians have used this model to explain some effects of United States foreign policy.   △

EXAMPLE 1.2   Suppose we now apply the principle to verify that for all $n \geq 1$,

$$\sum_{j=1}^{n} j = \frac{n(n+1)}{2}$$

Here we are dealing with the proposition

$$P(n): \qquad \sum_{j=1}^{n} j = \frac{n(n+1)}{2}$$

and the first integer under consideration is $k = 1$. Clearly

(i) $P(k) = P(1)$ is true, because

$$\sum_{j=1}^{n} j = 1 = \frac{1(2)}{2}$$

(ii) Now suppose that $P(n)$ is true for *some* $n \geq k = 1$. [Note that we do not assume that $P(n)$ is true for *all* $n \geq 1$, for that is what we are trying to prove. Furthermore, we know from (i) that $P(n)$ is true *for some* $n \geq 1$, namely $n = 1$, so that our supposition is indeed a reasonable one.] Can we then show that $P(n + 1)$ is true as a consequence? Yes, for the statement $P(n + 1)$ is simply

$$P(n + 1): \qquad \sum_{j=1}^{n+1} j = \frac{(n+1)(n+2)}{2}$$

and we can obtain it by means of the inductive hypothesis used on the second equality of the calculation:

$$\sum_{j=1}^{n+1} j = (n+1) + \sum_{j=1}^{n} j = (n+1) + \frac{n(n+1)}{2}$$

$$= \frac{2n+2+n^2+n}{2} = \frac{n^2+3n+2}{2}$$

$$= \frac{(n+1)(n+2)}{2}.$$ △

EXAMPLE 1.3   Lest the reader think that such identities (as in Example 1.2) are only mathematical curiosities, we quote from Publication 534, *Tax Information on Depreciation*, of the Internal Revenue Service as follows:

> *Sum of the years-digits method*   Under this method, as a general rule, you apply a different fraction each year to the cost or other basis of each single asset account reduced by estimated salvage value. The denominator (bottom number) of the fraction, which remains constant, is the total of the digits representing the years of estimated useful life of the property. For example, if the estimated useful life is 5 years, the denominator is 15, that is, the sum of $1 + 2 + 3 + 4 + 5$. To save time in arriving at the denominator, especially when an asset has a long life, square the life of the asset, add the life, and divide by 2. Thus, the asset with a 5-year life has the denominator $5 \times 5 + 5 \div 2 = 15$. The numerator (top number), ... $\triangle$

EXAMPLE 1.4   Oftentimes the proposition $P(n)$ will be an inequality. Still, the general outline of the inductive proof will be unchanged. As an illustration, suppose we verify that for all $n \geq 11$,

$$3n - 6 < \frac{n(n - 1)}{4}$$

With $n = 11$, we have 27 and $27\frac{1}{2}$ for the left- and right-hand sides, respectively. So we suppose that the inequality holds for some $n \geq 11$ and we proceed to compute:

$$3(n + 1) - 6 = (3n - 6) + 3 < \frac{n(n - 1)}{4} + 3 = \frac{n(n - 1) + 12}{4}$$

$$\leq \frac{n(n - 1) + (n + 1)}{4} < \frac{(n + 1)(n - 1) + (n + 1)}{4}$$

$$= \frac{(n + 1)((n - 1) + 1)}{4} = \frac{(n + 1)((n + 1) - 1)}{4} \qquad \triangle$$

A function or sequence $f: Z^+ \to Z^+$ is said to be *defined inductively* (or *by recursion*) if there is a given function $g: Z^+ \times Z^+ \to Z^+$ such that

$$f(1) = c \qquad \text{(constant element of } Z^+)$$

$$f(n + 1) = g(n, f(n)) \qquad (n \geq 1)$$

Ordinarily g is "more primitive" than f, that is to say, it is a function which is already well understood and generally computable by a given finite mechanical procedure. Of course, if this is the case, the inductive definition given earlier will provide a mechanical procedure or "algorithm" for computing each $f(n)$ in turn. Perhaps the best-known example is the *factorial function* (see Section 0.8):

$$f(1) = 1 \qquad\qquad 1! = 1$$
$$f(n + 1) = (n + 1) \cdot f(n) \qquad\qquad (n + 1)! = (n + 1) \cdot n!$$

that is

Here the function g is essentially multiplication. More precisely, we must have

$$g(n, m) = (n + 1) \cdot m$$

so that

$$f(n + 1) = (n + 1) \cdot f(n) = g(n, f(n))$$

as required.

It is easy to understand why statements that involve an inductively defined function may lend themselves to proofs by induction. After all, the whole trick in an inductive proof is to reduce statements about $n + 1$ to statements involving n (so that the inductive hypothesis can be used), and this much is provided immediately in $f(n + 1) = g(n, f(n))$. For instance, if we want to show that $n! \geq n$ for all $n \geq 1$, we will have

$$1! = 1 \geq 1$$

and assuming that $n! \geq n$ for some $n \geq 1$, we obtain

$$(n + 1)! = (n + 1) \cdot n! \geq (n + 1) \cdot n \geq n + 1$$

completing the inductive proof.

EXAMPLE 1.5   The meaning of the mathematical summation symbol ($\sum$) can be defined inductively (with $g = $ addition). Suppose that an infinite sequence of integers x_j is given. Then in order to define $\sum_{j=1}^{n} x_j (=f(n))$ inductively, we have only to write

$$\sum_{j=1}^{1} x_j = x_1 \quad \text{and} \quad \sum_{j=1}^{n+1} x_j = x_{n+1} + \sum_{j=1}^{n} x_j$$

As a matter of fact, this inductive definition was the whole key to the proof in Example 1.2.   △

At this point let us make appropriate preparations for our eventual discussion of algorithms by developing further properties of the integers. Most of these properties are related to the concept of divisibility. In a sense, the choice of material here is a matter of convenience; the main objective is to choose those topics that will lead to examples of only moderate difficulty.

Suppose $n, m \in Z$. We say that n *divides* m (evenly), written $n|m$, if there exists an integer q such that

$$m = q \cdot n$$

that is, if m is an integer multiple of n. Since this is a rather infrequent occurrence given any n and m, we cannot, strictly speaking, provide Z with a division operation; that is why the rational numbers were devised. If we step briefly into $Q = \{$rational numbers$\}$, however, we find that a useful substitute operation is available in Z. It goes by the name *integer division* and is found in most computer programming languages, ALGOL, FORTRAN, etc. For any $x \in Q$ we let tr(x) be that unique integer obtained by *truncation*, that is, by disregarding the fractional part of x. For instance, tr(3.14) = 3; similarly, tr(-3.14) = -3. Then for any pair of integers n, m where $n \neq 0$, we define the *integer quotient*

$$m \div n = \text{tr}(m/n)$$

EXAMPLE 1.6   If $m = 22$ and $n = 5$, we have

$$m \div n = \text{tr}(\tfrac{22}{5}) = \text{tr}(4.4) = 4$$

even though $n|m$ is false, that is, there does not exist an integer q such that

$$22 = q \cdot 5.$$

Note, however, that $q = 4$ comes very close.   △

   Closely related to this discussion is the so-called division algorithm that speaks of a quotient and remainder (q and r), when dividing a by b. If one takes $q = a \div b$ and $r = a - bq$ in the statement below, the justification should be clear. The remainder r is certainly less than b (otherwise, further division would be possible).

Division algorithm

For any integers $a, b \in Z^+$ there exist unique integers q and r such that

$$a = bq + r \quad \text{and} \quad 0 \leq r < b \qquad\qquad \square$$

EXAMPLE 1.7   In the preceding example ($a = 22$, $b = 5$) if we let

$$q = a \div b = 4$$
$$r = a - bq = 22 - 20 = 2$$

then certainly

$$22 = 5 \cdot 4 + 2 \quad \text{and} \quad 0 \leq 2 < 5 \qquad\qquad △$$

   As an application of the division algorithm, let us seek to determine the *greatest common divisor* gcd(a, b) of the integers $a, b \in Z^+$. By this we mean an integer $d = \text{gcd}(a, b)$ with the properties:

(a) $d|a, \; d|b$
(b) $c|a, \; c|b \Rightarrow c|d$

Note in particular that the adjective "greatest" here refers not primarily to d having a greater magnitude than any common divisor c, but instead, to d being a multiple of any such c.

   Given a, b whose gcd is sought, we write

$$a = bq_1 + r_1 \qquad 0 \leq r_1 < b$$

and as long as r_1 and the succeeding remainders are different from zero, we continue:

$$b = r_1 q_2 + r_2 \qquad 0 < r_2 < r_1$$
$$r_1 = r_2 q_3 + r_3 \qquad 0 < r_3 < r_2$$
$$\cdots$$
$$r_{k-2} = r_{k-1} q_k + r_k \qquad 0 < r_k < r_{k-1}$$
$$r_{k-1} = r_k q_{k+1} \qquad 0 = r_{k+1}$$

Observe that since the remainders decrease strictly and monotonically, one of them, r_{k+1}, say, must eventually be zero.

Now consider the first equation: $a = bq_1 + r_1$. We have

$$c|a, \; c|b \Rightarrow a = pc, \quad b = qc$$
$$\Rightarrow r_1 = (p - qq_1)c \Rightarrow c|r_1$$

that is, every common divisor of a, b also divides the remainder r_1. Conversely, the common divisors of b, r_1 will divide a as well. It follows that

$$\gcd(a, b) = \gcd(b, r_1) = \gcd(r_1, r_2) = \cdots = \gcd(r_{k-1}, r_k) = r_k$$

These observations form the basis for the *Euclidean algorithm* for computing greatest common divisors. A formal presentation of this algorithm later in the chapter will provide a few more details.

EXAMPLE 1.8   Suppose that $a = 252$ and $b = 180$. Then as before, we compute

$$252 = 180 \cdot 1 + 72$$
$$180 = 72 \cdot 2 + 36$$
$$72 = 36 \cdot 2 + 0$$

so that

$$\gcd(252, 180) = \gcd(180, 72) = \gcd(72, 36) = 36. \qquad \triangle$$

EXERCISES

1 Prove that for all $n \geq 1$

$$\sum_{j=1}^{n} j^2 = \frac{n(n + 1)(2n + 1)}{6}$$

2 Prove that for all $n \geq 1$

$$\sum_{j=1}^{n} j^3 = \left[\frac{n(n + 1)}{2} \right]^2$$

3 Prove by mathematical induction that

$$|\mathscr{P}(S)| = 2^{|S|}$$

for every finite set S. [*Hint*: In a set with $n + 1$ elements

$$S = \{x_1, \ldots, x_n, x_{n+1}\}$$

there are two kinds of subsets: those that include x_{n+1} and those that do not.]

4 Show that for all $n \geq 1$

$$\sum_{j=1}^{n} 2^{j-1} = 2^n - 1$$

5 Give an inductive proof that for all $n \geq 0$

$$\sum_{k=0}^{n} \binom{n}{k} = 2^n$$

6 Use the inductive definition of the summation symbol (Example 1.5) to show that for all $n \geq 1$

(a) $\displaystyle\sum_{j=1}^{n} x_j = x_1 + x_2 + \cdots + x_n$

(b) $\displaystyle\left(\sum_{j=1}^{n} x_j \right) y = \sum_{j=1}^{n} (x_j y)$

7 Prove that $3 | (n^3 - n)$ for all $n \geq 0$. Can you generalize this claim?

*8 According to the axioms of Giuseppe Peano (1858–1932), Z^+ can be viewed as an algebra with a *distinguished element*, $1 \in Z^+$, and a unary operation (see Section 1.2)

$$s: Z^+ \rightarrow Z^+$$

such that

(i) As a function, s is injective
(ii) $s(n) \neq 1$ for all $n \in Z^+$
(iii) Any subset $N \subseteq Z^+$ with the properties
    (a) $1 \in N$ and   (b) $n \in N \Rightarrow s(n) \in N$
    must be all of Z^+, that is, $N = Z^+$.

Show that the principle of mathematical induction follows from the Peano axioms. [*Hint*: s is called the *successor* function, and we agree to denote $s(1) = 2$, $s(2) = 3$, etc. We can then define addition and multiplication inductively by writing

$$\begin{array}{ll} 1 + m = s(m) & s(n) + m = s(n + m) \\ 1 \cdot m = m & s(n) \cdot m = n \cdot m + m] \end{array}$$

*9 Derive the properties of identity, associativity, commutativity, and distributivity (see Example 0.72) for Z^+; use inductive definitions for addition and multiplication. [*Hint*: See the hint for Exercise 8.]

10 Determine the quotients and remainders (q, r) of the division algorithm, given the following data:

(a) $a = 18, b = 5$ (b) $a = 121, b = 10$
(c) $a = 3512, b = 91$ (d) $a = 55, b = 5$
(e) $a = 63, b = 8$ (f) $a = 2764, b = 3$

11 Find the greatest common divisors for the following pairs:
(a) $a = 48, b = 42$ (b) $a = 154, b = 140$
(c) $a = 1350, b = 297$ (d) $a = 297, b = 256$
(e) $a = 1144, b = 351$ (f) $a = 8024, b = 412$

*12 Prove the second principle of mathematical induction: The two conditions

(i) $P(1)$ is true and
(ii) $P(j)$ is true for all $1 \le j \le n \Rightarrow P(n + 1)$ is true

imply that $P(n)$ is true for all $n \ge 1$. [*Hint*: Let

$$N = \{k: P(j) \text{ is true for all } 1 \le j \le k\}$$

in the Peano axioms (Exercise 8).]

*13 Show that the gcd function satisfies the properties
(a) $n > m \Rightarrow \gcd(n, m) = \gcd(n - m, m)$
(b) $\gcd(n, m) = \gcd(m, n)$
(c) $\gcd(n, n) = n$
for all n, m in Z^+.

14 It is claimed that Gauss proved the "IRS formula" by considering the scheme

$$1 + \quad 2 \quad + \cdots + n$$
$$n + (n - 1) + \cdots + 1$$

Complete Gauss' proof.

1.2 ALGEBRAS AND PROPOSITIONS

We used the term "proposition" informally in the preceding section and also in the preliminary chapter. The reader was left to interpret its meaning according to everyday usage. In some states, propositions are frequently placed on the ballot, and the voting population is asked to decide on each, yes or no. We hope that the reader is gradually adjusting to a more technical usage of the term, namely a proposition being a statement which may be true or false, depending on the values of the variables occurring in it. Our main interest in propositions, however, derives from their suitability as the decision-making ingredients of algorithms. The student who understands the use of diamond-shaped boxes for branching in flowcharts knows exactly where we are heading.

Formally, an *m*-ary *proposition* P is simply a subset of a corresponding product set $P \subseteq A_1 \times A_2 \times \cdots \times A_m$. The term *m*-ary is just a generalization of the words "unary," "binary," "ternary," etc., and indicates the number of *arguments* $a_i \in A_i$ needed for making the decision. Thus if $a_i \in A_i$ ($i = 1, 2, \ldots,$ m) we must have either

$$(a_1, a_2, \ldots, a_m) \in P \quad \text{or} \quad (a_1, a_2, \ldots, a_m) \notin P$$

And so we may think of P as representing a dichotomy: yes or no, true or false, relative to the question "$(a_1, a_2, \ldots, a_m) \in P$?" In the preceding section on induction we considered unary (1-ary) propositions $P \subseteq A_1 = Z^+$. A binary (2-ary) proposition $P \subseteq A_1 \times A_2$ is just a relation from A_1 to A_2 as introduced in Section 0.3. As with relations, we speak of P as an m-ary *proposition on the set A* if $P \subseteq A^m$, that is, if all the A_i are the same and equal to A. The actual description of a proposition can take many forms, but in most cases it will involve similar notations as in descriptions of sets. Ordinarily, we will use angle brackets $\langle \; \rangle$ rather than the braces $\{\;\}$ to distinguish propositions. This notation will also help to provide a link to the diamond-shaped boxes of the conditional statements in flowcharts, the main intended application for the propositions.

EXAMPLE 1.9 The 3-ary proposition

$$P = \langle (a, b, c): a^2 + b^2 = c^2 \rangle$$

considered as a subset of $Z^+ \times Z^+ \times Z^+$ is the set of *Pythagorean triples*. On $Z^+ \times Z^+ \times Z^+$ it establishes the dichotomy:

$$(3, 4, 5) \in P \qquad (3, 4, 6) \notin P$$
$$(5, 12, 13) \in P \qquad (2, 2, 2) \notin P, \text{ etc.} \qquad \triangle$$

EXAMPLE 1.10 Let E be the English dictionary (Example 0.29). On the product set $E \times E \times Z^+$ we may introduce the ternary proposition

$$P = \langle (x, y, n): |x| + |y| = n \rangle$$

Then we have

$$(\text{super, man, } 8) \in P \qquad (\text{a, the, } 5) \notin P$$
$$(\text{bye, bye, } 6) \in P \qquad (\text{mile, high, } 5280) \notin P, \text{ etc.}$$

This particular proposition can also be viewed as a function

$$E \times E \xrightarrow{f} Z^+$$

that is, for each $(x, y) \in E \times E$, there is one and only one $n \in Z^+$ such that $(x, y, n) \in P$, namely $n = |x| + |y|$. And so, if we set

$$f(x, y) = |x| + |y|$$

we may write

$$P = \langle (x, y, f(x, y)) : x, y \in E \rangle \qquad \triangle$$

EXAMPLE 1.11   Any relation R on a set S is a binary proposition on S. It follows that any of the familiar comparisons $<, \leq, =, \geq, >, \neq$ is to be considered a 2-ary proposition on Z say. (See Section 0.3.) Similarly (Example 0.22), the relation $\subseteq$ is a 2-ary proposition $\mathscr{P}(S)$ for any set S. $\triangle$

In order to set our propositions in a useful context, we now introduce the important notion of an algebra. The definition is completely flexible and allows for the introduction of arbitrary "arithmeticlike" operations on sets of our own choosing. After superimposing the decision capability afforded by certain "primitive" propositions, our algebras will begin to take on the appearance of machines. This is not the usual mathematical understanding of an algebra, but it does suit our purposes. For, we find in computer science that we often intend to simulate certain algebraic manipulations through the operations of an actual machine (computer). Then if we are sufficiently familiar with the details of the simulation, we can actually write computer programs that will answer questions of interest concerning the original algebraic structure.

Only the broadest outline of this simulation can be discussed here, but this much alone should be motivation enough. Consider the diagram in Figure 1.2 as an overview. A full discussion of the simulation process requires a knowledge of actual computer programming languages and their capabilities, some famil-

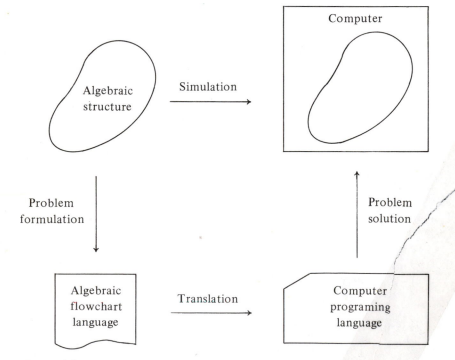

FIGURE 1.2

iarity with computer architecture, but more importantly, a detailed understanding of the computer data structures so crucial to the implementation. In a beginning course in discrete mathematical structures, we are in a position to make contact with this overall process only from the outside (the left-hand side of the diagram). It is left for the student to somehow complete the picture and make the necessary connections and associations in the course of studying programming languages, data structures, etc. On the other hand, our detatched point of view does offer certain advantages. We are able to discuss programming concepts in the abstract, without getting bogged down in the idiosyncrasies of a particular programming language or a particular machine capability. This abstraction will prove to be a very definite aid to clear thinking all around.

To proceed now with the definition of an algebra, we let A be any nonempty set. A function $f: A^n \rightarrow A$ for some integer $n \geq 1$ is called an *n-ary operation* on A. A system

$$A = (A, f_1, f_2, \ldots)$$

where the f_i are operations on A, is called an *algebra*. Note in particular that we follow the rather standard mathematical convention in referring to the algebra by the same name as the set on which its operations are given. Even allowing for this abuse of notation, the student should realize that the "equality" in $A = (A, f_1, f_2, \ldots)$ is only superficial—a kind of shorthand for saying that we would like to refer to the algebra $(A, f_1, f_2, \ldots)$ simply as A whenever this is convenient. Furthermore, it is temporarily convenient to let $\mathscr{F}_n = \mathscr{F}_n(A)$ denote the collection of *n*-ary operations in the algebra A.

EXAMPLE 1.12   Let $A = \{a, b, c, d, e, f\}$. We can describe a binary (2-ary) operation on A by means of a "multiplication table"

·	a	b	c	d	e	f
a	e	d	f	b	a	c
b	c	e	a	f	b	d
c	b	f	d	e	c	a
d	f	a	e	c	d	b
e	a	b	c	d	e	f
f	d	c	b	a	f	e

read from row to column, that is, $c \cdot b = f$ whereas $b \cdot c = a$. Here, the "infix" notation $x \cdot y$ replaces the rather awkward prefix notation $\cdot (x, y)$ that strict functional format would dictate. We have $\mathscr{F}_1 = \varnothing$, $\mathscr{F}_2 = \{\cdot\}$, and $\mathscr{F}_n = \varnothing$ for $n \geq 3$ in defining the algebra $A = (A, \cdot)$.

It should not be necessary to exhibit a definite application for each and every algebra that we introduce. At this stage we are more interested in illustrating the complete generality of the concept itself. We do note, however, that this particular algebra is rather special in that

$$x \cdot e = e \cdot x = x$$
$$x \cdot (y \cdot z) = (x \cdot y) \cdot z$$

for all $x, y, z \in A$. That is, the element e acts like a "one" or "identity" element in multiplication, and the operation itself is associative. We call such algebras *monoids*. They arise quite frequently in certain areas of computer science and will be discussed more thoroughly in Chapter 3.   △

EXAMPLE 1.13   Considering the ordinary arithmetic of addition and multiplication of integers as binary operations on Z^+, say, we see that the positive integers form an algebra

$$Z^+ = (Z^+, +, \cdot)$$

with $\mathscr{F}_2 = \{+, \cdot\}$.   △

EXAMPLE 1.14   If S is any set and $\mathscr{P}(S)$ is its power set, we obtain a corresponding algebra

$$\mathscr{P}(S) = (\mathscr{P}(S), \cup, \cap, \sim)$$

as in Section 0.2. Similarly, the set of all relations on S forms an algebra

$$\mathscr{R}(S) = (\mathscr{R}(S), \vee, \wedge, \neg)$$

as discussed in Section 0.6. Each of these algebras has two binary operations and one unary operation.   △

EXAMPLE 1.15   Let $B = \{\textbf{false}, \textbf{true}\}$ with the operations

$\vee$	false	true		$\wedge$	false	true
false	false	true		**false**	false	false
true	true	true		**true**	false	true

$$\neg \ \textbf{false} = \textbf{true}$$
$$\neg \ \textbf{true} \ = \textbf{false}$$

We have $\mathscr{F}_1 = \{\neg\}$ and $\mathscr{F}_2 = \{\vee, \wedge\}$ in the resulting algebra

$$B = (B, \vee, \wedge, \neg)$$

B is in fact a Boolean algebra as defined in Chapter 4, but for now, we refer to B simply as the algebra of *logical values* because its operations are consistent with the logical operations:

$\vee$ representing *or*
$\wedge$ representing *and*
$\neg$ representing *not*

Thus if your instructor says he will give you an *A* in a course if you hand in a perfect term paper *or* you do *not* score less than 90% on the final exam, what would be your fate if you should turn in a poor term paper but score 91% on the final? Is it true that you should receive an *A*? Yes, according to the computation:

$$\textbf{false} \ \vee \ \neg \ (\textbf{false}) \ = \ \textbf{false} \vee \textbf{true} \ = \ \textbf{true}.$$

In other contexts, the same algebra will appear with its elements simply renamed as *F*, *T* or sometimes 0, 1. Its operation symbols are often renamed as well. Electrical engineers most often use $+, \cdot, '$ in place of the $\vee, \wedge, \neg$ favored by logicians. In fact, we will switch to the former notation in Chapter 4 when more general Boolean algebras are studied. △

EXAMPLE 1.16 The same set or at least its synonym

$$B = \{0, 1\}$$

can be given a completely different algebraic structure if we define a binary operation $\oplus$ according to the following table:

$\oplus$	0	1
0	0	1
1	1	0

Note that this operation agrees with neither of those in the previous example. Those familiar with binary arithmetic, however, will recognize this as the addition operation performed by a computer at its hardware level. Thus in adding 5 and 4 (in binary arithmetic) we will have:

$$
\begin{array}{r}
1 \ 0 \ 1 \\
1 \ 0 \ 0 \\
\hline
1 \ 0 \ 0 \ 1
\end{array}
$$

with a "carry" into the fourth position. But "one plus one equals zero" in this system of arithmetic. We will find that there are other situations where we need this kind of addition. In any case, our set *B* is an entirely new algebra

$$B = (B, \oplus)$$

compared with that of the preceding example. △

One of the easiest ways to make new algebras out of old ones is to construct from $A = (A, f_1, f_2, \dots)$ the *product algebra*

$$A^n = (A^n, f_1, f_2, \dots)$$

in which the old operations are simply extended to *n*-tuples $(a_1, a_2, \dots, a_n)$ of A^n in coordinatewise fashion. Thus if *f* should happen to be a binary operation in A, then it can be extended to become a binary operation in A^n by writing

$$f((a_1, \ldots, a_n), (a_1', \ldots, a_n')) = (f(a_1, a_1'), \ldots, f(a_n, a_n'))$$

Other operations (unary, ternary, etc.) can be handled in exactly the same manner.

EXAMPLE 1.17   If $B = (B, \oplus)$ is the algebra of the preceding example, then for each integer $n \geq 1$ we can obtain the product algebras

$$B^n = (B^n, \oplus)$$

in which

$$(b_1, b_2, \ldots, b_n) \oplus (b_1', b_2', \ldots, b_n') = (b_1 \oplus b_1', b_2 \oplus b_2', \ldots, b_n \oplus b_n')$$

Thus in B^3 we would have

$$\begin{aligned}
(1, 1, 0) \oplus (0, 1, 1) &= (1, 0, 1) \\
(1, 1, 0) \oplus (1, 1, 0) &= (0, 0, 0), \quad \text{etc.}
\end{aligned}$$ △

As indicated in Figure 1.2, we intend to introduce an algebraic flowchart language for the design of algorithms in arbitrary algebras. But we will need to allow our algebras A to include propositions on A if the language is to be equipped with a capability for "branching" through the use of the various conditional statements. This is easily accomplished, by fiat if you wish. On the other hand, the reader has probably noticed that most of the algebras we have discussed seem to have associated propositions already. Think of $\leq$ for the integers or $\subseteq$ for set algebras. Therefore, it is entirely natural that we *augment our definition of an algebra* A to allow for both operations $\mathcal{F}_n$ and propositions $\mathcal{P}_m$ for $n, m \geq 1$. We may write

$$\mathcal{F} = \cup \mathcal{F}_n \quad \text{and} \quad \mathcal{P} = \cup \mathcal{P}_m$$

in referring to such algebras more succinctly as $A = (A, \mathcal{F}; \mathcal{P})$.

EXAMPLE 1.18   If we augment the algebra Z^+ of Example 1.13 by admitting the relation $\leq$, then

$$Z^+ = (Z^+, \mathcal{F}; \mathcal{P})$$

becomes an algebra with propositions in which

$$\mathcal{F} = \{+, \cdot\} \qquad \mathcal{P} = \{\leq\}.$$

We then have the capability for making comparisons. This means that in the algebraic flowchart language to be developed, we will be able to use this "primitive" proposition in constructing conditional statements of the general form:

if $n \leq m$, **then** . . .

if $n \leq 3$, **then** . . . **else** . . . ,   etc.   △

EXAMPLE 1.19   A similar decision-making capability will be afforded to the set algebras $\mathcal{P}(S)$ of Example 1.14 if we admit the binary proposition $\subseteq$ that tests whether one set is a subset of another. There will result for each set S an algebra

$$\mathcal{P}(S) = (P(S), \cup, \cap, \sim \; ; \subseteq).$$ $\triangle$

Again think of our algebras as machines. The primitive operations and propositions of the algebra can be thought of as representing the fundamental capabilities of the machine. Every real machine (computer) has its own rather limited repertoire of operations and decision-making capabilities. Whenever we write a program in a language comprehensible to a particular machine, we are limited to the repertoire of that machine, even if the capability is enlarged by a higher-level language. Therefore, a complex computational algorithm must be suitably decomposed into such elementary constructs as are permitted. We are simply mirroring this situation with our algebras.

EXERCISES

1 Give a formal description of the propositions

$$P \subseteq A_1 \times A_2 \times \cdots \times A_m$$

that serve to define the following.
(a) All pairs of positive integers in which the first is smaller than the second
(b) All prime numbers
(c) All triples (x, y, n) in which x is the last name of a postwar (World War II) United States president, y is the name of his vice president, and $n = |x| + |y|$
(d) All pairs of positive integers in which the first divides the second
(e) All quadruples (a, b, q, r) in which q is the quotient and r the remainder upon integer division of a by b
(f) All complex numbers of absolute value 1

2 Describe each of the following propositions in an English phrase; that is, do the reverse of Exercise 1:
(a) $\langle \text{HST, JFK, LBJ} \rangle$
(b) $\langle (n, m): n = m \rangle \subseteq Z \times Z$
(c) $\langle n: n = 2m \text{ and } m \in Z \rangle \subseteq Z$
(d) $\langle (a, b, d): d = \gcd(a, b) \rangle \subseteq Z^+ \times Z^+ \times Z^+$
(e) $\langle R: R \text{ is reflexive, symmetric, transitive} \rangle \subseteq \mathcal{R}(S)$
(f) $\langle (R, Q): Q = R^T \rangle \subseteq \mathcal{R}(S) \times \mathcal{R}(S)$

*3 For all $n \geq 1$ we have

$$A^n = A \times A \times \cdots \times A \qquad (n \text{ times})$$

and we usually define

$$A^0 = I \text{ (some set with one element)}$$

Why? Use this definition of A^0 to explain why a 0-ary operation on A is understood to be a "distinguished" (constant) element of A. [*Hint*: As motivation for the definition of A^0, consider an extension of the rule of the product $|A^n| = |A|^n$ to the case $n = 0$.]

4 Describe the following as algebras.
   (a) The arithmetic of the real numbers
   (b) The arithmetic of the rational numbers
   (c) The arithmetic of partitions (Section 0.6)
   (d) The arithmetic of the integers modulo n (Section 0.5)
   In each case, list any special algebraic properties that may hold.

1.3   ALGEBRAIC EXPRESSIONS

As mentioned at the outset, the algebraic expressions in any algebra become important as ingredients for developing algorithmic processes. The student may already be aware of this as a result of being familiar with arithmetic expressions in the assignment statements of FORTRAN, ALGOL, etc. If so, he or she may also be acquainted with the use of variables and constants according to their types, that is, integer, real, logical, etc., in these languages. Again, we simply try to imitate or model this situation in arbitrary algebras.

Let A be an algebra, and suppose that it has an associated collection $\mathscr{V}$ of *variables*. This suggestion is intended to be entirely harmless; you may imagine a collection of symbols to be used in that algebra in the same way that you ordinarily use x in forming expressions in high school algebra. In fact, this is exactly what we are trying to do here. We are building up expressions in the algebra A by using *constants* (elements of the set A), variables, and the given operations. These expressions will then be used to initiate computations in the algebra (machine?) in accordance with the rules of our flowchart language. Therefore, we define "inductively" the class $\mathscr{E}$ of *algebraic expressions* in A as follows:

(i) $A \subseteq \mathscr{E}$ and $\mathscr{V} \subseteq \mathscr{E}$
(ii) $x_1, \ldots, x_n \in \mathscr{E}, f \in \mathscr{F}_n \Rightarrow f(x_1, \ldots, x_n) \in \mathscr{E}$

That is, by (i), constants and variables (standing alone) are considered to be algebraic expressions in their own right; and if $x_1, \ldots, x_n$ are already known to be algebraic expressions, and if f is an n-ary operation of the algebra, then $f(x_1, \ldots, x_n)$ becomes, by (ii), an algebraic expression in good standing. In what sense is induction being used here?

Corresponding to each algebraic expression in A is a function (defined on an appropriate product set A^r) whose value can be computed upon substitution of constants for all the variables occurring in it, assuming that one has the means for performing all the operations in A. Of course, two different expressions may very well give rise to the same function in this correspondence. Briefly let us consider the rather special case in which the expression contains only constants! We then need to give a meaning to the idea of a product set A^0, because r is to denote the number of distinct variables occurring in the expression. The student is advised to ponder this question now, though we will approach the problem of defining A^0 only at a later point (Section 3.2). In this connection, the reader is advised to consult Exercise 3 of Section 1.2.

EXAMPLE 1.20 We may consider the natural numbers as an algebra

$$N = (N, +, \cdot)$$

much like that described for Z^+ in Example 1.13. Then

(a) 17 is an algebraic expression $[N \subseteq \mathscr{E}$ by (i)$]$
(b) n is an algebraic expression [by (i), assuming $n \in \mathscr{V}$]
(c) $n + 17$ is an algebraic expression [using (a), (b) in (ii) with the binary operation $+$]
(d) m is an algebraic expression (again, assuming $m \in \mathscr{V}$)
(e) $m \cdot (n + 17)$ is an algebraic expression [using (c), (d) in (ii) with the binary operation $\cdot$]

Continuing in this manner, we can verify that

$$(((m \cdot (n + 17)) + (m \cdot m)) \cdot n) + 4$$

is an algebraic expression in the algebra N. Observe that

$$4 + (n \cdot (((m \cdot m) + (m \cdot n)) + (17 \cdot m)))$$

is a different algebraic expression, but it will correspond to the same function $N^2 \to N$ upon substitution of constants for the variables n, m. Why? $\triangle$

EXAMPLE 1.21 In the same way, the set of all integers is also an algebra $Z = (Z, +, \cdot)$. In fact, we might admit subtraction as another binary operation here, if that were convenient. Then we would obtain algebraic expressions that are not permitted in the previous example. For instance, $(m \cdot (n - 17)) + n$ is an algebraic expression here, assuming of course that n, m are in the corresponding variable set. In this way, we see that the permissable algebraic expressions depend entirely on the structure of the algebra itself. $\triangle$

EXAMPLE 1.22 Such variables as we choose to employ in the algebra B (Example 1.15) are called *logical* variables, in accordance with the terminology used in the familiar computer programming languages. If K and L are logical variables, then

$$(L \lor (\neg K)) \land \textbf{false}$$

is a "logical expression," that is, it is an algebraic expression in the algebra B. To verify this, we simply proceed as before:

(a) $\neg K$ is an algebraic expression [using (ii) with the unary operation $\neg$]

(b) $L \vee (\neg K)$ is an algebraic expression [using (ii) with the binary operation $\vee$]

(c) **false** is an algebraic expression [$B \subseteq \mathscr{E}$ by (i)]

(d) $(L \vee (\neg K)) \wedge$ **false** is an algebraic expression [using (b), (c) with the binary operation $\wedge$]

The reader should see that a constant function is associated with this expression. $\triangle$

EXAMPLE 1.23   Consider the set $S = \{a, b, c\}$ in forming the algebra

$$\mathscr{P}(S) = (\mathscr{P}(S), \cup, \cap, \sim)$$

described in Example 1.14. If A, B are variables for this algebra, then

$$((A \cap (\sim B)) \cup ((\sim A) \cap B)) \cap \{a, c\}$$

is an algebraic expression. Note that $\{a, c\}$ is a constant. $\triangle$

When we first encounter algebraic expressions in high school algebra, we learn to write and to recognize them by example, without the benefit of a formal definition. It is only because we may intend eventually to present such expressions to a computer that the higher level of precision is required here. We have to be sure that the class of algebraic expressions has been clearly delineated in a mechanical fashion if a computer is to be programmed to recognize them by their form alone, and to process them appropriately. Surely we cannot afford to be less scrupulous in our abstract setting. For one thing, we want to return in Chapter 6 to this recognition problem and describe the algorithmic methods by which a computer may decide whether or not a given expression is indeed "well formed." Such methods must be entirely formal, and so a vague intuitive understanding of algebraic expressions will not suffice.

EXERCISES

1 Show that the following are algebraic expressions in the algebra Z^+ (Example 1.13):

(a) $((3 + n) \cdot m) \cdot n$ (b) $((m \cdot m) + 4) \cdot n$

(c) $n \cdot n \cdot n + 23$ (d) $3 + 2 + n$

(e) $(n + m + p) \cdot 3 + n \cdot m$ (f) $(m \cdot n + p) \cdot n + 2$

assuming that n, m, p are positive integer variables.

2 Show that the following are algebraic expressions in the algebra B (Example 1.15):

(a) $\neg(p \wedge (\neg q))$ (b) $(p \vee q) \wedge (\neg p)$

(c) $(p \vee (\neg \text{ true})) \vee \text{ false}$ (d) $\neg(p \wedge (\neg \text{ false}))$

assuming that p, q are logical variables.

3 Show that the following are algebraic expressions in the algebra A (Example 1.12):

(a) $(a \cdot x) \cdot b$ (b) $(x \cdot y) \cdot c$

(c) $((x \cdot x) \cdot (a \cdot b)) \cdot y$ (d) $(x \cdot y) \cdot (y \cdot x)$

assuming that x, y are variables of the algebra A.

4 Show that the following are algebraic expressions in the partition algebra [Exercise 4(c), Section 1.2] $\Pi(S)$ with $S = \{1, 2, 3, 4, 5\}$:

(a) $(\pi + \{\overline{1, 2}; \overline{3, 4}; \overline{5}\}) \cdot \gamma$ (b) $\pi \cdot \gamma + \{\overline{1, 2, 3}; \overline{4, 5}\}$

(c) $(\pi + \theta) \cdot \gamma$ (d) $(\pi + \gamma) \cdot \{\overline{1}; \overline{2}; \overline{3, 4}; \overline{5}\}$

assuming that π, γ are partition variables.

5 Show that the following are algebraic expressions in the subset algebra (Example 1.14) $\mathscr{P}(S)$ with $S = \{a, b, c\}$:

(a) $(A \cap (\sim B)) \cup \{a, c\}$ (b) $(\varnothing \cup B) \cap \{a\}$

(c) $(A \cup (\sim \{a, b\})) \cap B$ (d) $(S \cap A) \cup (\sim B)$

assuming that A, B are subset variables.

6 Show how the various algebraic expressions correspond to appropriate functions $f: A^r \to A$ by tabulating a few function arguments and values for the case of

(a) Exercise 1   (b) Exercise 2

(c) Exercise 3   (d) Exercise 4

(e) Exercise 5

1.4 EFFECTIVE ALGEBRAIC STRUCTURES

Whether or not the student has had prior programming experience, it should be clear that a single algebra A will rarely suffice as a backdrop for embedding a programming problem. Ordinarily, the design of an algorithm will require the use of several algebraic systems acting in concert. We may need integers for counting, logical values for decisions, sets for classification, etc., and there may be considerable interplay among the various algebras $A = (A, \mathscr{F}; \mathscr{P})$. In order to accommodate these complications, we need to develop an amalgamation of the material of the two preceding sections to allow for interacting algebras.

With these ideas in mind, we say that an *algebraic structure*

$$\mathscr{A} = (\mathscr{A}; \mathscr{F}; \mathscr{P}) = (A_1, A_2, \ldots; \mathscr{F}; \mathscr{P})$$

is a collection of sets (algebras), functions, and propositions. We write

$$\mathscr{F} = \bigcup_\alpha \mathscr{F}_\alpha$$

with the understanding that each $f \in \mathscr{F}_\alpha$ takes its values in the set A_α. The functions are no longer necessarily operations; instead, they may be defined on various products of the sets. Thus in the expression $f \in \mathscr{F}_\alpha$ the subscript α has a different meaning than that of n in Section 1.2. We only require that

$$f: A_{\alpha_1} \times A_{\alpha_2} \times \cdots \times A_{\alpha_n} \to A_\alpha$$

for certain sets $A_{\alpha_1}, A_{\alpha_2}, \ldots, A_{\alpha_n}$ in $\mathscr{A}$. Similarly, the propositions

$$P \subseteq A_{\alpha_1} \times A_{\alpha_2} \times \cdots \times A_{\alpha_m}$$

take arguments from some product set of $\mathscr{A}$.

The A_α may be thought of as constituting the individual algebras of the structure. As in Section 1.3, we may postulate the existence of a variable set $\mathscr{V}_\alpha$ for each A_α, and it is perhaps best to think of these sets as disjoint. For then we may speak unambiguously of variables of *type* α, and similarly with constants (elements of A_α). Then the *algebraic expressions* $\mathscr{E}_\alpha$ *of type* α in $\mathscr{A}$ can be defined much as before:

(i) $A_\alpha \subseteq \mathscr{E}_\alpha$ and $\mathscr{V}_\alpha \subseteq \mathscr{E}_\alpha$,

(ii) $x_{\alpha_1} \in \mathscr{E}_{\alpha_1}, \ldots, x_{\alpha_n} \in \mathscr{E}_{\alpha_n}$
$f \in \mathscr{F}_\alpha$ as above $\Rightarrow f(x_{\alpha_1}, \ldots, x_{\alpha_n}) \in \mathscr{E}_\alpha$.

Perhaps the most visible among the functions connecting the various algebras will be the *transfer functions*

$$f : A_\beta \to A_\alpha$$

so named because they merely convert expressions of one type into expressions of another type. They enter [in (ii)] into the determination of the total class $\mathscr{E} = \bigcup_\alpha \mathscr{E}_\alpha$ of expressions for the algebraic structure $\mathscr{A}$.

EXAMPLE 1.24   Consider the two algebras

$$\mathscr{P}(S) = (\mathscr{P}(S), \cup, \cap, \sim) \qquad N = (N, +, \cdot)$$

of Examples 1.14 and 1.20, respectively. The function

$$|\,| : \mathscr{P}(S) \to N$$

which delivers the cardinality of a set $A \in \mathscr{P}(S)$ is a transfer function. Accordingly,

$$|A \cup \{a, c\}| + 4 \cdot n$$

is an algebraic expression in any algebraic structure containing these two algebras and the indicated transfer function (assuming, of course, that n is a nonnegative integer variable, A is a set variable, and $a, c \in S$).   △

EXAMPLE 1.25   Suppose that we have an algebraic structure containing an algebra of sets and a partition algebra (Section 0.6)

$$\mathscr{P}(S) = (\mathscr{P}(S), \cup, \cap, \sim) \qquad \Pi(S) = (\Pi(S), +, \cdot)$$

for the same set S. Then the correspondence

$$A \to \{\overline{A}; \overline{\sim A}\}$$

is a transfer function from the set algebra into the partition algebra. For example,

$$\{3, 5\} \to \{\overline{3, 5}; \overline{1, 2, 4}\}$$

in case $S = 5^+$. If A is a set variable and π is a partition variable, then

$$(\{\overline{A}; \overline{\sim A}\} + \{\overline{1}; \overline{2, 3}; \overline{4, 5}\}) \cdot \pi$$

is an algebraic expression in the given structure. Note that $\{\overline{1}; \overline{2, 3}; \overline{4, 5}\}$ is a constant in $\Pi(S)$ for $S = 5^+$.   △

EXAMPLE 1.26   In the FORTRAN language, FLOAT and IFIX are transfer functions from the integers to real numbers, and vice versa. In ordinary algebra, the absolute value is a transfer function from Z to N.   △

EXAMPLE 1.27   If the set S is considered an unstructured algebra, that is, an algebra with no operations, then the singleton transfer function

$$\{\ \}: S \rightarrow \mathscr{P}(S)$$

associates with each $x \in S$ the subset $\{x\} \in \mathscr{P}(S)$.   △

So far we have deliberately avoided an important question about our simulation scheme (Figure 1.2): when is such a simulation feasible? Suppose that we have an ideal computer, one with infinite storage capacity (otherwise, we could not even contemplate an exact simulation of N). When can we decide "mechanically" as to the truth or falsity of one of our propositions, given definite arguments? When can our ideal computer make such a decision? In the same vein, when can we mechanically or, as we shall also say, "effectively" compute or otherwise determine the value of a function, given its arguments? Can our algebraic operations be implemented by the ideal machine?

These puzzling questions were considered by several famous mathematicians of the 1930s, notably Church, Turing, Post, Kleene, and Gödel. Unknowingly, they were, in a sense, the forerunners of today's computer scientists. It is rather surprising that such questions were even conceived before the dawn of the computer era. Equally amazing is the degree of consistency in the answers they each individually proposed. Unfortunately, a precise discussion of these questions is not possible here, though we will come back to them at a later point (Chapter 6).

For the time being, we are interested only in gaining some understanding of this delicate balance between what is known to be "effective" and what is not. When we are dealing with finite sets, there is no problem. Even if the set is large, such as encountered by the curator of the British Museum, it is possible to exhaustively survey all the items, and in a finite but perhaps very long time, the answer to each and every question that can be posed can be found. But what are we to do in the case of propositions and functions over (countably) infinite sets?

EXAMPLE 1.28   Consider the following unary proposition in Z^+:

$$P = \langle n: \text{there exists } x, y, z \text{ in } Z^+ \text{ with } x^n + y^n = z^n \rangle$$

We certainly have

$$1 \in P \quad \text{since} \quad 1^1 + 2^1 = 3^1$$
$$2 \in P \quad \text{since} \quad 3^2 + 4^2 = 5^2$$

But Fermat's last theorem would have us believe that $n \notin P$ for all $n > 2$. If this were true, then we could always decide P effectively; that is,

$$n \in P \Leftrightarrow n \leq 2$$

and we have no particular difficulty making the latter determination, given any $n \in Z^+$. Yet as history reports, Fermat wrote that he had a remarkable proof of the theorem but that there was not enough room for it in the margin of his notebook. Romantics like to think that this seventeenth-century "Prince of Amateurs" was correct, but to this day, his "last theorem" remains only a conjecture; and we can only say that there may or may not be a finite mechanical procedure for deciding the truth or falsity of P for a given n. Certainly no one knows of an appropriate algorithm or, indeed, whether there is one. Some propositions are known to be effectively undecidable, and we would surely want to avoid these. Here we do not know whether the given proposition is effectively decidable or not. Since no one knows for sure, it is best that we not admit P as a "primitive" proposition for one of our algebraic structures, for that would once again, frustrate the whole intention of the simulation symbolized by Figure 1.2. △

It seems that we would want somehow to restrict our algebraic structures so as to involve only *effectively decidable* propositions—those for which a finite mechanical procedure exists for determining the truth or falsity for every instance of its arguments. Similarly with the evaluation of functions, we would want to be sure that each such evaluation is *effectively computable* according to some finite sequence of mechanical calculations. We make these assumptions whenever we speak of an *effective* algebraic structure $\mathscr{A} = (\mathscr{A}; \mathscr{F}; \mathscr{P})$. Thus it is all right to include the factorial function (Section 1.1) or the binary proposition given by the comparison relation (Example 1.18), but as things stand, we are not prepared to allow the proposition P of Example 1.28. In Chapter 6 we will establish the existence of undecidable (noneffective) propositions, as well as functions that are not effectively computable. Even the existence of such propositions or functions may be difficult to imagine at this point, but rest assured that we are not whistling in the dark. Fortunately, the structures we will ordinarily encounter possess such "primitive" propositions $\mathscr{P}$ and functions $\mathscr{F}$ that their effective decidability and computability are unquestioned. Therefore, the feasibility of our intended simulation (Figure 1.2 again) is assured.

EXERCISES

1 Describe the following as transfer functions.
   (a) The absolute value of a complex number

(b) The absolute value of an integer

(c) The square root of a positive integer

(d) The number of blocks in a partition

(e) The congruence class (modulo n) of an integer [see Exercise 4(d) of Section 1.2]

2 Show that the following are algebraic expressions in the corresponding algebraic structures:

(a) $|A \cup \{a, c\}| + 4 \cdot n$ (Example 1.24)

(b) $|(A \cap B) \cup \{a\}| \cdot n$ (Example 1.24)

(c) $(\{\overline{A}; \overline{\sim A}\} + \{\overline{1}; \overline{2, 3}; \overline{4, 5}\}) \cdot \pi$ (Example 1.25)

Of what types are these expressions?

3 Of what types are the following expressions?

(a) $\#(\pi) + 3$

where $\#$ is the transfer function of Exercise 1(d)

(b) $(|z| + 17.6) \cdot x$

where $|\ |$ is the transfer function of Exercise 1(a)

(c) $|(A \cup \{a\}) \cap B| - 3 \cdot n$

where $|\ |$ is the transfer function of Example 1.24

4 Which of the following algebraic structures is "effective"?

(a) the algebra of real numbers $R = (R, +, \cdot)$

(b) the algebra of integers $Z = (Z, +, \cdot; \leq)$

(c) the algebra of positive integers $Z^+ = (Z^+, +, \cdot; \leq, P)$ where P is the proposition of Example 1.28

(d) the algebra of rational numbers $Q = (Q, +, \cdot; \leq)$

5 Show how the various algebraic expressions of the several exercises correspond to appropriate functions by tabulating a few function arguments and values.

(a) Exercise 2(a)   (b) Exercise 2(b)   (c) Exercise 2(c)

1.5 ALGEBRAIC FLOWCHART LANGUAGE

Given the intuitive idea of an "effective" algebraic structure

$$\mathscr{A} = (\mathscr{A}; \mathscr{F}; \mathscr{P}) = (A_1, A_2, \ldots; \mathscr{F}; \mathscr{P})$$

such as was just described, with associated constants A_α, variables $\mathscr{V}_\alpha$, and algebraic expressions $\mathscr{E}_\alpha$ of type α, we can now introduce a flowchart language for the design and communication of algorithms. In speaking to computer science students, it would probably suffice to say that a *flowchart* is simply a diagram of stylized "statement boxes" interconnected according to certain formation rules by directed *flowlines*. Each flowchart will have exactly one encircled "start" statement, and we may just as well assume similarly that there is but one "stop" statement. There is only one flowline emanating from the start statement, and owing to the remarks to be made concerning "collectors," we can similarly assume that there is only one flowline terminating at the stop statement. Naturally, all flowlines are provided with arrows to indicate the order in which the various statement boxes are encountered.

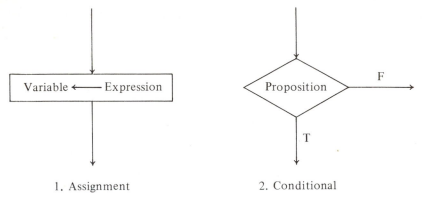

1. Assignment   2. Conditional

FIGURE 1.3

In the interest of simplicity, we use (other than "start" and "stop") only two kinds of flowchart statements (see Figure 1.3). In fact, on another level of precision we could use the language of Chapter 2 to say that a *flowchart* is a connected, labeled, and directed graph with only three kinds of nodes:

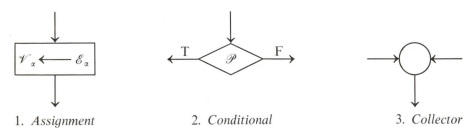

1. *Assignment*   2. *Conditional*   3. *Collector*

besides those that "orient" the graph by providing a start and a stop. The collector nodes are thought to embody only "dummy statements," and their sole purpose is to allow for the merging of two or more flowlines into one. In fact, in our flowcharts we will never represent the collectors explicitly; we prefer, instead, to freely merge flowlines whenever necessary.

The assignment statements will always be enclosed in a rectangular box, and (with the implicit use of collectors) they will have one entry and one exit flowline. It is assumed that the variable and the expression will be of the same type. If it is convenient, we may write more than one assignment statement in a single box. Then the statements are simply to be executed from top to bottom as described below. The conditional statements appear within diamond-shaped boxes with two exit flowlines, one marked F and one marked T.

The interpretation of these statements is straightforward. If we encounter the statement $V \leftarrow E$, we will think as before of the algebraic expression E as corresponding to a function. The value of this expression (function) computed according to the substitution of current values of all variables occurring in it, is assigned to the variable V, after which we proceed to the statement found upon following the exit flowline. If, on the other hand, we encounter the conditional statement $\langle P \rangle$, then the truth or falsity of P must first be decided, again according to the current values of all variables; thereafter we follow along the corresponding exit flowline (F if P is false, T if P is true). Actually we are to think of the algebra (machine) as performing the computations and making the decisions. The reference to an apparent human computer is made in the interest of clarity.

Most computer languages will provide input and output statements for communicating data to and from a machine, but it would detract from the readability of our algorithms if we incorporated these statements into our flowchart language. Instead, we include with each flowchart an input–output box that gives this information, thus making it easier to understand the overall performance of the algorithm. At the same time, we will *declare the type* of all variables appearing in the algorithm, according to the format

$$V, W, \ldots : \textbf{type } \alpha$$

Ordinarily we will provide a paragraph of commentary as well.

In whatever algebraic structure you find yourself involved, there will invariably be propositions, functions, or routines, of considerable and sometimes immediate interest that are not "primitive" to the structure at hand (otherwise, there would be no need for programmers)! Therefore, it is necessary that one be able to devise algorithms for their realization. The flowchart is a conceptual aid for formulating such algorithms.

EXAMPLE 1.29　In the algebra

$$Z = (Z, \mathscr{F} ; \mathscr{P}) = (Z, +, \cdot ; \leq)$$

there is no intrinsic capability for deciding whether or not two given integers are equal. It does not matter that you can tell at a glance. The fact remains that the given algebra (machine) does not have such a provision. If we recall, however, that

$$n = m \Leftrightarrow n \leq m \quad \text{and} \quad m \leq n$$

it is easy to describe an algorithm in the flowchart language for deciding the question. (See Figure 1.4.)

One may think of the pair of integers n, m as "inputs" to the algorithm. The output is the value of a logical variable L. Evidently at the conclusion of the

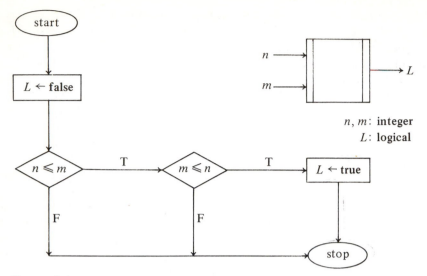

FIGURE 1.4

execution we will have

$$L = \textbf{false} \quad \text{if} \quad n \neq m$$
$$L = \textbf{true} \quad \text{if} \quad n = m$$

Thus our algorithm definitely decides on equality in the algebra Z.   △

This first example is not at all representative. In fact, it is pedantic in the extreme, but it does suffice for illustrating the flowchart symbolism and for demonstrating the associated notational conventions. Note that even in this simple flowchart, there were two algebras: Z and B. Thus we have an algorithm involving two types of variables: integer and logical. The type declarations, then, serve to clarify the roles of the variables occurring in the algorithm. The real necessity for such bookkeeping will become increasingly apparent as we graduate to examples of greater complexity.

EXERCISES

1 In the algebra of Example 1.29 describe an algorithm in flowchart language for deciding whether
(a) $n < m$ (b) $n = 2m$
(c) $n + m \neq 3$ (d) $n > m$
(e) $n - m < 0$ (f) $n < 0$
[*Hint*: In some cases it may be convenient to introduce "auxiliary" variables.]

2 In an algebra of subsets (Example 1.19) describe an algorithm in flowchart language for deciding whether

(a) $A = B$ (b) $A \cap B = \emptyset$
(c) $A \cup B = S$ (d) $A \neq \emptyset$

3 If we have the values $n = 3$, $m = 7$ before the execution of the flowchart segments shown in Figure 1.5, what will be their values after the execution?

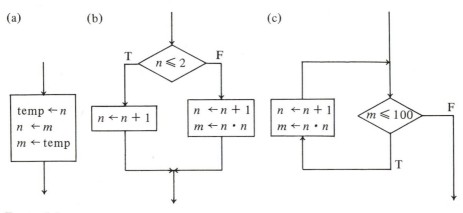

FIGURE 1.5

4 If we have the values $A = \{1, 3\}$, $B = \{2, 6\}$
   (with $S = 6^+$) before the execution of the flowchart segments shown in Figure 1.6, what will be their values after the execution?

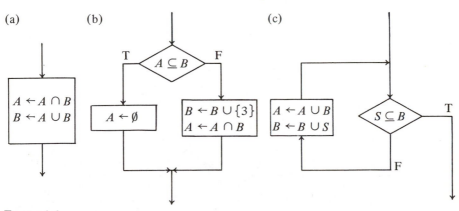

FIGURE 1.6

5 What is accomplished by the algorithms symbolized by the flowcharts shown in Figure 1.7? [*Hint*: Your answer in (c) may have something to do with triangles.]

(a)

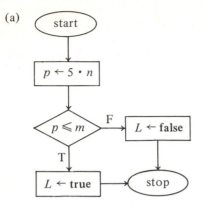

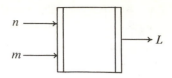

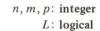

n, m, p: **integer**
L: **logical**

(b)

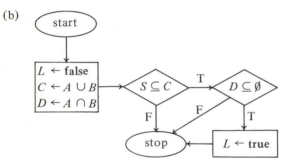

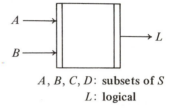

A, B, C, D: **subsets of** S
L: **logical**

(c)

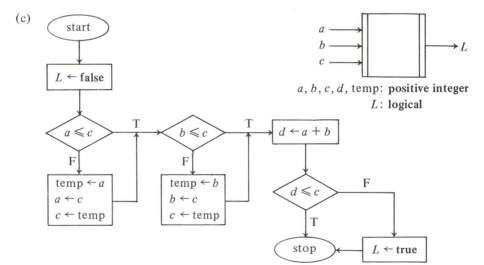

$a, b, c, d,$ temp: **positive integer**
L: **logical**

FIGURE 1.7

1.6 ALGORITHMIC CLOSURE

We would like to provide a formal characterization of some of the ideas concerning subroutines (FORTRAN) or procedures (ALGOL), and to incorporate these ideas into our flowchart language. As some readers may already know, these ideas basically involve the naming of an existing algorithm so that whenever the name is invoked, the corresponding algorithm will be activated. Suppose for definiteness that the algorithm decides a proposition or computes a function. We take the view that the actual naming of the proposition or function is not so important as the activation mechanism itself.

Consider Example 1.29 once more. If we were to write more algorithms in the same algebraic structure, we would no longer have to worry about equality. At the very least, we should be able to reproduce that flowchart whenever a decision on equality is required. Recognizing this fact, however, why do we not admit equality, say, as a proposition in good standing, after an algorithm deciding that question has been given?

To formalize this idea, let us assume again that we are given an effective algebraic structure $\mathscr{A} = (\mathscr{A}; \mathscr{F}; \mathscr{P})$ with its associated collections of variables $\mathscr{V}_\alpha$ and expressions $\mathscr{E}_\alpha$. And let us consider any proposition $P \subseteq A_{\alpha_1} \times A_{\alpha_2} \times \cdots \times A_{\alpha_m}$ whether in $\mathscr{P}$ or not. Now suppose that there exists an algorithm or flowchart with input variables $V_1, V_2, \ldots, V_m$ (V_i of type α_i) and a logical output variable L, having the following properties:

(1) The algorithm always halts, that is, leads to the "stop" instruction, regardless of the initial values of the input variables.
(2) At the conclusion of the execution of the algorithm, we have

$$L = \textbf{false} \quad \text{if} \quad P \text{ is false} \quad (V_1, V_2, \ldots, V_m) \notin P$$
$$L = \textbf{true} \quad \text{if} \quad P \text{ is true} \quad (V_1, V_2, \ldots, V_m) \in P$$

We will then say that the proposition P (ordinarily outside $\mathscr{P}$) is in the *algorithmic closure* $\overline{\mathscr{P}}$ of $\mathscr{P}$; that is, we write $P \in \overline{\mathscr{P}}$.

Similarly, if there exists an algorithm or flowchart with input variables $V_1, V_2, \ldots, V_n$ (V_i of type α_i) and output variable V of type α, such that

(1) the algorithm always halts (regardless of the values initially given to the input variables),
(2) at the conclusion of the execution of the algorithm, we have

$$V = f(V_1, V_2, \ldots, V_n)$$

then the function

$$f: A_{\alpha_1} \times A_{\alpha_2} \times \cdots \times A_{\alpha_n} \to A_\alpha$$

is said to be in the *algorithmic closure* $\overline{\mathscr{F}}$ of $\mathscr{F}$.

There are several important remarks that should be made to clarify all of this. First of all, it would perhaps be better if we spoke of the closure of the algebraic structure itself, because the functions in $\mathscr{F}$ participate in forming $\overline{\mathscr{P}}$, and conversely, propositions in $\mathscr{P}$ are involved in the formation of $\overline{\mathscr{F}}$. So let us admit this much when we speak in the future of the *algorithmic closure* $\overline{\mathscr{A}}$ of an algebraic structure $\mathscr{A} = (\mathscr{A}; \mathscr{F}; \mathscr{P})$ and write $P \in \overline{\mathscr{A}}$ or $f \in \overline{\mathscr{A}}$, as the case may be. Secondly, the student will very likely ask, "How do I know whether a given algorithm terminates?" The question seems important, since the idea of termination was crucial in the definitions. Indeed, it is a reasonable question. We will see, however, as the chapter unfolds that there are numerous specific examples where the "finiteness" of an algorithm can in fact be demonstrated. On the other hand, surprisingly perhaps, the general problem of determining whether a given algorithm terminates is itself effectively undecidable! Unfortunately, this fact cannot be demonstrated until we have a precise definition for the idea of an algorithm (Chapter 6). Nevertheless, it is highly important that the student develop an intuitive feeling for this "undecidability of the halting problem" at an early stage.

Finally, it is our hope that the reader somehow feels that we have not left the realm of the "effective" by extending our flowchart language to allow for corresponding statements (assignment and conditional) in the algorithmic closures. In any case, he or she will certainly be converted to this conviction as a variety of examples are presented.

EXAMPLE 1.30   Consider the positive integers as constituting an algebra

$$Z^+ = (Z^+, +, \cdot; \leq)$$

as in Example 1.18. Then, together with the algebra of logical values, we have an algebraic structure $\mathscr{A} = (\mathscr{A}; \mathscr{F}; \mathscr{P})$ comparable to that of Example 1.29. In fact, the same algorithm shows that equality is in $\overline{\mathscr{A}}$. The reader can provide similar flowcharts to show that $<, >, \geq$, and $\neq$ are also in $\overline{\mathscr{A}}$. According to the foregoing discussion, we are now permitted to use any of these comparisons in a conditional statement.

In Figure 1.8 we present a flowchart that shows that divisibility is in $\overline{\mathscr{A}}$. Note particularly that our convention for executing compound assignment statements by starting at the top is not affected by a flowline entering from the bottom. Thus we have $q \leftarrow q + 1$, then $p \leftarrow q \cdot n$ in Figure 1.8. In any event, we claim that at the conclusion of the execution of this algorithm, we have

$$L = \textbf{false} \quad \text{if} \quad n \nmid m$$
$$L = \textbf{true} \quad \text{if} \quad n \mid m$$

regardless of the initial values given to n and m. To see this, observe that the auxiliary variable q is first assigned the value $q = 1$, and successive multiples $q \cdot n$ of n are formed until $q \cdot n \geq m$ (as must eventually happen when m is fixed). If this last multiple is in fact equal to m, we will have $n \mid m$; otherwise, we will not.   △

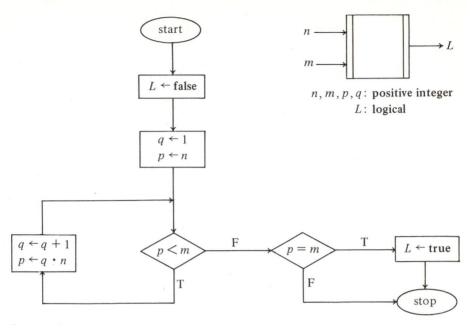

FIGURE 1.8

EXAMPLE 1.31   For any set S the power set algebra of Example 1.19

$$\mathscr{P}(S) = (\mathscr{P}(S), \cup, \cap, \sim; \subseteq)$$

can be embedded along with the algebra of logical values in an algebraic struc-
ture $\mathscr{A} = (\mathscr{A}; \mathscr{F}; \mathscr{P})$. Further, an algorithm analogous to that of Example 1.29
shows that (set) equality is a proposition in the closure $\overline{\mathscr{A}}$. So is the proposition that asks whether or not two given subsets are disjoint, as shown by the
flowchart of Figure 1.9.   △

EXAMPLE 1.32   Using the result just established, we can consider a larger
algebraic structure still, one that also includes the algebra of positive integers.
Then we can ask ourselves whether we can design an algorithm that will test
whether or not a given collection $A_1, A_2, \ldots, A_n$ of subsets of S constitutes a
partition of S. A clue as to how we might proceed is provided by conditions
I, II in the definition of a partition (Section 0.2). We can form the union of the
A_i sequentially to see whether we will obtain S, then test for disjointness exhaustively for all pairs A_i, A_j such that $i < j$. Actually, however, it is more
efficient to observe that disjointness can be checked while forming the union.
    Thus if $T_i = A_1 \cup A_2 \cup \cdots \cup A_i$, we must certainly have

$$A_1, A_2, \ldots, A_n \text{ a partition} \Leftrightarrow \begin{cases} \bigcup\limits_{i=1}^{n} A_i = S & \text{(that is, } T_n = S\text{)} \\ A_{i+1} \cap T_i = \varnothing & (1 \leq i < n) \end{cases}$$

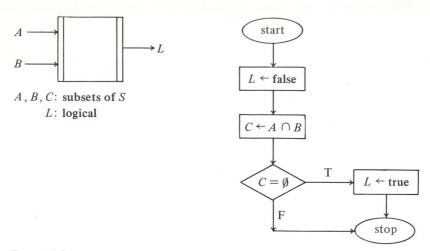

FIGURE 1.9

An algorithm based on this observation is shown in Figure 1.10. Note particularly how we invoke the algorithm of Example 1.31 as a subroutine. As a result of our simplification, this test for disjointness need be applied only n times. Actually the algorithm can be modified slightly so that we can get by with only $n - 1$ tests. In any case, this certainly compares favorably with the number of tests

$$1 + 2 + \cdots + (n - 1) = \frac{n(n - 1)}{2}$$

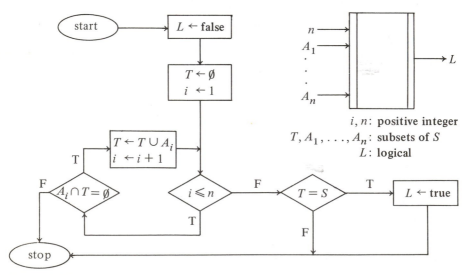

FIGURE 1.10

required according to the original formulation of the problem. Note our use once more of the IRS formula of Examples 1.2 and 1.3. $\triangle$

The last example illustrates several important principles concerning "programming philosophy," especially as it pertains to the aims of this chapter. In our algebraic algorithms no reference is made to the actual implementation. By design, the data representations and the coding into a real machine program have been deferred; in fact, they are outside the scope of this text. Moreover, one can make a very good case for claiming that this is the best way to proceed anyway. If somehow we should have delved immediately into the problem of physical implementation in the above example, quite possibly we would have overlooked our important simplification, one that could have been discovered only in the algebraic framework itself. The lesson here? Use your knowledge of the algebraic structure in which the problem arises as an aid in formulating an efficient algorithm! Almost without exception, the resulting efficiency will transfer to any eventual implementation.

One other point is suggested by the last two examples, particularly as regards the algorithms to be found later in the text. Ordinarily our use of the concept of closure, as applied to propositions and to expressions, will not proceed from the bottom up (as exemplified here) but from the *top down*. That is to say, we will often present algorithms that involve propositions and/or expressions which are not obviously primitive to the algebraic structure at hand. It is then left for the reader to assume that they are in the closure (otherwise, we would not have used them). And if a student is so inclined, he or she may stop to render an algorithm (subroutine) that shows that the assumption is well founded. This is the spirit in which this section has been presented.

EXERCISES

1 Consider the algebraic structure $\mathscr{A} = (\mathscr{A}; \mathscr{F}; \mathscr{P})$ of Example 1.30 and show that the following propositions are in $\overline{\mathscr{A}}$:

(a) $\langle (n, m, p): n, m, p$ are the (lengths of) sides of a triangle$\rangle \subseteq Z^+ \times Z^+ \times Z^+$

(b) $\langle (n, m, p): n, m, p$ are the sides of a right triangle$\rangle \subseteq Z^+ \times Z^+ \times Z^+$

(c) $\langle n: n$ is odd$\rangle \subseteq Z^+$

(d) $\langle n: n$ is prime$\rangle \subseteq Z^+$

Show that the following functions are in $\overline{\mathscr{A}}$:

(e) $f(m, n) = m^n$

(f) $f(n) = 2^n$

2 Consider the algebraic structure $\mathscr{A} = (\mathscr{A}; \mathscr{F}; \mathscr{P})$ of Example 1.31 and show that the following propositions are in $\overline{\mathscr{A}}$:

(a) $\langle (A, B): \{A, B\}$ is a partition of $S\rangle$

(b) $\langle (A, B, C): A \cap B \subseteq C\rangle$

(c) $\langle (A, B, C): ((A \cap (\sim B)) \cup ((\sim A) \cap B)) \cap C = \varnothing\rangle$

(d) $\langle (A, B): A \subseteq B$ and $A \neq B\rangle$

Show that the following functions are in $\overline{\mathscr{A}}$:

(e) $f(A, B) = (A \cap (\sim B)) \cup ((\sim A) \cap B)$

(f) $f(A, B, C) = \sim(A \cap B \cap C)$

3 Explain why you feel that the propositions (functions) in Exercise 1 and 2 are effectively decidable (computable).

4 Give a detailed account (including a flowchart) showing why you feel that the binary proposition

$$RP(n, m) = \langle (n, m): n \text{ and } m \text{ are relatively prime} \rangle \subseteq Z^+ \times Z^+$$

is effectively decidable. [*Hint*: Recall that n and m are said to be *relatively prime* if they have no common divisor (except 1).]

5 Repeat Exercise 4 for the unary proposition

$$P(n) = \langle n: n \text{ is prime} \rangle$$

6 Repeat Exercise 4 for the unary proposition

$$E(n) = \langle n: n \text{ is even} \rangle$$

7 Repeat Exercise 4 for the unary proposition

$$S(n) = \langle n: n \text{ is square free} \rangle$$

that is, there does not exist a square factor of n.

1.7 SYNTAX AND SEMANTICS

In the formal study of any complex subject drawn from our everyday experience, surely the technique of abstraction is the most powerful link in the chain of logical devices used to develop a theory. It is by this technique that we agree to concentrate on the similarities and ignore the irrelevant discrepancies from one situation to the next. Almost as important in the underlying logical framework, however, is the representation we choose for the real subject of investigation. If this representation is well chosen, we will almost identify the model with the abstract notion it replaces. Such may be the case with our highly stylized flowcharts, presented in the preceding sections, for the informal notion of an algorithm or finite mechanical procedure. However loosely we may have defined the rules of formation for flowcharts, the reader probably feels that he or she has grasped the nature of the permitted constructs so as to be able to recognize a "well-formed" flowchart when he or she sees one. But does a flowchart always represent an algorithm? To answer this question, we must realize that there is a distinction between form and substance, or as we prefer to say, between syntax and semantics. The latter has to do with meaning: just what is it that a given flowchart does?

If we begin with a fairly definite algorithmic process in mind, familiar from our everyday experience, and if we are reasonably adept in the use of the algebraic flowchart language, we can expect to be fairly successful in translating the process into a flowchart. Then the chances are quite good that the flowchart will do just what is intended. Even so, it would be useful to have some means for actually verifying that this is indeed the case. In the next few sections, we shall provide an introduction to certain techniques that are designed to attack this problem and also the converse problem. That is, suppose we start from the other direction. Suppose a flowchart is presented that apparently was written in accordance with our formal rules; that is, it is syntactically correct. Then can we say with any certainty what it is that the flowchart accomplishes? Can we learn its meaning? Can we even be sure that it represents an algorithm?

EXAMPLE 1.33   Look carefully at the flowchart of Figure 1.11 and try to decide what it accomplishes (no one knows for sure)! First of all, you will note the use of propositions ($\langle n$ odd$\rangle$) and expressions ($n \div 2$) that are not primitive to the algebra Z^+. It follows that the remarks made in the last paragraph of Section 1.6 are applicable here!

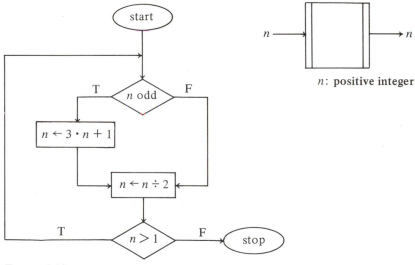

FIGURE 1.11

If this flowchart yields an output value for n, it must be $n = 1$; otherwise, we will never arrive at "stop." And it could be that this flowchart is simply an unusual algorithm for realizing the constant function $f(n) \equiv 1$, but no one knows for sure because it has never been established that the loop is always traversed only a finite number of times. (Recall that an algorithm is to be a finite procedure.) Therefore, we cannot say whether this is an algorithm or not.   △

While it is of no help in this particular example, the notion of the *trace* of an algorithm is useful in the informal analysis of flowcharts as an aid in gaining some understanding of their behavior. The idea is really nothing more than armchair "debugging." We simply draw up a table whose columns are headed by each of the variable names appearing in the flowchart. We then literally trace or step our way through the flowchart, recording all changes of values of the variables in the table as they occur. This is done for various test values for the inputs, and assuming that we arrive at "stop," the most recent values of the output variables give an indication as to whether or not the algorithm performs properly. Of course, this assumes that we know already the intent of the algorithm. Even then, we can only detect the existence of malfunction in this way; we can never *prove* the absence of malfunction by examining only selected test cases, finite structures notwithstanding.

EXAMPLE 1.34   The divisibility algorithm of Example 1.30 provides a good illustration of the concept of a trace table. (See Table 1.1.) Input, auxiliary, and output variables are listed from left to right. Suppose that we try the test case $n = 3$, $m = 12$, where of course we expect a final value $L = $ **true**. First, we have to enter the input values of the test case:

TABLE 1.1

	n	m	q	p	L	
Start	3	12	1	3	**false**	
			2	6		
			3	9		
			4	12	**true**	Stop

Then we trace through the flowchart, changing the values of the variables accordingly.

We first encounter the assignment $L \leftarrow$ **false**; this value is entered at the appropriate column. Then we arrive at the pair of assignments $q \leftarrow 1$; $p \leftarrow n$; these values are entered. Note that the current value of n is 3, so we enter $p = 3$. Then we find a conditional statement $\langle p < m \rangle$. Since p is 3 and m is 12, we exit along the flowline marked T. This causes us to arrive at the pair of assignments $q \leftarrow q + 1$; $p \leftarrow q \cdot n$. The current value of q is 1 and so it gets updated to $q = 2$, after which we set $p \leftarrow q \cdot n = 2 \cdot 3 = 6$. Still $p < m$, and so we traverse the loop once again. After three circulations, we will have $p = 12$, so that $p < m$ is false (m has not changed). Then, since $p = m$ is true, we set $L \leftarrow$ **true** and stop.

The preceding analysis may give no firm assurance that the correct performance will result if we let $n = 93$ and $m = 9034$. But if the behavior of the algorithm is clearly understood in examining the test case, then we are certainly justified in our faith.   △

So far, our selection of algorithms has been guided mainly by pedagogical motives. We have sought to develop a basic understanding of the algebraic flowchart language and a general appreciation for the concepts of effective decidability and computability. Hereafter we may regard the flowchart language as a tool for the design and communication of effective procedures. We shall become interested in such procedures either because they illuminate a particular algebraic structure or because they solve an important problem arising in the structure. It is therefore appropriate that we stop to document these procedures by way of a formal algorithm. In this spirit we conclude this section with a more detailed presentation of the division algorithm and the Euclidean algorithm, first encountered in Section 1.1. They are the first of a sequentially numbered series of algorithms sprinkled throughout this text.

EXAMPLE 1.35   In the division algorithm (Algorithm I) the quotient q is found by forming successive multiples of b, much as was done in the case of Example 1.30. As an illustration, suppose that we trace the behavior of the algorithm for the input data $(a = 22, b = 5)$ of Example 1.7. In the trace table (Table 1.2), we see that the correct result $(q = 4, r = 2)$ is obtained for the quotient and the remainder. It seems that this is the first algorithm we have presented that computes a function (actually two functions) rather than deciding a proposition. In accordance with the discussion of Section 1.6, the function $q(a, b) = a \div b$ may now be used as a "subroutine" in succeeding algorithms.   △

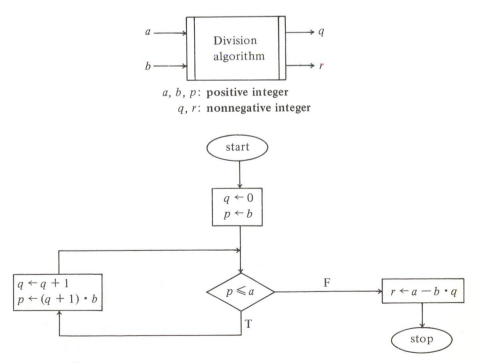

a, b, p: **positive integer**
q, r: **nonnegative integer**

ALGORITHM I

TABLE 1.2

a	b	p	q	r
22	5	5	0	
		10	1	
		15	2	
		20	3	
		25	4	2

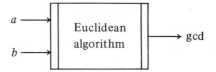

a, b, gcd, temp: **positive integer**

r: **nonnegative integer**

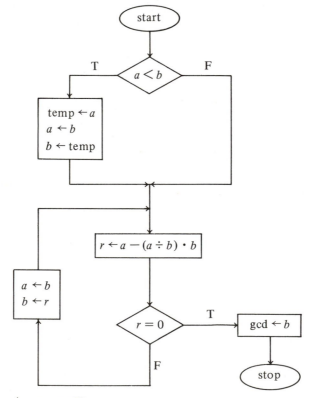

ALGORITHM II

We recall that the Euclidean algorithm determines the greatest common divisor of positive integers a, b. This is a very important function in number theoretic investigations. More and more, these investigations are carried out with the aid of computers, and so a number of algorithms have been devised for computing the greatest common divisors. The one given here (Algorithm II) is perhaps the best known. You will notice that a portion of the division algorithm is used as a "subroutine." Otherwise, the procedure is in strict accordance with the observations of Section 1.1, once a has been arranged to be the larger of the two inputs a, b.

EXAMPLE 1.36 Suppose that we test Algorithm II with the input data of Example 1.8, but with the values interchanged ($a = 180, b = 252$). (See Table 1.3.) Then we can illustrate how the box with three assignment statements will interchange a, b if necessary to achieve $a \geq b$. Observe the shifting of the values

TABLE 1.3

a	b	r	temp	gcd
180	252		180	
252	180	72		
180	72	36		
72	36	0		36

of b, r into a, b, as is characteristic of the Euclidean greatest common divisor algorithm. When $r = 0$, the current divisor (in this case, 36) is the greatest common divisor of a and b. △

EXERCISES

1 Perform a trace of the flowchart of Example 1.33 (Figure 1.11) with the following input data:
(a) $n = 7$ (b) $n = 9$ (c) $n = 11$ (d) $n = 12$ (e) $n = 13$

2 Perform a trace of the flowchart of Example 1.30 (Figure 1.8) with the following input data:
(a) $n = 3, m = 18$ (b) $n = 3, m = 19$
(c) $n = 6, m = 25$ (d) $n = 6, m = 30$
(e) $n = 7, m = 30$ (f) $n = 7, m = 42$

3 Perform a trace of the flowchart of Example 1.32 (Figure 1.10) with the following input data ($S = 7^+$):
(a) $n = 4, A_1 = \{1, 3, 7\}, A_2 = \{2\}, A_3 = \{6\}, A_4 = \{4, 5\}$
(b) $n = 4, A_1 = \{1, 3\}, A_2 = \{2\}, A_3 = \{6\}, A_4 = \{4, 5\}$
(c) $n = 4, A_1 = \{1, 3\}, A_2 = \{2\}, A_3 = \{6\}, A_4 = \{3, 4, 5\}$

4 Perform a trace of Algorithm I with the data in Exercise 10 of Section 1.1.

5 Perform a trace of Algorithm II with the data in Exercise 11 of Section 1.1.

6 Perform a trace of the algorithm in Exercise 5(a) of Section 1.5 with the data:
   (a) $n = 3, m = 14$ (b) $n = 4, m = 22$

7 Perform a trace of the algorithm in Exercise 5(b) of Section 1.5 with the data ($S = 7^+$):
   (a) $A = \{1, 3, 7\}, B = \{2, 4, 5, 6\}$ (b) $A = \{1, 3, 4, 7\}, B = \{2, 4, 5, 6\}$

8 Perform a trace of the algorithm in Exercise 5(c) of Section 1.5 with the data:
   (a) $a = 7, b = 5, c = 4$ (b) $a = 7, b = 5, c = 1$

*1.8 STRUCTURED ALGORITHMS

In the preceding section we began to make reference to the meaning of an algorithm. More specifically, we sought to describe the performance of an algorithm or flowchart through the trace induced by specific input values. It is clear that this technique can illuminate many of the important features of an algorithm, and can even be used to detect logical errors in the formulation of the algorithm. But it will rarely suffice for proving that such errors are totally nonexistent. It may occasionally happen that only finitely many input values are intended and that these are sufficiently few in number for an exhaustive testing to be feasible. However, this is clearly the exception and not the rule.

Since the early 1970s, a significant advance in computer science has been to provide the means for actually proving the correctness of algorithms. It can be shown that these techniques are most successful when the algorithm (the flowchart, say) is *structured*. By this we mean that the flowchart is block decomposable into nested elementary flowchart constructs, the latter being chosen from the now standard repertoire first introduced or at least popularized by Dijkstra* and his colleagues.

EXAMPLE 1.37   Consider Algorithm I, the division algorithm. We have already run a test (Example 1.35) of this procedure for specific input values to illustrate the concept of the trace of an algorithm. On the basis of this test, or perhaps 100 similar tests on randomly selected data, the student might be tempted to conclude that the algorithm indeed does what it was intended to do; this would not be a valid conclusion.

Instead, one might argue that because an actual computer is only capable of representing a finite (but large) number of the integers, say up to 2^{48}, we could test all such pairs a, b as would ever be encountered. But even if the

* See Reference 1.5

entire algorithm could be executed in a microsecond (10^{-6} second), we would
still need

$$2^{48} \ 2^{48} \ 10^{-6} \text{ seconds} \doteq 2.5 \times 10^{15} \text{ years}$$

to complete the testing!   △

For the purpose of analyzing algorithms, we introduce three basic rules,
called the *rules of algorithmic verification*. They will seem to be entirely natural,
almost obvious on the face of it. Nevertheless, they are extremely powerful.
The rules are in direct correspondence with the statements and interconnections
permitted in our flowchart language. Each rule refers to the propositions
$Q_1, Q_2, \ldots$, variously called *assertions* or conditions that are satisfied (true) at
specific points in the flowchart. The Q_i are entered literally as *annotations* to
the flowchart.

Rule 1   (Junction rule)

If two (or more) flowlines merge into one at a junction, and we make the annotated assertions Q_1, Q_2, Q at the points indicated in Figure 1.12, then we
must have the logical implications:

$$Q_1 \Rightarrow Q \qquad Q_2 \Rightarrow Q$$

Similarly, $Q_i \Rightarrow Q$ for all i if more than two flowlines merge into the junction.

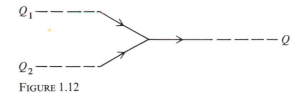

FIGURE 1.12

EXAMPLE 1.38   It is logically consistent to have the annotated flowchart junction shown in Figure 1.13, whereas the labeling in Figure 1.14, is inconsistent
with Rule 1 because $x \geq 0 \nRightarrow x > 0$.   △

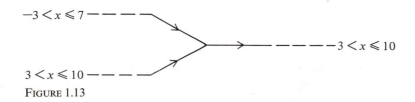

FIGURE 1.13

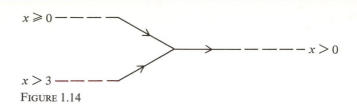

FIGURE 1.14

EXAMPLE 1.39   The annotated junction shown in Figure 1.15 meets the conditions of Rule 1, that is, we only have to apply (a) in Example 0.10 to obtain

$$A \subseteq B \Rightarrow A \subseteq B \cup C$$
$$A \subseteq C \Rightarrow A \subseteq B \cup C.$$

△

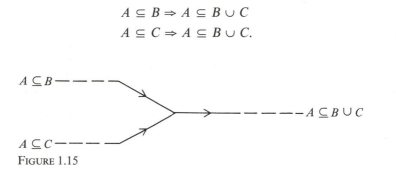

FIGURE 1.15

Rule 2   (Conditional rule)

If the entry point to a conditional statement $\langle P \rangle$ is annotated with the assertion Q, then the T, F exit points are to be annotated $Q \wedge P$ and $Q \wedge \neg P$, respectively, as shown in Figure 1.16.

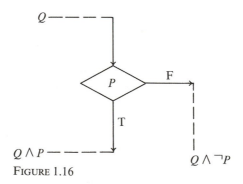

FIGURE 1.16

EXAMPLE 1.40   Suppose that we have the annotation $-3 < x \leq 3$ at the entry to the conditional $\langle x > 0 \rangle$. Then the exit points should be labeled as in Figure 1.17. Note, for example, that "$-3 < x \leq 3$ *and* $x > 0$" is the same as "$0 < x \leq 3$."

△

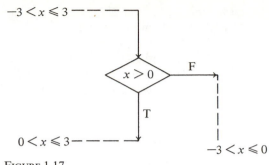

FIGURE 1.17

EXAMPLE 1.41   If we enter the conditional $\langle A = \varnothing \rangle$ with the assertion $A \cup B = C$, then the labels of Figure 1.18 are appropriate for the exit flowlines. Note particularly that we use property (P6a) of Section 0.2 to rewrite the condition "$A \cup B = C$ *and* $A = \varnothing$" as "$B = C$."   △

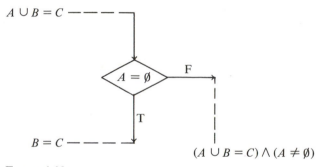

FIGURE 1.18

Rule 3   (Assignment rule)

The annotations for the entry and exit points of the assignment statement $V \leftarrow E$ are related by $Q(E)$ and $Q(V)$, respectively. (See Figure 1.19.) That is,

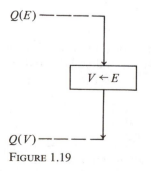

FIGURE 1.19

whatever one asserted for the expression E before the assignment, can be asserted for the variable V afterward, and conversely.

EXAMPLE 1.42   The consistency of the two assertions in Figure 1.20 can be seen by substituting $a - b \cdot q$ for r in the exit assertion. The result of the substitution is $0 \leq a - b \cdot q < b$, and after adding qb to each expression, we obtain the entry assertion.   △

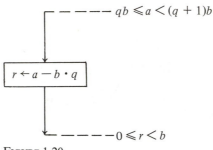

FIGURE 1.20

EXAMPLE 1.43   It seems obvious that if we can assert that $q = 6$ before the incrementation assignment $q \leftarrow q + 1$, then we should certainly be able to say that $q = 7$ after the assignment. Therefore, it is reassuring to know that the particular labeling shown in Figure 1.21 is consistent with Rule 3.   △

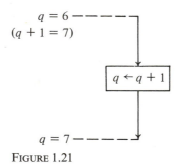

FIGURE 1.21

These rules can be used to prove the correctness of algorithms in the following way. We recall that each algorithm is provided with an input-output box, as shown in Figure 1.22. We generally claim that the final values of the output variables are somehow related to the initial values of the input. If we regard these claims as propositions

antecedent $Q(x_1, x_2, \ldots, x_n)$
consequent $R(y_1, y_2, \ldots, y_m)$

then the correctness of the algorithm is assured if $Q \rightarrow R$ (*Q leads to R*) by virtue

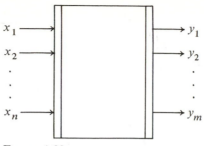

FIGURE 1.22

of a consistent annotation of flowlines, "start" being labeled by Q and "stop" by R. The consistency is measured against the three rules of algorithmic verification.

With the initial and terminal assertions Q, R given in advance, so that we can move backward and forward through the flowchart according to the rules, establishing further assertions is a reasonably straightforward procedure, except for the problem created by loops—that is, the question, "Which came first, the chicken or the egg?" Perhaps the best approach, given a simple loop, is to assume a conjecture C at some point of the circuit and then follow the rules around the loop to see whether we can return to the same condition C, or at least to a condition $C' \Rightarrow C$ (in the case of a forward traversal). If this procedure is not successful, then we must regretably begin again with another conjecture. The situation is not usually so bleak as the preceding remarks might suggest. We are usually guided in the making of a conjecture by the claims we have made for the algorithm. The person who developed the algorithm should have a fair idea as to its characteristics and intended behavior, so that loop conjectures (*loop-invariant assertions*, if they are chosen properly) should spring quickly to mind. It isn't as though they are selected at random.

EXAMPLE 1.44   Suppose that we are presented with the flowchart of Figure 1.23 and the claim that the algorithm computes $z = a^b$ for $a > 0$ and $b \geq 0$. Let us try to determine a likely candidate for the loop-invariant assertion Q. We should have

$$Q_0 = \langle a > 0, b \geq 0 \rangle \to Q$$

in order to use Rule 3 to pass through the three initial assignment statements. At the same time, we want

$$Q \wedge \neg \langle y > 0 \rangle = \langle z = a^b \rangle$$

according to Rule 2. The proponent of the algorithm would surely know enough about its performance to suggest

$$Q = \langle zx^y = a^b, y \geq 0, x > 0 \rangle$$

given the preceding clues.

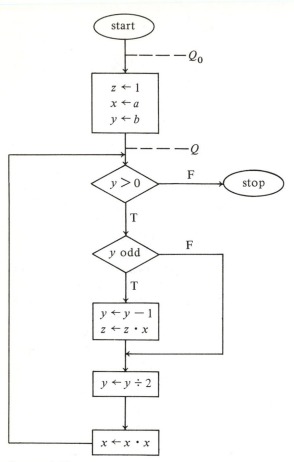

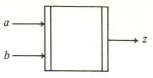

a, x, z: positive integer
b, y: nonnegative integer

FIGURE 1.23

Note that with this choice of Q, we have

$$Q \wedge \neg \langle y > 0 \rangle = \langle zx^y = a^b, y = 0, x > 0 \rangle$$
$$= \langle z = a^b \rangle$$

as required. Furthermore, using Rule 3 three times, we obtain

$$Q: zx^y = a^b, y \geq 0, x > 0$$
$$zx^b = a^b, b \geq 0, x > 0 \quad \text{(substituting } b \text{ for } y\text{)}$$
$$za^b = a^b, b \geq 0, a > 0 \quad \text{(substituting } a \text{ for } x\text{)}$$
$$Q_0: b \geq 0, a > 0 \qquad \qquad \text{(substituting 1 for } z\text{)}$$

In the next section we shall verify that the assertion Q is indeed an invariant for the loop.   △

We have enough of a challenge with a simple loop, but the situation becomes much more difficult when the loop complexity increases. Imagine a wildly entangled flowchart with loops going every which way! If nothing else, we can at least predict that such an algorithm would be quite difficult to understand. Can we be more specific? How about the hopes for analyzing such an algorithm, for proving its correctness? Again the chances for success in these endeavors are minimal. It is in response to such obvious difficulties that the concept of *structured programming* has surfaced, representing a vital force and hopeful trend in computer science.

As with most nested or recursive constructions (for instance, the algebraic expressions of Section 1.3), the easiest way to introduce the concept of *structure* for algorithms is by means of an inductive definition. We begin with a precondition:

(0) an assignment statement $V \leftarrow E$ standing alone is structured.

Then if S_1 and S_2 are structured flowchart segments (each has only one entry and one exit flowline), so are the segments obtained by

(1) *concatenation*
(2) *selection*
(3) *repetition*

as defined in Figure 1.24. The main thing to observe is that the three constructs or *compound statements* (as well as the assignment statement itself) have only one entry and one exit point. That is what allows for the substitution of any of these for the S_i as often and as variously as desired. There results a nesting of the same elementary constructs from the *top* level *down*, finally to the levels of the assignment statements. In this way, it should be clear that the structured

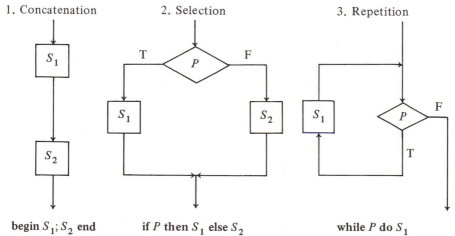

1. Concatenation   2. Selection   3. Repetition

begin S_1; S_2 end   if P then S_1 else S_2 while P do S_1

FIGURE 1.24

flowcharts are quite restricted in their form, especially in their loop complexity. Nevertheless, and somewhat surprisingly, it has been shown* that every algorithm has a structured realization!

EXAMPLE 1.45   Algorithm I has the structured decomposition of Figure 1.25, where the small boxes at the bottom levels are the assignment statements. Note

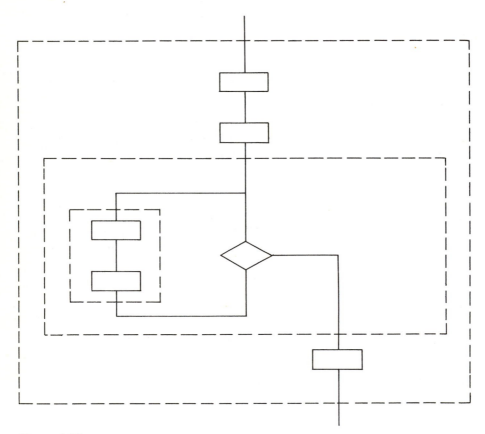

FIGURE 1.25

that it is convenient to generalize the concatenation construct to permit any finite number of components

$$\textbf{begin } S_1; S_2; \dots ; S_n \textbf{ end}$$

as was done in the figure for $n = 4$.   △

---

* Corrado Böhm and Guiseppe Jacopini, "Flow Diagrams, Turing Machines and Languages with only Two Formation Rules," *Comm. A.C.M.*, **9**, 366–371 (May 1966).

EXAMPLE 1.46   The exponentiation algorithm of Figure 1.23 has the structured decomposition shown in Figure 1.26. Note that we allow the selection construct

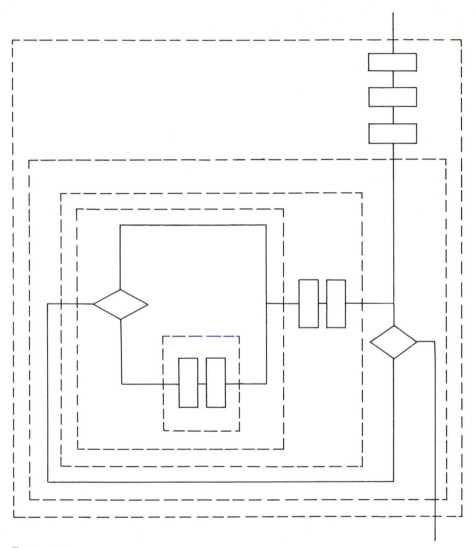

FIGURE 1.26

with one of its alternatives empty: the so called "if-then" instruction denoted by

if *P* **then** *S*   △

EXERCISES

1 Decide whether the annotations shown in Figure 1.27 are consistent with Rule 1.

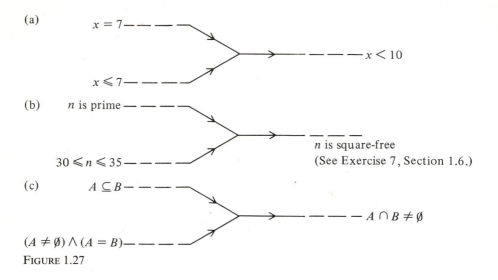

(a)

$x = 7$

$x \leqslant 7$

$x < 10$

(b) n is prime

$30 \leqslant n \leqslant 35$

n is square-free
(See Exercise 7, Section 1.6.)

(c) $A \subseteq B$

$(A \neq \emptyset) \wedge (A = B)$

$A \cap B \neq \emptyset$

FIGURE 1.27

2 Decide whether the annotations shown in Figure 1.28 are consistent with Rule 2.

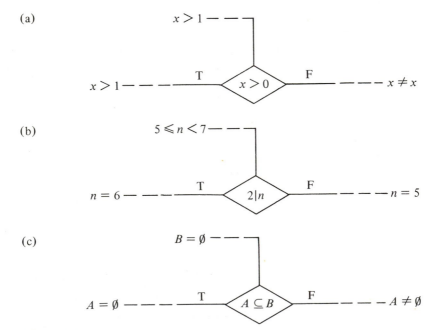

(a)

$x > 1$

$x > 1$ ———— T ———— $x > 0$ ———— F ———— $x \neq x$

(b)

$5 \leqslant n < 7$

$n = 6$ ———— T ———— $2 \mid n$ ———— F ———— $n = 5$

(c)

$B = \emptyset$

$A = \emptyset$ ———— T ———— $A \subseteq B$ ———— F ———— $A \neq \emptyset$

FIGURE 1.28

3 Decide whether the annotations shown in Figure 1.29 are consistent with Rule 3.

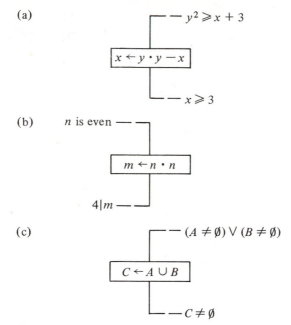

(a)

$$y^2 \geqslant x + 3$$

$$x \leftarrow y \cdot y - x$$

$$x \geqslant 3$$

(b) n is even

$$m \leftarrow n \cdot n$$

$$4 \mid m$$

(c)

$$(A \neq \emptyset) \vee (B \neq \emptyset)$$

$$C \leftarrow A \cup B$$

$$C \neq \emptyset$$

FIGURE 1.29

4 Supply the (consistent) missing annotations on the flowchart segments shown in Figure 1.30.

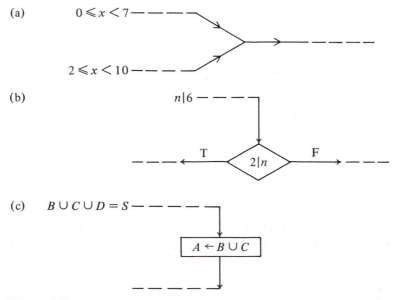

(a) $0 \leqslant x < 7$

$2 \leqslant x < 10$

(b) $n \mid 6$

T   F

$2 \mid n$

(c) $B \cup C \cup D = S$

$$A \leftarrow B \cup C$$

FIGURE 1.30

(d)

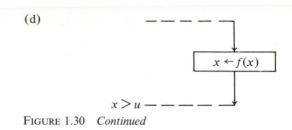

FIGURE 1.30   *Continued*

5 In each of the flowchart segments shown in Figure 1.31 supply the missing annotations consistent with Rule 3.

(a) $(p = n^2) \wedge (m = 3p - 2)$

$$m \leftarrow m + 3$$

$$p \leftarrow p + m$$

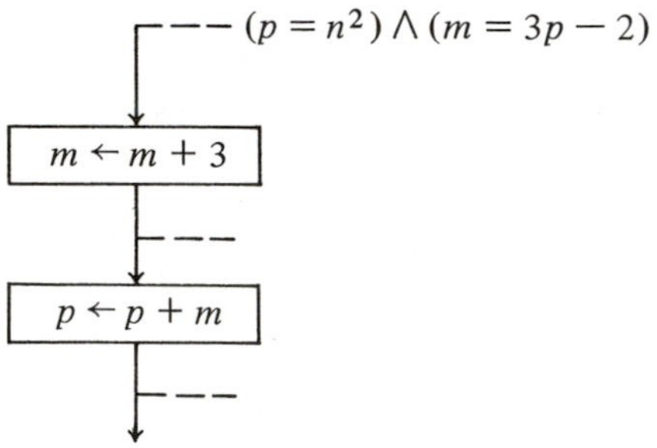

(b)

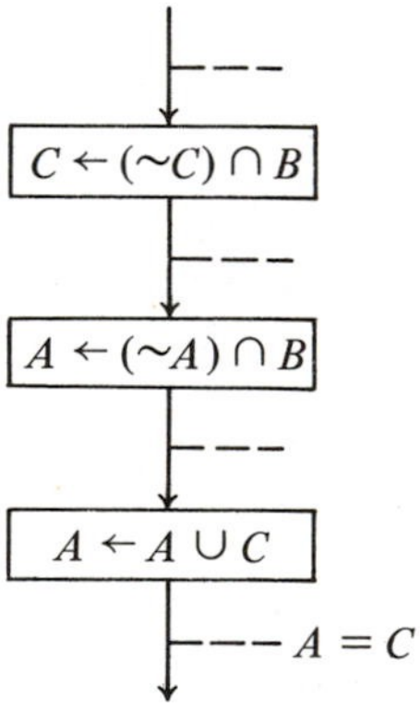

$$C \leftarrow (\sim C) \cap B$$

$$A \leftarrow (\sim A) \cap B$$

$$A \leftarrow A \cup C$$

$A = C$

FIGURE 1.31

6 Which of the drawings in Figure 1.32 are structured flowcharts? Indicate, as in Figures 1.25 and 1.26, the structural decomposition for those that are structured. Do the same for the flowcharts of

(a)

(b)

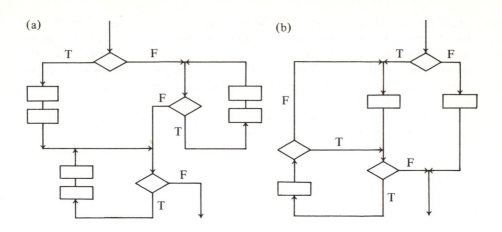

(c)

(d)

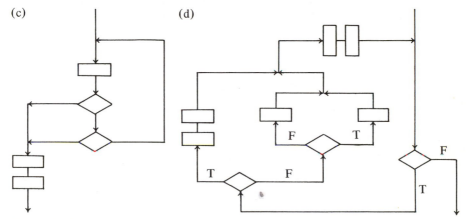

FIGURE 1.32

(e) Figure 1.42

(f) Exercise 1 of Section 1.9

(g) Exercise 4 of Section 1.9

*1.9   ANALYSIS OF ALGORITHMS

It would be very easy for the student to misread our intentions in providing these few details for verifying the correctness of algorithms. We certainly do not mean to propose that these techniques be applied every time one writes a program. Otherwise, he or she would never be able to write a program of any appreciable size. It would be akin to asking mathematicians to supply a proof for the validity of induction for each and every instance in which they use the principle. On most occasions they probably *know* that the induction will go through, without actually proving it. In the same way, a programmer generally *knows* that a loop will terminate with the proper conditions on the variables without carrying out a formal verification. Nevertheless, as computer scientists, we cannot altogether avoid the obligation to use our unique understanding of a program to make assertions about its computational behavior and to communicate these assertions to others.

We saw earlier that the analysis and correctness proofs of algorithms depend on finding an annotation of flowlines consistent with the rules of algorithmic verification. If this can be done with

$$Q_0 = Q_0(x_1, x_2, \ldots, x_n)$$
$$R_0 = R_0(y_1, y_2, \ldots, y_m)$$

as labels for the "start" and "stop", respectively, then we have demonstrated that Q_0 leads to R_0, the one being the *antecedent* and the other being *consequent* for the algorithm as a whole.

In the case of a structured algorithm, the search is more orderly. It is enough to find a Q, R labeling of the entry and exit to each constituent block of the decomposition consistent with the following set of *structured rules*:

S.R. 0   (Assignment)

The assignment statement $V \leftarrow E$ is to be annotated exactly as before (the old Rule 3), that is, with $Q(E)$ as antecedent and $Q(V)$ as consequent.

Then if Q_i, R_i are respectively the antecedent and consequent to the structured segment S_i, we should have annotations for the three compound statements in accordance with the structured rules given in Figure 1.33.

EXAMPLE 1.47   Consider Algorithm I once more, along with its structured decomposition (Example 1.45). At the top level of the decomposition we have a concatenation of four statements; all except the third are just simple assignment statements. We would expect to verify that the algorithm as a whole has the antecedent and consequent

$$Q_0: a \geq 0, b > 0$$
$$R_0: a = bq + r, \qquad 0 \leq r < b$$

Using S.R. 1, we see that R_0 is also the desired consequent of the assignment statement $r \leftarrow a - b \cdot q$. We use S.R. 0 (the old Rule 3) to obtain its antecedent by substituting $a - bq$ for r, as in Example 1.42. We obtain $qb \leq a < (q + 1)b$. By S.R. 1, this inequality should be the consequent

$$R: qb \leq a < (q + 1)b$$

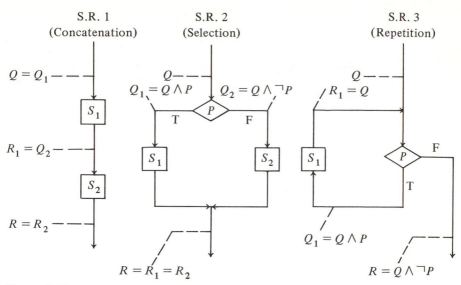

FIGURE 1.33

of the repetition, which is the third block of the concatenation. At the same time we should have

$$R = Q \wedge \neg P$$

by S.R. 3, where P is the proposition $\langle p \le a \rangle$. However, what is Q?
    Note that

$$\neg P: a < p$$

strongly suggests that

$$p = (q + 1)b$$
$$Q = \langle p = (q + 1)b \wedge qb \le a \rangle$$
$$= \langle (q + 1)b = p \le a + b \rangle$$

Then we have

$$Q \wedge \neg P = \langle p = (q + 1)b \wedge qb \le a \wedge a < p \rangle$$
$$= \langle qb \le a < (q + 1)b \rangle = R$$

as required. In any case, according to S.R. 3, Q is to be the antecedent of the repetition block, and at the same time it is to be the consequent of the two assignments $q \leftarrow 0$, $p \leftarrow b$. Therefore, suppose that we try to derive Q_0 from Q by these assignment statements. We find that we need the additional statement $b > 0$ in the assertion Q, for then we have

$$Q: b > 0, (q + 1)b = p \le a + b$$
$$b > 0, (q + 1)b = b \le a + b \qquad \text{(substituting } b \text{ for } p)$$
$$0 < b = b \le a + b \qquad \text{(substituting 0 for } q)$$

and the latter is equivalent to

$$Q_0: a \geq 0, b > 0$$

Observe that the earlier analysis remains valid when we use the new Q.

Suppose that we now check to see whether Q is indeed a loop-invariant assertion for the third block. We need to examine the concatenation shown in Figure 1.34. According to S.R. 3 and S.R. 1, Q should be the consequent of the

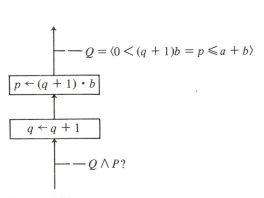

FIGURE 1.34

assignment statement $p \leftarrow (q + 1) \cdot b$. Its antecedent, by S.R. 0, would be

$$b > 0, \qquad (q + 1)b = (q + 1)b \leq a + b$$

or

$$b > 0, \qquad qb \leq a.$$

Taking this assertion as the consequent of $q \leftarrow q + 1$, we obtain

$$b > 0, \qquad (q + 1)b \leq a$$

as antecedent. This conclusion agrees with $Q \wedge P$. The completely annotated flowchart is shown in Figure 1.35. Being consistent with the structured rules, these annotations demonstrate the correctness of Algorithm I.   △

EXAMPLE 1.48 Suppose that we return to the exponentiation algorithm of Example 1.44. To verify that

$$Q: zx^y = a^b, \qquad y \geq 0, \quad x > 0$$

is a loop-invariant assertion, we begin by using S.R. 0 to proceed backward through the assignment statements $y \leftarrow y \div 2, x \leftarrow x \cdot x$. Then we have

$$zx^{2y} = a^b, \qquad y \geq 0, \qquad x > 0, \quad 2y \text{ even} \qquad \text{(substituting } x \cdot x \text{ for } x\text{)}$$
$$zx^y = a^b, \qquad y \geq 0, \qquad x > 0, \quad y \text{ even} \qquad \text{(substituting } y \div 2 \text{ for } y\text{)}$$

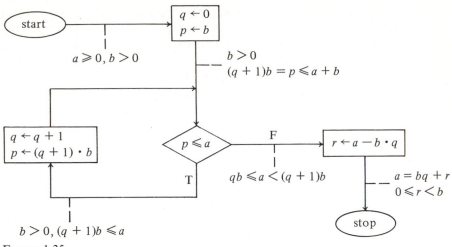

FIGURE 1.35

It remains for us only to check the performance of the selection block shown in Figure 1.36. Note that with an if-then construction one has only to verify that $Q_2 \Rightarrow R$ for Rule I (the junction rule) to be satisfied. Beginning with R and moving backward through the two assignment statements, we obtain

$zxx^y = a^b,$	$y \geq 0, \quad x > 0, \quad y$ even	(substituting zx for z)
$zxx^{y-1} = a^b,$	$y - 1 \geq 0, \quad x > 0, \quad y - 1$ even	(substituting $y - 1$ for y)

or

$$Q_1: zx^y = a^b, \qquad y > 0, \quad x > 0, \quad y \text{ odd}$$

as required.   △

$$Q_1 = \langle zx^y = a^b, y > 0, x > 0, y \text{ odd} \rangle$$

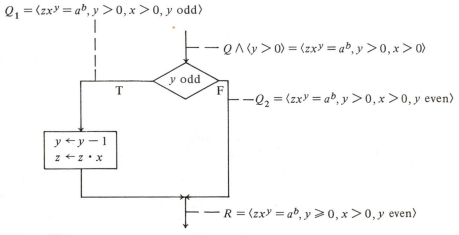

$$Q \wedge \langle y > 0 \rangle = \langle zx^y = a^b, y > 0, x > 0 \rangle$$

$$Q_2 = \langle zx^y = a^b, y > 0, x > 0, y \text{ even} \rangle$$

$$R = \langle zx^y = a^b, y \geq 0, x > 0, y \text{ even} \rangle$$

FIGURE 1.36

After considering these examples of structured algorithms, we would like to make one observation in summary. At a particular level of investigation the details at a lower level are to a certain extent irrelevant. For instance, if a repetition statement is encountered in a concatenation, the number of times that the loop is repeated is of no particular concern; all that matters is that it is executed the proper number of times. It is only the net effect that is important at the given level of investigation. Let us study this phenomenon a bit closer, again as it applies to loops.

We mentioned earlier that the principle of mathematical induction somehow plays a role in the verification of algorithms. That is so, but the induction takes place over the number (n) of times that a loop is executed. As we just said, this variable n is ordinarily suppressed; but the proposition $P(n)$ is the invariant assertion of the loop. On various occasions, we checked that an assertion held when we entered the loop from the outside ($n = 0$) and then proceeded to verify that it held again after one pass through the loop (if it had held at the outset). Surely it is the principle of mathematical induction at work when we claim that $P(n)$ therefore holds however many passes are made.

A further consideration, however, must accompany the inductive reasoning when it is applied to the analysis of algorithms. You may recall that it was implicit in the formulation of the whole idea of an algorithm that the process be finite. Surely our flowcharts are finite, and we expect each of their seperate activities to be accomplished in some finite time. But what about the presence of loops? How do we know that a loop is not traversed infinitely often? Without some further analysis, the answer is, we don't. In the case of a structured algorithm, however, again the required analysis is reasonably straightforward.

If we have a structured flowchart, then we can at least claim that computational loops arise only in the form of the compound repetition statement (Figure 1.24): **while** P **do** S. That is, while P is true, we continue to repeat the (perhaps compound) statement S. It is clear that a necessary condition for preventing an infinite loop traversal is that S must change the value of some variable contained in the proposition P. More definitely, we can say that the finiteness of the repetition is assured if we can exhibit an integer-valued function F of some of the variables of the algorithm, for which

(i) P is true $\Rightarrow F > 0$

(ii) $F_{n+1} < F_n$ (the subscript referring to the number of loop traversals)

Thus we have, by (ii), a strictly decreasing integer-valued function, whose value must eventually be nonpositive. At that point, by (i), P becomes false and we exit, terminating the repetition.

EXAMPLE 1.49   In Algorithm I, we look at the proposition

$$P = \langle p \leq a \rangle$$

and condition (i) to decide that it is reasonable to choose

$$F = a - p + 1$$

because we then have

$$P \text{ is true} \Rightarrow p \le a \Rightarrow F \ge 1 \Rightarrow F > 0$$

as required by (i). Furthermore,

$$F_n = (a - p + 1)_n = (a - (q + 1)b + 1)_n$$

by the invariant assertion of Example 1.47. But

$$q_{n+1} = q_n + 1 > q_n$$

by virtue of the assignment $q \leftarrow q + 1$, and this causes $F_{n+1} < F_n$, as required by (ii). It follows that the loop is traversed only a finite number of times.   △

We have already seen several advantages for structured algorithms. They are generally easier to understand and lend themselves more easily to the construction of correctness proofs. They also facilitate the demonstration of finiteness arguments. Finally, we will show in the next section that the structured algorithms permit a direct translation into "sequential form." It is no wonder that structured programming represents such a hopeful trend in computer science.

EXERCISES

1 The flowchart shown in Figure 1.37, while quite different from that of Algorithm II, nevertheless computes the greatest common divisor of two given positive integers a, b.

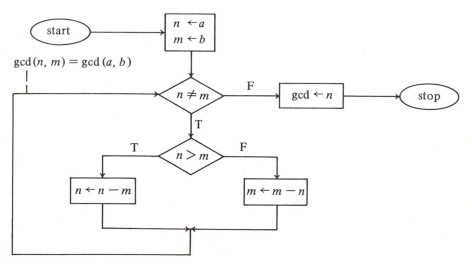

FIGURE 1.37

Use the given loop-invariant assertion (after showing that it is indeed an invariant assertion) to verify the correctness of the algorithm. [*Hint*: Determine the necessary annotations through the use of the properties given in Exercise 13 of Section 1.1.]

2 Show that the flowchart of Exercise 1 is structured.

3 Verify the finiteness of the repetition for the loop in
(a) Exercise 1   (b) Figure 1.23
(c) Algorithm II   (d) Exercise 4
(e) Exercise 5   (f) Exercise 6

4 What is accomplished by the flowchart shown in Figure 1.38? Verify the correctness of your assertion. [*Hint*: Try $f \cdot n^m = n^n$ as an invariant assertion.]

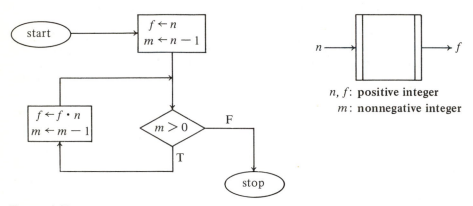
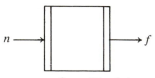

n, f: **positive integer**
m: **nonnegative integer**

FIGURE 1.38

5 Repeat Exercise 4 for the flowchart in Figure 1.39.

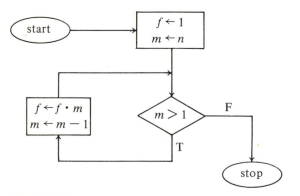

n, m, f: **positive integer**

FIGURE 1.39

*6 Repeat Exercise 4 for the flowchart in Figure 1.40.

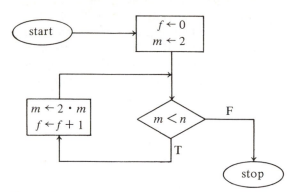
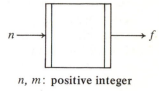

n, m: **positive integer**
f: **nonnegative integer**

FIGURE 1.40

1.10 SEQUENTIAL FORM

Flowcharts are inherently two dimensional (at best) and therefore not suitable as a program input to a computer. With the most common input devices, the program must be in the form of a linear sequence of symbols from the alphabet of a suitable programming language. The required translation is a likely source of errors, unless the errors are mechanical, following as it does very definite rules of transcription. If we can think of a flowchart as *block decomposed* into just a few basic structured flowchart constructs, then an automatic translation to a linear or *sequential form* is possible.

Consider the three basic structured flowchart constructs shown in Figure 1.24. Each has its own characteristic linear rendition:

(1) Concatenation   **begin** S_1; S_2; ... ; S_n **end**
(2) Selection   **if** P **then** S_1 **else** S_2
(3) Repetition   **while** P **do** S

In the first case we are simply to execute two or more flowchart segments S_i in sequence (one after the other). In the second case we make a decision: if P is true, we execute the segment S_1; and if P is false, the flowchart segment S_2 is executed instead. We permit one of these alternatives to be empty, as in the instruction

if P **then** S

encountered in Example 1.46. The third instruction involves a computational loop: we repeat the flowchart segment S as long as P is true. Each of the segments S_i or S is to have only one entry and one exit point. Each may consist of a simple assignment statement or, more importantly, one of the basic structured constructs 1, 2, or 3 all over again. And so it goes. As long as we are able to so decompose a flowchart, however, a direct translation to *sequential form* is immediately available. Note in particular that the assignment statements at the bottom levels of the decomposition are already in linear form.

In the case of algorithms in certain restricted algebraic structures, the reader who is familiar with ALGOL will see immediately that the resulting sequential form of the algorithm could be called "ALGOL-like." With the more general algebraic structures, there remains the problem of choosing the appropriate data structures, etc., but this problem must be faced eventually in any case.

EXAMPLE 1.50   Reviewing Algorithm I and its structured decomposition in Figure 1.25, we note that a "top-down" development of the sequential form would start at the top level with the concatenation

$$\textbf{begin } S_1 \,;\, S_2 \,;\, S_3 \,;\, S_4 \textbf{ end}$$

Here S_3 is the only compound block, having the form

$$\textbf{while } p \leq a \textbf{ do } S$$

where S itself is the concatenation $q \leftarrow q + 1; p \leftarrow (q + 1) \cdot b$. Altogether, when accounting for the blocks (assignment statements) S_1, S_2, and S_4, we have the sequential form

$$
\begin{aligned}
&\textbf{begin } q \leftarrow 0; p \leftarrow b; \\
&\quad \textbf{while } p \leq a \textbf{ do} \\
&\quad \textbf{begin } q \leftarrow q + 1; p \leftarrow (q + 1) \cdot b \\
&\quad \textbf{end}; \\
&\quad r \leftarrow a - b \cdot q \\
&\textbf{end}
\end{aligned}
$$

$\triangle$

The syntax of the sequential form permits a strict linear notation, but for purposes of enhancing readability, we shall employ certain indentation conventions. Unless the concatenation is quite short, we usually line up each paired **begin** . . . **end** vertically and indent as the decompositional level increases. At the same time we will ordinarily left-justify each **if** or **while** within its own level so that it is more easily noticed.

EXAMPLE 1.51   Considering the flowchart of Figure 1.23 and its structured decomposition (Figure 1.26), we are led to the following sequential form:

$$
\begin{aligned}
&\textbf{begin } z \leftarrow 1; x \leftarrow a; y \leftarrow b; \\
&\quad \textbf{while } y > 0 \textbf{ do} \\
&\quad \textbf{begin if } y \text{ odd } \textbf{then} \\
&\qquad \textbf{begin } y \leftarrow y - 1; z \leftarrow z \cdot x \textbf{ end}; \\
&\qquad y \leftarrow y \div 2; x \leftarrow x \cdot x \\
&\quad \textbf{end} \\
&\textbf{end}
\end{aligned}
$$

$\triangle$

The concept of structured programming has sometimes been erroneously designated as "go-to-less programming." This oversimplification is usually a result of someone's misreading of the classic note by E. W. Dijkstra,* in which we are warned of the dangers inherent in an unbridled use of "go to" statements (as found in the familiar programming languages). It is true that such excesses will lead to programs with a complex loop structure, such as we have sought to avoid. But that is only part of the story. We hope we were able to show in the two previous sections that much more is involved in the idea of structured algorithms than the mere avoidance of "go-to" as an end in itself.

EXAMPLE 1.52   In fact, most proponents of structured programming are prefectly willing to permit the *alarm exit* as a legitimate construct, that is,

if *P* then exit

Yet this is a restricted use of the "go to" statement (that is, go to "stop"). In effect, we used this construction in Example 1.32 (the flowchart of Figure 1.10).   △

The reader should observe that the algorithmic sequential form retains the same algebraic generality as our original flowchart language. In the interest of brevity and clarity, we intend to propose the use of the sequential form for the longer algorithms to be presented in the remainder of the text. For the same reason, we will need to introduce additional repetition constructs, similar to the DO loops of FORTRAN, or the **for** statements of ALGOL.

Assume that V is a variable of an *ordered* data type A. That is, suppose that we can imagine the distinct elements of A as being arranged in a sequence:

$$A = \langle \ldots, a_{i-1}, a_i, a_{i+1}, \ldots \rangle$$

with the convention that

$$a_i < a_j \Leftrightarrow i < j.$$

(The integers are the best-known example.) Then we may speak of *successor* and *predecessor* functions

$$\text{succ}(a_i) = a_{i+1} \qquad \text{pred}(a_i) = a_{i-1}$$

In the case of the integers, $\text{succ}(n) = n + 1$ and $\text{pred}(n) = n - 1$. Now we may wish to repeat the execution of a segment S, once for each value of V in the range from a to b. Ordinarily $a \leq b$, where a and b are allowed to be any algebraic expressions of the same type as V. For this purpose, we write

for $V \leftarrow a$ to b do S

---

* E. W. Dijkstra, " Go to Statement Considered Harmful," *Comm. A.C.M.*, **11**, 147–148 (March 1968).

This statement is to have the same effect as the sequence:

$$V \leftarrow a; S; V \leftarrow \text{succ}(a); S; \ldots ; V \leftarrow b; S$$

Alternatively, we may think of the new repetition as being composed of the statements

$$V \leftarrow a;$$
while $V \leq b$ **do**
begin $S; V \leftarrow \text{succ}(V)$ **end**

as shown in Figure 1.41(a).

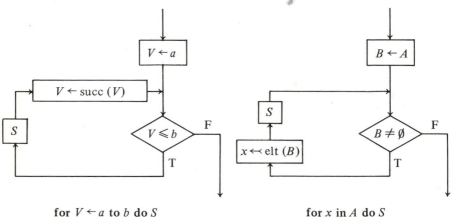

for $V \leftarrow a$ to b do S for x in A do S

(a)   (b)

FIGURE 1.41

Next suppose that A is just a finite set without any particular structure, but we still wish to repeat a segment S, once for each element x in A. We could introduce a new kind of assignment statement

$$x \leftarrow \text{elt}(A)$$

which has the effect of choosing (to assign to the variable x the value of) an element from A more or less at random. Repeated choices, however, might result in the same element being selected more than once, contrary to our plans. A solution is given in Figure 1.41(b), where the assignment arrow with the "tail"

$$x \twoheadleftarrow \text{elt}(B)$$

has the following meaning:

(1) Assign x the value of an arbitrary element of B, as in the assignment $x \leftarrow \text{elt}(B)$ just described.

(2) At the same time delete this element from the set B.

The tail on the arrow is a reminder that B is modified as well as x. The whole construction of Figure 1.42(b) can be decomposed much as before:

$$B \leftarrow A;$$
$$\textbf{while } B \neq \emptyset \textbf{ do}$$
$$\textbf{begin } x \leftarrowtail \text{elt}(B); S \textbf{ end}$$

so that it is easy to see why we prefer the abbreviation

$$\textbf{for } x \textbf{ in } A \textbf{ do } S$$

Note by way of comparison that if $A = n^+$, then the only difference between the two statements:

$$\textbf{for } x \textbf{ in } n^+ \textbf{ do } S \qquad \textbf{for } x \leftarrow 1 \textbf{ to } n \textbf{ do } S$$

is that in the first case the elements x may not be selected in their natural order.

EXAMPLE 1.53   In Figure 1.42 is shown a flowchart for deciding the proposition $\langle n$ is prime$\rangle$ in an algebraic structure containing the positive integers and the

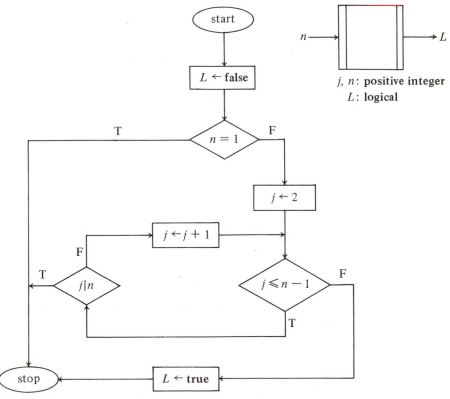

FIGURE 1.42

algebra of logical values. Note the two alarm exits, one for $n = 1$ and the other in case some integer in the range $2 \leq j \leq n - 1$ should happen to divide n. (In neither case is n a prime number.) Evidently the loop can be expressed with a **for** statement, as in the following sequential form of the algorithm:

$$\begin{aligned} &\textbf{begin } L \leftarrow \textbf{false}; \\ &\quad \textbf{if } n = 1 \textbf{ then exit}; \\ &\quad \textbf{for } j \leftarrow 2 \textbf{ to } n - 1 \textbf{ do} \\ &\quad \textbf{if } j|n \textbf{ then exit}; \\ &\quad L \leftarrow \textbf{true} \\ &\textbf{end} \end{aligned}$$

$\triangle$

In both of our **for** statements, the iteration variable (V or x in Figure 1.41) is called a *controlled* variable because it is given a value only "locally." Its name should not be invoked outside the loop; furthermore, its value is not allowed to change within the iterated statement S. A violation of either of these rules would be contrary to the whole purpose of the **for** statement. A **for** statement is appropriate only when the number of necessary repetitions is known ahead of time. If it is determined only while the repetitions are being performed, then the **while** statement is the one to use.

We find the greatest use of the **for** statements in connection with the manipulation of "arrays." It is well to briefly review the notion of an array structure at this time, even though it is quite likely that the student has already encountered similar ideas in an elementary programming course. We suppose once more that we are given an algebraic structure

$$\mathscr{A} = (\mathscr{A}; \mathscr{F}; \mathscr{P}) = (A_1, A_2, \ldots; \mathscr{F}; \mathscr{P}).$$

When we introduce an *array type*, as in the declaration

$$M: \textbf{array } A_{\alpha_1} \times A_{\alpha_2} \times \cdots \times A_{\alpha_n} \textbf{ of } A_\alpha$$

we are not quite saying that M is a function

$$M: A_{\alpha_1} \times A_{\alpha_2} \times \cdots \times A_{\alpha_n} \rightarrow A_\alpha$$

which would be a "constant" array or *matrix*. Instead, we wish to stress that:

1. M is a collection of component variables, each of type α.
2. Each component is directly accessible (*indexed*) by means of the denotation $M[a_1, a_2, \ldots, a_n]$, where the a_i are algebraic expressions of type α_i.

As an array is declared, the types α_i of its *indices* are defined as well as that of its component variables.

EXAMPLE 1.54   If we make the declaration

$$M: \textbf{array } n^+ \times n^+ \textbf{ of } B$$

then M is a collection of n^2 logical variables, indexed by ordered pairs of integers from 1 to n. A particular component $M[i, j]$ is a variable in its own right, but the whole collection goes by one name, M. $\triangle$

EXAMPLE 1.55   In order to illustrate the use of **for** statements in connection with arrays, suppose that we consider a simple algorithm for finding the minimum y of n integers $x_1, x_2, \ldots, x_n$. We may think of these integers as the initial values (input) of an array, such as in Figure 1.43. Then our algorithm might read as follows:

$$\textbf{begin } i \leftarrow 1;$$
$$\textbf{for } j \leftarrow 2 \textbf{ to } n \textbf{ do}$$
$$\textbf{if } x[j] < x[i] \textbf{ then } i \leftarrow j;$$
$$y \leftarrow x[i]$$
$$\textbf{end} \hspace{8cm} \triangle$$

x: **array** n^+ **of** Z
y, i, j, n: **integer**

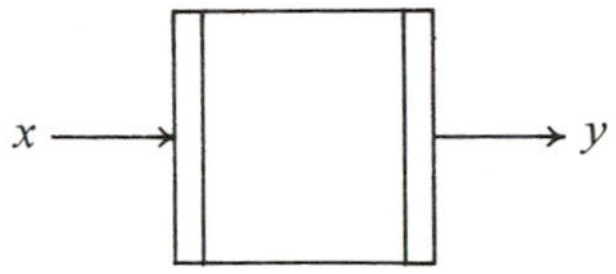

FIGURE 1.43

We hesitate at this time to present more examples merely for the purpose of illustrating the use of array structures. Such examples will occur so frequently as we proceed that the student will have every occasion to reinforce his basic understanding of their purpose and application.

To conclude this discussion, we propose the following useful exercise to the student. Recall our earlier remark as to how we plan to use the sequential form for describing the larger algorithms to be encountered in the remainder of the text. Until that form of presentation becomes second nature to the student, it is perhaps safe to say that the flowchart language will seem to be the clearer of the two. Yet we have described only a method for conversion one way, from a block-decomposed flowchart to the sequential form. Certainly the process must be reversible; otherwise, the sequential form would be ambiguous! To be convinced of this, it would be a good idea for the student to perform such reverse conversions on the algorithms presented thus far, and to continue this practice until he or she feels more comfortable with the sequential notation.

SUGGESTED REFERENCES

1.1 Birkhoff, G., and S. MacLane. *A Survey of Modern Algebra*, 3rd ed. Macmillan, New York, 1965.

1.2 Dean, R. A. *Elements of Abstract Algebra*. Wiley, New York, 1968.

1.3 Cohn, P. M. *Universal Algebra.* Harper and Row, New York, 1965.

1.4 Wells, M. B. *Elements of Combinatorial Computing.* Pergamon Press, New York, 1971.

1.5 Dahl, O. J., E. W. Dijkstra, and C. A. R. Hoare. *Structured Programming.* Academic Press, New York, 1972.

1.6 Wirth, N. *Systematic Programming: An Introduction.* Prentice-Hall, Englewood Cliffs, N.J., 1973.

The author knows of no useful elementary book on algebraic structures, but it is possible to gain an understanding of some of the same ideas encountered here by consulting a good book on general abstract algebra. Two such books are listed above. The first is a classic text, the forerunner of all the others. The book by Dean has been equally well received, however. The student would profit by gaining some familiarity with either of these. The book by Cohn is an excellent but advanced treatment of algebraic structures, suitable for the graduate student.

The fourth reference is unusual. It has a built-in algorithmic language, designed with much the same purpose as the present book. The style of the language here, however, is influenced to a greater degree by the last two references, either of which will provide an excellent and extensive introduction to the concept of structured programming. The latter is more of a textbook than the former, but both should be consulted.

2

Graphs and Digraphs

The subject of graph theory may be considered a part of combinatorial mathematics. It has its origins in a 1736 paper by the celebrated mathematician Leonard Euler, but this now famous paper remained a neglected and isolated contribution for nearly a century. Interest in graphs began to stir only gradually, as the same mathematical structure began to arise in seemingly unrelated disciplines, one after another. Nevertheless, because a number of popular puzzles could be formulated directly in the language of graph theory, it was still some time before the subject came to be taken seriously.

The relevance of graph theory to applications is no longer a subject of debate. The theory has greatly contributed to our understanding of programming, communication theory, electrical networks, switching circuits, biology, chemistry, and psychology. From the standpoint of applications, it is safe to say that graph theory has become the most important part of combinatorial mathematics. It follows that anyone interested in computer applications should be aware of the more important concepts and a few of the most typical algorithms having to do with graphs. Since the applications are found in the more recently emerging sciences, present-day students have a distinct advantage over their predecessors. Ordinarily there should be no difficulty at all in perceiving the relevance of graph theory, even at a first exposure.

For the uninitiated, however, one word of caution is in order. As we use the term here, graphs have only the slightest connection with the graphs of functions such as discussed in calculus. Here a graph is simply a particular kind of relation on a set, and so the subject falls within the scope of general algebraic structures. Conceptually they are the simplest of the algebras we shall study, undoubtedly because there are no operations at all, only the proposition (relation) of adjacency among vertices.

139

2.1 GRAPHS AND SUBGRAPHS

Whenever we encounter a network or diagram whose nodes or vertices are joined by various arcs, it is likely that a graphical formalism is appropriate to the analysis. This is true of highway maps, electrical circuits, flowcharts, transportation networks, and a variety of situations. If the arcs or edges have an associated orientation or direction, then we enter the realm of the *directed graphs* to be discussed later in the chapter. For the time being, we do not assume an orientation, and so our edges are merely connecting links for the various nodes or vertices.

Formally, a *graph* $G = (V, E)$ is an algebraic structure in which V is a non-empty set of *vertices* and E is a relation on V satisfying the properties (as introduced in Section 0.3):

(i') *irreflexive* $v \not\mathrel{E} v$
(ii) *symmetric* $v \mathrel{E} w \Rightarrow w \mathrel{E} v$

for all v and w in V. We recall that it is customary to view a relation also as a subset, in which case $E \subseteq V \times V$, and we write $(v, w) \in E$ just in case $v \mathrel{E} w$. In view of the properties (i'), (ii), however, it is more convenient here to consider *un*ordered pairs $\{v, w\}$ such that

$$E \subseteq V \otimes V = \{ \{v, w\} : v \neq w \text{ in } V \}$$

Thus we write $\{v, w\} \in E$ iff $v \mathrel{E} w$. Such unordered pairs $\{v, w\}$ are called *edges* of the graph G. Note that $\{v, w\} = \{w, v\}$ and also that we are already deviating from the convention established in Chapter 1, which was to denote an algebra by the same name as the set on which the operations and/or propositions were defined. It seems that every rule must have its exceptions.

In this text we shall consider only finite graphs, that is, graphs $G = (V, E)$ where in each case the set of vertices is a finite set, say $|V| = n$. When the number of vertices is finite, so too is the number of edges. Considering E as the subset $E \subseteq V \otimes V$, we ordinarily denote its size by $|E| = m$.

EXAMPLE 2.1   Let $V = \{v_1, v_2, v_3, v_4, v_5\}$ and suppose that we present E in the form of a relation matrix $E = (e_{ij})$, as in Section 0.3. Then, as in our characterization of irreflexive and symmetric relations, the two required conditions translate into the following matrix conditions:

(i') $e_{ii} = 0$
(ii) $e_{ij} = e_{ji}$

It follows that the following matrix at the left represents a graph $G = (V, E)$. Owing to conditions (i'), (ii), the lower triangular portion ($e_{ij} : i > j$) at the right still retains all of the necessary information.

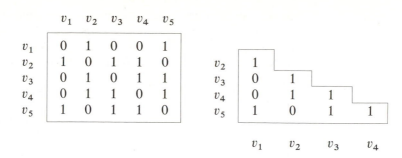

In any case, considering E as a subset $E \subseteq V \otimes V$, we have

$$|V| = n = 5 \qquad |E| = m = 7$$

and

$$E = \{\{v_1, v_2\}, \{v_1, v_5\}, \{v_2, v_3\}, \{v_2, v_4\}, \{v_3, v_4\}, \{v_3, v_5\}, \{v_4, v_5\}\}.$$

Note that $|E| = m$, the number of edges in a graph, is always given by the number of ones in the lower-triangular matrix representation.

A *picture* of a graph $G = (V, E)$ is obtained in the following way. We draw a point for each vertex and join the points v and w by an arc or line segment just in case $v \, E \, w$. The lines need not be straight, and the points can be arranged at will. Consequently, both of the pictures of Figure 2.1 are valid descriptions of the graph just introduced.

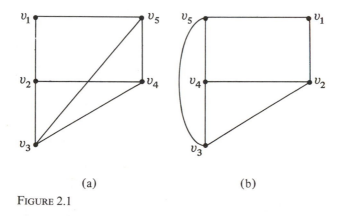

(a)   (b)

FIGURE 2.1

We must bear in mind, however, that a graph is not a picture, but an algebraic structure. Thus the same graph can be pictorially represented in various ways, as we just illustrated. Also erroneous conclusions might be drawn from a particular picture. Thus it might appear, on the basis of Figure 2.1(a), that our graph has six vertices, which is not true. The apparent intersection of the edges

$\{v_2, v_4\}$ and $\{v_3, v_5\}$ does not represent a vertex. It is simply a peculiarity of the particular way in which the graph was represented. Nevertheless, we will find it convenient to introduce examples of graphs by means of pictures. As long as we heed these few words of warning, no confusion will arise.   △

Unfortunately, graph theory is one of those areas of mathematics in which the number of definitions seems almost to outweigh the number of significant results. We will try not to overburden the reader with definitions, but a certain number of them is unavoidable. Thus we shall present a few relating to the concept of a subgraph. Obviously one should expect such an idea to be somehow connected with the notion of subset, and that is indeed the case. The graph $H = (V_1, E_1)$ is said to be a *subgraph* of the graph $G = (V, E)$ if $V_1 \subseteq V$ and $E_1 \subseteq E$. Note also that $E_1 \subseteq V_1 \otimes V_1$ by virtue of the fact that $H = (V_1, E_1)$ is a graph. If $V_1 = V$, then we say that H is a *spanning* subgraph of G. Now supposing that $\varnothing \neq U \subseteq V$ in any graph $G = (V, E)$, the subgraph *generated by* U is defined to be

$$\langle U \rangle = (U, E_U)$$

where

$$\{v, w\} \in E_U \quad \text{iff} \quad v, w \in U \quad \text{and} \quad \{v, w\} \in E.$$

Thus we include in $\langle U \rangle$ all edges that were already joining some vertices of U in the original graph G. In this sense $\langle U \rangle$ is completely determined by the subset U (and the graph G). It follows that $\langle U \rangle$ is never a spanning subgraph unless $U = V$; but then $\langle U \rangle = G$.

EXAMPLE 2.2   Among the subgraphs of the graph $G = (V, E)$ of Example 2.1, three are pictured in Figure 2.2. Both H_2 and H_3 are spanning subgraphs, but H_1 is not. On the other hand, $H_1 = \langle U \rangle$ where $U = \{v_1, v_2, v_4, v_5\}$. Of course, each subgraph is a graph in its own right. Thus

$$H_1 = (V_1, E_1)$$

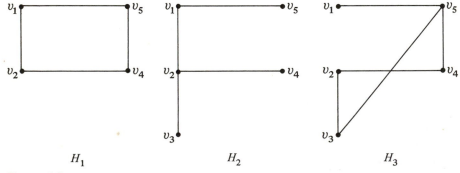

H_1 H_2 H_3

FIGURE 2.2

with

$$V_1 = \{v_1, v_2, v_4, v_5\}$$

$$E_1 = \{ \{v_1, v_2\}, \{v_1, v_5\}, \{v_2, v_4\}, \{v_4, v_5\} \}$$

and from this representation we see clearly that $V_1 \subseteq V$ and $E_1 \subseteq E$ as required by the definition of a subgraph. △

Continuing with our imitation of the algebra of sets, suppose that we have an arbitrary collection of subgraphs $H_\alpha = (V_\alpha, E_\alpha)$ of the graph $G = (V, E)$. Then we can define *unions* and *intersections* in an obvious way:

$$\bigcup_\alpha H_\alpha = \left(\bigcup_\alpha V_\alpha, \bigcup_\alpha E_\alpha \right)$$

$$\bigcap_\alpha H_\alpha = \left(\bigcap_\alpha V_\alpha, \bigcap_\alpha E_\alpha \right)$$

again viewing the E_α as subsets $E_\alpha \subseteq V_\alpha \otimes V_\alpha \subseteq V \otimes V$. In the case of the union $\bigcup_\alpha H_\alpha$, we write $\sum_\alpha H_\alpha$ and speak of a *direct union* if $V_\alpha \cap V_\beta = \emptyset$ whenever $\alpha \neq \beta$.

Every graph is a spanning subgraph of some complete graph, one in which all possible edges appear. More precisely, $G = (V, E)$ is said to be a *complete graph* on $|V| = n$ vertices if

$$E = V \otimes V = \{ \{v, w\} : v \neq w \text{ in } V \}.$$

Ordinarily the complete graph on n vertices is denoted by the symbol K_n. Now if we fix V in considering only the spanning subgraphs $G = (V, E)$ of the complete graph $K = (V, V \otimes V)$, then we may introduce for each G a *complementary* graph

$$G' = (V, E') \quad \text{where} \quad \{v, w\} \in E' \quad \text{iff} \quad \{v, w\} \notin E.$$

Thus for the same set of vertices, G' has precisely those edges that are not found in G. Its relation matrix will have ones (zeros) where the matrix for G has zeros (ones), except on the main diagonal.

A graph G is said to be *bipartite* (two parts) if its vertices can be partitioned into two separate blocks, U and V, so that the only edges of G are those of the form $\{u, v\}$ with $u \in U$ and $v \in V$. All such edges need not appear, but if in fact they do, then G is called a *complete* bipartite graph. By analogy, the complete bipartite graphs are designated by $K_{m, n}$ when $|U| = m$ and $|V| = n$. In order to normalize the situation, it is usually assumed that U and V have been chosen so that $m \leq n$. It is easily seen that the general bipartite graphs $G = (U \cup V, E)$ are appropriate for the study of matching problems, such as applicants to jobs, as discussed in Section 0.3. In abstract terms, they simply provide a graphical representation for relations from one set to another.

EXAMPLE 2.3   Figure 2.3 shows the complete graphs K_n for $1 \le n \le 6$. We recall that the interior intersections do not represent vertices for these graphs in the case of K_4, K_5, and K_6.   △

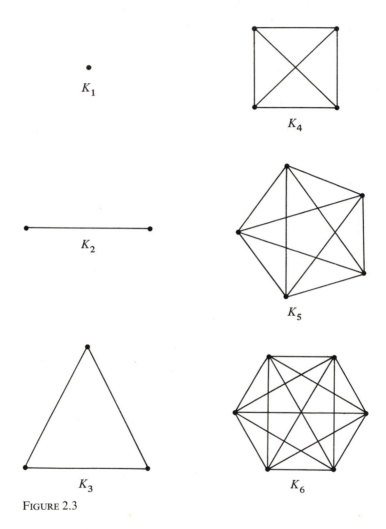

FIGURE 2.3

EXAMPLE 2.4   A few of the smaller complete bipartite graphs $K_{m,n}$ are shown in Figure 2.4.   △

EXAMPLE 2.5   To illustrate the concepts of union and intersection of subgraphs, consider H_1 and H_2 as shown in Figure 2.2. Their union and intersection are shown in Figure 2.5.   △

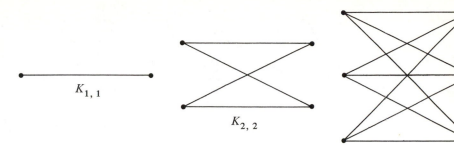

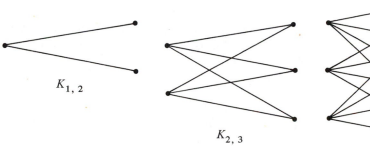

FIGURE 2.4

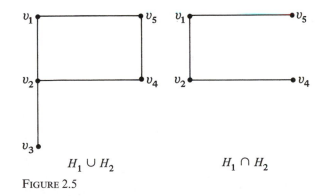

$$H_1 \cup H_2 \qquad\qquad H_1 \cap H_2$$

FIGURE 2.5

EXAMPLE 2.6 Consider once more the graph $G = (V, E)$ of Example 2.1. The complementary graph G' has the same set of vertices

$$V = \{v_1, v_2, v_3, v_4, v_5\}$$

but the complementary set of edges:

$$E' = \{\{v_1, v_3\}, \{v_1, v_4\}, \{v_2, v_5\}\}$$

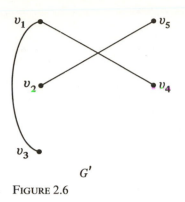

FIGURE 2.6

A picture of $G' = (V, E')$ is shown in Figure 2.6. As a graph in its own right, G' has subgraphs H_1' and H_2', as shown in Figure 2.7. Since the vertex sets are disjoint, their union is direct and we write

$$H_1' \cup H_2' = H_1' + H_2' = G'$$

with the addition sign in order to emphasize the decomposition of a graph into its separate "connectivity components." The latter concept has to do with the "paths" in a graph, as discussed in the next section. △

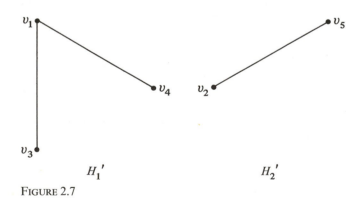

FIGURE 2.7

EXAMPLE 2.7   The complete graph K_5 has 10 edges, as shown in Figure 2.3, and K_6 has 15 edges. Can we obtain a general formula for the number of edges in K_n? The combinatorial nature of the problem allows for several determinations. If we recall the manner in which the lower-triangular matrix representation was derived (taking half of the matrix after deleting the diagonal), we arrive at the figure

$$\frac{n^2 - n}{2} = \frac{n(n-1)}{2}$$

because the relation matrix of K_n consists entirely of ones. Alternatively, we may add the number of entries in successive rows of the triangular representation. According to the IRS formula of Examples 1.2 and 1.3, we obtain

$$1 + 2 + \cdots + (n - 1) = \frac{n(n - 1)}{2}$$

the same result again. As a third approach, consider the number $(n - 1)$ of edges emerging at each of the n vertices. Multiplying $n(n - 1)$ would count each edge twice, and so we would have to divide this result by two, again leading us to the same conclusion. △

The last calculation makes use of the notion of the degree of a vertex. Let $G = (V, E)$ be any graph and consider any vertex $v \in V$. The number of edges $\{v, w\} \in E$ *incident* to or emerging from this fixed vertex v is called the *degree* of v and is denoted by $\delta(v)$. As was demonstrated earlier, these figures will often enter into a graphical enumeration problem. We would like to state two elementary consequences of the definition of degree, each of which is proved by a simple counting argument.

Lemma

$\sum_{v \in V} \delta(v) = 2m$, where m is the number of edges in $G = (V, E)$.

Proof As in Example 2.7, every edge contributes twice to the summation. □

Theorem 2.1

In any graph the number of vertices of odd degree is even.

Proof Suppose that $V = V_e \cup V_o$ is the decomposition into vertices of even and odd degrees. Obviously $\sum_{v \in V_e} \delta(v)$ is even; and so is $\sum_{v \in V} \delta(v)$, according to the lemma. Then the difference

$$\sum_{v \in V_o} \delta(v) = \sum_{v \in V} \delta(v) - \sum_{v \in V_e} \delta(v)$$

must also be even. Now a sum of odd numbers can be even only if the number of summands is even. It follows that $|V_o|$ is even. □

EXAMPLE 2.8 In Example 2.1 we have $\delta(v_1) = 2$, but otherwise,

$$\delta(v_2) = \delta(v_3) = \delta(v_4) = \delta(v_5) = 3$$

which is odd. Therefore, we have an even number (4) of vertices of odd degree. △

EXERCISES

1 Consider the graphs $G = (V, E)$ represented by the following matrices. In each case, draw a picture of the graph.

(a)

v_2	0			
v_3	1	1		
v_4	1	1	1	
v_5	1	1	0	1

$\qquad v_1 \quad v_2 \quad v_3 \quad v_4$

(b)

v_2	0				
v_3	0	0			
v_4	0	1	0		
v_5	1	1	0	1	
v_6	1	0	0	1	1

$\qquad v_1 \quad v_2 \quad v_3 \quad v_4 \quad v_5$

(c)

v_2	1				
v_3	1	1			
v_4	1	1	0		
v_5	1	1	1	1	
v_6	0	1	1	1	1

$\qquad v_1 \quad v_2 \quad v_3 \quad v_4 \quad v_5$

2 Determine the degree of each vertex in the graphs of Exercise 1, and show that Theorem 2.1 holds.

3 Consider the graphs $G = (V, E)$ shown in Figure 2.8. In each case determine the (lower-triangular) matrix representation.

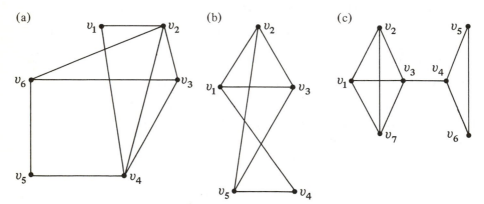

FIGURE 2.8

4 Determine the degree of each vertex in the graphs of Exercise 3, and show that Theorem 2.1 holds.

5 Draw (pictures of) the graphs G' complementary to those of Exercise 3.

6 Draw (pictures of) the subgraphs $\langle U \rangle$ generated by
   (a) $U = \{v_2, v_3, v_4, v_6\}$ in Exercise 3(a)
   (b) $U = \{v_1, v_2, v_3, v_5\}$ in Exercise 3(b)
   (c) $U = \{v_1, v_2, v_3, v_5, v_6\}$ in Exercise 3(c)

7 Obtain a formula for the number of edges in $K_{m, n}$.

8 For each of the following graphs G, list two subgraphs H, I where neither is a subgraph of the other. Then draw (pictures of) $H \cup I$ and $H \cap I$.

    (a) the graph of Exercise 3(a)   (b) Exercise 3(b)   (c) Exercise 3(c)

2.2 PATHS AND CIRCUITS

The real utility of graphs becomes apparent when we begin to discuss routes or itineraries as if the edges were to represent links of a shipping, distribution, or transportation network. If we look at a road map of the continental United States, we see that it always seems possible to arrange a route between any two given points. The lost traveler is not always so convinced of this fact, however; and if we were to look at a world road map, then there are certainly points, San Francisco and London for instance, that cannot be joined by an automobile route. Obviously the *connected* graphs are of special importance—those graphs for which connecting routes can always be found for any pair of vertices. It is necessary to take all of these various situations into account in a general treatment of graph theory. Each problem has its own peculiarities when expressed in graph-theoretic terms, and we must allow for every eventuality. Consider the special requirements of the traveling salesman. He would like to visit a certain group of cities without duplications and then finally return to his point of origin. In graph-theoretic terms, such a route is called an *elementary circuit*. We find that all such situations can be described in one way or another in terms of the various types of paths found in the underlying graph.

    In any graph $G = (V, E)$ a *path* from v to w is a sequence $\langle v_0, v_1, \ldots, v_k \rangle$ of (not necessarily distinct) vertices $v_i \in V$ with the properties

$$v_0 = v \quad \text{and} \quad \{v_{i-1}, v_i\} \in E$$
$$v_k = w$$

for all $i = 1, 2, \ldots, k$. If $v_0 = v_k$, then the path is called a *circuit*. If all the edges $\{v_{i-1}, v_i\}$ are different, then we have a *simple* path or circuit. It is an *elementary path* if none of the vertices is repeated; in an *elementary circuit* only v_0 is repeated (as v_k). [*Note.* elementary $\Rightarrow$ simple.]

    A graph is said to be *connected* if every pair of vertices $v \neq w$ is joined by a path. A *tree* is a connected graph without circuits. The latter concept has numerous applications in computer science, so we shall give a fairly extensive account of this special class of graphs in Section 2.4.

    The property of *being joined by a path* is an obvious equivalence relation on the set of vertices of a graph $G = (V, E)$. According to our characterization of equivalence relations (Section 0.5), V is partitioned into nonoverlapping subsets of vertices,

$$V = V_1 \cup V_2 \cup \cdots \cup V_p$$

for which every pair of vertices in a particular V_i is joined by a path (related) in the original graph G. We call p, the number of equivalence classes for this relation, the *connectivity number* of G. The subgraphs $\langle V_i \rangle$ generated by the various equivalence classes are called the *connected components* of G. Furthermore, according to Section 2.1, we may write

$$G = \sum_{i=1}^{p} \langle V_i \rangle$$

We use the notation $|G| = p$ in speaking of this decomposition of G into its p connected components.

EXAMPLE 2.9   In the graph G of Figure 2.1,

$v_1 v_5 v_3 v_4 v_5$ is a simple path from v_1 to v_5, but it is not elementary

$v_1 v_5 v_3 v_4$ is an elementary path from v_1 to v_4

$v_1 v_5 v_3 v_4 v_2 v_1$ is an elementary circuit

$v_1 v_5 v_4 v_1$ is not a circuit

In the last instance $\{v_4, v_1\}$ is not an edge of the graph G, so that $v_1 v_5 v_4 v_1$ is not a path in G. In the first instance v_5 is repeated, which is the reason the path $v_1 v_5 v_3 v_4 v_5$ is not elementary.   △

EXAMPLE 2.10   In Figure 2.2, H_2 is a tree. Although they are connected, H_1 and H_3 are not trees because they have circuits. H_1 has the circuit $v_1 v_5 v_4 v_2 v_1$ and H_3 has the circuit $v_5 v_3 v_2 v_4 v_5$.   △

EXAMPLE 2.11   A few more examples of trees are presented in Figure 2.9. Fortunately, we have a case of well-chosen terminology here. They actually look like trees.   △

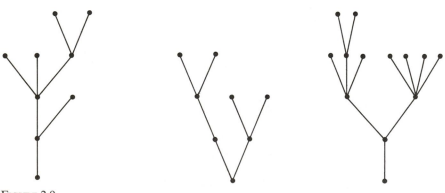

FIGURE 2.9

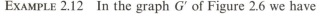

EXAMPLE 2.12   In the graph G' of Figure 2.6 we have

$$|G'| = p = 2$$

as discussed in Example 2.6. Its connected components H'_1 and H'_2 are displayed in Figure 2.7.   △

If we have a picture of a small graph G (as in Example 2.12), then in most cases it is easy enough for a human observer to determine the connectivity number $|G| = p$. The situation is quite different however, for larger graphs or computer simulations. In the latter case the graph is not generally represented pictorially, and the computer is limited in its perceptive capabilities, in any case. What we need is an algorithm for determining this important graphical parameter p. A simple procedure for this purpose is presented in Algorithm III.

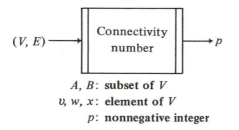

A, B: **subset of** V

v, w, x: **element of** V

p: **nonnegative integer**

begin $p \leftarrow 0; B \leftarrow \emptyset;$
   while $B \neq V$ **do**
      begin $x \leftarrow$ elt $(\sim B); A \leftarrow \{x\};$
         while $(A, \sim A) \neq \emptyset$ **do**
            for $v \in A$ **do**
            for $w \notin A$ **do**
            if $v E w$ **then**
               $A \leftarrow A \cup \{w\};$
         $B \leftarrow B \cup A;$
         $p \leftarrow p + 1$
      end
   end
ALGORITHM III

The procedure is quite straightforward and should be intuitively clear. We make use of a new notation [see Figure 2.10(a)]

$$(U, W) = \{ \{u, w\} \in E : u \in U, w \in W \}$$

for representing the set of all edges in E joining U to W for disjoint $U, W \subseteq V$ in any graph $G = (V, E)$. In Algorithm III, this notation is used in the interior **while** statement to represent the set of all edges emerging from the set A. [See Figure 2.10(b).] A is the set of vertices in a growing connected component. We say "growing" because we continually adjoin vertices w corresponding to edges $\{v, w\}$ with $v \in A$ and $w \notin A$. B represents the set of vertices so far selected in

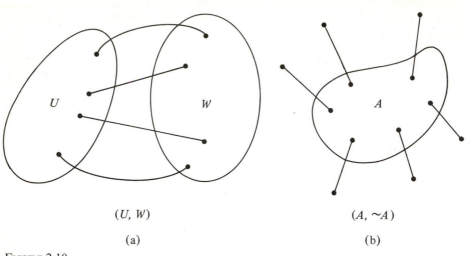

(U, W)   (A, ~A)

(a)   (b)

FIGURE 2.10

forming all these components. Until $B = V$, we continue to choose an (unselected) element x as the starting point for building a new component. Note that we begin with $p = 0$ and increment p in the assignment statement $p \leftarrow p + 1$ every time a new component is exhausted. So we cannot help but arrive at the connectivity number of G at the conclusion of the execution of the algorithm.

EXAMPLE 2.13   Consider the graph $G = (V, E)$ given by the following lower-triangular matrix representation:

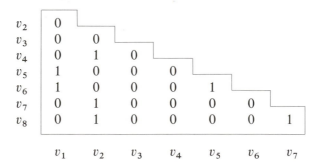

	v_1	v_2	v_3	v_4	v_5	v_6	v_7
v_2	0						
v_3	0	0					
v_4	0	1	0				
v_5	1	0	0	0			
v_6	1	0	0	0	1		
v_7	0	1	0	0	0	0	
v_8	0	1	0	0	0	0	1

Partial intermediate results from a trace of the algorithm can be given as follows:

$x = v_1$ $A = \{v_1, v_5, v_6\}$ $B = \{v_1, v_5, v_6\}$ $p = 1$

$x = v_7$ $A = \{v_7, v_2, v_8, v_4\}$ $B = \{v_1, v_2, v_4, v_5, v_6, v_7, v_8\}$ $p = 2$

$x = v_3$ $A = \{v_3\}$ $B = V$ $p = 3$

Note that the arbitrary selection of the element $x \notin B$ will have an effect on the sequence of computations, but not on the final result $p = 3$.   △

If a graph with n vertices is given, say, by its relation matrix or by the lower-triangular portion, then the reader can devise a truly elementary algorithm for computing m, the number of edges in the graph. We have just provided an algorithm for determining p, the connectivity number of the graph. The three parameters n, m, and p are important in the detailed analysis of the circuit structure of a graph, as shall be presented in the following section.

EXERCISES

1 Give an example of a simple and a nonsimple path in the graph of
   (a) Exercise 3(a) of Section 2.1   (b) Exercise 3(b) of Section 2.1
   (c) Exercise 3(c) Section 2.1

2 Repeat Exercise 1 for circuits.

3 Give an example of a simple nonelementary path in the graph of
   (a) Exercise 3(a) of Section 2.1   (b) Exercise 3(b) of Section 2.1
   (c) Exercise 3(c) of Section 2.1

4 Give an example of a simple nonelementary circuit in the graph of Exercise 1(c) of Section 2.1.

5 Use Algorithm III to compute the connectivity number p for each of the following graphs:
   (a) Exercise 3(b) of Section 2.1
   (b) The complement of the graph of Exercise 3(b) of Section 2.1
   (c) The graph given by the matrix

$$
\begin{array}{c|ccccc}
v_2 & 0 \\
v_3 & 0 & 0 \\
v_4 & 1 & 0 & 0 \\
v_5 & 0 & 1 & 0 & 0 \\
v_6 & 1 & 0 & 0 & 1 & 0 \\
\hline
 & v_1 & v_2 & v_3 & v_4 & v_5
\end{array}
$$

6 Devise a formal algorithm for determining m, the number of edges in a graph $G = (V, E)$ given by its (lower-triangular) matrix.

2.3   CIRCUIT RANK

An electrical network as usually conceived has an underlying graph in which every edge lies on a circuit. Each of the edges is labeled with the value of an associated electrical component (battery, resistor, etc.), but the graph itself is of primary importance in any comprehensive electrical analysis. In particular, the circuit structure of the graph plays a central role in the various methods for determining the electrical currents flowing in the various branches (edges) of the network.

In one popular method for studying electrical networks (loop current analysis), one assumes that certain unknown and fictitious loop currents are flowing simultaneously around various circuits of the underlying graph. Using Ohm's law and the familiar Kirchhoff voltage law from high school physics, one is able to write linear algebraic (loop) equations involving these unknown currents. Provided that the resulting system of equations is properly determined, that is *nonsingular* in the sense of linear algebra, we are able to solve the system and determine the unknown currents.

EXAMPLE 2.14   Readers who are not overly familiar with electrical networks and/or linear algebra should not try to follow this example too closely. If they simply try to gain an overview of the situation without paying much attention to the details, it will surely be sufficient. We are only trying to motivate the definition of the circuit rank of a graph. It happens that the most natural setting for such a presentation is in the area of electrical networks.

Consider the electrical network of Figure 2.11(a) together with the underlying graph as shown in Figure 2.11(b). We have assumed names $e_1, e_2, \ldots, e_{12}$ and directions for the unknown voltage drops across each of the branches. Suppose that we imagine loop currents I_1, I_2, I_3, I_4 corresponding to the circuits

$$c_1 : v_6 v_5 v_8 v_9 v_6$$
$$c_2 : v_5 v_4 v_7 v_8 v_5$$
$$c_3 : v_4 v_1 v_2 v_3 v_6 v_5 v_4$$
$$c_4 : v_1 v_4 v_5 v_8 v_9 v_6 v_3 v_2 v_1$$

respectively. Note that the circuits imply a definite direction of traversal, and these directions in turn determine the algebraic sign of the voltage drops e_j to be taken in applying Kirchhoff's law.

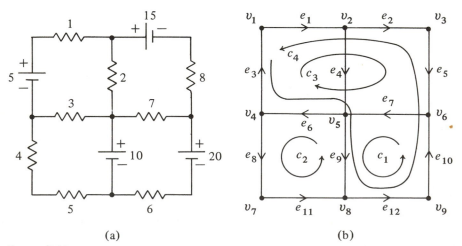

(a)   (b)

FIGURE 2.11

The application of Kirchhoff's voltage law to each of the four loops will lead to the following system of linear equations:

$$L_1: e_7 + e_9 + e_{12} + e_{10} = 0$$
$$L_2: e_6 + e_8 + e_{11} - e_9 = 0$$
$$L_3: e_3 + e_1 + e_2 + e_5 + e_7 + e_6 = 0$$
$$L_4: -e_3 - e_6 + e_9 + e_{12} + e_{10} - e_5 - e_2 - e_1 = 0$$

If we use Ohm's law across the resistive branches, then the system becomes:

$$L_1: 7(I_1 + I_3) + 10 + 6(I_1 + I_4) - 20 = 0$$
$$L_2: 3(I_2 + I_3 - I_4) + 4I_2 + 5I_2 - 10 = 0$$
$$L_3: -5 + 1(I_3 - I_4) + 15 + 8(I_3 - I_4) + 7(I_1 + I_3)$$
$$+ 3(I_2 + I_3 - I_4) = 0$$
$$L_4: 5 + 3(-I_2 - I_3 + I_4) + 10 + 6(I_1 + I_4) - 20 + 8(-I_3 + I_4)$$
$$- 15 + 1(-I_3 + I_4) = 0$$

The important thing to observe in all of this, however, is the identity:

$$L_1 - L_3 - L_4 \equiv 0$$

In the sense of linear algebra, we say that the four equations are *not* linearly independent, and it is well known that one cannot solve such a system for the unknowns I_1, I_2, I_3, I_4, at least in a way that provides a unique solution. If we had written a system of equations that could be solved uniquely, then we would have been able to obtain the individual branch currents by superposition of loop currents, that is,

$I_1 + I_4$ would have been the current from v_8 to v_9

$I_4 - I_2 - I_3$ would have been the current from v_4 to v_5, etc.

In order to obtain a system of equations that admits a unique solution, we need to choose a maximum number of loops or circuits $c_1, c_2, \ldots, c_r$ such that *dependencies*

$$\pm L_{\alpha_1} \pm L_{\alpha_2} \pm \cdots \pm L_{\alpha_s} \equiv 0$$

do not occur among any of the corresponding loop equations. In view of the dependency $L_1 - L_3 - L_4 \equiv 0$, it might be thought that $r < 4$ in the present example, but that is not the case. In fact, if we simply change our fourth circuit only slightly, taking

$$c_4: v_2 v_5 v_8 v_9 v_6 v_3 v_2$$

then we will obtain a new (maximal) system of equations without dependencies. Its solution will then be obtainable according to standard methods of linear algebra. △

As a by-product of our analysis of the circuit structure of graphs, we will be able to describe an algorithmic process for the determination of a proper collection $c_1, c_2, \ldots, c_r$ of circuits on which to define loop currents in the electrical network problem. Any ad hoc procedure would become hopelessly difficult as the complexity of the network increases, but graph theory comes to the rescue and finally offers a most orderly and systematic approach.

We shall no longer consider electrical networks but, instead, return to a discussion of arbitrary graphs $G = (V, E)$. If we designate the set of edges by the notation

$$E = \{e_1, e_2, \ldots, e_m\}$$

then there are at least two ways to look at circuits. Instead of the sequence $v_0, v_1, \ldots, v_k = v_0$ of vertices encountered in traversing the circuit, we may consider the circuit c as the associated collection of edges encountered; that is,

$$c = \{e_{i_1}, e_{i_2}, \ldots, e_{i_k}\}$$

as a subset of E. Evidently we restrict our attention to simple circuits by this convention and we lose track of the direction of traversal by this agreement, but otherwise the circuits are quite adequately described. In fact, we can go one step further in representing c as a vector of the set B^m (see Example 0.18.), that is,

$$c = (c_1, c_2, \ldots, c_m)$$

with the components determined by the rule

$$c_j = \begin{cases} 1 & \text{if } e_j \text{ is an edge of the circuit} \\ 0 & \text{otherwise} \end{cases}$$

EXAMPLE 2.15   In Example 2.1 let us name the edges

$$E = \{e_1, e_2, e_3, e_4, e_5, e_6, e_7\}$$

as in Figure 2.12. Then we can list the circuits and their corresponding vectors

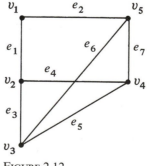

FIGURE 2.12

in the new notation as follows:

$$\{e_1, e_2, e_4, e_7\} \qquad c^1 = (1, 1, 0, 1, 0, 0, 1)$$
$$\{e_3, e_4, e_5\} \qquad c^2 = (0, 0, 1, 1, 1, 0, 0)$$
$$\{e_5, e_6, e_7\} \qquad c^3 = (0, 0, 0, 0, 1, 1, 1)$$
$$\{e_1, e_2, e_4, e_5, e_6\} \qquad c^4 = (1, 1, 0, 1, 1, 1, 0)$$
$$\{e_3, e_4, e_6, e_7\} \qquad c^5 = (0, 0, 1, 1, 0, 1, 1)$$
$$\{e_1, e_2, e_3, e_5, e_7\} \qquad c^6 = (1, 1, 1, 0, 1, 0, 1)$$
$$\{e_1, e_2, e_3, e_6\} \qquad c^7 = (1, 1, 1, 0, 0, 1, 0) \qquad \triangle$$

EXAMPLE 2.16 If we label the edges of the network of Example 2.14 as in Figure 2.13, then the circuits discussed in that example become identified with the vectors

$$c^1 = (0, 0, 0, 0, 0, 0, 1, 0, 1, 1, 0, 1)$$
$$c^2 = (0, 0, 0, 0, 0, 1, 0, 1, 1, 0, 1, 0)$$
$$c^3 = (1, 1, 1, 0, 1, 1, 1, 0, 0, 0, 0, 0)$$
$$c^4 = (1, 1, 1, 0, 1, 1, 0, 0, 1, 1, 0, 1)$$

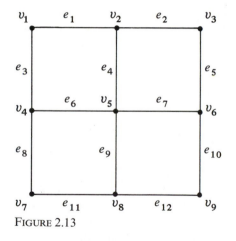

FIGURE 2.13

respectively. On the other hand, the corresponding loop equations can be identified as vectors on the "directed" edges of Figure 2.11(b) as follows:

$$L_1: (0, 0, 0, 0, 0, 0, 1, 0, 1, 1, 0, 1)$$
$$L_2: (0, 0, 0, 0, 0, 1, 0, 1, -1, 0, 1, 0)$$
$$L_3: (1, 1, 1, 0, 1, 1, 1, 0, 0, 0, 0, 0)$$
$$L_4: (-1, -1, -1, 0, -1, -1, 0, 0, 1, 1, 0, 1)$$

Here our dependency

$$L_1 - L_3 - L_4 \equiv 0$$

is all the more apparent. Now observe that we would also have

$$c^1 \oplus c^3 \oplus c^4 = 0 \qquad [= (0, 0, 0, 0, 0, 0, 0, 0, 0, 0, 0, 0)]$$

in an algebra for which

$$0 \oplus 0 = 0$$
$$0 \oplus 1 = 1 \oplus 0 = 1$$
$$1 \oplus 1 = 0$$

but this is exactly the behavior of the algebra discussed in Examples 1.16 and 1.17!   △

Motivated by the discussions of Examples 2.14 and 2.16, we say that a collection $\{c^\alpha\}$ of circuits in a graph G is *independent* if there does not exist a nonempty subset summing to zero

$$c^{\alpha_1} \oplus c^{\alpha_2} \oplus \cdots \oplus c^{\alpha_s} = 0$$

in the algebra $B^m = (B^m, \oplus)$ of Example 1.17. Otherwise, the collection is said to be *dependent*. At the same time the maximum number of independent circuits is called the *circuit rank* $\rho(G)$ of the graph G. Eventually we will see that the circuit rank can be computed by an arithmetic formula, but for the time being, we shall simply appeal to the definition.

EXAMPLE 2.17   In Example 2.15 the collection $\{c^1, c^2, c^6\}$ is dependent because

$$\begin{aligned} c^1 \oplus c^2 \oplus c^6 &= (1, 1, 0, 1, 0, 0, 1) \oplus (0, 0, 1, 1, 1, 0, 0) \\ &\qquad \oplus (1, 1, 1, 0, 1, 0, 1) \\ &= (1 \oplus 0 \oplus 1, 1 \oplus 0 \oplus 1, 0 \oplus 1 \oplus 1, \\ &\qquad 1 \oplus 1 \oplus 0, 0 \oplus 1 \oplus 1, 0 \oplus 0 \oplus 0, \\ &\qquad 1 \oplus 0 \oplus 1) \\ &= (0, 0, 0, 0, 0, 0, 0) \end{aligned}$$

It follows that any collection of circuits that contains $\{c^1, c^2, c^6\}$ as a subset will also be dependent.

Any collection $\{c^\alpha\}$ consisting of only one or two circuits will always be independent. (Why?) As a consequence, the circuit rank of the present graph (Example 2.1 or 2.15) is at least two. On the other hand, the collection $\{c^1, c^2, c^4\}$ is independent because $c^1 \neq 0$, $c^2 \neq 0$, $c^4 \neq 0$, $c^1 \oplus c^2 \neq 0$, $c^1 \oplus c^4 \neq 0$, $c^2 \oplus c^4 \neq 0$, and also

$$c^1 \oplus c^2 \oplus c^4 = (0, 0, 1, 1, 0, 1, 1) \neq 0.$$

Therefore, the circuit rank is at least three. We leave it to the readers to convince themselves that no larger independent sets exist, so that the circuit rank is precisely three.   △

Soon we will see that the circuit rank of a graph $G = (V, E)$ can be computed according to the simple formula

$$\rho(G) = m - n + p$$

whenever

$$|G| = p \qquad |V| = n \qquad |E| = m$$

Let us temporarily denote $r(G) = m - n + p$ and make the following observation. If we augment $G = (V, E)$ by adding an edge to obtain $G^+ = (V, E \cup \{e\})$, where $e = \{v, w\} \notin E$, then we have two cases to consider:

1. v and w are joined by a path in G;
2. They are not.

In the first case, $r(G^+) = r(G) + 1$ because m increases by one while p remains unchanged. In the second case, $r(G^+) = r(G)$ since m increases by one while p decreases by one. This simple observation is important in obtaining the following fundamental result.

Theorem 2.2

In any graph $G = (V, E)$ such that

$$|G| = p \qquad |V| = n \qquad |E| = m$$

the circuit rank is given by the formula

$$\rho(G) = m - n + p$$

Proof   We will show how to find $m - n + p$ independent circuits in G. If $E = \{e_1, e_2, \ldots, e_m\}$, we consider the sequence of subgraphs

$$G_0, G_1, \ldots, G_m = G$$

each having the vertex set V but in which

$$E_0 = \varnothing \quad \text{and} \quad E_i = E_{i-1} \cup \{e_i\}$$

Thus we simply add one edge at a time, so that each $G_i = G_{i-1}^+$ relative to the addition of $e_i = \{v_i, w_i\}$.

First, we compute the initial and final values for the function r. We evidently have

$$r(G_0) = 0 - n + n = 0 \qquad r(G_m) = r(G) = m - n + p$$

Then we notice that v_i and w_i are joined by a path in G_{i-1} if and only if the addition of the edge e_i closes a new circuit c^i. It follows by our preliminary observation that r increases by one with each c^i. Altogether we must obtain $r(G_m) - r(G_0) = m - n + p$ increases and, accordingly, $m - n + p$ circuits c^i in succession. They are independent because each c^i has an edge (namely e_i) not found in any of the preceding.

The argument that shows that it is not possible to find more than $m - n + p$ independent circuits requires some knowledge of linear algebra, so it will not be presented.   □

This result has important theoretical implications; but the proof was constructive and in fact provides an algorithm for the systematic determination of a proper collection of circuits on which to base the loop current method for solving electrical network problems. Other techniques will be suggested as we proceed. Especially in the case of graphs of a particular structure (for instance, the planar graphs of Section 2.5), these techniques may be more appropriate, more efficient than the one presented here, but the present approach is reasonably effective in the general case. Note, however, that in a machine implementation, we must be able to decide whether or not two given vertices (v_i and w_i here) are joined by a path in a graph (the graph G_{i-1} in this case). We need to know the nature of such a path in order to form the circuit c^i. The reader can devise his own algorithm for answering these questions. Otherwise, he can wait until the algorithms for solving the various path problems are discussed later in the chapter.

EXAMPLE 2.18   In Example 2.1 or 2.15 we have

$$p = 1 \qquad n = 5 \qquad m = 7$$

and accordingly,

$$\rho(G) = m - n + p$$
$$= 7 - 5 + 1 = 3$$

which agrees with the calculations of Example 2.17.

In order to illustrate the construction in the proof of Theorem 2.2, let G_i have the edges $E_i = \{e_1, e_2, \ldots, e_i\}$ as in Figure 2.14. The calculations

$$r(G_0) = 0 - 5 + 5 = 0$$
$$\vdots$$
$$r(G_3) = 3 - 5 + 2 = 0$$
$$r(G_4) = 4 - 5 + 1 = 0$$
$$r(G_5) = 5 - 5 + 1 = 1$$
$$r(G_6) = 6 - 5 + 1 = 2$$
$$r(G_7) = 7 - 5 + 1 = 3$$

show that r increases by one every time that the augmenting edge closes a new circuit.

The circuits obtained by this procedure are not always uniquely determined because of the possible existence of more than one path joining the vertices v_i and w_i in G_{i-1}. They are always independent, however, regardless of the choice of paths, because each circuit has an edge (e_i) that is not found in any of the

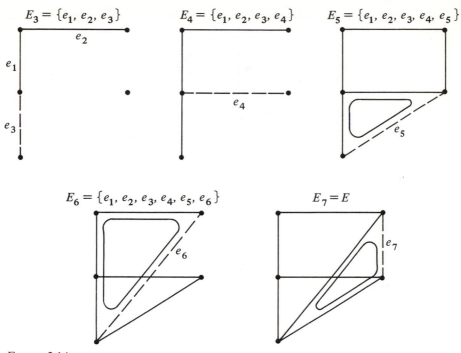

FIGURE 2.14

preceding circuits. In the present example, we have chosen paths that lead to the circuits

$$\{e_3, e_4, e_5\} \qquad \{e_1, e_2, e_3, e_6\} \qquad \{e_5, e_6, e_7\}$$

as shown in Figure 2.14.   △

EXERCISES

1 Show that the following collections of circuits are dependent in the graph of Example 2.15 (Figure 2.12):
   (a) $\{c^1, c^3, c^4\}$ (b) $\{c^1, c^2, c^3, c^4\}$
   (c) $\{c^4, c^5, c^6, c^7\}$ (d) $\{c^2, c^3, c^4, c^5\}$
   (e) $\{c^2, c^3, c^5\}$ (f) $\{c^1, c^2, c^6, c^7\}$

2 Find a collection of three independent circuits in the graph of Example 2.15 other than $\{c^1, c^2, c^4\}$. Show that your collection is independent.

3 Find a collection of four independent circuits in the graph of Example 2.14 and 2.16 (Figure 2.13). Show that your collection is independent.

4 In addition to the circuits listed in Example 2.16 (for Figure 2.13), suppose we name

$$c^5 = (1, 0, 1, 1, 0, 0, 0, 1, 1, 0, 1, 0)$$
$$c^6 = (0, 1, 0, 1, 1, 1, 0, 1, 0, 1, 1, 1)$$
$$c^7 = (0, 0, 0, 0, 0, 1, 1, 1, 0, 1, 1, 1)$$
$$c^8 = (1, 1, 1, 0, 1, 0, 0, 1, 0, 1, 1, 1)$$

Show that the following collections of circuits are dependent.
(a) $\{c^1, c^5, c^6, c^7, c^8\}$ (b) $\{c^1, c^2, c^3, c^5, c^8\}$
(c) $\{c^1, c^2, c^3, c^8\}$ (d) $\{c^1, c^3, c^5, c^7, c^8\}$

5 Use Theorem 2.2 to compute the circuit rank of the graphs of:
   (a) Exercise 1(a) of Section 2.1   (b) Exercise 1(c) of Section 2.1
   (c) Exercise 3(a) of Section 2.1   (d) Exercise 3(b) of Section 2.1
   (e) Exercise 3(c) of Section 2.1   (f) Exercise 5(c) of Section 2.1
   (g) Example 2.16 (Figure 2.13)   (h) The cube with its (four) diagonals

6 Suppose that an ordered collection $\{c^1, c^2, \ldots, c^k\}$ of circuits in a graph G has the property that each c^i has an edge not found in any of the preceding. Prove that the collection is independent.

*7 (For students having a familiarity with linear algebra) Prove that there are $n - p$ independent vectors d^j orthogonal to all the circuit vectors c^i in a graph G. [*Hint*: First suppose that G is connected and find $n - 1$ such vectors by considering a sequence of subsets

$$V_1 \subseteq \cdots \subseteq V_{n-1}$$

of V, each of increasing size ($|V_j| = j$). Eventually associate a vector d^j with each V_j representing the edges of $(V_j, V \sim V_j)$, to use the notation introduced in discussing Algorithm III. Then use an argument analogous to Exercise 6 to obtain independence, and use the fact that each d^j has an even number of edges in common with any circuit to obtain the orthogonality.]

*8 (Again, for students having a familiarity with linear algebra) Use the result of Exercise 7 to prove that there cannot be more than $m - n + p$ independent circuits in a graph G. [*Hint*: The proof hinges on an important theorem in linear algebra.]

9 As in Example 2.18, illustrate the construction in the proof of Theorem 2.2 for each of the graphs in Exercise 5.

*10 Considering circuits as sets of edges, let c^1, c^2 be circuits in any graph G, and let $e \in c^1 \cap c^2$. Prove that there exists a circuit c such that $e \notin c \subseteq c^1 \cup c^2$.

2.4 TREES

Recall that an acyclic (without circuits) connected ($p = 1$) graph is called a tree. As graphs, trees are quite special, but they have immense importance in applications, especially in computer science. The reader will be easily convinced of this importance as we proceed, but first we wish to establish a number of convenient characterizations of trees. We shall do this in the following theorem, where we state six logically equivalent propositions. When we use the term *edge maximality* (*minimality*) with respect to a certain property, we simply mean that the graph G has the property in question, but if an edge is added (deleted), then the property will no longer hold. See (4) and (5) in the theorem.

Theorem 2.3

The following statements are equivalent for a graph on n vertices, with $n \geq 2$:

(1) G is connected and acyclic (that is, G is a tree).
(2) G is acyclic and has $n - 1$ edges.
(3) G is connected and has $n - 1$ edges.
(4) G is edge maximal with respect to being acyclic.
(5) G is edge minimal with respect to being connected.
(6) Every pair of vertices in G is joined by exactly one path.

Proof   As discussed in Section 0.10, we seek to prove a cycle of implications in order to establish the equivalence. In this case, we try (1) $\Rightarrow$ (2) $\Rightarrow$ (3) $\Rightarrow$ (4) $\Rightarrow$ (5) $\Rightarrow$ (6) $\Rightarrow$(1). It turns out that each individual implication is entirely straightforward, and most of them follow rather easily from the circuit rank formula (Theorem 2.2).

(1) $\Rightarrow$ (2).   Since $p = 1$ and $\rho(G) = m - n + p = m - n + 1 = 0$ because of the assumptions in (1), we have $m = n - 1$ immediately.

(2) $\Rightarrow$ (3).   Since (2) implies $\rho(G) = m - n + p = (n - 1) - n + p = 0$,   we obtain $p = 1$ by transposition.

(3) $\Rightarrow$ (4).   With $p = 1$ and $m = n - 1$,

$$\rho(G) = m - n + p = (n - 1) - n + 1 = 0$$

which shows that G is acyclic. If, however, we were to add an edge, then m would be increased by one, but p and n would remain unchanged. As a result, $\rho(G)$ would increase to one.

(4) $\Rightarrow$ (5).   Suppose (4) is true. Then if G were not connected, there would exist vertices v, w not joined by a path. So adjoining the edge $\{v, w\}$ would not form a circuit. This contradicts the assumption (4), so we must have $p = 1$. Still assuming (4), we would have

$$\rho(G) = m - n + p = m - n + 1 = 0$$

which shows that $m = n - 1$. Now if we were to delete an edge, we would obtain a graph G' with

$$m' = m - 1 = n - 2 = n' - 2$$
$$\rho(G') = m' - n' + p' = 0$$

the latter because we had no circuits to begin with. It follows that $p' = n' - m' = n' - (n' - 2) = 2$, which shows that G is edge minimal with respect to being connected.

(5) $\Rightarrow$ (6). Assuming (5), we see that G is connected, so that every pair of vertices v, w is joined by a path. If there were more than one path from v to w, then the deletion of an edge belonging to a second one and not the first would not disconnect G, which is a contradiction of (5).

(6) $\Rightarrow$ (1). Clearly (6) implies that G is connected. But if, additionally, G had a circuit, then some pair of vertices would be joined by two paths, again a contradiction.   □

Corollary

Every connected graph has a spanning tree.

Proof   When G is connected, choose an edge which may be deleted (assuming there is one) without disconnecting G. Continue this process until it is no longer possible to delete an edge without disconnecting G. Since according to Theorem 2.3, (5) is equivalent to (1), we then have a (spanning) tree.   □

EXAMPLE 2.19   Let us use this method to find a spanning tree for the graph shown in Figure 2.15(a). We may remove in succession the edges af, ag, ef, bc. Then we will have the subgraph shown in Figure 2.15(b). Now, we cannot remove fg or cf, for example, because either way we would disconnect the graph. But we can delete de, say, followed by ad and ab, leaving the subgraph shown in Figure 2.15(c). Now it is impossible to continue. The removal of any more edges would disconnect the graph. The resulting subgraph is a spanning tree.   △

A computer implementation of the procedure suggested by the proof of the corollary is quite straightforward. For one thing, we can modify Algorithm III (Section 2.2) slightly so as to decide the question of connectivity of the graphs obtained by a deletion. We can even determine ahead of time just how many edges need to be removed! Note in Example 2.19 that we removed

$$7 = 13 - 7 + 1 = m - n + p = \rho(G)$$

edges altogether. Is this a coincidence? In agricultural terms, is the circuit rank "the number of dikes one must remove in order to flood all the rice fields?" To answer this question, the reader should observe that the reinsertion of the deleted edges, one at a time, will produce a (unique!) circuit. That is because of the equivalence of (4) and (1) in Theorem 2.3. At the same time, the collection of circuits so obtained is independent, because each has an edge—the previously deleted one—not found in any of the others.

The last remark suggests an alternative procedure for finding a maximal independent collection of circuits in any graph G:

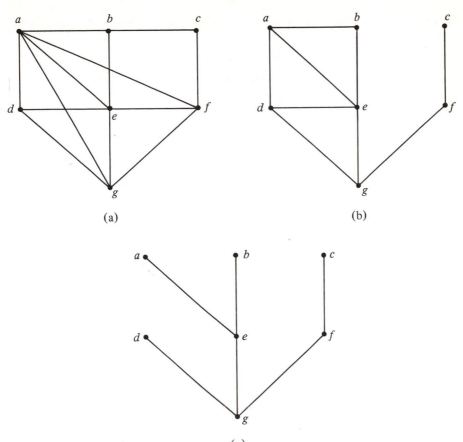

(a)   (b)

(c)

FIGURE 2.15

1. Obtain a *spanning forest* T, a spanning tree in each of the p connectivity components, and let the edges of G that are not in T be called *chords*. The latter will be $\rho(G)$ in number.
2. The graphical union of T with each individual chord gives rise to a unique circuit in G. The total collection of circuits will be independent and $\rho(G)$ in number.

Once again, the procedure gives a solution to the electrical network problem which motivated the discussion in Section 2.3, and it should be compared with the process based on the proof of Theorem 2.2.

EXAMPLE 2.20   In Example 2.19 we have the chords

$$af, ag, ef, bc, de, ad, ab$$

When reinserted, one at a time, these chords give rise to circuits as follows (see Figure 2.16):

$$af: aegfa$$
$$ag: agea$$
$$ef: egfe$$
$$bc: begfcb$$
$$de: dged$$
$$ad: aegda$$
$$ab: aeba$$

The reader should check that this collection of circuits is independent.   △

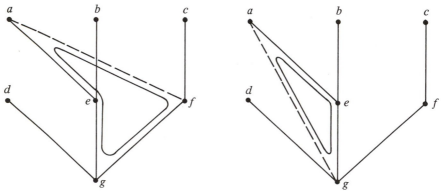

FIGURE 2.16

The proof of the corollary gave an outline of an algorithm for finding a spanning tree in any connected graph $G = (V, E)$, but the whole construction can be approached from the other direction. If $|V| = n$, we can build a sequence of subgraphs

$$G_0, G_1, \ldots, G_{n-1}$$

each having the vertex set V, but in which $|E_i| = i$. We simply add one edge (from E) after another, being careful that our selections do not produce a circuit. This time it is the equivalence of (4) with (1) in Theorem 2.3 that is applicable. It shows that we again obtain a spanning tree when the augmentation can no longer be carried out. If we try to implement this algorithm, we will have to be able to detect the presence of a circuit, but we have seen already that this is only a matter of arithmetic when m, n, and p are known.

EXAMPLE 2.21   In the graph of Example 2.19 [Figure 2.15(a)] suppose we have already selected the edges ab, bc, ad, be, cf. Then the subgraph G_5 is as in

Figure 2.17(a). At this stage we cannot select *de*, say, since that would give rise to the circuit *adeba*. We can select *ag*, however, obtaining the subgraph G_6 shown in Figure 2.17(b). Now no further edges can be added without producing a circuit, and the subgraph G_6 is a spanning tree. △

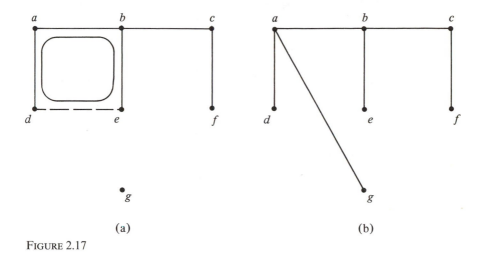

(a)   (b)

FIGURE 2.17

The applications of the concept of spanning trees are many and varied. In order to gain some appreciation for this fact, suppose that the vertices of the set *V* represent cities, and the edges denote pipelines, telephone cables, railways, or any such form of communication or conveyance. In any graph $G = (V, E)$ whose edges represent the feasible or proposed connections, we may be able to associate a mileage or cost with each edge in order to represent the length or expense in joining the various cities. In a *minimal connector problem*, we seek to find the most economical network linking all the cities. The solution must surely be a spanning tree. Otherwise, if there were circuits, links could be removed to lower the cost.

To approach the minimal connector problem, we have only to make a slight change in the procedure just introduced. We choose e_1 to be a cheapest edge, and having already determined the subgraph G_{i-1}, we choose e_i to be the most economical link such that $G_i = G_{i-1} \cup \{e_i\}$ is without circuits. The resulting spanning tree G_{n-1} is called an *economy tree*. We will soon show that they are in fact *minimal spanning trees*, that is, solutions to the minimal connector problem.

EXAMPLE 2.22 Suppose that we want to design a minimum-length rail network connecting the seven cities shown in the mileage chart in Figure 2.18. To build

Salt Lake City	512				
Albuquerque	422	611			
Oklahoma City	615	1108	525		
Cheyenne	102	462	522	706	
Omaha	540	955	895	454	493
Dodge City	301	682	425	212	355	336

Denver   Salt Lake City   Albuquerque   Oklahoma City   Cheyenne   Omaha

FIGURE 2.18

an economy tree, we first choose e_1 to be the link between Denver and Cheyenne (102 miles), the smallest distance between any two cities in the table. Then in succession we choose

e_2: Dodge City to Oklahoma City (212 miles)

e_3: Denver to Dodge City (301 miles)

e_4: Dodge City to Omaha (336 miles)

e_5: Denver to Albuquerque (422 miles)

e_6: Cheyenne to Salt Lake City (462 miles)

Finally, we are left with the economy tree of Figure 2.19. Note that e_5 was chosen over a cheaper link, from Cheyenne to Dodge City, but the latter would have produced a circuit.   △

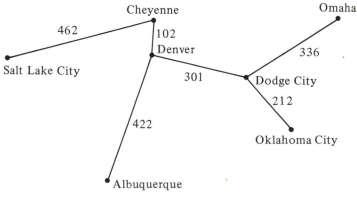

FIGURE 2.19

Theorem 2.4

Every economy tree is a minimal spanning tree.

Proof Suppose that we are given an economy tree G_{n-1}. Among all the minimal spanning trees, let T be one that has a maximum number of edges in common with G_{n-1}. We are to compare the *cost* $c(T)$ with that for G_{n-1}. Each of these trees has the same number $(n - 1)$ of edges, so we can let e be an edge (*of minimum cost*) that is in G_{n-1} but not in T. Otherwise, $T = G_{n-1}$ and we would be finished. According to Theorem 2.3, the graph $T \cup \{e\}$ has a unique circuit, and there must be an edge f in this circuit that is not an edge of G_{n-1}, simply because G_{n-1} has no circuits.

Suppose that we denote by T' the spanning tree obtained by exchanging e for f in T. Then the inequality $c(f) > c(e)$ would contradict the fact that T was a minimal spanning tree [compare $c(T') < c(T)$], and an equality would contradict the fact that among all minimal spanning trees, T had a maximum number of edges in common with G_{n-1} (T' has one more). So we have to conclude that $c(f) < c(e)$.

Just as before, the graph $G_{n-1} \cup \{f\}$ has a unique circuit, and in this circuit we can find an edge e' that is not an edge of T. An inequality $c(f) < c(e')$ would contradict the whole construction of the economy tree. So we are left to conclude that

$$c(e') \le c(f) < c(e)$$

but this is in opposition to the choice of e to be of minimum cost among the edges in G_{n-1} but not in T! Therefore, there is no alternative to having $T = G_{n-1}$ to begin with. $\qquad\square$

EXERCISES

1 Find spanning trees by the method of the proof of the corollary for each of the following graphs:

(a) Exercise 1(a) of Section 2.1   (b) Exercise 1(c) of Section 2.1

(c) Exercise 3(a) of Section 2.1   (d) Exercise 3(b) of Section 2.1

(e) Exercise 3(c) of Section 2.1   (f) Figure 2.12

(g) Example 2.16 (Figure 2.13)   (h) The cube with its (four) diagonals

2 For the same graphs (as in Exercise 1), again obtain spanning trees, this time by approaching the problem from the other end (as in Example 2.21).

3 Having obtained a spanning tree for each graph of Exercise 1, use the method of Example 2.20 to obtain a maximal independent collection of circuits.

4 Suppose that the entries in the following matrices represent feasible pipeline routes among six cities. If the costs or distances are as shown, design a minimal-cost (or length) pipeline system connecting the cities. List your edges in their order of selection.

(a)

	a	b	c	d	e
b	130				
c	95	140			
d		170	175		
e	120	100		160	
f	90	125	75	150	110

(b)

	a	b	c	d	e
b	105				
c	100	150			
d	210	210	210		
e	225		230	250	
f		200	210	205	230

*5 Prove that exactly $\rho(G)$ edges must be removed from a graph G to obtain a spanning tree (forest).

*6 Characterize all connected graphs having the same number of vertices as edges. That is, what must such a graph look like and why?

2.5   PLANARITY

A graph G is said to be *planar* if it can be drawn or pictured in the plane in such a way that its edges intersect only at the vertices of the graph. Intuitively, we see that in identifying a planar graph G with such a picture $G = (V, E, R)$, we will partition the plane into a collection R of *regions*, one of which is infinite or unbounded. The boundaries or *contours* of the various regions will be formed by the edges of the graph. We may suppose that a definite planar (pictorial) representation has been chosen, and we let $r = |R|$ denote the number of regions, including the unbounded one. We will find that r is not dependent on the choice of the planar representation. Except perhaps in the case of the infinite region, the boundaries or contours of each region will correspond to circuits of the underlying graph G.

EXAMPLE 2.23   The graph of Example 2.1 is planar because it can be drawn as in Figure 2.20. Here we have $r = |R| = 4$ because

$$R = \{R_1, R_2, R_3, R_4\}$$

the fourth region being infinite.   △

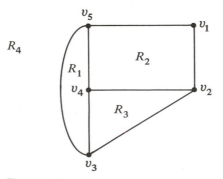

FIGURE 2.20

EXAMPLE 2.24   The graph of Example 2.19 [Figure 2.15(a)] is also planar. It can be drawn without intersecting edges, as shown in Figure 2.21. Here we have $r = 8$. △

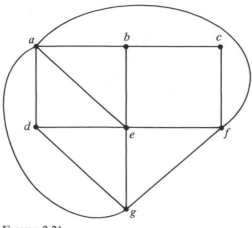

FIGURE 2.21

It is not always so easy to demonstrate that a graph is nonplanar, however compelling the evidence. That no matter how the graph is drawn, there will always be intersecting edges, can be quite a difficult thing to establish. Fortunately, certain necessary and even sufficient conditions have been developed for planarity, and we will touch on a few of these later in the section.

EXAMPLE 2.25   In the arid southwestern corner of the United States three houses have been built side by side. Each has its own well at some distance to the rear of the house. Due to lack of rainfall, one or more of the wells frequently runs dry. So it is essential that each family have access to all three wells. As time goes on, the families develop rather strong dislikes for one another. They decide that it would be best for all concerned if they were to construct paths to all three wells in such a way that they would never have to meet as they walked to and from the wells. But is this possible?

Evidently, the proposed network of walkways is the bipartite graph $K_{3,3}$ of Figure 2.4, where there are intersections at every turn. We try to improve the picture in Figure 2.22, but try as we may, we always seem to fall one path short of the objective. Later we will be able to *prove* that this graph $K_{3,3}$ is nonplanar. It is no wonder that we are having difficulty. △

Questions of planarity or near planarity do arise in graphical applications to computer science, though we cannot say that they are always of central concern. Let us mention briefly the printed circuit problem and the flowcharting problem merely to give an indication.

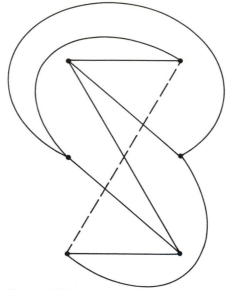

FIGURE 2.22

A printed circuit is an electrical network, usually mass fabricated by an electroplating process, the "wires" being deposited on a nonconducting surface. Such a network can be printed on a single surface if and only if the underlying graph of the network is planar. Actually, it is just as common to use both sides of the printed circuit board, in which case the possibilities for network realization depend on the nature of the interconnections permitted from side to side. Suppose that we allow all vertices to appear on both sides, with conducting wires joining the node pairs through holes in the nonconducting board. Even then we will find that there are nonrealizeable networks.

Flowcharts are prepared on a two-dimensional grid so as to allow statements executed sequentially to be placed in a reasonable proximity. In the interest of clarity, it is desirable that there be as few overlapping or crossing flowlines as possible. Thus planarity or at least near planarity becomes an objective here as well.

As we develop a few of the notions surrounding planar graphs, the reader will gain broader insights into the general theory of graphs. Our first result, Euler's formula, may be known to the student already, particularly when phrased in the language of geometric polyhedra. As with most of our results thus far, the Euler formula follows quite easily from the circuit rank formula, but only after a preliminary result has first been established. We call this result a lemma, but actually it is quite important in its own right. We should not have to point out its obvious utility in connection with electrical network analysis.

Lemma

In any planar graph $G = (V, E, R)$, the contours of the finite regions represent a largest collection of independent circuits.

Proof We use induction on the number of finite regions. If there is only one finite region (or no finite region), then the statement of the lemma is obviously true. So suppose that the statement holds for all planar graphs with r regions altogether $(r \geq 2)$, and let $G^+ = (V, E^+, R^+)$ be planar with $|R^+| = r + 1$. Then we can delete an edge $e = \{v, w\}$ separating a finite region from the infinite region (See Figure 2.23) to obtain a graph $G = (V, E, R)$ such that

$$E = E^+ \sim \{e\}$$
$$|R| = r$$

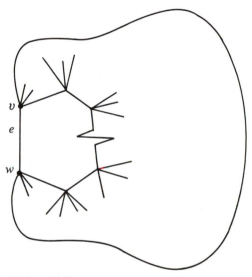

FIGURE 2.23

By the inductive hypothesis, the contours of the finite regions of G form a largest number $(r - 1)$ of independent circuits for G and $\rho(G) = r - 1$. In replacing e to recover the graph G^+, we note that a new finite circuit is created, independent of the finite contours of G, because none of these circuits involves the edge e. Since v and w were already joined by a path in G,

$$\rho(G^+) = \rho(G) + 1 = (r - 1) + 1 = r$$

Since this coincides with the number of finite contours of G^+, our proof is complete. $\square$

Theorem 2.5   (Euler)

In a connected planar graph $G = (V, E, R)$ where

$$|V| = n \qquad |E| = m \qquad |R| = r$$

we have the formula:

$$n - m + r = 2$$

Proof   According to the lemma, the number of finite regions of G is the same as the circuit rank $\rho(G)$. Using Theorem 2.2 for the case $p = 1$, we obtain

$$r = \rho(G) + 1 = (m - n + 1) + 1 = m - n + 2$$

which is equivalent to the Euler formula.   □

EXAMPLE 2.26   Checking this formula against the data of Examples 2.23 and 2.24, we have

$$n - m + r = 5 - 7 + 4 = 2$$
$$n - m + r = 7 - 13 + 8 = 2$$

respectively. Of course, more often we make use of the formula when only two of the quantities n, m, r are known.   △

EXAMPLE 2.27   In geometry Theorem 2.5 is known as Euler's polyhedral formula. Consider the stereographic projection of a planar graph onto a sphere, where we imagine the sphere to be resting on the plane at its south pole [Figure 2.24(a)]. From each vertex v of the graph (and from points along the edges as well) a ray is drawn to the north pole. The point of intersection with the sphere is taken as the *image v'* of the point v. We will speak of stereographic projection in either direction, of the graph onto the sphere or of a polyhedron into a planar graph. In the first case, we see on the sphere how there is no difference between the infinite region and any of the others. In fact, the regions become the faces of a somewhat distorted polyhedron. In this context, Euler's polyhedral formula is usually written:

number of vertices − number of edges + number of faces = 2

for any polyhedron.   △

EXAMPLE 2.28   Consider the cube, shown both as a planar graph and (its stereographic projection) as a polyhedron in Figures 2.24(b) and 2.24(c), respectively. Note in particular that the infinite region for the graph is identified as one of the six faces after projection onto a sphere. In any case, we have

number of vertices − number of edges + number of faces = 8 − 12 + 6 = 2

as predicted by Euler's polyhedral formula.   △

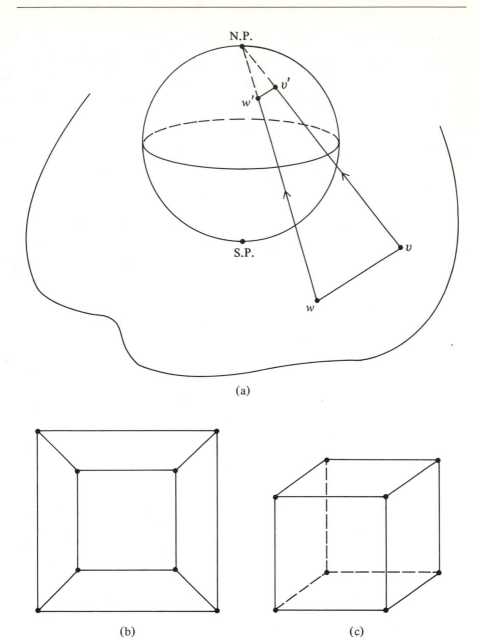

(a)

(b)

(c)

FIGURE 2.24

Theorem 2.6

In a planar graph $G = (V, E, R)$ where

$$|E| = m \qquad |R| = r$$

we always have

$$3r \leq 2m$$

If every region has at least four edges as contour, then

$$4r \leq 2m$$

Proof   Each region has at least three edges as contour, but when counting the edges that appear in these contours, we count each edge twice.   ☐

Corollary

The so-called *Kuratowski graphs* (Figures 2.3 and 2.4)

$$K_5 \quad \text{and} \quad K_{3,3}$$

are nonplanar.

Proof   If K_5 were planar, then according to Theorem 2.5, we would have

$$r = m - n + 2 = 10 - 5 + 2 = 7$$

But Theorem 2.6 asserts that

$$3r \leq 2m \qquad \text{(or } 21 \leq 20)$$

which is an obvious contradiction. Again, if $K_{3,3}$ were planar, the number of regions would be

$$r = m - n + 2 = 9 - 6 + 2 = 5$$

But since no region of $K_{3,3}$ can be triangular, Theorem 2.6 yields

$$4r \leq 2m \qquad \text{(or } 20 \leq 18)$$

again a contradiction.   ☐

EXAMPLE 2.29   The last result shows that there is no solution to the dilemma of the three southwestern neighbors (Example 2.25).   △

EXAMPLE 2.30   Note that if we insert vertices of degree two along an edge of a nonplanar graph (say $K_{3,3}$), as in Figure 2.25, then the resulting graph is just as nonplanar as before. The same is true with the removal of such vertices of degree two. These constructions have no effect on the planarity or nonplanarity of a graph. This is the sense of the parenthetical insertion in the following remarkable theorem of Kuratowski.   △

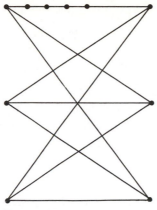

FIGURE 2.25

Theorem 2.7   (Kuratowski)

A graph is planar if and only if it does not contain K_5 or $K_{3,3}$ (perhaps modified by the insertion of vertices of degree two) as a subgraph.

Proof   Our corollary establishes the necessity of the condition. The proof of sufficiency is rather involved, and will not be given here.   □

Until the appearance of Kuratowski's paper* in 1930, the problem of characterizing the planar graphs had not been solved. Now there are several known criteria, but none more striking than this (small) catalog of "forbidden" subgraphs. Still, the conditions of Kuratowski's theorem can be quite difficult to apply. A test for planarity, based on these conditions has been devised,† but more efficient methods‡ are available. Furthermore, some simpler necessary conditions for planarity are provided in the following results.

Theorem 2.8

Let $G = (V, E, R)$ be a connected planar graph. Then

$$(1) \ m \le 3n - 6 \qquad (2) \ \partial(v) \le 5 \text{ for some } v \in V$$

Proof   For (1) we use Theorems 2.5 and 2.6 as follows:

$$2 = n - m + r \le n - \tfrac{3}{2}r + r = n - \tfrac{1}{2}r \qquad (r \le 2n - 4)$$
$$2 = n - m + r \le n - m + (2n - 4) \qquad (m \le 3n - 6)$$

As for (2), we again have $3r \le 2m$ because of Theorem 2.6. Now, if we had $\partial(v) \ge 6$ for every vertex v, then an easy counting argument would give $6n \le 2m$,

* K. Kuratowski, "Sur le Problème des Courbes Gauches en Topologie," *Fund. Math.*, **15**, 271–283 (1930).
† P. S. Mei and N. E. Gibbs, "A Planarity Algorithm Based on the Kuratowski Theorem," *Proc. AFIPS* 1970 *Spring Joint Computer Conf.*, pp. 91–95 (1970).
‡ J. Hopcroft and R. Tarjan, "Efficient Planarity Testing," *J.A.C.M.*, **21**, 549–568 (1974).

and according to Euler's formula, we would obtain

$$2 = n - m + r \leq \tfrac{1}{3}m - m + \tfrac{2}{3}m = 0$$

an obvious contradiction. □

Corollary

Let G and G' be complementary graphs on n vertices. If $n < 8$, then at least one of them is planar, whereas if $n > 8$, then at least one of them is nonplanar.

Proof Since $G \cup G'$ is the complete graph K_n, we may use the result of Example 2.7 to write

$$m + m' = \tfrac{1}{2}n(n - 1)$$

and it follows that one of the two numbers m, m' must be greater than or equal to $\tfrac{1}{4}n(n - 1)$. But Example 1.4 shows that

$$\tfrac{1}{4}n(n - 1) > 3n - 6 \qquad (n \geq 11)$$

so we simply use (1) in Theorem 2.8 when $n \geq 11$.

On the other hand, one of the two numbers is less than or equal to $\tfrac{1}{4}n(n - 1)$. Furthermore, the reader can check that

$$\tfrac{1}{4}n(n - 1) < 9 \qquad (n \leq 6)$$

whereas the Kuratowski graphs K_5 and $K_{3,3}$ have at least nine edges. So we can apply Kuratowski's theorem in case $n \leq 6$.

The remaining cases $n = 7, 9, 10$ have been analyzed exhaustively, and the results are found to be consistent with the statement of the corollary. □

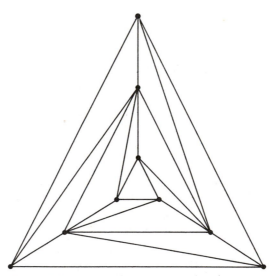

FIGURE 2.26

EXAMPLE 2.31   The graph on nine vertices in Figure 2.26 is planar. We can use the corollary to assert that its complement is nonplanar. On the other hand, the corollary says nothing about the complement of the graph in Figure 2.24 (since it has eight vertices). So far as we know, it may or may not be planar.   △

EXAMPLE 2.32   Consider the two-sided printed circuit board problem mentioned earlier. Suppose we have an electrical network whose underlying graph is K_9. Then no matter how we separate the graph into $G \cup G'$, our corollary shows that one of them will be nonplanar. It follows that we cannot realize this circuit with the physical constraints described.   △

EXERCISES

1 An ancient religion, Cabalism (Kabbalah), persists even today in certain regions of the Near East. It represents one of the world's oldest systems of mystical thought, believed by some to hold the key to all the mysteries of the universe. The essence of the Kabbalah is symbolized in the graph shown in Figure 2.27 on 10 vertices and 22 edges, depicting the emanations of God. Is the graph of the Kabbalah a planar graph? Either prove that it is nonplanar or find a planar representation.

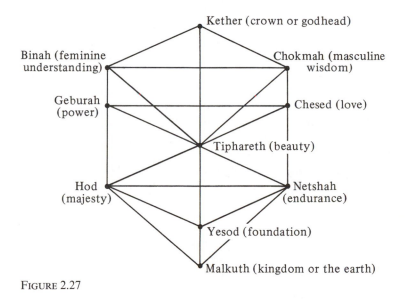

FIGURE 2.27

2 Which of the following graphs are planar?
  (a) The graph of Exercise 1(c) of Section 2.1
  (b) The graph of the cube with its (four) diagonals
  (c) The complement of the graph in (b)
  (d) The graph of Exercise 3(a) of Section 2.1
  (e) $K_{2,4}$

(f) The complement of the graph in Figure 2.24(b)

In each case, either prove that the graph is nonplanar or exhibit a planar representation.

3 Without using Kuratowski's theorem, prove that

(a) K_n is nonplanar for all $n > 4$

(b) $K_{m,n}$ is nonplanar for all $m, n > 2$

4 Verify Euler's formula for the graph of the Kabbalah (Exercise 1).

5 Verify Euler's formula for each of the following planar graphs.

(a) The tetrahedron (four triangular faces)

(b) The dodecahedron (12 pentagonal faces)

(c) The octahedron (eight triangular faces)

(d) A chess board (81 vertices)

(e) A bingo card (36 vertices)

(f) The star of David (12 vertices)

6 Sometimes we are able to say that a graph is nonplanar simply because there are "too many" edges. Explain.

2.6   ISOMORPHISM AND INVARIANTS

How many graphs are there on n vertices? We need to have such information in the event that an exhaustive analysis is required. How many graphs need to be considered in analyzing the cases $n = 7, 9, 10$ in the corollary to Theorem 2.8? Evidently the answer has something to do with the number of entries in the lower-triangular matrix representation. Since Example 2.7 shows this figure to be $\frac{1}{2}n(n - 1)$, and we have two choices for each entry, Rule 3 of Section 0.8 would lead us to conclude that there are $2^{n(n-1)/2}$ graphs on n vertices. When $n = 7$, this number is already more than two million!

The number can be greatly reduced by considering two graphs to be *essentially the same* when they differ only in the naming or labeling of their vertices. Thus we say that the graphs $G_1 = (V_1, E_1)$ and $G_2 = (V_2, E_2)$ are *isomorphic* if there exists a

$$\left.\begin{array}{l}\text{bijection}\\ \text{surjective and injective function}\\ \text{one-to-one correspondence}\\ \text{one-to-one and onto mapping}\end{array}\right\} \quad \varphi\colon V_1 \to V_2$$

(whichever you choose to call it) such that for all v, w

$$v\, E_1\, w \Leftrightarrow (\varphi v)\, E_2\, (\varphi w)$$

Here as elsewhere, we allow a slight abuse of the functional notation, omitting the parentheses in writing φv rather than $\varphi(v)$. We will do this whenever it seems to improve the readability. In any case, the mapping φ is called an *isomorphism* of graphs. Quite obviously, *is isomorphic to* is an equivalence relation on the set of all graphs with n vertices. On the other hand, the reduced number, that of the distinct isomorphism (equivalence) classes, is a rather complicated function of n, which we do not wish to pursue here. To the mathematician, the theory of graphs is very much concerned with enumeration questions such as the foregoing, which is part of the reason that graph theory falls within the realm of combinatorial mathematics. We are mainly interested in making the student aware of the existence of such questions, and in reminding the student that the information sought is of the utmost importance in the event that a problem requires exhaustive analysis.

EXAMPLE 2.33   Let $G_1 = (V_1, E_1)$ be the graph of Example 2.1, that is,

$$V_1 = \{v_1, v_2, v_3, v_4, v_5\}$$
$$E_1 = \{ \{v_1, v_2\}, \{v_1, v_5\}, \{v_2, v_3\},$$
$$\{v_2, v_4\}, \{v_3, v_4\}, \{v_3, v_5\}, \{v_4, v_5\} \}$$

and suppose that $G_2 = (V_2, E_2)$ has

$$V_2 = \{w_1, w_2, w_3, w_4, w_5\}$$
$$E_2 = \{ \{w_1, w_2\}, \{w_1, w_4\}, \{w_1, w_5\},$$
$$\{w_2, w_3\}, \{w_2, w_5\}, \{w_3, w_4\}, \{w_4, w_5\} \}$$

as shown in Figure 2.28. Then we can check that the mapping

$$\varphi: V_1 \rightarrow V_2$$

given by

$$\varphi(v_1) = w_3$$
$$\varphi(v_2) = w_2$$
$$\varphi(v_3) = w_5$$
$$\varphi(v_4) = w_1$$
$$\varphi(v_5) = w_4$$

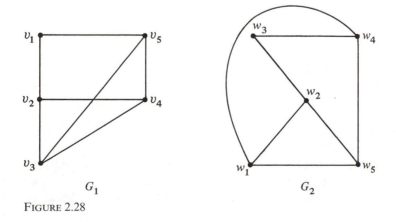

G_1 G_2

FIGURE 2.28

is an isomorphism of graphs. First of all, it is obviously a bijection. Moreover, we can check that

$$v_i \, E_1 \, v_j \Leftrightarrow (\varphi v_i) \, E_2 \, (\varphi v_j)$$

for all v_i and v_j in V_1. For instance,

$$v_1 \, E_1 \, v_2 \quad \text{and also} \quad w_3 = (\varphi v_1) \, E_2 \, (\varphi v_2) = w_2$$
$$v_1 \, E_1 \, v_5 \quad \text{and also} \quad w_3 = (\varphi v_1) \, E_2 \, (\varphi v_5) = w_4$$
$$v_1 \, \not{E}_1 \, v_3 \quad \text{and also} \quad w_3 = (\varphi v_1) \, \not{E}_2 \, (\varphi v_3) = w_5, \text{ etc.}$$

After an appropriate relabeling, we see that the two graphs are *abstractly equivalent*. △

EXAMPLE 2.34   The two graphs in Figure 2.29 are not isomorphic, even though they each have five vertices and six edges. Any candidate for an isomorphism φ must map the two vertices v, w of degree three onto the vertices of degree three in the other graph, say $\varphi(v) = v'$ and $\varphi(w) = w'$, as labeled in Figure 2.29. Nevertheless, we have

$$v \, E \, w \quad \text{whereas} \quad v' \, \not{E}' \, w' \qquad \qquad △$$

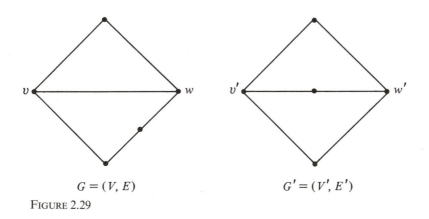

$$G = (V, E) \qquad\qquad\qquad G' = (V', E')$$

FIGURE 2.29

Reexamining Example 2.33, we find one facet of the argument particularly bothersome. It would appear that to test two graphs for isomorphism, we must examine $n!$ mappings, n being the number of vertices in each of the graphs. We cannot have an isomorphism (or even a bijection) unless the number of vertices agree. But then there are $n!$ separate bijections to consider, according to Rule 4 of Section 0.8. So again we come up against a rather large number as a real obstacle to the investigation.

Conceptually there is no particular difficulty in testing for graphical isomorphism, but on purely technical grounds we must reject the straightforward approach of examining all the bijections. One is in a much better position to undertake such an investigation if a few of the graphical "invariants" can be used as tools in the quest. Suppose that we write $G_1 \simeq G_2$ if G_1 and G_2 are isomorphic. Then a *graphical invariant* is a function η from graphs to integers (or sequences of integers) for which

$$G_1 \simeq G_2 \Rightarrow \eta(G_1) = \eta(G_2).$$

We already have a few elementary invariants at our disposal:

$$n = n(G) = \text{the number of vertices in } G$$
$$m = m(G) = \text{the number of edges in } G$$
$$p = p(G) = \text{the number of components in } G$$

Since isomorphism involves a one-to-one correspondence of vertices, a bijection is also established between the edges as well as the connected components for the two graphs in question. It follows that two graphs cannot be isomorphic unless they agree in each of these respects. At the same time Theorem 2.2 shows that

$$\rho = \rho(G) = \text{the circuit rank of } G$$

is likewise a graphical invariant. When we actually employ a graphical invariant η, it is usually used in the contrapositive form, as

$$\eta(G_1) \neq \eta(G_2) \Rightarrow G_1 \not\simeq G_2$$

If G_1 and G_2 disagree on a particular invariant, then they cannot possibly be isomorphic, and we don't need to examine all the possible bijections only to come to the same conclusion.

If a graphical invariant is to be useful, it should be fairly easy to compute, and we should describe algorithms for the computation. Certainly these conditions have been met for the invariants introduced thus far. Now consider the row sums (the number of ones in a particular row) of the relation matrix of a graph. Evidently these figures provide the degrees $\delta(v)$ of the corresponding vertices, which in turn allow for the computation of the *degree spectrum*

$$d = d(G) = \langle d_0, d_1, \ldots, d_{n-1} \rangle$$

of a graph G. This invariant is a sequence of integers where

$$d_i = \text{the number of vertices of degree } i$$

in G. In proceeding to define further graphical invariants, our hope is to find a collection $\{\eta_1, \eta_2, \ldots, \eta_N\}$ such that

$$G_1 \simeq G_2 \Leftrightarrow \eta_1(G_1) = \eta_1(G_2), \ldots, \eta_N(G_1) = \eta_N(G_2)$$

Such a collection is called a *complete set* of invariants because it will allow a complete resolution of the isomorphism question in terms of the separate computations. Unfortunately, a useful complete set of graphical isomorphism invariants has not yet been found. Still, the utility of the easily computed invariants cannot be denied.

EXAMPLE 2.35   The row sums of the relation matrix of Example 2.1 are 2, 3, 3, 3, 3, respectively. Therefore, the graph of that example has the degree spectrum $\langle 0, 0, 1, 4, 0 \rangle$. Since the graph G in Figure 2.29 has the spectrum $\langle 0, 0, 3, 2, 0 \rangle$, we know that these two graphs are not isomorphic. Can you find a way to obtain the same result more easily?   △

EXAMPLE 2.36   The two graphs G and G' in Figure 2.29 agree on each of the invariants n, m, p, d, but still they are not isomorphic (as shown in Example 2.34). It is necessary to introduce additional invariants to distinguish the two graphs.   △

In spite of the fact that there exist graphs with the same degree spectrum that are not isomorphic, the spectrum itself is extremely useful in reducing the number of bijections that may have to be considered. Corresponding vertices must have the same degree in any graphical isomorphism. Therefore, in the case of graphs of the same degree spectrum

$$d = \langle d_0, d_1, \ldots, d_{n-1} \rangle$$

we need not consider all the $n!$ bijections as candidates for an isomorphism. The vertices of degree i must be mapped onto vertices of degree i, and according to Rules 1 and 4 of Section 0.8, only $d_0! \, d_1! \cdots d_{n-1}!$ mappings will meet this requirement. Unless all the vertices have the same degree, this number will be smaller, often considerably smaller, than $n!$

EXAMPLE 2.37   Suppose that we return once again to the two graphs of Example 2.34. Each has the degree spectrum $\langle 0, 0, 3, 2, 0 \rangle$, but in looking for an isomorphism, only $3!2! = 12$ mappings need to be considered, which is considerably less than the total number ($5! = 120$) of bijections.   △

In proceeding to introduce further numerical invariants, we now have a new purpose in mind, beyond their obvious implications for the isomorphism problem. Some of the graphical invariants—for example, the "domination," "covering," and "independence" numbers of a graph—are important in applications because they express an optimality of one form or another. Accordingly, their definitions involve various maxima or minima on the size of certain collections of vertices in the graph. These rather special collections may be introduced as follows. If $G = (V, E)$, we say that the vertices $\{v_1, v_2, \ldots, v_s\}$ constitute

a *dominating set* if

$$v \in V \Rightarrow v = v_j \text{ or } v \, E \, v_j \quad (\text{for some } j, 1 \leq j \leq s)$$

a *covering set* if

$$\{v, w\} \in E \Rightarrow v = v_j \text{ or } w = v_j \quad \text{(for some } j)$$

an *independent set* if

$$\{v_i, v_j\} \notin E \quad \text{(for all } i, j)$$

In a dominating set every vertex of the graph is, or is adjacent to, a vertex of the set. A covering set "covers" the edges in that every edge has at least one of its extremities in the set. All of the vertices of an independent set are mutually nonadjacent. In each of these cases, we can introduce corresponding graphical invariants:

$$\delta = \delta(G) = \min_{\delta} \{v_1, v_2, \ldots, v_\delta : \text{dominating set}\}$$

$$\alpha = \alpha(G) = \min_{\alpha} \{v_1, v_2, \ldots, v_\alpha : \text{covering set}\}$$

$$\beta = \beta(G) = \max_{\beta} \{v_1, v_2, \ldots, v_\beta : \text{independent set}\}$$

which are called the *domination number, covering number,* and *independence number,* respectively. Thus δ is the size of a smallest dominating set, α is the size of a smallest covering set, and β is the size of a largest independent set.

Actually, we always get two invariants for the price of one, for we may speak of any of these concepts relative to the complementary graph $G' = (V, E')$ as if they were properties of G itself. For instance, we may define

$$\beta' = \beta'(G) = \beta(G')$$

This invariant is called the *clique number* of G because of its sociological meaning. When graphs are used in certain sociological contexts, the relation $v\, E\, w$ is taken to mean that v and w are friends, acquaintances, or whatever. All the vertices in a *clique* $\{v_1, v_2, \ldots, v_s\}$ are mutually adjacent in the graph. Consequently, a clique in G is an independent set in G' (and vice versa). It follows that the clique number of G is the size of a largest clique in G (independent set in G').

EXAMPLE 2.38   The graph of Example 2.1 is reproduced in Figure 2.30(a). The collection $\{v_2, v_3, v_4\}$ is a clique for this graph G, since

$$v_2\, E\, v_3 \qquad v_2\, E\, v_4 \qquad v_3\, E\, v_4$$

Evidently the clique number of a graph is the size (n) of the largest complete subgraph (K_n) to be found. Thus the clique number is $\beta'(G) = 3$ in this case.

The complementary graph G' is shown in Figure 2.30(b). Here the same set $\{v_2, v_3, v_4\}$ becomes an independent set of vertices, none of which is joined by an edge. In this way we have an illustration of the relationship $\beta'(G) = \beta(G')$.   △

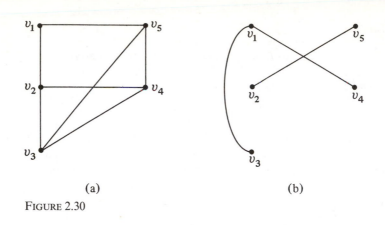

(a)   (b)

FIGURE 2.30

EXAMPLE 2.39   Consider a communications channel for transmitting symbols from an alphabet Σ. Due to distortion in the channel, certain symbols may be confused; that is, various pairs of symbols may received as if they were the same. In such a situation it would be advisable not to use the entire alphabet, but to select instead a largest subalphabet having the property that no two letters can be confused. Such a selection would be a largest independent set of vertices in a graph having edges that depict all the confusable pairs of symbols.   △

EXAMPLE 2.40   On a chessboard as shown in Figure 2.31 it is possible to place as many as eight queens in such positions that none is threatened; but if there are nine queens or more, then no matter how they are placed, there will always be a threatened queen. It follows that $\beta(G) = 8$ for the graph (on 64 vertices) representing the queen's dominating relation.   △

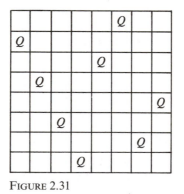

FIGURE 2.31

EXAMPLE 2.41   Suppose that the vertices of a graph $G = (V, E)$ represent various trouble spots in a prison complex, where the edges might correspond to corridors, stairways, etc. The fewest number of guards necessary to control the complex would then be obtained if we could find a smallest dominating set of vertices. △

EXAMPLE 2.42   Suppose, instead, that the edges of $G = (V, E)$ are interpreted as streets in a city, and suppose further that it is desired that every street have a fire hydrant at one corner or another at the ends of the street. Then a minimum number of hydrants needed for the purpose is obtained by determining a smallest covering set of vertices. △

EXAMPLE 2.43   Again with the graph $G = (V, E)$ of Figure 2.30(a), suppose that we want to see if we can determine the domination and covering numbers. Consider the collection $\{v_2, v_5\}$, which is a dominating set of vertices because

$$v_1 \, E \, v_5 \qquad v_4 \, E \, v_5 \qquad v_3 \, E \, v_2$$

Obviously there is no single vertex that dominates, so we must have $\delta(G) = 2$.

In a graph without isolated vertices, it is clear that every covering set of vertices is also a dominating set. The converse, however, is not true. Here the dominating set $\{v_2, v_5\}$ is not a covering set, because the edge from v_3 to v_4 is not covered. In fact, there is not any pair of vertices that covers all the edges of this graph, as the reader can easily see. Since $\{v_2, v_3, v_5\}$ is a covering set, we conclude that $\alpha(G) = 3$. △

EXAMPLE 2.44   Consider the two graphs G and G' in Figure 2.29. We recall (Examples 2.34 and 2.36) that these graphs are nonisomorphic and yet agree on all of our earlier numerical invariants. Since $\alpha(G) = 3$ but $\alpha(G') = 2$, we finally have an invariant that distinguishes between the two. △

As more sophisticated computational techniques are developed in subsequent chapters, we will present formal algorithms for determining these various graphical invariants. For the time being, however, we risk straining the reader's patience by introducing yet another facet of these investigations. It so happens that the literature is full of equalities and inequalities relating the several graphical invariants. Here we can afford to present only two of these just to give an indication of the overall interdependence.

Theorem 2.9

$\beta \geq \delta$ in any graph $G = (V, E)$.

Proof   We will show that an independent set is maximal if and only if it is a dominating set. The inequality will then follow quite easily, since a largest independent set will definitely dominate, though it may not be smallest set with this property.

When we speak of $\{v_1, v_2, \ldots, v_s\}$ as a maximal independent set, we mean that we cannot adjoin an additional vertex without losing the independence.

Such a set must surely dominate because $v \neq v_j$ and $v \not\mathrel{E} v_j$ for all $j = 1, 2, \ldots, s$ would contradict the maximality. Conversely, suppose $\{v_1, v_2, \ldots, v_s\}$ is a dominating independent set. For any $v \in V$ such that $v \neq v_j$ (all j) we must have $v \mathrel{E} v_j$ for some j, so that $\{v, v_1, v_2, \ldots, v_s\}$ cannot be independent.   □

Theorem 2.10

For any graph $G = (V, E)$ with n vertices,

$$\alpha + \beta = n$$

Proof   This time we show that if $V = A \cup B$ and $A \cap B = \varnothing$, then A is a covering set iff B is an independent set. It will then follow immediately that $\alpha + \beta = n$.

With the assumption that $V = A \cup B$ and $A \cap B = \varnothing$, suppose that A is a covering set. Then consider any two elements $v, w \in B$. If it were true that $v \mathrel{E} w$, then we must have $v \in A$ or $w \in A$. But either of these would contradict the disjointness of A and B. So we have $v \not\mathrel{E} w$ for all $v, w \in B$, and B is independent. Conversely, if B is an independent set and $v \mathrel{E} w$, then we cannot have both vertices in B. It follows that one of them is in A, and A is a covering set of vertices.   □

EXAMPLE 2.45   Given the graph of Figure 2.30(a), Theorem 2.10 tells us that we should have an independence number

$$\beta = n - \alpha = 5 - 3 = 2$$

by the result of Example 2.43. Sure enough, the largest independent sets to be found are $\{v_1, v_3\}$, $\{v_1, v_4\}$, and $\{v_2, v_5\}$.   △

EXAMPLE 2.46   Reviewing Example 2.40 in connection with Theorem 2.9, we see that clearly eight queens (or perhaps fewer?) will dominate a chess board. Putting one queen on each row will do that, although actually five queens will suffice, as shown in Figure 2.32. In any case, we have $\beta \geq \delta$ as claimed.   △

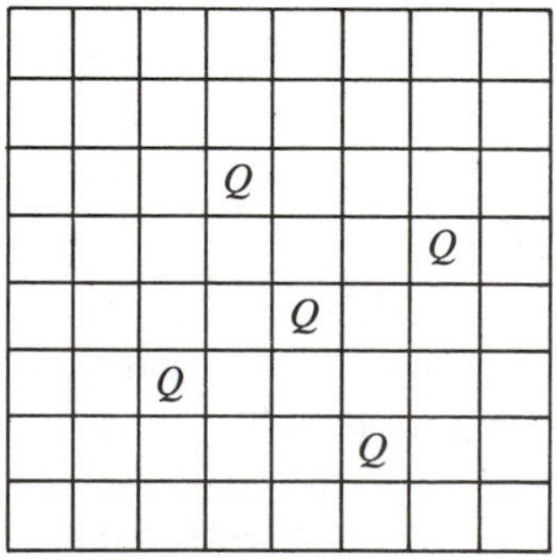

FIGURE 2.32

We have tried to give vivid illustrations to help demonstrate the meaning of the domination, covering, and independence numbers of a graph. The meaning of the next and last of our numerical invariants is suggested by its name— "chromatic" number. In fact, the illustration is in technicolor! Imagine that we have a box of crayons for the purpose of coloring the vertices of a graph $G = (V, E)$. We add the stipulation that adjacent vertices must be given different colors. What is the fewest number of colors that will suffice? This number is the *chromatic number* $\kappa = \kappa(G)$ for the graph G. Phrased somewhat differently, let a *proper coloring* of G be a partition

$$V = \bigcup_{i=1}^{s} V_i \qquad V_i \cap V_j = \varnothing \; (i \neq j)$$

of the vertices subject to the condition

$$v, w \in V_i \Rightarrow v \not\!E w$$

Then $\kappa = \kappa(G)$ is the fewest number of blocks (colors) to be found among all the proper colorings of G.

Regardless of its applications, we would not have been able to exclude the chromatic number from our sampling of the graphical invariants because this number is the subject of perhaps the most celebrated unsolved problem in all of mathematics: the *four-color problem*. The task is to determine the fewest number of colors that will serve in a proper coloring of every *planar* graph. It is known that three colors are not enough and that five colors will always suffice. But the question remains, are four colors always sufficient? No one knows. This question, though tantalizingly simple to state, has survived the onslaught of the world's most capable mathematicians. Indeed, many important results in graph theory have been discovered in the course of generations of unsuccessful pursuit.

EXAMPLE 2.47   The complete graph K_4 is planar but obviously requires four colors, that is, $\kappa(K_4) = 4$. This shows that three colors are not enough for all the planar graphs. We often hear of the *four-color conjecture*: $\kappa(G) \leq 4$ for every planar graph G. Exhaustive computer analyses of all the planar graphs up to 40 vertices have revealed none with the chromatic number of 5, and there is a proof that 5 colors will always suffice. All these results lend credence to the conjecture, but the question still remains an open one.   △

EXAMPLE 2.48   In scheduling students for classes at a university, consider a graph whose vertices represent the classes and let there be an edge from v to w if there is a time conflict, for example when a student must schedule both classes. Such classes must be arranged to meet at different times. A proper coloring will then correspond to a schedule without conflicts.   △

EXAMPLE 2.49   Consider the graph of Figure 2.30(a) once more. Suppose that we color the vertex v_5 red. Then we would have to choose a different color, say white, for v_4. Yet a third color, say blue, would have to be selected for vertex v_3, because it is adjacent to both v_4 and v_5. Now v_2 cannot be white or blue, but we can color it red. We would then merely have to choose white or blue as a color for v_1 in order to achieve a proper coloring with only three colors. Obviously $\kappa \neq 2$ because of the clique $\{v_3, v_4, v_5\}$. It follows that $\kappa = 3$.   △

EXERCISES

1 Find all the nonisomorphic graphs having
   (a) Three vertices   (b) Four vertices   (c) Five vertices

2 Use a different invariant (than that of Example 2.35) to show that the graphs of Figures 2.1 and 2.29 are not isomorphic.

3 Use various numerical invariants to show that no two of the following graphs are isomorphic:
   (a) $K_{2,3}$ (b) Figure 2.12
   (c) Exercise 1(a) of Section 2.1   (d) Exercise 1(b) of Section 2.1

4 Repeat Exercise 3 for the following graphs:
   (a) $K_{3,3}$ (b) Exercise 1(b) of Section 2.1
   (c) Exercise 1(c) of Section 2.1   (d) Exercise 3(a) of Section 2.1

5 Repeat Exercise 3 for the following graphs:
   (a) $K_{3,4}$ (b) Exercise 3(c) of Section 2.1
   (c) Figure 2.15(a)   (d) K_7

6 Determine the degree spectrum of the following graphs:
   (a) Exercise 1(a) of Section 2.1   (b) Exercise 3(b) of Section 2.1
   (c) Exercise 3(c) of Section 2.1   (d) Exercise 1(b) of Section 2.1
   (e) Exercise 1(c) of Section 2.1   (f) Exercise 3(a) of Section 2.1
   (g) Figure 2.15(a)   (h) Exercise 1 of Section 2.5
   (i) Exercise 5(h) of Section 2.3

7 Determine the independence number β of each graph of Exercise 6 and exhibit a maximal independent set.

8 Determine the domination number δ of each graph of Exercise 6 and exhibit a minimal dominating set.

9 Determine the covering number α of each graph of Exercise 6 and exhibit a minimal covering set.

10 Determine the chromatic number κ of each graph of Exercise 6 and exhibit a minimal proper coloring of the vertices.

11 Determine the clique number β' of each graph of Exercise 6 and exhibit a maximal clique.

*12 (For chess players only) Show directly that $\beta \geq \delta$ for the graph of the knight's dominating relation.

*13 Prove that in any graph with n vertices

$$n/\beta \leq \kappa \leq n - \beta + 1$$

*14 In analogy to a set of vertices that "covers all the edges" and an "independent" set of vertices, we may consider for connected graphs:
  (i) A covering set of edges (they cover all the vertices)
  (ii) An independent set of edges (no two of them share a vertex)
  and corresponding minimal and maximal numbers α_e and β_e, respectively. These are the *edge-covering* and *edge independence numbers*. Prove that
  (a) $\alpha_e + \beta_e = n$ (b) $\frac{1}{2}n \leq \alpha_e \leq n - 1$

15 Determine the edge-covering number α_e (see Exercise 14) of each graph of Exercise 6 and exhibit a minimal covering set of edges.

16 Determine the edge independence number β_e (see Exercise 14) of each graph of Exercise 6 and exhibit a maximal independent set of edges.

2.7 DIRECTED GRAPHS

In many of the important applications to computer science, it is more natural to deal with a graph model in which a direction is associated with each edge. Actually, there is direction whenever we are talking about a flow of information, the transport of some commodity, a transfer of control, etc. Though the definition of the directed graphs or *digraphs* is straightforward, the theory that comes with it becomes somewhat cluttered as compared with that for undirected models. This is the main reason for our first focusing on the graphs with undirected edges.

EXAMPLE 2.50   The daily schedule offered by a particular airline will provide connections between various cities as illustrated in Figure 2.33 but if we want to travel from A to B, a direct flight may not be available. On the other hand, there may be a series of flights that will accomplish the same thing. It can be seen that a route from A (through C, then D) to B can be arranged, provided that the connections are feasible, that is, the connecting flight does not depart before we have landed. Even so, there is also the path from A through C, then E, to B to be considered. If we had to choose between these two routes, it would be nice to know the various flight times, distances, fares, or some other criteria for helping us to make a decision. On the other hand, if we wished to travel from G to C, then we had better consult another airline or wait until another day. As they say in Vermont, "You can't get there from here."   △

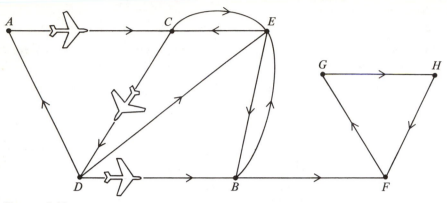

FIGURE 2.33

In such overly simplified settings, many of the important problems in the theory of directed graphs have already surfaced. The distinction between the digraphs and the graphs previously studied is immediately apparent. The edges here are oriented with arrows to indicate a direction. Now, it may appear that these digraphs return us to the theory of relations all over again—relations on a set (of vertices). Abstractly, that is true enough, but the geometric point of view is emphasized here. Also we shall eventually superimpose enough additional structure (such as distances or other labels for the edges and vertices) to render our new subject worthy of separate treatment. At the same time we wish to show how the two theories borrow from each other while maintaining their separate identities. Then after a suitable generalization of the concepts in the theory of undirected graphs (paths, connectedness, etc.), we will be led to a series of problems (minimum-distance paths, reachability, etc.) not ordinarily encountered in the abstract theory of relations, but which are the problems of greatest concern in applications, as can be easily understood.

In contrast with the definition of Section 2.1, a *directed graph* (or *digraph* as it is abbreviated) is a system $G = (V, E)$ in which E is *any* relation on V. Then either of the alternative notations $(v, w) \in E$ or $v \, E \, w$ will indicate a corresponding directed edge, from v to w, in the picture of the digraph. As before, we make dual use of the symbol E to stand for the relation and the relation matrix. The context will always eliminate any possibility of confusion.

But for one simple change in notation—the representation of edges as (v, w) rather than $\{v, w\}$—all our previous definitions of paths and circuits go through exactly as before. For instance, a *path* from v to w (of *length* k) in the digraph $G = (V, E)$ is a sequence $\langle v_0, v_1, \ldots, v_k \rangle$ of vertices $v_i \in V$ such that:

$$\begin{aligned} v_0 &= v \\ v_k &= w \end{aligned} \quad \text{and} \quad (v_{i-1}, v_i) \in E$$

for all $i = 1, 2, \ldots, k.$

EXAMPLE 2.51   As in the case of undirected graphs, a relation matrix can serve
to define a directed graph $G = (V, E)$. If

$$V = \{v_1, v_2, v_3, v_4, v_5\}$$

and E is the following relation:

	v_1	v_2	v_3	v_4	v_5
v_1	0	1	0	1	0
v_2	0	1	1	0	0
v_3	1	1	0	0	1
v_4	0	0	1	0	1
v_5	1	0	0	0	1

then G is the digraph shown in Figure 2.34. Note that a reverse edge $w \, E \, v$ may
or may not accompany an edge $v \, E \, w$. Note also that we may or may not have
edges $v \, E \, v$. Such edges (as $v_2 \, E \, v_2$ and $v_5 \, E \, v_5$ here) are called *loops*.

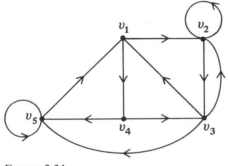

FIGURE 2.34

In this digraph the sequence $\langle v_4, v_3, v_1, v_4, v_5, v_5, v_1 \rangle$ is a path of length 6
from v_4 to v_1. On the other hand, $\langle v_2, v_3, v_4, v_1, v_2 \rangle$ is not a circuit, even though
such a circuit does appear in the undirected graph obtained by removing the
arrows. It just happens that (v_3, v_4) and (v_4, v_1) are not edges of the digraph.   △

EXAMPLE 2.52   In the digraph of Example 2.50, $\langle A, C, D, B \rangle$ is a path of length
3 from A to B. On the other hand, there is no path from G to C. The sequence
$\langle G, H, F, B, E, C \rangle$ is not a path from G to C because (F, B) is not an edge of the
digraph; the arrow points the other way. In the terminology of the next section,
we say that C is not "reachable" from G.   △

EXAMPLE 2.53   The sequences $\langle A, C, D, A \rangle$ and $\langle G, H, F, G \rangle$ are circuits in the
digraph of Example 2.50. Similarly, $\langle v_1, v_2, v_3, v_1 \rangle$ and $\langle v_1, v_4, v_3, v_5, v_1 \rangle$ are
circuits in the digraph of Figure 2.34.   △

In a *labeled digraph* the vertices or edges (or both) may carry information that supplements or replaces the usual identifying names. Ordinarily these labels will themselves be elements of other algebraic systems, reflecting the particular use being made of the digraph model. It may be helpful at times to think of this labeling as being accomplished by functions

$$l: V \rightarrow A_V \qquad l: E \rightarrow A_E$$

from the vertices and edges of the digraph $G = (V, E)$ into sets from an appropriate algebraic structure. More often than not, however, we simply indicate the labels on the pictures as if they were to replace the usual identifying names for the vertices or edges.

EXAMPLE 2.54   In the digraph shown in Figure 2.35 we have labeled the edges with numbers to represent perhaps the distances along various highways. In this way it is evident that the labeled digraphs are an appropriate formalism for treating the various path problems, for instance the problem of finding the shortest path between two given points. We shall return to a study of these problems in Section 2.9.

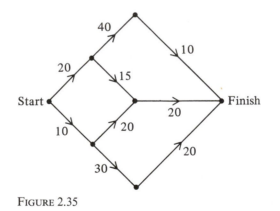

FIGURE 2.35

On the other hand, the same digraph might be used to depict a complex manufacturing process from start to finish. Then the edges would correspond to activities or processes, with the understanding that an activity originating at a given node could not begin until all the activities terminating at that node had been completed. Thus, if the labels designated the duration of a process, it would be a longest path from start to finish that is important, that is, a "critical path" holding back the completion time. Such situations are discussed in Section 2.10.   △

EXAMPLE 2.55   The skeleton of a flowchart is a digraph. The arrows on the edges indicate the flow of the algorithmic process from one statement to the next. With given initial values for the input variables, the trace of a computation (Section 1.7) will correspond to a path in the diagraph from "start" to "stop" We may think of the digraph as having its vertices labeled by instructions and its edges labeled with annotations, as in Sections 1.8 and 1.9, or we may simply say that certain of its edges are labeled with the symbols T or F associated with the various conditional statements.   △

Before embarking on a study of these problems, we would like to show how ideas are borrowed from the abstract theory of relations. They have to do mainly with the notion of composition of relations, and so a review of the idea of functional composition (Section 0.7) would be helpful here. As a matter of fact, we have purposely sought to avoid any mention of relational composition until now, because the idea is most easily understood in the context of the intended applications—applications to questions of reachability and path problems connected with digraphs. The whole concept, however, is simple enough. If you can appreciate that "is an uncle of" is obtained by composing the two relations, "is a brother of" and "is a parent of," then you are prepared for the forthcoming presentation.

Suppose that two relations are given:

$$R \text{ from the set } U \text{ to the set } V$$

$$S \text{ from the set } V \text{ to the set } W$$

Note that the range of R coincides with the domain of S, since this is important to the eventual construction, just as it was for functional composition. In these circumstances we can define a *composite relation* $R \circ S$ from U to W:

$$u(R \circ S)w \Leftrightarrow \text{ there exists } v \text{ in } V \text{ with } u \, R \, v \quad \text{and } v \, S \, w$$

Observe particularly that if R and S are functions, then our definition coincides with the idea of functional composition. The interesting question, however, is this: How do we construct $R \circ S$ when R and S are given? That is, can we obtain its graphical representation if we have the graphical representations of R and S? And can we obtain the relation matrix for $R \circ S$ if the matrices for R and S are at hand? We can give affirmative answers to both questions, but we will find that the second one is somewhat more difficult than the first.

If graphical representations for R and S are given as directed edges connecting certain vertices of U to V, and similarly V to W, then the definition of composition states that a particular u is joined by an arc to a paricular w if we can find an intermediate point $v \in V$ with edges (u, v) and (v, w). There may be more than one way in which this can come about, but that is all right. Of course, if there is none, if no such v exists, then we have $u(R \not\circ S)w$ and we do not join u to w.

EXAMPLE 2.56   Two relations R and S are depicted graphically in Figure 2.36(a). The relation R is from U to V and S is from V to W, where $U = \{u_1, u_2, u_3, u_4\}$, $V = \{v_1, v_2, v_3, v_4, v_5\}$, and $W = \{w_1, w_2, w_3\}$. We recall that in the composite relation $R \circ S$ we are to draw an arc from u to w if there is a path $\langle u, v, w \rangle$ from u to w through an intermediate vertex $v \in V$. Thus we draw the arc (u_1, w_3) in Figure 2.36(b) because of the path $\langle u_1, v_1, w_3 \rangle$ in Figure 2.36(a). There is also a path $\langle u_1, v_2, w_3 \rangle$, but that is all right. All we need is one. On the other hand, we do not draw an edge (u_1, w_1) in Figure 2.36(b) because there is no intermediate vertex $v \in V$ on a path $\langle u_1, v, w_1 \rangle$. In this way, the graphical representation of the composite relation $R \circ S$ is given by Figure 2.36(b).   △

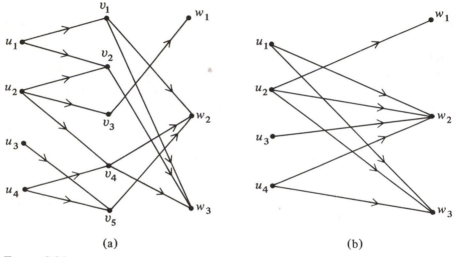

(a)   (b)

FIGURE 2.36

As for the equivalent matrix problem, we have to make use of the algebra B (Example 1.15) with its elements renamed 0, 1. For, suppose R and S denote the corresponding relation matrices, $R = (r_{ij})$ and $S = (s_{ij})$. How are we to tell whether or not there exists a path $\langle u_i, v, w_j \rangle$? We observe that the ith row of the matrix R gives the entry $r_{ik} = 1$ corresponding to the edge (u_i, v_k). Similarly, the jth column of S has an entry $s_{kj} = 1$ if (v_k, w_j) is an edge in the graphical description of the relation S. So what we need to look for is a common entry:

"1" in position k of row $r_i = (r_{i1}, r_{i2}, \ldots, r_{in})$
"1" in position k of column $s_j = (s_{1j}, s_{2j}, \ldots, s_{nj})$

which will signify the existence of an appropriate intermediate vertex v_k. Using the algebra B, we compute

$$r_{i1} \wedge s_{1j} = 1?$$
$$r_{i2} \wedge s_{2j} = 1?$$
$$\vdots$$
$$r_{in} \wedge s_{nj} = 1?$$

(Recall that only $1 \wedge 1 = 1$ in the algebra B.) If more than one of these calculations yields a 1, that is all right; we still have $u_i(R \circ S)w_j$. Accordingly, when we take the sum of these results we obtain the correct answer because $1 \vee 1 = 1$ in the algebra B. Altogether we have proved the following:

Theorem 2.11

The composite relation $R \circ S$ has the matrix

$$RS = T = (t_{ij})$$

in which

$$t_{ij} = \bigvee_{k=1}^{n} (r_{ik} \wedge s_{kj})$$

(The reader who is familiar with ordinary matrix multiplication will note that our formula is completely analogous after a change in algebras from the real numbers to B.)   □

EXAMPLE 2.57   In Example 2.56, the relations R and S have the matrices

$$R = \begin{bmatrix} 1 & 1 & 0 & 0 & 0 \\ 0 & 1 & 1 & 1 & 0 \\ 0 & 0 & 0 & 0 & 1 \\ 0 & 0 & 0 & 1 & 1 \end{bmatrix} \qquad S = \begin{bmatrix} 0 & 1 & 1 \\ 0 & 0 & 1 \\ 1 & 0 & 0 \\ 0 & 1 & 1 \\ 0 & 1 & 0 \end{bmatrix}$$

According to Theorem 2.11, the entry t_{13} in the matrix for $R \circ S$ is

$$t_{13} = (r_{11} \wedge s_{13}) \vee (r_{12} \wedge s_{23}) \vee (r_{13} \wedge s_{33}) \vee (r_{14} \wedge s_{43}) \vee (r_{15} \wedge s_{53})$$
$$= (1 \wedge 1) \vee (1 \wedge 1) \vee (0 \wedge 0) \vee (0 \wedge 1) \vee (0 \wedge 0)$$
$$= 1 \vee 1 \vee 0 \vee 0 \vee 0$$
$$= 1$$

Note that this calculation involves only the first row of R and the third column of S, also that the two 1's in the expression $1 \vee 1 \vee 0 \vee 0 \vee 0$ correspond to the two paths from u_1 to w_3 in Figure 2.36(a), one through v_1 and the other through v_2.

If the formula of Theorem 2.11 is used in this way to compute each of the entries t_{ij}, we arrive finally at the matrix

$$RS = T = \begin{bmatrix} 0 & 1 & 1 \\ 1 & 1 & 1 \\ 0 & 1 & 0 \\ 0 & 1 & 1 \end{bmatrix}$$

The reader can check that this is the proper relation matrix for the composite relation of Figure 2.36(b).   △

EXAMPLE 2.58   When we specialize to the case of relations on a single set V, the very definition of relational composition leads to the familiar properties

(a) *associativity*: $R \circ (S \circ T) = (R \circ S) \circ T$
(b) *identity*: $R \circ 1 = 1 \circ R = R$

where 1 is the relation

$$1 = \{(v, v): v \in V\}$$

What can you say about the corresponding relation matrices?

Suppose that we demonstrate the associativity of relational composition. If $u(R \circ (S \circ T))v$, then there exists an element x such that $u R x$ and $x (S \circ T) v$. The latter implies that there is an element y such that $x S y$ and $y T v$. A reconstitution of these relations gives $u(R \circ S)y$ and finally $u((R \circ S) \circ T)v$. Obviously the converse goes through in a similar fashion.   △

How do we use the notion of relational composition and the computational scheme of Theorem 2.11 in the theory of directed graphs? We specialize to the case where $U = W = V$, the set of vertices of a digraph $G = (V, E)$, and merely compose the edge relation E with itself. The composite $E \circ E$ will give us all the paths of length 2 in G; for longer paths, we just keep going. Thus we have Theorem 2.12.

Theorem 2.12

There is a path (of length k) from v to w in the digraph $G = (V, E)$ if and only if

$$v(\underbrace{E \circ E \circ \cdots \circ E}_{k})w$$

Proof   Note first of all that we can write the composites of E without parentheses because of the associativity discussed in Example 2.58. Then we use induction on k. Clearly, the theorem holds for $k = 1$ because an edge is a path of length 1. So suppose the result is true for some $k - 1 \geq 0$. Then if we have a path of length k from v to w, there is a sequence $\langle v_0, v_1, \cdots, v_k \rangle$ such that

$$\begin{aligned} v_0 &= v \\ v_k &= w \end{aligned} \quad \text{and} \quad (v_{i-1}, v_i) \in E$$

for $i = 1, 2, \ldots, k$. It follows that $\langle v_0, v_1, \ldots, v_{k-1} \rangle$ is a path of length $k - 1$ from v to v_{k-1}, and by the inductive hypothesis, we have $v\, E^{k-1}\, v_{k-1}$. But also $v_{k-1}\, E\, w$, so that $v\,(E^{k-1} \circ E)\, w$, that is, $v\, E^k\, w$. Conversely, if $v\, E^k\, w$, then by the definition of composition, there is a vertex u such that $v\, E^{k-1}\, u$ and $u\, E\, w$. By the inductive hypothesis again, there is a path of length $k - 1$ from v to u, and we use it to construct a path of length k from v to w in the obvious way. $\square$

Corollary

If $V = \{v_1, v_2, \ldots, v_n\}$ is the set of vertices for the digraph $G = (V, E)$, then there is a path from v_i to v_j (of length k) if and only if

$$e_{ij}{}^k = 1 \qquad \text{(that is, the } i, j \text{ entry in matrix } E^k \text{ is 1)} \qquad \square$$

EXAMPLE 2.59   The digraph $G = (V, E)$ of Example 2.51 (Figure 2.34) has the edge relation matrix

$$E = \begin{bmatrix} 0 & 1 & 0 & 1 & 0 \\ 0 & 1 & 1 & 0 & 0 \\ 1 & 1 & 0 & 0 & 1 \\ 0 & 0 & 1 & 0 & 1 \\ 1 & 0 & 0 & 0 & 1 \end{bmatrix}$$

Suppose that we use the formula of Theorem 2.11 to compute

$$E^2 = \begin{bmatrix} 0 & 1 & 1 & 0 & 1 \\ 1 & 1 & 1 & 0 & 1 \\ 1 & 1 & 1 & 1 & 1 \\ 1 & 1 & 0 & 0 & 1 \\ 1 & 1 & 0 & 1 & 1 \end{bmatrix} \qquad E^3 = \begin{bmatrix} 1 & 1 & 1 & 0 & 1 \\ 1 & 1 & 1 & 1 & 1 \\ 1 & 1 & 1 & 1 & 1 \\ 1 & 1 & 1 & 1 & 1 \\ 1 & 1 & 1 & 1 & 1 \end{bmatrix}$$

In somewhat more detail, we have

$$E^2 = EE = \begin{bmatrix} 0 & 1 & 0 & 1 & 0 \\ 0 & 1 & 1 & 0 & 0 \\ 1 & 1 & 0 & 0 & 1 \\ 0 & 0 & 1 & 0 & 1 \\ 1 & 0 & 0 & 0 & 1 \end{bmatrix} \begin{bmatrix} 0 & 1 & 0 & 1 & 0 \\ 0 & 1 & 1 & 0 & 0 \\ 1 & 1 & 0 & 0 & 1 \\ 0 & 0 & 1 & 0 & 1 \\ 1 & 0 & 0 & 0 & 1 \end{bmatrix} = \begin{bmatrix} 0 & 1 & 1 & 0 & 1 \\ 1 & 1 & 1 & 0 & 1 \\ 1 & 1 & 1 & 1 & 1 \\ 1 & 1 & 0 & 0 & 1 \\ 1 & 1 & 0 & 1 & 1 \end{bmatrix}$$

$$E^3 = E^2E = \begin{bmatrix} 0 & 1 & 1 & 0 & 1 \\ 1 & 1 & 1 & 0 & 1 \\ 1 & 1 & 1 & 1 & 1 \\ 1 & 1 & 0 & 0 & 1 \\ 1 & 1 & 0 & 1 & 1 \end{bmatrix} \begin{bmatrix} 0 & 1 & 0 & 1 & 0 \\ 0 & 1 & 1 & 0 & 0 \\ 1 & 1 & 0 & 0 & 1 \\ 0 & 0 & 1 & 0 & 1 \\ 1 & 0 & 0 & 0 & 1 \end{bmatrix} = \begin{bmatrix} 1 & 1 & 1 & 0 & 1 \\ 1 & 1 & 1 & 1 & 1 \\ 1 & 1 & 1 & 1 & 1 \\ 1 & 1 & 1 & 1 & 1 \\ 1 & 1 & 1 & 1 & 1 \end{bmatrix}$$

with the sample calculations:

$$e_{42}{}^2 = \bigvee_{k=1}^{5} (e_{4k} \wedge e_{k2})$$

$$= (e_{41} \wedge e_{12}) \vee (e_{42} \wedge e_{22}) \vee (e_{43} \wedge e_{32}) \vee (e_{44} \wedge e_{42}) \vee (e_{45} \wedge e_{52})$$

$$= (0 \wedge 1) \vee (0 \wedge 1) \vee (1 \wedge 1) \vee (0 \wedge 0) \vee (1 \wedge 0)$$

$$= 0 \vee 0 \vee 1 \vee 0 \vee 0 = 1$$

$$e_{14}{}^3 = \bigvee_{k=1}^{5} (e_{1k}{}^2 \wedge e_{k4})$$

$$= (e_{11}{}^2 \wedge e_{14}) \vee (e_{12}{}^2 \wedge e_{24}) \vee (e_{13}{}^2 \wedge e_{34}) \vee (e_{14}{}^2 \wedge e_{44}) \vee (e_{15}{}^2 \wedge e_{54})$$

$$= (0 \wedge 1) \vee (1 \wedge 0) \vee (1 \wedge 0) \vee (0 \wedge 0) \vee (1 \wedge 0)$$

$$= 0 \vee 0 \vee 0 \vee 0 \vee 0 = 0$$

Then E^2 tells us whether there is a path of length 2 connecting the various vertices. Evidently there is a path of length 2 from v_1 to v_2, from v_1 to v_3, and from v_1 to v_5, but none from v_1 to v_1 or from v_1 to v_4. Note that the path from v_1 to v_2 uses a loop. Similarly E^3 shows that there are paths of length 3 connecting every pair of vertices except v_1 to v_4. If we were interested in paths of length 4, then we would have to compute the matrix E^4, and so on. In the next section however, we will show how to bring an end to this seemingly endless computation when we are only interested in the existence or nonexistence of paths.   △

EXERCISES

1 Draw pictures of the digraphs represented by the following matrices:

(a)

	v_1	v_2	v_3	v_4	v_5
v_1	0	0	1	0	0
v_2	0	0	1	0	1
v_3	1	1	0	0	1
v_4	1	0	0	0	1
v_5	1	1	0	0	0

(b)

	v_1	v_2	v_3	v_4	v_5
v_1	0	1	1	0	1
v_2	0	0	0	1	1
v_3	0	0	0	0	0
v_4	1	0	1	0	1
v_5	0	0	1	0	0

(c)

	v_1	v_2	v_3	v_4	v_5	v_6
v_1	0	0	1	1	0	0
v_2	1	1	0	1	0	1
v_3	1	1	1	0	1	1
v_4	0	0	1	0	0	1
v_5	1	1	1	0	1	1
v_6	0	0	1	1	0	0

2 Obtain matrices that represent the digraphs shown in Figure 2.37.

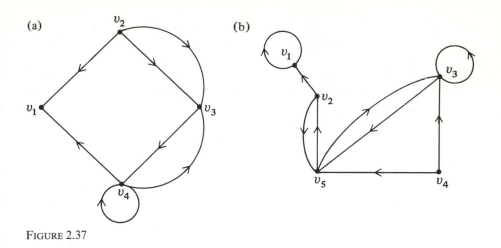

FIGURE 2.37

3 Draw pictures of the composite relations $R \circ S$ shown in Figure 2.38.

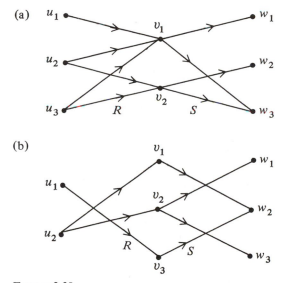

FIGURE 2.38

4 Compute the matrices corresponding to $R \circ S$ by means of Theorem 2.11:

(a)

$$R = \begin{pmatrix} 1 & 0 & 1 & 1 \\ 0 & 1 & 1 & 0 \\ 0 & 1 & 0 & 1 \\ 0 & 0 & 1 & 0 \end{pmatrix} \qquad S = \begin{pmatrix} 1 & 1 & 0 \\ 0 & 1 & 1 \\ 1 & 0 & 0 \\ 0 & 1 & 1 \end{pmatrix}$$

(b)

$$R = \begin{pmatrix} 1 & 1 & 1 \\ 0 & 1 & 1 \\ 0 & 1 & 0 \\ 1 & 0 & 1 \end{pmatrix} \qquad S = \begin{pmatrix} 1 & 1 & 1 & 0 & 1 \\ 0 & 1 & 0 & 0 & 0 \\ 1 & 1 & 0 & 0 & 1 \end{pmatrix}$$

5 Use the corollary to Theorem 2.12 to determine all the paths of length two in the digraph of:

(a) Exercise 1(a)   (b) Exercise 1(b)
(c) Exercise 2(a)   (d) Exercise 2(b)

6 Repeat Exercise 5 for paths of length three.

2.8 REACHABILITY

In any transportation network or a directed graph indicating the flow of some commodity, it is of interest to know whether one can travel from one point to another, or whether the goods from point A can be transported to point B. Once the means for answering these existence questions are understood, we can move on to the more detailed optimality considerations, how do we find a shortest route or a most economical one in case several choices are available.

We say that a vertex w is *reachable* from v in the directed graph $G = (V, E)$ if there is at least one path from v to w. There may be more than one such path, but that does not matter. The question is simply, Can I get there from here? A feasible, though not particularly efficient procedure for answering these questions simultaneously (for all $v, w \in V$) is immediately suggested by the corollary of the preceding section. The matrix E^k gives an explicit accounting of all the vertices joined by paths of length k. Since we are not interested in the length at this stage, we might try to form the *reachability matrix*

$$E^* = I \vee E \vee E^2 \vee E^3 \vee \cdots$$

in which corresponding elements of the various matrices are joined in the algebra $B = (B, \vee, \wedge, \neg)$ of Example 1.15. The *identity matrix* I has ones on the diagonal and zeros off the diagonal so as to account for paths of length 0 from any vertex to itself. In this way, the definition of E^* can be phrased as follows:

$$e_{ij}^* = \delta_{ij} \vee e_{ij} \vee e_{ij}{}^2 \vee e_{ij}{}^3 \vee \cdots$$

where

$$\delta_{ij} = \begin{cases} 0 & \text{if} \quad i \neq j \\ 1 & \text{if} \quad i = j \end{cases}$$

Since $1 \vee 0 = 0 \vee 1 = 1 \vee 1 = 1$ in the algebra B, we have

$$e_{ij}^* = \begin{cases} 0 & \text{if} \quad v_j \text{ is not reachable from } v_i \\ 1 & \text{if} \quad v_j \text{ is reachable from } v_i \end{cases}$$

This is all well and good, but it seems that we are locked into an infinite procedure here, unless we observe the following elementary fact:

Lemma

$$v \, E^* \, w \quad \text{if and only if} \quad v(I \vee E \vee E^2 \vee \cdots \vee E^{n-1})w$$

Proof   As usual, n denotes the number of vertices in the digraph $G = (V, E)$. In one direction, the proof of the lemma is obvious. So suppose $v \, E^* \, w$. This only means that there is some path from v to w in G. Now suppose that its length is k so that the path can be described as a sequence $\langle v_0, v_1, \ldots, v_k \rangle$ with

$$\begin{aligned} v_0 &= v \\ v_k &= w \end{aligned} \quad \text{and} \quad (v_{i-1}, v_i) \in E$$

for $i = 1, 2, \ldots, k$. If $k > n - 1$, then according to the pigeonhole principle (Section 0.7), some vertex v_j must be encountered more than once in traversing this path. This shows that the path includes a circuit

$$\langle v_0, v_1, \ldots; v_k \rangle = \langle v_0, \ldots, v_j, \ldots, v_j, \ldots, v_k \rangle$$

(see Figure 2.39.) It follows that we can substitute a shorter path if the circuit

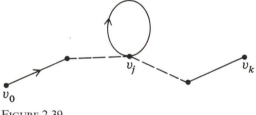

FIGURE 2.39

is by-passed. Since we can continue to do this as long as the path length is greater than $n - 1$, we must arrive eventually at a path from v to w of length less than or equal to $n - 1$. Then $v(I \vee E \vee E^2 \vee \cdots \vee E^{n-1})w$, as claimed.   □

Theorem 2.13

The reachability matrix E^* is given by the finite computation

$$E^* = I \vee E \vee E^2 \vee \cdots \vee E^{n-1}$$

in any digraph with n vertices.   □

EXAMPLE 2.60   Let $G = (V, E)$ be the digraph shown in Figure 2.40. Then we have the relational matrix

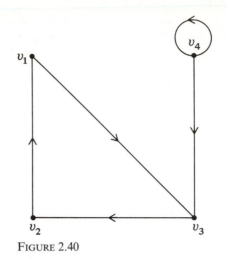

FIGURE 2.40

$$E = \begin{pmatrix} 0 & 0 & 1 & 0 \\ 1 & 0 & 0 & 0 \\ 0 & 1 & 0 & 0 \\ 0 & 0 & 1 & 1 \end{pmatrix}$$

Using the method of Theorem 2.11, we compute

$$E^2 = \begin{pmatrix} 0 & 1 & 0 & 0 \\ 0 & 0 & 1 & 0 \\ 1 & 0 & 0 & 0 \\ 0 & 1 & 1 & 1 \end{pmatrix} \qquad E^3 = \begin{pmatrix} 1 & 0 & 0 & 0 \\ 0 & 1 & 0 & 0 \\ 0 & 0 & 1 & 0 \\ 1 & 1 & 1 & 1 \end{pmatrix}$$

Then, according to Theorem 2.13,

$$E^* = I \vee E \vee E^2 \vee E^3$$

$$= \begin{pmatrix} 1 & 0 & 0 & 0 \\ 0 & 1 & 0 & 0 \\ 0 & 0 & 1 & 0 \\ 0 & 0 & 0 & 1 \end{pmatrix} \vee \begin{pmatrix} 0 & 0 & 1 & 0 \\ 1 & 0 & 0 & 0 \\ 0 & 1 & 0 & 0 \\ 0 & 0 & 1 & 1 \end{pmatrix} \vee \begin{pmatrix} 0 & 1 & 0 & 0 \\ 0 & 0 & 1 & 0 \\ 1 & 0 & 0 & 0 \\ 0 & 1 & 1 & 1 \end{pmatrix} \vee \begin{pmatrix} 1 & 0 & 0 & 0 \\ 0 & 1 & 0 & 0 \\ 0 & 0 & 1 & 0 \\ 1 & 1 & 1 & 1 \end{pmatrix}$$

$$= \begin{pmatrix} 1 & 1 & 1 & 0 \\ 1 & 1 & 1 & 0 \\ 1 & 1 & 1 & 0 \\ 1 & 1 & 1 & 1 \end{pmatrix}$$

Our result indicates that v_4 is not reachable from v_1, v_2, or v_3, which is consistent with the figure. △

EXAMPLE 2.61   For the digraph of Example 2.59 (Figure 2.34) we evidently obtain

$$E^* = \begin{bmatrix} 1 & 1 & 1 & 1 & 1 \\ 1 & 1 & 1 & 1 & 1 \\ 1 & 1 & 1 & 1 & 1 \\ 1 & 1 & 1 & 1 & 1 \\ 1 & 1 & 1 & 1 & 1 \end{bmatrix} = 1$$

Such a digraph is said to be *strongly connected*, because each vertex is reachable from every other vertex. These are useful in the design of communications networks where every pair of stations needs a path of communication. △

A more efficient method for calculating the reachability matrix E^* is provided by an algorithm due to Warshall.* The approach has many useful generalizations, but we merely summarize the basic technique in Algorithm IV. The main idea is fairly easy to grasp if we think in terms of the relations $E^{(k)}$:

$$v \, E^{(k)} w \iff \quad \text{there is a path from } v \text{ to } w \text{ using only} \\ \text{intermediate vertices from among } \{v_1, \ldots, v_k\}$$

Thus we begin with an array

$$E^{(0)} = I \vee E \quad \text{(paths with no intermediate vertices)}$$

and because k runs up to n, we eventually obtain

$$E^{(n)} = E^* \quad \text{(paths having arbitrary intermediate vertices)}$$

Evidently the interior assignment statement

$$e_{ij}^{(k)} \leftarrow e_{ij}^{(k-1)} \vee e_{kj}^{(k-1)}$$

i, j, k, n: **positive integer**
I, E^*, E: **array** $n^+ \times n^+$ **of** B

```
begin E* ← I ∨ E;
    for k ← 1 to n do
    for i ← 1 to n do
        if (i ≠ k) ∧ (E*[i, k] = 1) then
        for j ← 1 to n do
            E*[i, j] ← E*[i, j] ∨ E*[k, j]
end
```

ALGORITHM IV

---

* S. Warshall, "A Theorem on Boolean Matrices," *J.A.C.M.*, **9**, 11–12 (1962).

(rewritten slightly here) shows how $E^{(k)}$ is derived from $E^{(k-1)}$. The **if** statement signifies the existence of a path from v_i to v_k through intermediate vertices of the set $\{v_1, \ldots, v_{k-1}\}$ whenever $e_{ik}^{(k-1)} = 1$. (See Figure 2.41.) Then obviously there is a path from v_i to v_j involving intermediate vertices from among $\{v_1, \ldots, v_k\}$ in just the two cases:

Case 1.   There is a path from v_i to v_j using intermediate vertices from among $\{v_1, \ldots, v_{k-1}\}$;

Case 2.   There is a path from v_k to v_j using intermediate vertices from among $\{v_1, \ldots, v_{k-1}\}$. (See Figure 2.41 once more.)

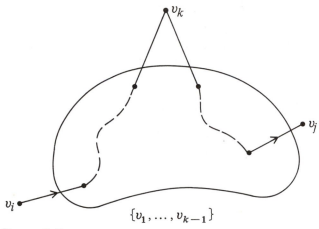

FIGURE 2.41

The assignment statement simply accounts for either of these. An example will clarify any remaining difficulties that the student may have.

EXAMPLE 2.62   Suppose that we examine once more the digraph of Example 2.60 (Figure 2.40). The algorithm begins with the assignment

$$E^* \leftarrow \begin{pmatrix} 1 & 0 & 1 & 0 \\ 1 & 1 & 0 & 0 \\ 0 & 1 & 1 & 0 \\ 0 & 0 & 1 & 1 \end{pmatrix} = \begin{pmatrix} 1 & 0 & 0 & 0 \\ 0 & 1 & 0 & 0 \\ 0 & 0 & 1 & 0 \\ 0 & 0 & 0 & 1 \end{pmatrix} \vee \begin{pmatrix} 0 & 0 & 1 & 0 \\ 1 & 0 & 0 & 0 \\ 0 & 1 & 0 & 0 \\ 0 & 0 & 1 & 1 \end{pmatrix}$$

From this point on a trace of the algorithm would appear as shown in Table 2.1.

When $k = 1$ and $i = 2, j = 3$, the interior assignment statement sets

$$e_{23}^* \leftarrow e_{23}^* \vee e_{13}^* = 0 \vee 1 = 1$$

[see Figure 2.42(a)] because the statement is operating only under the assump-

TABLE 2.1

n	i	j	k	$e_{11}^* e_{12}^* e_{13}^* e_{14}^*$	$e_{21}^* e_{22}^* e_{23}^* e_{24}^*$	$e_{31}^* e_{32}^* e_{33}^* e_{34}^*$	$e_{41}^* e_{42}^* e_{43}^* e_{44}^*$
4	1		1	1 0 1 0	1 1 0 0	0 1 1 0	0 0 1 1
	2	1			1		
		2			1 ↓		
		3			1		
		4			0		
	3						
	4						
	1		2				
	2				↓		
	3	1				1	
		2				1	
		3				1	
		4				0	
				. . .			
				1 1 1 0	1 1 1 0	1 1 1 0	1 1 1 1

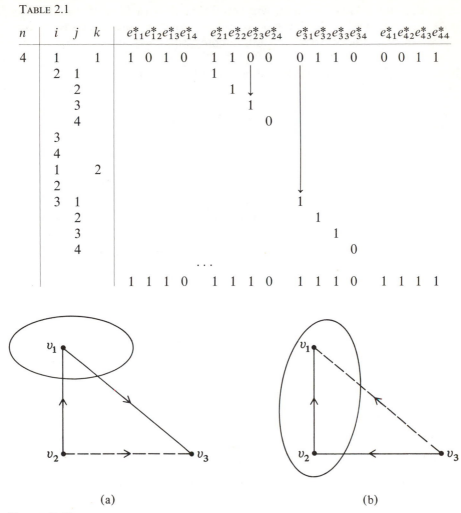

(a)   (b)

FIGURE 2.42

tion that $e_{ik}^* = e_{21}^* = 1$. Later on, with $k = 2$ and $i = 3, j = 1$, we have $e_{ik}^* = e_{32}^* = 1$, so we set

$$e_{31}^* \leftarrow e_{31}^* \lor e_{21}^* = 0 \lor 1 = 1$$

in accordance with Figure 2.42(b).   △

Algorithm IV is extremely versatile. Sometimes we see it in a slightly modified form with the initial assignment $E^* \leftarrow E$ rather than $E^* \leftarrow I \lor E$. Then it computes all the paths of length 1 or more. To see how pursuits of directed graphs can reflect back on the theory of abstract relations, we make the following observation: The modified Warshall algorithm computes the transitive closure of any relation E. You will recall that this was one of our items of unfinished business in Section 0.6. It follows from the discussions there that we can use this algorithm as one link in computing the sum of two partitions.

EXAMPLE 2.63   Referring back to Example 0.44, we write

$$E = r(\pi) \lor r(\gamma) = \begin{bmatrix} 1 & 1 & 0 & 0 & 0 & 0 & 0 \\ 1 & 1 & 1 & 0 & 0 & 0 & 0 \\ 0 & 1 & 1 & 0 & 0 & 0 & 0 \\ 0 & 0 & 0 & 1 & 1 & 1 & 0 \\ 0 & 0 & 0 & 1 & 1 & 1 & 1 \\ 0 & 0 & 0 & 1 & 1 & 1 & 1 \\ 0 & 0 & 0 & 0 & 1 & 1 & 1 \end{bmatrix}$$

Warshall's algorithm computes the transitive closure:

$$E^* = \overline{r(\pi) \lor r(\gamma)} = \begin{bmatrix} 1 & 1 & 1 & 0 & 0 & 0 & 0 \\ 1 & 1 & 1 & 0 & 0 & 0 & 0 \\ 1 & 1 & 1 & 0 & 0 & 0 & 0 \\ 0 & 0 & 0 & 1 & 1 & 1 & 1 \\ 0 & 0 & 0 & 1 & 1 & 1 & 1 \\ 0 & 0 & 0 & 1 & 1 & 1 & 1 \\ 0 & 0 & 0 & 1 & 1 & 1 & 1 \end{bmatrix}$$

$\triangle$

EXERCISES

1 Use Theorem 2.13 to compute the reachability matrix E^* for the digraph of
   (a) Exercise 1(a) of Section 2.7   (b) Exercise 1(b) of Section 2.7
   (c) Exercise 2(a) of Section 2.7   (d) Exercise 2(b) of Section 2.7

2 Repeat Exercise 1, but use Warshall's algorithm.

3 Use the modified Warshall algorithm as an aid in computing
   (a) $\{\overline{1, 2, 4}; \overline{3, 5}; \overline{6, 7, 8}\} + \{\overline{1, 3}; \overline{2, 5}; \overline{4}; \overline{6, 7}; \overline{8}\}$
   (b) $\{\overline{1, 5, 6}; \overline{2, 7}; \overline{3, 8}; \overline{4}\} + \{\overline{1, 3}; \overline{2, 4}; \overline{5}; \overline{6, 8}; \overline{7}\}$

4 A digraph is *strongly connected* if for each pair of vertices v, w there is a path from v to w, and vice versa. Give a matrix characterization of strongly connected graphs.

5 Devise a formal algorithm for testing whether a given digraph is strongly connected (See Exercise 4).

6 A digraph is *unilaterally connected* if for each pair of vertices v, w there is a path joining them (in one direction or the other). Give a matrix characterization of unilaterally connected graphs.

7 Devise a formal algorithm for testing whether a given digraph is unilaterally connected (See Exercise 6).

*2.9   PATH PROBLEMS

We have already alluded to the existence of various path problems in the theory of directed graphs. Actually the undirected graphs offer path problems of their own. So it is perhaps well to begin with these as a kind of warmup exercise. It turns out that the most famous problems for undirected graphs can either be solved quite easily or not at all. Thus they can all be discussed in fairly short order.

In the Sunday supplement of the major newspapers you will often find a puzzle section to help you while away the weekend. It is surprising how many of these puzzles have an underlying graphical interpretation. Sometimes we are asked to find our way out of a maze. Another favorite puzzle asks us to trace just once around the edges of a certain line drawing without lifting the pencil from the page. This last example is simply a variation of a problem considered by Euler in 1736: the Königsberg bridge problem, which was the start of the whole subject of graph theory. In the city of Königsberg there were seven bridges connecting the banks of the Pregel River with its two islands, as shown in Figure 2.43(a). Euler wondered whether it was possible to take a stroll in which one would start and end at the same point while crossing each bridge just once. A similar problem arises in the design of parade routes.

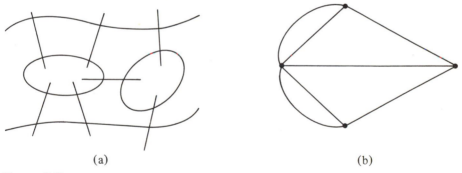

(a)   (b)

FIGURE 2.42

In the case of the Königsberg bridge problem we draw a corresponding *multigraph*, "multi" meaning that there may be more than one edge joining a given pair of vertices. [See Figure 2.43(b).] Then it is clear that we have returned to the pencil-pushing problem of the Sunday newspaper. In any case, we say that an *Euler path* (*circuit*) in a graph of multigraph is a path (circuit) that traverses each edge exactly once.

In the event that an Euler path can be traced, we will have transformed our graph or multigraph G into a digraph or multidigraph simply by observing the direction of traversal of the edges. Now suppose that in any digraph or multidigraph we generalize the notion of the degree of a vertex as follows. We say that v has the *in-degree* and *out-degree*

$$\partial^-(v) = \text{number of directed edges entering } v$$

$$\partial^+(v) = \text{number of directed edges leaving } v$$

respectively. In some applications a vertex with zero in-degree is called a *source* and one with zero out-degree is called a *sink*. In the case of a digraph resulting from the tracing of an Euler circuit, the in-degree and out-degree at each vertex must coincide. Every time we enter a vertex along one edge, we must leave that vertex along a different edge. When phrased in terms of the degrees of the vertices in the original undirected graph or multigraph, this observation leads to the following conclusion.

Theorem 2.14

A connected (multi-) graph has an Euler circuit iff every vertex has an even degree.

Proof   We have already shown the necessity of the condition. For sufficiency, we use the fact (which is easily shown) that every connected graph without vertices of odd degree has a circuit. In supposing that G has only vertices of even degree, we may therefore choose a circuit C_1. If $G = C_1$, we are done. Otherwise, we introduce an obvious temporary notation in claiming that $G - C_1$ will again have vertices only of even degree, because C_1 uses a pair of edges at each of its vertices. Continuing in this way, we can construct an inductive argument to show that we will eventually have an Euler circuit. Note however that this construction must be done with some care, since there is the possibility that intermediate graphs $G - C_1 - C_2 - \cdots - C_k$ will not be connected. In addition, we must argue that the resulting sequence of circuits may yet be combined into a single Euler circuit.   □

Corollary

A connected (multi-) graph has an Euler path, not a circuit, if and only if it has exactly two vertices of odd degree.   □

EXAMPLE 2.64   The Königsberg multigraph does not have an Euler circuit because it has vertices of odd degree. In fact, all four of its vertices are of odd degree, so according to the corollary, there does not even exist an Euler path.   △

EXAMPLE 2.65   The pair of spectacles (Figure 2.44) has an Euler path, not a circuit, because it has just two vertices of odd degree.   △

EXAMPLE 2.66   In the digraph of Figure 2.35 the vertex labeled "start" is a source and the vertex labeled "finish" is a sink. The vertex in the center has the in-degree two and out-degree one.   △

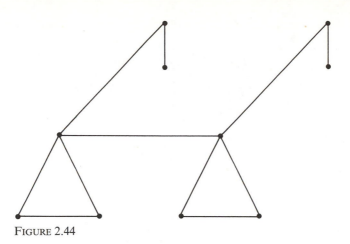

FIGURE 2.44

EXAMPLE 2.67   The graph of Figure 2.26 has an Euler circuit because every vertex has even degree. Can you trace around the figure, arriving finally at your point of origin, without lifting your pencil from the page?   △

If we revise the Euler type problem by asking instead that every vertex be encountered once and only once, we arrive at the *Hamiltonian paths* (*circuits*) of an undirected graph. W. R. Hamilton was a famous Irish mathematician of the nineteenth century. Being somewhat more ambitious than Euler, Hamilton sought to "go around the world" on a dodecahedron whose vertices were labeled by 20 cities of his choosing, and of course, the tour should encounter each city just once, excepting the origin to which one eventually returns. If this particular polyhedron is projected onto the plane, according to the method described in Example 2.27, we will obtain the planar graph of Figure 2.45. Showing that Hamilton met with more success than Euler, we have indicated a Hamiltonian circuit for this graph.

Though the problem of Hamiltonian paths bears a superficial resemblance to the Euler problem, no simple criterion is known for deciding whether or not such paths exist. This is unfortunate, because the problem of arranging itineraries is quite real. The traveling salesman, the circus, any traveling road company is faced with a problem not unlike that considered by Hamilton. What has come to be called the traveling salesman problem, however, is even more difficult. The salesman must visit a number of cities, and he knows the distance between any two. He wishes to arrange a minimum distance itinerary that includes a visit to each city just once, returning to his point of origin. It seems that these problems are all quite difficult, and except in very special cases, no efficient solutions are known. In the case of a Hamiltonian path, most of the special solutions are to the effect that "if the graph has enough edges," then a Hamiltonian path exists. That is true of the following result, which the reader can verify inductively by "extending the path."

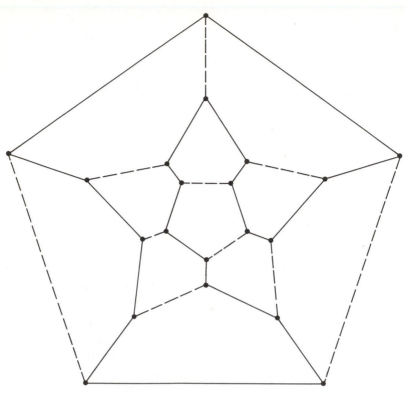

FIGURE 2.45

Theorem 2.15

Every complete graph has a Hamiltonian path. □

We now describe (Algorithm V) a "propagation procedure" for determining a shortest path connecting two given vertices u, v in an undirected graph $G = (V, E)$. This algorithm should be studied fairly closely, since it can then be taken as a model for understanding the subsequent shortest-path algorithms in labeled digraphs. In the latter case each directed edge $(x, y) \in E$ will be labeled with a *length* $\Delta(x, y)$. In the case of undirected graphs we simply assume that each edge is of length $\Delta = 1$.

Treating the vertex u as the origin, our algorithms actually define a (minimum) *distance* $d(w)$ from u, for each vertex w on the shortest path from u to v. The path itself must then be obtained by a "backtrack" algorithm. Let us explain. We let the eventual shortest path be denoted by the sequence $\langle v_0, v_1, \ldots, v_k \rangle$, where as usual,

$$v_0 = u \quad \text{and} \quad \text{each } \{v_{i-1}, v_i\} \in E$$
$$v_k = v$$

$[(v_{i-1}, v_i) \in E$ in the case of a directed graph]. Assuming that there is a path from u to v, the shortest path algorithms will eventually determine $d(v) = d(v_k)$; that is, we choose $v_k = v$. Supposing that $v_k, v_{k-1}, \ldots, v_i$ have already been chosen, we look for a vertex w with the properties

$$w \; E \; v_i \quad \text{and} \quad d(w) = d(v_i) - \Delta(w, v_i)$$

and we choose $v_{i-1} = w$, continuing to "backtrack" until we reach vertex u. We have not incorporated this backtrack procedure in our shortest-path algorithms, since that would tend to obscure the main ideas involved. It is left for the reader to formalize this technique and to tie the procedure in tandem with each of our shortest-path algorithms.

The idea behind Algorithm V is really quite simple. Like the ripples in a stream that emanate from the point at which a pebble is dropped, we imagine a wave W which has its origin at u. The wave systematically traces all the paths of increasing distance from u, marking (with the subset M) the distance of each vertex as it is determined. In this way, circuits can be avoided; obviously a shortest path would not contain a circuit. In the final block of statements, we extend the wavefront to a new set of unmarked vertices, while those already

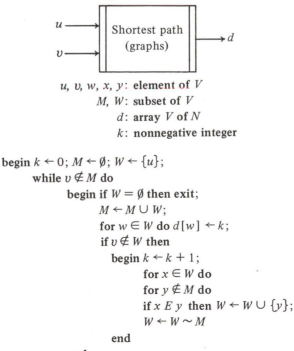

u, v, w, x, y: **element of** V

M, W: **subset of** V

d: **array** V **of** N

k: **nonnegative integer**

```
begin k ← 0; M ← ∅; W ← {u};
    while v ∉ M do
        begin if W = ∅ then exit;
            M ← M ∪ W;
            for w ∈ W do d[w] ← k;
            if v ∉ W then
                begin k ← k + 1;
                    for x ∈ W do
                    for y ∉ M do
                    if x E y then W ← W ∪ {y};
                    W ← W ∼ M
                end
        end
end
```

ALGORITHM V

marked are removed from W. With the process continuing until $v \in M$ [until v is marked and its distance $d(v)$ is determined], we eventually mark a sequence of vertices w on a shortest path from u to v.

EXAMPLE 2.68   Consider the graph of Figure 2.15(a), redrawn in Figure 2.46(a). Suppose that we use Algorithm V to determine a shortest path from c to d. A trace of the algorithm would then appear as shown in Table 2.2. In this particular example all the vertices are marked, and their minimum distances

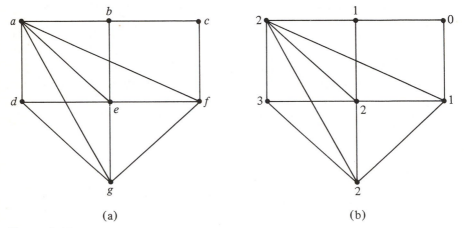

(a)   (b)

FIGURE 2.46

TABLE 2.2

u	v	k	M	W	d
c	d	0	$\varnothing$	$\{c\}$	
			$\{c\}$		$d[c] = 0$
		1		$\{b, c, f\}$	
				$\{b, f\}$	
			$\{b, c, f\}$		$d[b] = 1$
					$d[f] = 1$
		2		$\{a, b, e, f, g\}$	
				$\{a, e, g\}$	
			$\{a, b, c, e, f, g\}$		$d[a] = 2$
					$d[e] = 2$
					$d[g] = 2$
		3		$\{a, d, e, g\}$	
				$\{d\}$	
			$\{a, b, c, d, e, f, g\}$		$d[d] = 3$

from c are determined as in Figure 2.46(b). A shortest path from c to d is then found by means of the backtrack algorithm. △

EXAMPLE 2.69   To illustrate the backtrack procedure applied to the results of Example 2.68, we let $v_3 = d$ be the last vertex in our path $\langle v_0, v_1, v_2, v_3 \rangle$ from c to d. Then we have three choices for v_2, namely $v_2 = a, e,$ or g, because each satisfies the conditions

$$v_2 \ E \ d \quad \text{and} \quad d(v_2) = d(d) - 1 = 2$$

Suppose that we let $v_2 = g$. Then we have to look for a vertex v_1 such that

$$v_1 \ E \ g \quad \text{and} \quad d(v_1) = d(g) - 1 = 1$$

Here only $v_1 = f$ is appropriate. Finally, we obtain $v_0 = c$ such that

$$v_0 \ E \ f \quad \text{and} \quad d(v_0) = d(f) - 1 = 0$$

as required. All together this gives us the shortest path $\langle c, f, g, d \rangle$. △

Now we are ready to consider shortest paths in the labeled digraphs. By intention we have phrased Algorithm VI in much the same language as Algorithm V. Nevertheless, there are a few necessary distinctions. We first set all

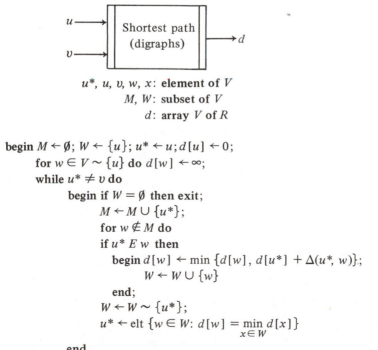

$$u^*, u, v, w, x: \textbf{element of } V$$
$$M, W: \textbf{subset of } V$$
$$d: \textbf{array } V \textbf{ of } R$$

```
begin M ← ∅; W ← {u}; u* ← u; d[u] ← 0;
      for w ∈ V ∼ {u} do d[w] ← ∞;
      while u* ≠ v do
            begin if W = ∅ then exit;
                  M ← M ∪ {u*};
                  for w ∉ M do
                  if u* E w then
                        begin d[w] ← min {d[w], d[u*] + Δ(u*, w)};
                              W ← W ∪ {w}
                        end;
                  W ← W ∼ {u*};
                  u* ← elt {w ∈ W: d[w] = min d[x]}
                                              x∈W
            end
end
ALGORITHM VI
```

$d(w) = \infty$ for $w \neq u$. In practice, we can use the largest number of the machine as infinity. Here W is only a "candidate" set from which the next marked element u^* is selected. The last statement of the algorithm ensures that u^* is a vertex whose (minimum) distance from u is known. Otherwise, the algorithm is quite similar to the preceding. An example will clarify the few subtle distinctions.

EXAMPLE 2.70   In Figure 2.47 we have a digraph whose edges are labeled with various lengths. If $u = v_1$ and $v = v_8$, then a trace of the algorithm will produce Table 2.3.

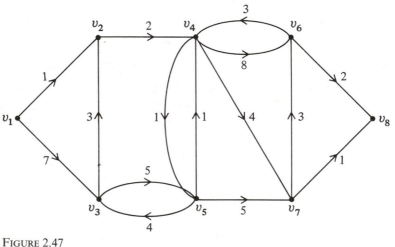

FIGURE 2.47

We must bear in mind that for this algorithm, the distances $d(v_i)$ are not final, that is, not known to be minimal until v_i has been marked. Thus we had $d(v_6) = 11$, then $d(v_6) = 10$, and it is only a coincidence that this is the minimum distance from $u = v_1$ to v_6. Suppose there had been an edge of length one from v_8 to v_6!

To provide a few more details, let us pick up the action at the point where u^* has just been assigned the value v_5. Then we set

$$M \leftarrow M \cup \{u^*\} = \{v_1, v_2, v_4\} \cup \{v_5\} = \{v_1, v_2, v_4, v_5\}$$

For $w = v_3, v_7$ we compute

$$d(v_3) \leftarrow \min \{d(v_3), d(v_5) + \Delta(v_5, v_3)\} = \min \{7, 4 + 4\}$$
$$= 7$$

$$d(v_7) \leftarrow \min \{d(v_7), d(v_5) + \Delta(v_5, v_7)\} = \min \{7, 4 + 5\}$$
$$= 7$$

and adjoin v_3, v_7 to W. In this case they are already in W. Then we delete

TABLE 2.3

u	v	u^*	M	W	d
v_1	v_8	v_1	$\varnothing$	$\{v_1\}$	$d[v_1] = 0$
			$\{v_1\}$	$\{v_1, v_2, v_3\}$	$d[v_2] = 1$
					$d[v_3] = 7$
		v_2	$\{v_1, v_2\}$	$\{v_2, v_3\}$ $\{v_2, v_3, v_4\}$ $\{v_3, v_4\}$	$d[v_4] = 3$
		v_4	$\{v_1, v_2, v_4\}$	$\{v_3, v_4, v_5, v_6, v_7\}$	$d[v_5] = 4$
					$d[v_6] = 11$
					$d[v_7] = 7$
		v_5	$\{v_1, v_2, v_4, v_5\}$	$\{v_3, v_5, v_6, v_7\}$ $\{v_3, v_5, v_6, v_7\}$	$d[v_3] = 7$
					$d[v_7] = 7$
		v_7	$\{v_1, v_2, v_4, v_5, v_7\}$	$\{v_3, v_6, v_7\}$ $\{v_3, v_6, v_7, v_8\}$	$d[v_6] = 10$
					$d[v_8] = 8$
		v_3	$\{v_1, v_2, v_3, v_4, v_5, v_7\}$	$\{v_3, v_6, v_8\}$ $\{v_6, v_8\}$	
		v_8			

$u^* = v_5$ from W and select an element $w \in W$ of minimum (tentative) distance $d(w)$ to be the new u^*. In the present situation this selection involves a comparison of

$$d(v_3) = 7 \qquad d(v_6) = 11 \qquad d(v_7) = 7$$

and we have chosen $u^* \leftarrow v_7$. At this point we are ready to begin another iteration.

From here on the reader should be able to follow the trace of the algorithm to its conclusion. The termination comes as soon as $u^* = v = v_8$. The student should also check that the backtracking procedure gives rise to the path $\langle v_1, v_2, v_4, v_7, v_8 \rangle$, a shortest path from v_1 to v_8. △

A somewhat more efficient algorithm for finding a shortest path has been described by E. W. Dijkstra,* but the algorithm does not easily convert to a longest-path algorithm. This objection is not too serious, because, the "longest-path problem" would arise only in connection with a digraph without directed circuits. In Figure 2.47 there are several circuits, for instance $v_4 v_7 v_6 v_4$, so that looking for the longest path would hardly make sense. One could traverse the

---

* E. W. Dijkstra, "A Note on Two Problems in Connection with Graphs," *Numerische Matematik,* **1**, 269–271 (1959).

given circuit as often as desired, increasing the length by 10 each time. Therefore there is no upper bound on the lengths of paths from v_1 to v_8. We need to restrict our attention to acyclic digraphs in order to properly discuss the problem of longest paths, in which case, however, as we will see, a much simpler algorithm can be given for the shortest as well as the longest path.

EXERCISES

1 Determine all the in-degrees and out-degrees of the following digraphs:
  (a) Exercise 1(a) of Section 2.7   (b) Exercise 1(b) of Section 2.7
  (c) Exercise 2(a) of Section 2.7   (d) Exercise 2(b) of Section 2.7
  List all the sources and sinks.

2 Decide whether or not the following graphs have an Euler circuit and explain your answers.
  (a) Exercise 1(a) of Section 2.1   (b) Exercise 1(c) of Section 2.1
  (c) $K_{2,4}$ (d) Figure 2.13
  (e) K_6 (f) Exercise 5(h) of Section 2.3

3 Give an inductive proof of Theorem 2.15.

4 Use Algorithm V and the backtrack procedure to find a shortest path from u to v in each of the following situations:
  (a) Exercise 1(c) of Section 2.1, $u = v_1, v = v_6$
  (b) Exercise 3(c) of Section 2.1, $u = v_1, v = v_6$
  (c) Exercise 1 of Section 2.5, u = Malkuth, v = Kether
  (d) Exercise 5(f) of Section 2.5, u and v two opposite points
  (e) Figure 2.12, $u = v_1, v = v_3$
  (f) Figure 2.13, $u = v_1, v = v_9$

*5 Devise a formal algorithm for implementing the backtrack procedure
  (a) For undirected graphs, all edge lengths being equal to 1
  (b) For labeled digraphs

6 Use Algorithm V and the backtrack procedure to find a shortest path from start to finish in each of the mazes (the distance between adjacent dots being one unit) shown in Figure 2.48.

7 Use Algorithm VI and the backtrack procedure to find shortest paths from u to v in the following digraphs (with unit distance along each edge):
  (a) Exercise 1(a) of Section 2.7, $u = v_1, v = v_5$
  (b) Exercise 1(c) of Section 2.7, $u = v_2, v = v_5$
  (c) Exercise 2(b) of Section 2.7, $u = v_4, v = v_1$

8 Use Algorithm VI and the backtrack procedure to find shortest paths from u to v in the labeled digraphs shown in Figure 2.49.

(a)

(b) Start at center

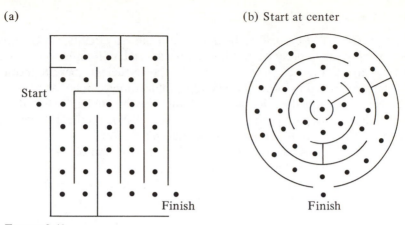

FIGURE 2.48

(a)

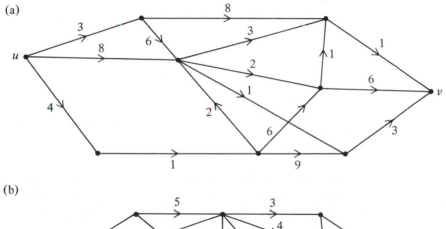

(b)

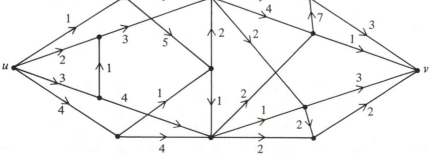

FIGURE 2.49

*2.10   ACYCLIC DIGRAPHS

A digraph $G = (V, E)$ is said to be *acyclic* if it has no circuits, in the directed sense. Even if the underlying graph is connected, such digraphs need not look like trees at all. Figure 2.35 is an example. Nevertheless, these digraphs are sufficiently restricted in form that several simplifications can be made. For one thing, as we mentioned earlier, here the longest-path problem will make good sense. Indeed, both path problems, the longest and shortest, lend themselves to simplified solutions in the case of acyclic digraphs.

That acyclic digraphs can be easily handled derives from the fact that we can find for each of them a *consistent labeling* of the vertices. By consistent labeling we actually mean a relabeling or resequencing of the vertices as $v_1, v_2, \ldots, v_n$ such that

$$v_i \, E \, v_j \Rightarrow i < j$$

We can state this property as a theorem, which can then serve to characterize the acyclic digraphs.

Theorem 2.16

The digraph $G = (V, E)$ is acyclic if and only if its vertices can be given a consistent labeling.

Proof   It is clear that if G has a directed circuit $u = v_{i_1} v_{i_2} \cdots v_{i_k} = u$, then we will not be able to find a consistent labeling, for we will have

$$j_1 < j_2 < \cdots < j_k = j_1$$

that is, $j_1 < j_1$, regardless of how the vertices v_i are relabeled as v_j. On the other hand, if G is acyclic, then we can give an algorithm (Algorithm VII) for achieving a consistent labeling. $\qquad\square$

Algorithm VII is due to R. B. Marimont,* and given the way it is phrased here, it can even be used to test whether or not a given digraph is acyclic. Thus we include as output a logical variable L, whose value at the conclusion of execution gives the decision:

$$L = \textbf{true} \quad \text{if } G \text{ is acyclic}$$

$$L = \textbf{false} \quad \text{otherwise}$$

For each vertex w the algorithm makes use of a set of vertices $A(w)$, which might be called the set of *antecedents* of w; that is

$$A(w) = \{v \in V : (v, w) \in E\}$$

We evidently have $A(w) = \varnothing$ iff w is a source. It is easy to show that every

---

* R. B. Marimont, "A New Method for Checking the Consistency of Precedence Matrices," *J.A.C.M.*, **6**, 164 (1959).

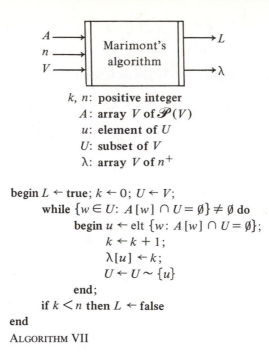

$$k, n: \text{ positive integer}$$
$$A: \text{ array } V \text{ of } \mathscr{P}(V)$$
$$u: \text{ element of } U$$
$$U: \text{ subset of } V$$
$$\lambda: \text{ array } V \text{ of } n^+$$

begin $L \leftarrow$ **true**; $k \leftarrow 0$; $U \leftarrow V$;
  while $\{w \in U: A[w] \cap U = \emptyset\} \neq \emptyset$ **do**
      begin $u \leftarrow$ **elt** $\{w: A[w] \cap U = \emptyset\}$;
          $k \leftarrow k + 1$;
          $\lambda[u] \leftarrow k$;
          $U \leftarrow U \sim \{u\}$
      end;
    if $k < n$ **then** $L \leftarrow$ **false**
end

ALGORITHM VII

acyclic digraph has at least one source, so that if G is acyclic, then the proposition

$$\{w \in U : A(w) \cap U = \emptyset\} \neq \emptyset$$

is true at the start of the algorithm, when $U = V$. The whole procedure can then be described as successive removal of sources. If the proposition becomes false before $k = n$, that is before all the vertices have been removed, then we must have come up against a circuit in the digraph. Obviously there are no sources in a circuit.

EXAMPLE 2.71   If we apply Marimont's algorithm to the digraph of Example 2.70 (Figure 2.47), we will expect it to tell us that the digraph is not acyclic. We first determine that

$$A(v_1) = \emptyset \quad (v_1 \text{ is a source})$$
$$A(v_2) = \{v_1, v_3\}$$
$$A(v_3) = \{v_1, v_5\}$$
$$A(v_4) = \{v_2, v_5, v_6\}$$
$$A(v_5) = \{v_3, v_4\}$$
$$A(v_6) = \{v_4, v_7\}$$

$$A(v_7) = \{v_4, v_5\}$$

$$A(v_8) = \{v_6, v_7\}$$

Then a trace of the algorithm will begin as follows:

V	n	k	u	U	L	λ
$\{v_1, v_2, \ldots, v_8\}$	8	0		$\{v_1, v_2, \ldots, v_8\}$	**true**	
		1	v_1	$\{v_2, \ldots, v_8\}$		$\lambda[v_1] = 1$
					false	

At this point, we have

$$A(v_2) \cap U = \{v_3\}$$

$$A(v_3) \cap U = \{v_5\}$$

and otherwise $A(v_i) \cap U = A(v_i)$. It follows that the proposition in the **while** statement is false. We have come up against a circuit. No new sources have been created by the removal of v_1 (and edges emanating from v_1). Consequently, the whole operation is aborted and L will be set to **false**.   △

EXAMPLE 2.72   Suppose that we remove the edges (v_5, v_3), (v_5, v_4), and (v_6, v_4) in the digraph of Example 2.70. Then there won't be any circuits. To apply Marimont's algorithm again, we have to change the antecedent sets accordingly. Thus we have

$$A(v_1) = \varnothing$$

$$A(v_2) = \{v_1, v_3\}$$

$$A(v_3) = \{v_1\}$$

$$A(v_4) = \{v_2\}$$

$$A(v_5) = \{v_3, v_4\}$$

$$A(v_6) = \{v_4, v_7\}$$

$$A(v_7) = \{v_4, v_5\}$$

$$A(v_8) = \{v_6, v_7\}$$

Now if we run a trace of Algorithm VII on the new digraph, we will obtain the results shown in Table 2.4.

Thus at the same point that we examined the original graph (Example 2.71) we now have, with $U = \{v_2, \ldots, v_8\}$:

$$A(v_2) \cap U = \{v_3\}$$

$$A(v_3) \cap U = \varnothing$$

$$\vdots$$

TABLE 2.4

V	n	k	u	U	L	λ
$\{v_1, \ldots, v_8\}$	8	0		$\{v_1, \ldots, v_8\}$	**true**	
		1	v_1	$\{v_2, \ldots, v_8\}$		$\lambda[v_1] = 1$
		2	v_3	$\{v_2, v_4, v_5, v_6, v_7, v_8\}$		$\lambda[v_3] = 2$
		3	v_2	$\{v_4, v_5, v_6, v_7, v_8\}$		$\lambda[v_2] = 3$
		4	v_4	$\{v_5, v_6, v_7, v_8\}$		$\lambda[v_4] = 4$
		5	v_5	$\{v_6, v_7, v_8\}$		$\lambda[v_5] = 5$
		6	v_7	$\{v_6, v_8\}$		$\lambda[v_7] = 6$
		7	v_6	$\{v_8\}$		$\lambda[v_6] = 7$
		8	v_8	$\varnothing$		$\lambda[v_8] = 8$

A new source has been created by the removal of v_1 and its two emanating edges. Accordingly, we assign $u \leftarrow v_3$ and continue. The appearance of new sources persists all the way up to v_8, at which point the labeling function λ provides a new consistent labeling of the vertices as shown in Figure 2.50. △

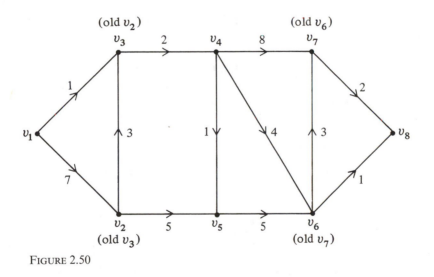

FIGURE 2.50

We can give a simplified procedure (Algorithm VIII) for determining a shortest path in an acyclic diagraph after the vertices have been relabeled according to Marimont's algorithm. We assume that the consistent labeling is now $V = \{v_1, v_2, \ldots, v_n\}$, and there is no loss of generality in supposing that it is a shortest path from v_1 to v_n that is desired. As before, we begin evaluating the distance by setting $d(v_1) = 0$ and all other distances equal to infinity. Then the actual distances $d(v_2), \ldots, d(v_n)$ are determined in succession by iterating the last line in Algorithm VIII. If, instead, we want a longest path, then we

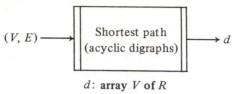

$$d: \text{ array } V \text{ of } R$$
$$j: \text{ nonnegative integer}$$
$$v: \text{ element of } V$$

begin $d[v_1] \leftarrow 0$;
  for $v \in V \sim \{v_1\}$ **do** $d[v] \leftarrow \infty$;
  for $j \leftarrow$ **to** n **do**
    $d[v_j] \leftarrow \min \{d[v_j], d[v_i] + \Delta(v_i, v_j)\}$
     $v_i \, E \, v_j$
     $i < j$

end

ALGORITHM VII

simply have to change the "minimum" in this last statement to "maximum finite."

EXAMPLE 2.73   Consider the consistent labeling of Figure 2.50, and suppose that we are trying to find a longest path from v_1 to v_8. Then the modified Algorithm VIII determines in succession:

$$d(v_1) = 0$$
$$d(v_2) = \max \text{ fin } \{\infty, 0 + 7\} = 7$$
$$d(v_3) = \max \text{ fin } \{\infty, 0 + 1, 7 + 3\} = 10$$
$$d(v_4) = \max \text{ fin } \{\infty, 10 + 2\} = 12$$
$$d(v_5) = \max \text{ fin } \{\infty, 7 + 5, 12 + 1\} = 13$$
$$d(v_6) = \max \text{ fin } \{\infty, 13 + 5, 12 + 4\} = 18$$
$$d(v_7) = \max \text{ fin } \{\infty, 12 + 8, 18 + 3\} = 21$$
$$d(v_8) = \max \text{ fin } \{\infty, 18 + 1, 21 + 2\} = 23$$

The backtrack procedure will produce the longest path $v_1 v_2 v_3 v_4 v_5 v_6 v_7 v_8$. Needless to say, this striking sequence is only a coincidence.   △

A possible application of the longest-path problem was discussed briefly in Example 2.54. Most complex modern design or production processes are composed of many separate activities, and each product is created by the successful completion of each of these activities according to some prearranged schedule. Sometimes it is possible to arrange some of these activities to be carried out simultaneously. Nevertheless, some of the activities cannot begin until others have been completed. All such features, which are common to most complicated processes, can be incorporated into a special kind of labeled digraph called a *scheduling network*. We could design an illustration to show the reader how to save x number of working hours on a Detroit assembly line, but the moral of the story will be brought home just as effectively (and perhaps more interestingly) if, instead, we discuss the planning of a picnic.

There was once a popular song with a refrain along the following lines: "Did you bring the pickles, pickles, pickles? Did you bring the mustard, mustard, mustard? . . ." Its composer was acutely aware that the planning of a picnic must be given very careful consideration if we are not to forget an important item. Suppose that the family is at home on the morning that the plan is hatched, and the following menu is proposed (and agreed to by the children):

(a) hamburgers
(b) potato salad
(c) coffee, (iced) tea, or milk

together with the usual condiments. Then someone notices that there is no tea in the cupboard. So somebody must go shopping. If one parent goes, then the other parent can start preparing the potato salad. Meanwhile, the children are told to gather all the picnic supplies so that they will be in order when it is time to pack. A similar splitting of the chores can take place at the picnic grounds. One person can gather wood for the fire, another find a picnic table, and a third look for a source of water. All together the activities can be diagramed in Figure 2.51. Note that the iced tea can't be made until the shopping has been

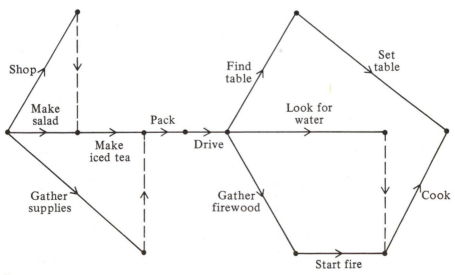

FIGURE 2.51

done. So we must introduce a "dummy" activity (with a dashed line) to indicate this dependence of the start of one activity on the completion of another.

The next stage in the analysis of a schedule is the assignment of estimated time for the completion of each of the various activities. This assignment and the (consistent) labels of the vertices are shown in Figure 2.52. The dummy activities are assigned a completion time of zero. Such a labeled digraph is called a *scheduling network*. As we proceed with the formal definition and analysis of scheduling networks, there is one main convention that the reader should constantly keep in mind: an activity (edge) emanating from a given vertex cannot begin until all the activities terminating at that vertex have been completed.

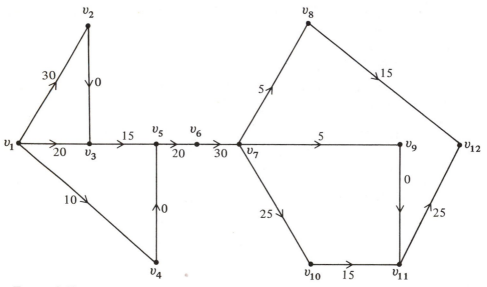

FIGURE 2.52

Formally, a *scheduling network* is an acyclic labeled digraph $G = (V, E)$ having exactly one source and one sink. We may assume that the vertices are consistently labeled so that v_1 is the source and v_n is the sink. We will still assume that the labels on the edges are denoted by $\Delta(v_i, v_j)$, but of course these now represent completion times and not distances. For each vertex $v \in V$ we define an *earliest starting time* $E(v)$ for all the activities emanating from v as

$$E(v_1) = 0$$

$$E(v_j) = \max_{v_i \, E \, v_j} \{E(v_i) + \Delta(v_i, v_j)\} \qquad (j = 2, \ldots, n)$$

and similarly, the *latest completion time* $L(v)$ for all activities terminating at v as

$$L(v_n) = E(v_n)$$

$$L(v_i) = \min_{v_i \, E \, v_j} \{L(v_j) - \Delta(v_i, v_j)\} \qquad (i = n - 1, \ldots, 1)$$

All of these quantities can be determined in succession because of the consistent labeling of the vertices. Finally, the *float time* of the edge or activity $e = (v, w)$ is defined by

$$f(e) = f(v, w) = L(w) - E(v) - \Delta(v, w)$$

It represents the maximum delay that can be tolerated in the corresponding activity without the overall completion time of the process being upset. An activity is said to be *critical* if it has a zero float time. Similarly, a *critical path* is one all of whose edges represent critical activities.

EXAMPLE 2.74   Let us determine these various quantities for the picnic network of Figure 2.52. We can obtain Tables 2.5 and 2.6 for the earliest starting times, latest completion times, and float times. Note that the order of the

TABLE 2.5

v	$E(v)$	$L(v)$
v_1	0	0
v_2	30	30
v_3	30	30
v_4	10	45
v_5	45	45
v_6	65	65
v_7	95	95
v_8	100	145
v_9	100	135
v_{10}	120	120
v_{11}	135	135
v_{12}	160	160

TABLE 2.6

$e = (v, w)$	$\Delta(v, w)$	$f(v, w)$
(v_1, v_2)	20	0
(v_1, v_3)	20	10
(v_1, v_4)	10	35
(v_2, v_3)	0	0
(v_4, v_5)	0	35
(v_3, v_5)	15	0
(v_5, v_6)	20	0
(v_6, v_7)	30	0
(v_7, v_8)	5	45
(v_7, v_9)	5	35
(v_7, v_{10})	25	0
(v_{10}, v_{11})	15	0
(v_9, v_{11})	0	35
(v_8, v_{12})	15	45
(v_{11}, v_{12})	25	0

computations is from top to bottom in the column for $E(v)$, but from bottom to top in the column for $L(v)$. As sample calculations, we cite

$$E(v_3) = \max \{0 + 20, 30 + 0\} = 30$$

$$L(v_5) = \min \{65 - 20\} = 45$$

$$f(v_3, v_5) = L(v_5) - E(v_3) - \Delta(v_3, v_5) = 45 - 30 - 15 = 0$$

The last line shows that there is "no time to spare" in making the iced tea; it is a critical activity. On the other hand, the fact that $f(v_1, v_4) = 35$ shows that the children have time to watch a half-hour television program without upsetting the schedule. The float times show that $v_1, v_2, v_3, v_5, v_6, v_7, v_{10}, v_{11}, v_{12}$ constitute a critical path. They are the only activities whose expedition can improve the picnic schedule. Having the children work faster in gathering the picnic supplies will not help matters a bit.   △

The whole purpose of critical-path analysis is to isolate those activities whose efficiency must be improved in order to reduce the overall project completion time. A knowledge of the critical paths in a project allows the production supervisor or manager to decide on the most effective means of reorganizing the project. He may determine a way of arranging more activities to be done in parallel or reallocate the manpower to improve the efficiency of certain critical links in the process. Since its discovery in the 1950s, the approach has met with phenomenal success in any number of industrial situations.

We conclude this section and our chapter on graphs with a result, offered without proof, that should agree with the reader's intuitions.

Theorem 2.17

A critical path in any scheduling network is a longest path from source to sink.   □

EXERCISES

1 Prove that every acyclic digraph has a source.

2 Use Algorithm VII to determine a consistent labeling of the vertices in the digraph of
   (a) Exercise 8(a) of Section 2.9   (b) Exercise 8(b) of Section 2.9
   (c) Figure 2.51

3 Use Algorithm VIII with a consistent labeling of the vertices to find a shortest path from u to v in the digraph of
   (a) Exercise 8(a) of Section 2.9   (b) Exercise 8(b) of Section 2.9

4 Repeat Exercise 3 with the modified Algorithm VIII for finding a longest path.

5 Use Algorithm VII to detect the existence of a directed circuit in
   (a) Exercise 1(a) of Section 2.7   (b) Exercise 1(b) of Section 2.7

6 Use the modified Algorithm VIII to find a longest path (from v_1 to v_{12}) in the picnic network of Figure 2.52.

*7 Prove Theorem 2.17.

8 Tabulate the earliest starting times, latest completion times, and float times for the activities characterized by
   (a) Figure 2.49(a)   (b) Figure 2.49(b)

Begin by providing a consistent labeling of the vertices for each. Then find a critical path from u to v.

*9 Prove that there is always a critical path in any scheduling network.

SUGGESTED REFERENCES

2.1 Ore, O. *Graphs and Their Uses.* Random House, New York, 1963.

2.2 Ore, O. *Theory of Graphs.* A.M.S. Colloquium Publications, Providence, R.I., 1962.

2.3 Harary, F. *Graph Theory.* Addison-Wesley, Reading, Mass., 1969.

2.4 Wilson, R. *Introduction to Graph Theory.* Oliver and Boyd, Edinburgh, 1972.

2.5 Busacker, R. G., and T. L. Saaty. *Finite Graphs and Networks.* McGraw-Hill, New York, 1965.

2.6 Even, S. *Algorithmic Combinatorics.* Macmillan, New York, 1973.

There are now any number of fine books on graph theory. The two books by Ore are excellent, though at different levels of presentation. The first belongs to the SMSG (School Mathematics Study Group) series, designed for advanced high school students. The book by Harary reveals the depth of the subject at its advanced levels. Wilson's book gives an enjoyable account of the basic concepts of graph theory.

The book by Busacker and Saaty has an excellent chapter on the many applications of graph theory. Overall, it is somewhat more concerned with digraphs than any of the references mentioned previously. We have included the interesting book by Even because of its wealth of graphical algorithms.

3

Monoids and Machines

As we mentioned earlier, high school algebra is really only the study of a few specific algebraic structures: the arithmetic of the integers, the algebra of the real numbers, perhaps some polynomial arithmetic, but that is about all. In computer science, as the student must be beginning to learn, quite a diversity in algebraic structures is encountered. A systematic study of algebraic structures can be made somewhat easier if a few basic algebraic systems are singled out as central to the overall algebraic theory. To do exactly this was one of our goals in the preceding chapter. It remains the goal of this chapter. Traditionally in undergraduate mathematics curricula, group theory is presented as the core of algebra. For the study of computer science, however, we have to include the more general monoids because of their application to finite state machines, formal languages, etc.

The important notion of homomorphisms for algebras appears in its most crystalline form in the theory of monoids. As we will see, the concept of homomorphism is only slightly more general than that of isomorphism as illustrated in Chapter 2 for graphs. The characterization of homomorphic images by means of congruence relations on the domain is nowhere more transparent than in connection with monoids, and the related concepts of simulation and realization are aptly illustrated through the machine design techniques deriving from homomorphism theory. The latter topics, taken up toward the end of the chapter, will help the student develop a feel for the range of application of the earlier material.

3.1 MONOIDS AND SUBMONOIDS

No doubt the student has come to realize that the binary operations

$$\circ: A \times A \to A$$

play a prominent role in algebra. If not, we shall nevertheless try to convey this impression in our next several chapters. Temporarily we shall use the generic symbol ∘ to represent an arbitrary binary operation and recall that we prefer to use the "infix" notation $x \circ y$ [rather than $\circ(x, y)$] in writing

$$(x, y) \to x \circ y$$

in accordance with conventional arithmetic notation. Then the reason for the prominence of binary operations should be easily understood: a useful algebraic theory would naturally seek to imitate and generalize those features that are found in standard arithmetics, that is, in high school algebra, computer arithmetics, etc. Indeed, the term "arithmetic" already suggests an overall concern with binary operations. Additions, multiplications, subtractions, and divisions, each in its proper arithmetic setting, constitute four binary operations; thus we add *two* numbers to obtain a third.

Certainly there are unary operations all around: absolute value, squaring, complementation in the algebra of sets, etc. What about 0-ary (*nullary*) operations? What would such a terminology mean? What could we understand by an operation

$$f: A^0 \to A$$

in an algebra A? A satisfactory answer can be obtained by first clarifying the idea of a product set A^0. If we insist on extending the rule of the product

$$|A^n| = |A|^n$$

to the case $n = 0$, then we will evidently obtain $|A^0| = |A|^0 = 1$, which implies that for any set A, the product A^0 is a singleton set, say $A^0 = \{1\}$ for convenience. Thus a nullary operation simply selects or *distinguishes* a particular element in A. Rather than explicitly admitting a nullary operation in our algebras, we will ordinarily say, "let e be a distinguished element." Usually this will be quite a special element, as in the case of 0 or 1 in standard arithmetics.

The theory of monoids is rather like the "tip of the iceberg," where the iceberg refers to the theory of algebraic structures. Frankly, very few mathematicians are interested in monoids because monoids are so lacking in structure. On the other hand, that is precisely the reason that it may be most sensible for us to approach the whole subject of algebraic structures by way of monoids— to try to grasp the essential algebraic notions at a level where the operational machinery is not yet overly burdensome. Then, because the monoid axioms are so general, you will first get the impression that "everything in God's world is a monoid." Certainly all of the familiar arithmetic systems are monoids, and it is this generality that makes the study of the monoids a convenient point of departure.

EXAMPLE 3.1   In Example 1.55 we designed a simple algorithm for computing the minimum of n numbers $x_1, x_2, \ldots, x_n$. A closer inspection of that procedure reveals that in effect the algorithm allows us to compare only two numbers at a time. If $n = 4$, for example, then the algorithm actually gives us

$$\min (\min (\min (x_1, x_2), x_3), x_4)$$

as if min were a binary operation: $(((x_1 \circ x_2) \circ x_3) \circ x_4)$. Should we expect a different answer with

$$(x_1 \circ x_2) \circ (x_3 \circ x_4) \quad \text{or} \quad x_1 \circ ((x_2 \circ x_3) \circ x_4)?$$

No, we are well aware that the final result does not depend on the order in which the repeated minima are taken. When expressed for arbitrarily many elements and all possible ways of inserting parentheses, we have here an instance of a generalized associativity principle. $\triangle$

Returning to our original convention of designating an algebra by the same symbol as the set on which its operations are defined, we will say that a *monoid* $M = (M, \cdot, e)$ consists of a set M with a distinguished element e and a binary operation

$$(x, y) \rightarrow x \cdot y$$

such that the following axioms hold:

(1) *associativity* $\qquad x \cdot (y \cdot z) = (x \cdot y) \cdot z$
(2) *identity* $\qquad\qquad x \cdot e = e \cdot x = x$

for all x, y, z in M. There is no particular reason for using a multiplication sign here. Any other operational symbol could just as well have been used. With this choice, however, we say that the distinguished element e acts like a "one" element [in light of axiom (2)] in multiplication. On the other hand, axiom 1 assures us that repeated "products" can be computed without regard to parentheses, for it is easy to show that the generalized associative law follows, by induction, from axiom (1). In any case, it is certainly clear that the monoids are among the more general algebras that one could imagine. Still more elementary, however, are the *semigroups* $S = (S, \cdot)$ in which only axiom (1) is satisfied. It happens that most of the semigroups that one finds in computer science will have an identity element, so it is just as well to begin with monoids.

We say that a *submonoid* N of a monoid $M = (M, \cdot, e)$ is a subset $N \subseteq M$ that is itself a monoid with respect to the same operation and identity as those of M itself. Thus, if N is a submonoid of $M = (M, \cdot, e)$, then $e \in N$. Apart from this requirement, we have only to check the *closure condition*

$$x \in N \quad \text{and} \quad y \in N \Rightarrow x \cdot y \in N \qquad (\text{for } N \subseteq M)$$

in order to guarantee that N is indeed a submonoid, for the associativity in N will follow from that in M. The closure condition is verified only to make sure that $\cdot$ is an operation on N. (That M is closed with respect to the multiplication follows from the definition of $\cdot$ as a binary operation on M.)

EXAMPLE 3.2   Evidently the algebra $Z = (Z, \cdot, 1)$ is a monoid with the ordinary multiplication of integers as the binary operation. We know that the axioms are satisfied because

(1) $n \cdot (m \cdot p) = (n \cdot m) \cdot p$
(2) $n \cdot 1 = 1 \cdot n = n$

are familiar laws of arithmetic, as listed in Example 0.72. Similarly, the real numbers $R = (R, \cdot, 1)$ form a monoid. △

EXAMPLE 3.3   The algebra $Z = (Z, +, 0)$ is also a monoid. This time the operation is the addition of integers. Since

(1) $n + (m + p) = (n + m) + p$
(2) $n + 0 = 0 + n = n$

for all n, m, and p, the axioms are satisfied. It is only a matter of convenience to use the generic symbol $(\cdot)$ in speaking of monoids. In the present example the actual operation is addition. The same thing as the preceding can be said for the additive monoid $R = (R, +, 0)$ of real numbers.

The natural numbers $N = (N, +, 0)$ form a submonoid of Z; and of course the nonnegative real numbers form a submonoid of $R = (R, +, 0)$. At the same time, these submonoids can be considered as monoids in their right. As submonoids of Z and R, respectively, however, each will contain the identity element 0, and each satisfies the required closure condition. If we add two nonnegative integers, we will again obtain a nonnegative integer. The same is true of the nonnegative reals. △

EXAMPLE 3.4   Other examples of monoids from the previous chapters are:

$(\mathscr{P}(S), \cup, \varnothing)$ and $(\mathscr{P}(S), \cap, S)$	(Section 0.2)
$(\mathscr{R}(S), \vee, 0)$ and $(\mathscr{R}(S), \wedge, 1)$	(Section 0.6)
$(B, \vee, 0)$ and $(B, \wedge, 1)$	(Example 1.15)
$(B, \oplus, 0)$	(Example 1.16)
$(B^n, \oplus, (0, 0, \ldots, 0))$	(Example 1.17)

All of these taken together serve to emphasize the extreme generality of the monoid axioms. △

EXAMPLE 3.5   According to the discussion in Example 3.1, $N = (N, \min)$ is a semigroup but not a monoid because there is no identity element for the min operation. Note that if there were an (identity) element e such that

$$\min(n, e) = \min(e, n) = n$$

for all $n \in N$, then $n \le e$ for all $n \in N$. Obviously there is no such "largest" integer e. On the other hand, in light of Example 3.3, we see that the same set may form a semigroup in different ways. The same, however, may be said of monoids.   △

EXAMPLE 3.6   Consider the set $M = \{e, a, b\}$ with the operations

·	e	a	b		⊙	e	a	b
e	e	a	b		e	e	a	b
a	a	b	e		a	a	a	e
b	b	e	a		b	b	e	a

Since

$$b \odot (b \odot a) = b \odot e = b$$

$$(b \odot b) \odot a = a \odot a = a$$

the second operation does not give rise to a monoid; the operation is not associative. The first operation, however, does, as the reader may verify. In fact, $M = (M, \cdot, e)$ is a *group*. By this we mean that a third axiom is satisfied, beyond (1), (2), namely: to each element $x \in M$ there corresponds an element x^{-1} (the *inverse* of x) such that

(3) $x \cdot x^{-1} = x^{-1} \cdot x = e$

In the present example, we have $e^{-1} = e$, $a^{-1} = b$, $b^{-1} = a$. The study of such more restricted algebras is deferred until Chapter 5.   △

EXAMPLE 3.7   By an exhaustive check of the condition

$$x \cdot (y \cdot z) = (x \cdot y) \cdot z$$

the reader can verify that $M = \{e, a, b, c\}$ with the "multiplication" table:

·	e	a	b	c
e	e	a	b	c
a	a	e	c	b
b	b	b	b	b
c	c	c	c	c

represents a monoid. To look for the identity element in the multiplication table is quite an easy matter. Do not look for an element named e: there may not be one. Look for an element whose row and column simply repeat the respective headings. In this example the identity element *happens* to be named e.

The subset $N = \{e, a\}$ is a submonoid. It contains the identity element of M, as required, and another examination of the multiplication table reveals that the closure condition is also satisfied. At the intersection of the rows and columns corresponding to elements of N, we again find elements only from the subset N.

Observe, finally, that

$$a \cdot b = c \neq b = b \cdot a$$

showing that there exist *noncommutative* monoids. Is this monoid a group? $\triangle$

EXAMPLE 3.8 Trivially, the subsets $\{e\}$ and M are submonoids of any monoid M. The student should verify that in the case of the monoid of Example 3.6, these are the only submonoids. $\triangle$

EXAMPLE 3.9 A certain primitive tribe can count no higher than three. They use the monoid $M = (\{0, 1, 2, 3, \infty\}, +, 0)$ with the addition table shown below:

+	0	1	2	3	∞
0	0	1	2	3	∞
1	1	2	3	∞	∞
2	2	3	∞	∞	∞
3	3	∞	∞	∞	∞
∞	∞	∞	∞	∞	∞

To this tribe, infinity means large, that is, bigger than three. $\triangle$

In the last few examples, we saw several instances of finite monoids $M = (M, \cdot, e)$. In finite monoids the cardinality $|M|$ of the set M is called the *order* of the monoid. It may be a useful exercise for the student to find all the *essentially different* monoids of small order, say $|M| = 1, 2, 3, 4$. To do so, however, it will be necessary to use the formal algebraic meaning of "essentially different," namely nonisomorphic.

In light of our discussion of isomorphic graphs (Section 2.6), it should be clear how we can proceed in similar fashion for monoids. A bijective mapping $\varphi: M \to M'$ from one monoid to another is called a *monoid isomorphism* (and we write $M \simeq M'$) if

(i) $\varphi(e) = e$
(ii) $\varphi(x \cdot y) = \varphi(x) \cdot \varphi(y)$

for all x, y in M. Is (i) necessary or does (ii) $\Rightarrow$ (i) here? Technically speaking, we should denote the identity element of M' somewhat differently, say by e'. Similarly, the operation should be distinguished from that in M. Then our two conditions would read:

(i) $\varphi(e) = e'$
(ii) $\varphi(x \cdot y) = \varphi(x) \cdot' \varphi(y)$

but that would be carrying the formality a bit too far. The context will always reveal which identity element and which operation we are currently considering.

The main thing is to see that two isomorphic monoids are abstractly indistinguishable. All that we need is the key φ to the one-to-one renaming process; once we have that, anything that could be learned about one monoid could immediately be inferred for the other. Consider Figure 3.1 in which we treat the monoid operations as if they were performed by actual computational modules. Examining the computation of $u \cdot v$ in the monoid M', we see that it is only an imitation of a computation in M. Since φ is surjective, there exist elements x, y in M such that $\varphi(x) = u$ and $\varphi(y) = v$; these elements x, y are uniquely determined because φ is injective as well. The condition (ii) shows that the "transform" of the product $x \cdot y$ agrees with the computation in M'; that is,

$$\varphi(x \cdot y) = \varphi(x) \cdot \varphi(y) = u \cdot v$$

The same is true in the reverse direction, as we can see by using the inverse mapping φ^{-1} where we just used φ. In short, every computation in M has its analog in M', and conversely.

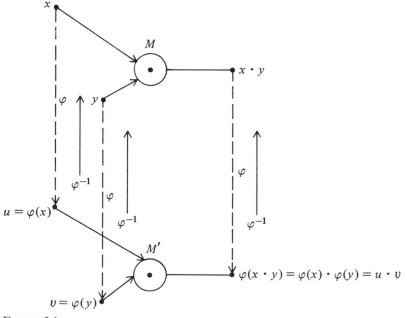

FIGURE 3.1

EXAMPLE 3.10   Let $Z = (Z, +, 0)$ be the monoid of Example 3.3, and suppose that $E = (E, +, 0)$ is the additive monoid of even integers. Then the mapping

$$\varphi(n) = 2n$$

is an isomorphism $\varphi \colon Z \to E$. Checking the two required conditions, we obtain

(i) $\varphi(0) = 2 \cdot 0 = 0$
(ii) $\varphi(n + m) = 2(n + m) = 2n + 2m = \varphi(n) + \varphi(m)$

At the same time φ is obviously a bijection. For instance,

$$\varphi(n) = \varphi(m) \Rightarrow 2n = 2m \Rightarrow n = m$$

shows that φ is injective.

Intuitively it is clear that any computation in Z could be "scaled up," so as to be performed in the monoid E. As monoids, Z and E are abstractly equivalent.   △

EXAMPLE 3.11   As with graphs, two finite monoids cannot be isomorphic unless they are of the same order, and it is almost a combinatorial problem to decide whether an isomorphism exists. Consider the two multiplication tables shown below, where M is the monoid of Example 3.7:

		M						M'		
·	e	a	b	c	·	w	x	y	z
e	e	a	b	c	w	z	z	w	w
a	a	e	c	b	x	z	y	x	w
b	b	b	b	b	y	w	x	y	z
c	c	c	c	c	z	w	w	z	z

If we use the technique described in Example 3.7 in order to find an identity element for M', we will conclude that it must be the element y. The reader can verify that the operation is associative, so that M' is indeed a monoid. But are the two monoids isomorphic?

Because of condition (i), we necessarily have

$$\varphi(e) = y$$

in any isomorphism φ. The fact that $a \cdot a = e$ helps to determine the image of a in an isomorphism. We must have

$$\varphi(a) \cdot \varphi(a) = \varphi(a \cdot a) = \varphi(e) = y$$

because of condition (ii). Examining the multiplication table for M', we see that there are only two elements whose "squares" are y, namely x and y. We cannot let $\varphi(a) = y$, for then the mapping φ would not be injective. So we are left with the partial definition

$$e \xrightarrow{\varphi} y$$
$$a \xrightarrow{\varphi} x$$

for the proposed isomorphism. It can be completed so as to become a bijection in just two ways:

$$b \xrightarrow{\varphi} w \qquad b \xrightarrow{\varphi} z$$
$$\text{or}$$
$$c \xrightarrow{\varphi} z \qquad c \xrightarrow{\varphi} w$$

In the first case, we would have

$$w = \varphi(b) = \varphi(b \cdot b) = \varphi(b) \cdot \varphi(b) = w \cdot w = z$$

In the second case, we would obtain

$$z = \varphi(b) = \varphi(b \cdot c) = \varphi(b) \cdot \varphi(c) = z \cdot w = w$$

Since both cases lead to a contradiction, we conclude that there can be no isomorphism. $\triangle$

EXERCISES

1 Verify that the algebras of Example 3.4 are monoids.

2 Which of the monoids of Exercise 1 are groups?

3 Find all the submonoids of the monoid in Example 3.9.

4 Find all the essentially different monoids M of order $|M| = 3$.

5 Devise a formal algorithm for determining whether or not a given multiplication table represents a monoid on the finite set $M = \{x_1, x_2, \ldots, x_n\}$.

6 Which of the following multiplication tables represent monoids?

(a) ·	1	2	3	4
1	2	1	4	3
2	1	2	3	4
3	4	3	2	1
4	3	4	1	2

(b) ·	1	2	3	4
1	1	2	3	4
2	2	1	4	4
3	3	3	2	1
4	4	3	1	2

(c) ·	1	2	3	4
1	1	2	3	4
2	2	1	4	3
3	3	3	3	3
4	4	4	4	4

Find all the submonoids of those that are monoids.

7 Which of the following are monoids?
(a) Even integers, ordinary addition
(b) Odd integers, ordinary addition
(c) Nonnegative integers, ordinary addition
(d) Positive integers, ordinary multiplication

8 Exhibit an isomorphism between a monoid of Exercise 6 and the monoid M of Example 3.11.

9 Show that in any monoid, the identity element is unique, that is, there cannot be more than one identity element.

10 Show that there do not exist two isomorphic monoids in Exercise 6.

*11 Prove that the following monoids are isomorphic: positive real numbers with multiplication and all real numbers with addition. [*Hint*: Use a logarithmic isomorphism.]

12 Devise a formal algorithm for determining whether a given monoid is a group.

3.2 TRANSFORMATION MONOIDS

We have established that there is apparently quite a variety of monoids. Some are finite, others are not. Some of them are groups. Some are commutative, others are not. In this section, however, our objective is to show that there is an essential similarity among all the examples of the last section. This is done by means of a characterization theorem, usually attributed to the nineteenth-century English mathematician Arthur Cayley. The theorem refers to transformations—functions or mappings from a set into itself—as discussed briefly in Section 0.7. It happens that there is an obvious way to construct a whole class of monoids whose elements are transformations; and "up to isomorphism," Cayley's theorem shows that this much is characteristic of monoids generally.

Let S be any set and suppose that $T = \{T_\alpha\}$ is a collection of mappings $T_\alpha: S \to S$ which contains the identity transformation 1_S and is *closed* with respect to composition of mappings, that is,

$$T_\alpha \in T \quad \text{and} \quad T_\beta \in T \quad \Rightarrow \quad T_\alpha \circ T_\beta \in T$$

Then T is called a *transformation monoid* $T = (T, \circ, 1_S)$ when we take composition as the monoid operation; that is to say,

$$(T_\alpha \circ T_\beta)(s) = T_\beta(T_\alpha(s))$$

is taken as the product $T_\alpha \circ T_\beta$ in the monoid T. In order to be sure that T is indeed a monoid, one should verify the axioms

(1) $T_\alpha \circ (T_\beta \circ T_\gamma) = (T_\alpha \circ T_\beta) \circ T_\gamma$
(2) $T_\alpha \circ 1_S = 1_S \circ T_\alpha = T_\alpha$

Recalling the definition for the equality of functions (Section 0.7), we compute for any $s \in S$:

$$\begin{aligned}
T_\alpha \circ (T_\beta \circ T_\gamma)(s) &= (T_\beta \circ T_\gamma)(T_\alpha(s)) \\
&= T_\gamma(T_\beta(T_\alpha(s))) \\
&= T_\gamma((T_\alpha \circ T_\beta)(s)) \\
&= (T_\alpha \circ T_\beta) \circ T_\gamma(s)
\end{aligned}$$

which verifies the associativity axiom (1). Axiom (2) follows from the properties of the identity transformation derived in Section 0.7.

EXAMPLE 3.12   We list in Figure 3.2 eight transformations of the set $S = \{0, 1, 2, 3\}$. We have $1_S = T_1$, and the collection

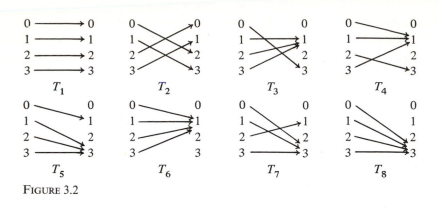

FIGURE 3.2

$$T = \{T_1, T_2, T_3, T_4, T_5, T_6, T_7, T_8\}$$

is closed with respect to functional composition, that is, $T_3 \circ T_2 = T_5$, etc. (Figure 3.3). It follows that T is a transformation monoid. Just as before, the entire multiplication table for $T = (T, \circ, 1_S)$ may be computed as shown in Table 3.1.

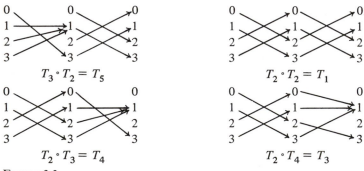

FIGURE 3.3

TABLE 3.1

$\circ$	T_1	T_2	T_3	T_4	T_5	T_6	T_7	T_8
T_1	T_1	T_2	T_3	T_4	T_5	T_6	T_7	T_8
T_2	T_2	T_1	T_4	T_3	T_7	T_6	T_5	T_8
T_3	T_3	T_5	T_6	T_6	T_8	T_6	T_8	T_8
T_4	T_4	T_7	T_6	T_6	T_8	T_6	T_8	T_8
T_5	T_5	T_3	T_6	T_6	T_8	T_6	T_8	T_8
T_6	T_6	T_8	T_6	T_6	T_8	T_6	T_8	T_8
T_7	T_7	T_4	T_6	T_6	T_8	T_6	T_8	T_8
T_8	T_8	T_6	T_6	T_6	T_8	T_6	T_8	T_8

Of course, these are but a small fraction of the total number of transformations of the set S. According to Rule 3 of Section 0.8, there are $4^4 = 256$ transformations in all. △

Apparently Rule 3 of Section 0.8 shows that there are n^n transformations of the set n^+. Surely they are closed with respect to composition and include the identity transformation. It follows that

$$T_n = \{f : f \text{ is a transformation of } n^+\}$$

is itself a transformation monoid. We call it the *full transformation monoid* on n symbols. It is clear that any transformation monoid on a finite set S will be isomorphic to a submonoid of some T_n.

EXAMPLE 3.13　The transformation monoid of Example 3.12 is (isomorphic to) a submonoid of T_4. T_4 is the full transformation monoid on four symbols. △

EXAMPLE 3.14　Suppose that somehow we came across a monoid $M = (M, \cdot, 1)$ with the multiplication table shown in Table 3.2.

TABLE 3.2

$\cdot$	1	2	3	4	5	6	7	8
1	1	2	3	4	5	6	7	8
2	2	1	4	3	7	6	5	8
3	3	5	6	6	8	6	8	8
4	4	7	6	6	8	6	8	8
5	5	3	6	6	8	6	8	8
6	6	8	6	6	8	6	8	8
7	7	4	6	6	8	6	8	8
8	8	6	6	6	8	6	8	8

Then, because of our prior experience with Example 3.12, we would be entirely justified in saying that M is isomorphic to a transformation monoid. The interesting thing, however, as Cayley's theorem asserts, is that this can be said of *any* monoid! △

Theorem 3.1　(Cayley)

Every monoid is isomorphic to a transformation monoid.

Proof　Let $M = (M, \cdot, e)$ be any monoid and $T_M = \{T_g : g \in M\}$ be the collection of transformations $T_g : M \to M$ given by

$$T_g(x) = x \cdot g \qquad (x \in M)$$

We intend to show that the mapping $g \to T_g$ is a monoid isomorphism $(M \simeq T_M)$.

Our mapping is obviously surjective, but also

$$T_g = T_h \Rightarrow T_g(x) = T_h(x) \qquad \text{(all } x \in M)$$
$$\Rightarrow x \cdot g = x \cdot h \qquad \text{(all } x \in M)$$
$$\Rightarrow \quad g = h \qquad \text{(taking } x = e)$$

showing that our mapping is injective as well. Next, we observe that the mapping transforms the identity element of M into the identity element of T_M:

$$e \to T_e = 1_M \qquad [T_e(x) = x \cdot e = x = 1_M(x)]$$

This is condition (i) in the definition of a monoid isomorphism. But the calculation

$$T_{gh}(x) = x(gh) = (xg)h = T_h(T_g(x)) = (T_g \circ T_h)(x)$$

shows that condition (ii) is also satisfied. □

EXAMPLE 3.15 Let us illustrate the construction in the proof of Cayley's theorem by considering the monoid $M = (M, \cdot, e)$ of Example 3.7. The transformations T_g are given by the columns of the multiplication table, as shown in Figure 3.4. Equivalently, they may be expressed as in Figure 3.5. Paying

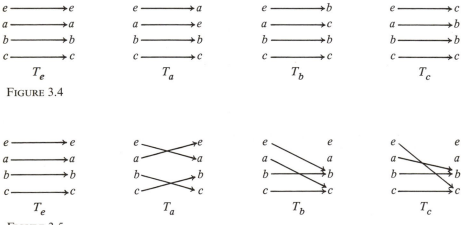

FIGURE 3.4

FIGURE 3.5

no further attention to the multiplication table, we can construct a new table for the transformation monoid

$$T_M = (\{T_e, T_a, T_b, T_c\}, \circ, T_e = 1_M)$$

by composing the various transformations. Thus we have $T_a \circ T_a = T_e$

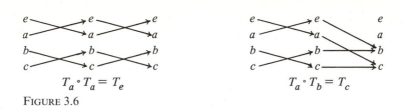

$$T_a \circ T_a = T_e \qquad\qquad T_a \circ T_b = T_c$$

FIGURE 3.6

(Figure 3.6), etc. Altogether, we are led to the following table:

$\circ$	T_e	T_a	T_b	T_c
T_e	T_e	T_a	T_b	T_c
T_a	T_a	T_e	T_c	T_b
T_b	T_b	T_b	T_b	T_b
T_c	T_c	T_c	T_c	T_c

From here on the isomorphism $g \rightarrow T_g$ of M and T_M is abundantly clear.   △

EXERCISES

1 Show that the set of transformations shown in Figure 3.7 forms a (transformation) monoid by deriving its multiplication table.

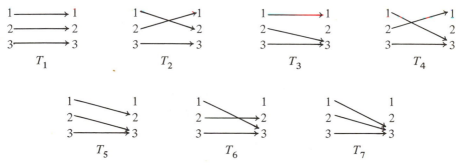

FIGURE 3.7

2 Find all the submonoids of the monoid in Exercise 1.

3 Illustrate the construction in the proof of Cayley's theorem by deriving a transformation monoid that is isomorphic to the monoid of
(a) Exercise 6(a) of Section 3.1   (b) Example 3.9   (c) Example 3.6

4 Find the smallest transformation monoid containing each of the mappings shown in Figure 3.8. In each case obtain the resulting multiplication table.

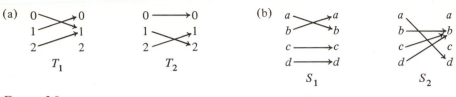

FIGURE 3.8

5 Derive the multiplication table for the full transformation monoid T_2.

3.3   HOMOMORPHISMS OF ALGEBRAS

One of the most important ideas in modern mathematics concerns structure-preserving mappings from one algebra into another; these mappings may or may not be bijective. Ordinarily we want the two algebras to be *similar*, that is, possess corresponding operations and propositions, each of the same "arity." Further, any special property (associativity, commutativity, etc.) enjoyed by one algebra should be shared by the other. For instance, the two algebras may be monoids, in which case each will have a binary and a nullary operation, and the properties of associativity and identity will be common to both algebras. In such a situation, it may be that one algebra *simulates* the other, exhibiting some of the same characteristics, though perhaps not reflecting every aspect in the behavior of the other. On the other hand, one algebra may have all of the capability of the other, and more, whereupon we say that the one algebra *realizes* the other. If we have somehow managed to impart a "machinelike" flavor to our terminology, so much the better. Still, the fact remains that each of these ideas finds its most natural expression in the homomorphism theory for algebras.

Computer arithmetics are inherently finite, so there cannot be an exact correspondence (isomorphism) between these arithmetics and the actual number systems they attempt to imitate. Nevertheless, for many purposes, the imitation or simulation is quite acceptable, because the correspondence of the actual numbers and their computer representations has been designed so as to preserve certain important algebraic features. The homomorphism theory will isolate these features and reveal the precise conditions under which one algebra can simulate or realize another.

For the time being it will be advantageous for us to operate in full generality. As we move on to particular instances, all of the general ideas will crystallize. We recall (Section 1.2) that our generalized algebras $A = (A, \mathscr{F} ; \mathscr{P})$ allow for n-ary operations ($n \geq 0$) and m-ary propositions ($m \geq 1$) on the set products of A. When we are given two similar algebras

$$A = (A, \mathscr{F}_A; \mathscr{P}_A) \qquad B = (B, \mathscr{F}_B; \mathscr{P}_B)$$

the "arities" of the various operations ($f_A \leftrightarrow f_B$) and propositions ($P_A \leftrightarrow P_B$)

will be in complete agreement, after an appropriate pairing. Then, without assuming either injectivity or surjectivity, we call a mapping $\varphi: A \to B$ a *homomorphism* if

$$\varphi f_A(a_1, a_2, \ldots, a_n) = f_B(\varphi a_1, \varphi a_2, \ldots, \varphi a_n)$$

for every operation f and

$$(a_1, a_2, \ldots, a_m) \in P_A \Leftrightarrow (\varphi a_1, \varphi a_2, \ldots, \varphi a_m) \in P_B$$

for every proposition P. (See Figure 3.9.) In words, the definition amounts to

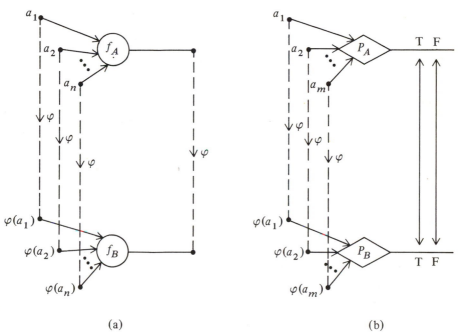

(a)   (b)

FIGURE 3.9

the following. If we perform a given n-ary operation $f : A^n \to A$ in the algebra A, then transform the value into B, we will obtain the same result as we would by first transforming all of the arguments into B, then performing the corresponding operation in the algebra B. And again, the same yes or no answer determined by the proposition $P \subseteq A^m$ should also be obtained in transforming the arguments for presentation to the corresponding proposition in B.

A homomorphism $\varphi: A \to B$ is said to be *injective* (respectively, *surjective*) if the underlying mapping is injective (surjective). Furthermore, a bijective (injective and surjective) homomorphism is called an *isomorphism*, and we write $A \simeq B$. The representations in Figure 3.10 may help in visualizing the effect of

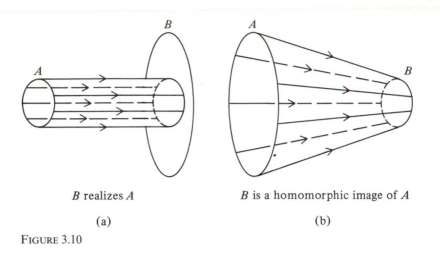

B realizes A   B is a homomorphic image of A

(a)   (b)

<small>FIGURE 3.10</small>

the injective and surjective homomorphisms. In connection with homomorphisms we introduce one further bit of terminology. We say that *B realizes A* when the homomorphism $\varphi: A \to B$ is injective; when the homomorphism is surjective, we say that *B is a homomorphic image of A*.

EXAMPLE 3.16   In the case of graphs $G = (V, E)$, there is no operation and just one (binary) proposition: the edge relation E. Therefore, our definition of an isomorphism for graphs (Section 2.6) requiring a bijective mapping $\varphi: G_1 \to G_2$ subject to the condition

$$v \, E_1 \, w \Leftrightarrow (\varphi v) \, E_2 \, (\varphi w)$$

is entirely consistent with the ideas presented here.   △

EXAMPLE 3.17   A monoid has no propositions, but it has two operations. There is the binary operation, multiplication if you wish, around which the algebra is designed, and there is the nullary operation that distinguishes the identity element. So if $M = (M, \cdot, e)$ and $M' = (M', \cdot, e)$ are monoids, it is appropriate to say that a mapping

$$\varphi: M \to M'$$

is a *monoid homomorphism* provided that

(i) $\varphi(e) = e$
(ii) $\varphi(x \cdot y) = \varphi(x) \cdot \varphi(y)$

This idea is consistent with the general idea of a homomorphism for similar algebras.   △

EXAMPLE 3.18   As discussed in Examples 3.3 and 3.4,

$$Z = (Z, +, 0) \quad \text{and} \quad B = (B, \oplus, 0)$$

are monoids, and we can show that the mapping $\varphi: Z \to B$, namely

$$\varphi(n) = \begin{cases} 0 & \text{if } n \text{ is even} \\ 1 & \text{if } n \text{ is odd} \end{cases}$$

is a surjective monoid homomorphism. Obviously the mapping is surjective, because there are both even and odd integers. And since 0 is even, we are assured of condition (i). The condition (ii)

$$\varphi(n + m) = \varphi(n) \oplus \varphi(m)$$

can be verified by a simple exhaustive consideration of the four cases:

n	m
even	even
even	odd
odd	even
odd	odd

For instance, if n and m are both even, then $n + m$ is also even. Therefore,

$$\varphi(n + m) = 0 = 0 \oplus 0 = \varphi(n) \oplus \varphi(m)$$

as required.

Evidently we may say that B is a homomorphic image of Z, but notice how "weak" this particular simulation is: B only implements the arithmetic rules

$$\text{even} + \text{even} = \text{even}$$
$$\text{even} + \text{odd} = \text{odd} + \text{even} = \text{odd}$$
$$\text{odd} + \text{odd} = \text{even}$$

△

EXAMPLE 3.19   Consider the monoid $M = (M, \cdot, e)$ of Example 3.7 whose multiplication table is repeated below, and let $M' = (M', \cdot, E)$ have the multiplication table shown at the right:

$\cdot$	e	a	b	c
e	e	a	b	c
a	a	e	c	b
b	b	b	b	b
c	c	c	c	c

$\cdot$	E	B
E	E	B
B	B	B

We see that the mapping

$$e \xrightarrow{\varphi} E$$
$$a \xrightarrow{\varphi} E$$
$$b \xrightarrow{\varphi} B$$
$$c \xrightarrow{\varphi} B$$

is a (surjective) monoid homomorphism $\varphi \colon M \to M'$, for we have

$$\varphi(e) = E$$

$$\varphi(ab) = \varphi(c) = B = EB = \varphi(a)\varphi(b)$$

$$\varphi(ba) = \varphi(b) = B = BE = \varphi(b)\varphi(a), \quad \text{etc.}$$

Note that we often omit the multiplication signs in a product, as in ordinary algebra.

For purposes of later discussions, we observe that the kernel (Section 0.7) of the mapping φ is represented by the partition $\{\overline{e, a} \; ; \overline{b, c}\}$ of the set M; and we notice that in multiplying elements in M, the equivalence class of the product depends only on the equivalence classes from which the two factors are chosen. Such equivalence relations will be called monoid *congruences*. They play an important role in characterizing the homomorphic images of a given monoid.   △

EXAMPLE 3.20   The four entries in the upper left-hand corner of the multiplication table for the monoid M of Example 3.19:

$\cdot$	e	a
e	e	a
a	a	e

show that M realizes the monoid $B = (B, \oplus, 0)$. The mapping

$$0 \overset{\varphi}{\nearrow} \begin{matrix} e \\ a \end{matrix}$$
$$1 \nearrow \begin{matrix} b \\ c \end{matrix}$$

is an injective monoid homomorphism $\varphi \colon B \to M$.   △

EXAMPLE 3.21   The discussions of Examples 3.13 and 3.14 show that T_4 (the full transformation monoid on four symbols) realizes the monoid of Example 3.14.   △

EXAMPLE 3.22 In view of the identities

$$T_1 = T_2 \circ T_2$$
$$T_4 = T_2 \circ T_3$$
$$T_5 = T_3 \circ T_2$$
$$T_6 = T_3 \circ T_3$$
$$T_7 = T_2 \circ T_5 = T_2 \circ T_3 \circ T_2$$
$$T_8 = T_3 \circ T_5 = T_3 \circ T_3 \circ T_2$$

in the transformation monoid T of Example 3.12, we are lead to say that T is *generated by* the subset $X = \{T_2, T_3\}$. Every element in T has a representation as an algebraic expression (in this case, a product or composition) in T_2 and T_3. In fact, this is the way that the monoid T was constructed. Beginning with the two transformations T_2 and T_3, a succession of transformations was composed until closure was obtained. Each time a new transformation was obtained, it was added to the multiplication table. The procedure eventually came to an end because there are only finitely many (n^n) transformations of an n-set.

A monoid homomorphism $\varphi \colon T \to M$ is completely determined by the images $\varphi(T_2)$ and $\varphi(T_3)$, just because T_2, T_3 generate T. Thus

$$\varphi(T_1) = \varphi(T_2 \circ T_2) = \varphi(T_2)\varphi(T_2)$$

$$\varphi(T_4) = \varphi(T_2 \circ T_3) = \varphi(T_2)\varphi(T_3)$$

$$\varphi(T_5) = \varphi(T_3 \circ T_2) = \varphi(T_3)\varphi(T_2), \quad \text{etc.}$$

On the other hand, if we assign images

$$T_2 \to \varphi(T_2)$$

$$T_3 \to \varphi(T_3)$$

in a given monoid M, then there may or may not be an extension to a homomorphism $\varphi \colon T \to M$. Suppose M is the monoid of Example 3.19, and we try to assign

$$T_2 \overset{\varphi}{\to} b$$

Then we should have

$$b = b \cdot b = \varphi(T_2)\varphi(T_2) = \varphi(T_2 \circ T_2) = \varphi(T_1) = e$$

which is a contradiction. The identity $T_2 \circ T_2 = T_1$ prevents any extension to a homomorphism, however T_3 may be assigned. △

In a *free* algebra, one that is free of any identities except those implied by the axioms, any mapping of the generators into a similar algebra will admit a unique extension to a homomorphism. In the light of the preceding example, we can easily anticipate the precise formulation of this important principle. We say that the algebra A is *freely generated* by the elements of the finite subset (the set of *generators*)

$$X = \{x_1, x_2, \ldots, x_n\} \subseteq A$$

provided that

(a) every $x \in A$ has a representation as an algebraic expression in the elements $x_1, x_2, \ldots, x_n$;
(b) every mapping $\varphi: X \to B$ of the generators into a similar algebra B has an extension to a homomorphism $\varphi: A \to B$.

By an *extension*, we mean that the images $\varphi(x_i)$ in the homomorphism are exactly those assigned in the mapping of X into B.

EXAMPLE 3.23   The natural numbers $N = (N, +, 0)$ are an example of a free monoid—free on one generator. Taking $X = \{1\}$, we can verify the two conditions:

(a) Every $n \in N$ has a representation as an algebraic expression $n = 1 + 1 + \cdots + 1$ (n times). Note that we consider 0 to be an empty expression of this type.
(b) If M is any monoid and we assign $\varphi(1)$ in M, then the extension $\varphi: N \to M$ is given by

$$\varphi(n) = \varphi(1) + \varphi(1) + \cdots + \varphi(1) \qquad (n \text{ times})$$
$$= n \cdot \varphi(1)$$

assuming that M is denoted additively, $M = (M, +, 0)$, for we can then verify the conditions for a monoid homomorphism (Example 3.17):

(i) $\varphi(0) = 0 \cdot \varphi(1) = 0$
(ii) $\varphi(n + m) = (n + m)\varphi(1) = n\varphi(1) + m\varphi(1)$
$$= \varphi(n) + \varphi(m)$$

It is abundantly clear that our homomorphism is an extension of the generator assignment. $'\triangle$

EXERCISES

1 Show that every finite monoid is realized by a full transformation monoid.

2 Exhibit a surjective monoid homomorphism $\varphi: M \to M'$, where M is the table of Exercise 6(a) of Section 3.1 and M' is given by the table

·	e	a
e	e	a
a	a	e

3 Exhibit an injective monoid homomorphism $\varphi: M' \to M$ for the monoids of Exercise 2.

4 Find a minimal set of generators for the transformation monoid of Exercise 1 of Section 3.2. Show that the monoid is not freely generated by this generating set.

5 Consider the transformations of Exercise 4 of Section 3.2 as generators of transformation monoids. Show that the mappings

(a) $T_1 \to c$
$\quad\ T_2 \to b$ into the monoid M of Example 3.19

(b) $S_1 \to 1$
$\quad\ S_2 \to 3$ into the monoid of Exercise 6(a) of Section 3.1

do not extend to monoid homomorphisms.

6 Given a pair of monoids M, M', define the *direct product*

$$M \times M' = \{(x, x'): x \in M \quad \text{and} \quad x' \in M'\}$$

to be the monoid with the operation

$$(x, x') \cdot (y, y') = (x \cdot y, x' \cdot y')$$

Show that

(a) $M \times M'$ is a monoid
(b) M and M' are realized by $M \times M'$
(c) Each is a homomorphic image of $M \times M'$

7 Construct the direct product (Exercise 6) of M, the monoid of Exercise 8, and M' as given in Exercise 2. Exhibit the homomorphisms that show that M and M' are realized by $M \times M'$ and that each is a homomorphic image of $M \times M'$.

8 Let M be the monoid with the multiplication table

$\cdot$	0	1
0	0	0
1	0	1

Show that there is a surjective homomorphism $\varphi: T_2 \to M$. Then generalize to T_n.

9 Show that the real numbers with addition is a homomorphic image of the monoid of complex numbers with complex addition. Explain the connection with Exercise 6.

3.4 FREE MONOIDS

In Section 0.4 we introduced the notion of an alphabet

$$\Sigma = \{\sigma_1, \sigma_2, \ldots, \sigma_n\}$$

of symbols for use in composing "words":

$$x = \sigma_{i_1} \sigma_{i_2} \cdots \sigma_{i_k}$$

(including the "null word" ϵ, a word of length zero). The notation Σ^* was used for the collection

$$\Sigma^* = \{x: x \text{ is a word on the alphabet } \Sigma\}$$

which is the set of all words generable from the given alphabet. We are now going to introduce an interesting binary operation, evidently quite elementary, relative to which this collection becomes a monoid. The operation is usually called *concatenation* (or *juxtaposition*) of words. With x given as above, and $y = \sigma_{j_1}\sigma_{j_2} \cdots \sigma_{j_l}$, we merely string the symbols of the second word after those of the first in forming the product:

$$x \cdot y = \sigma_{i_1}\sigma_{i_2} \cdots \sigma_{i_k}\sigma_{j_1}\sigma_{j_2} \cdots \sigma_{j_l}$$

If it is understood that

$$x \cdot \epsilon = \epsilon \cdot x = x$$

for every word x, then $\Sigma^* = (\Sigma^*, \cdot, \epsilon)$ apparently becomes a monoid because the operation of concatenation is obviously associative. In view of our next lemma, we call Σ^* the *free monoid* generated by the alphabet Σ.

EXAMPLE 3.24   If $\Sigma = \{a, b, c, d\}$, then we may illustrate the concatenation operation in Σ^* with the following examples:

$$abba \cdot dabba = abbadabba$$
$$da \cdot da = dada$$
$$cab \cdot \epsilon = cab$$

If Σ is the entire lowercase English alphabet of Example 0.25, then we have the further examples:

$$\text{single} \cdot \text{ton} = \text{singleton}$$
$$\text{water} \cdot \text{gate} = \text{watergate}$$
$$\epsilon \cdot \text{water} = \text{water} \qquad \triangle$$

EXAMPLE 3.25   If $\Sigma = \{0, 1\}$, then

$$\Sigma^* = \{\epsilon, 0, 1, 00, 01, 10, 11, 000, 001, \ldots\}$$

and as concatenations, we cite

$$01 \cdot 101 = 01101$$
$$101 \cdot 01 = 10101$$
$$01 \cdot \epsilon = 01$$

It is clear that unless Σ is a one-letter alphabet, the monoid Σ^* is noncommutative; but the monoids are always infinite, even in the case of a one-letter alphabet, say $\Sigma = \{1\}$. Examining a few concatenations in the alphabet just mentioned

$$11 \cdot 1111 = 111111$$
$$111 \cdot 111 = 111111$$
$$1111 \cdot 1 = 11111$$

we are tempted to say that the monoid is for adding natural numbers with a primitive style of notation. In fact, the reader is invited to conclude that $N = (N, +, 0)$ is isomorphic to Σ^* whenever Σ is a one-letter alphabet. In view of the fact that we just verified (Example 3.23) that N is a free monoid, the preceding observation rests well with the following lemma. △

Lemma

If M is any monoid and Σ is an alphabet, then every mapping $\varphi: \Sigma \to M$ has a (unique) extension to a monoid homomorphism $\varphi: \Sigma^* \to M$.

Proof Suppose that we are given an arbitrary mapping $\varphi: \Sigma \to M$ relative to an arbitrary monoid $M = (M, \cdot, e)$. For each word $x = \sigma_{i_1}\sigma_{i_2} \cdots \sigma_{i_k}$ in Σ^* we define [with the understanding that $\varphi(\epsilon) = e$]

$$\varphi(x) = \varphi(\sigma_{i_1}) \cdot \varphi(\sigma_{i_2}) \cdots \varphi(\sigma_{i_k})$$

where the multiplication of course takes place in M. Since

$$\begin{aligned}
\varphi(x \cdot y) &= \varphi(\sigma_{i_1}\sigma_{i_2} \cdots \sigma_{i_k}\sigma_{j_1}\sigma_{j_2} \cdots \sigma_{j_l}) \\
&= \varphi(\sigma_{i_1}) \cdot \varphi(\sigma_{i_2}) \cdots \varphi(\sigma_{i_k}) \cdot \varphi(\sigma_{j_1}) \cdot \varphi(\sigma_{j_2}) \cdots \varphi(\sigma_{j_l}) \\
&= \varphi(x) \cdot \varphi(y)
\end{aligned}$$

we have succeeded in extending φ to a monoid homomorphism $\varphi: \Sigma^* \to M$. We leave the reader the task of showing that such an extension is unique, and we note that φ is not necessarily surjective. □

Theorem 3.2

Every finitely generated monoid is a homomorphic image of a free monoid.

Proof Suppose that M is generated by the finite subset

$$\{\mu_1, \mu_2, \ldots, \mu_n\} \subseteq M$$

Then we choose any alphabet Σ with n symbols and assign

$$\varphi(\sigma_i) = \mu_i \qquad (i = 1, 2, \ldots, n)$$

According to the lemma, φ extends to a monoid homomorphism $\varphi: \Sigma^* \to M$. But this homomorphism is surjective, because it is onto a set of generators for M. Explain! □

Corollary

Every finite monoid is a homomorphic image of a free monoid.   □

We now have two representation theorems for finite monoids. Applying Cayley's theorem, we see that every finite monoid is realized by (thus isomorphic to a submonoid of) some full transformation monoid. On the other hand, the corollary says that each finite monoid is a homomorphic image of some free monoid. These observations impart a certain degree of universality to the full transformation monoids and the free monoids, which is the reason that they occupy a position of such prominence in the theory of monoids.

EXAMPLE 3.26   The monoid of Example 3.7 is generated by the pair a, b because $a \cdot b = c$ and $a \cdot a = e$. So we choose any alphabet with two symbols, say $\Sigma = \{\sigma, \tau\}$, and according to the proof of Theorem 3.2, we are to assign

$$\varphi(\sigma) = a \qquad \varphi(\tau) = b$$

or vice versa. Then the proof of the lemma shows how we can extend this assignment to a monoid homomorphism $\varphi \colon \Sigma^* \to M$:

$$\varphi(\sigma\sigma) = \varphi(\sigma) \cdot \varphi(\sigma) = a \cdot a = e$$
$$\varphi(\sigma\tau) = \varphi(\sigma) \cdot \varphi(\tau) = a \cdot b = c$$
$$\varphi(\tau\sigma) = \varphi(\tau) \cdot \varphi(\sigma) = b \cdot a = b$$
$$\varphi(\tau\tau) = \varphi(\tau) \cdot \varphi(\tau) = b \cdot b = b$$

$$\varphi(\sigma\sigma\sigma) = \varphi(\sigma) \cdot \varphi(\sigma) \cdot \varphi(\sigma) = a \cdot a \cdot a = a$$
$$\varphi(\sigma\sigma\tau) = \varphi(\sigma) \cdot \varphi(\sigma) \cdot \varphi(\tau) = a \cdot a \cdot b = b$$
$$\varphi(\sigma\tau\sigma) = \varphi(\sigma) \cdot \varphi(\tau) \cdot \varphi(\sigma) = a \cdot b \cdot a = c$$
$$\varphi(\sigma\tau\tau) = \varphi(\sigma) \cdot \varphi(\tau) \cdot \varphi(\tau) = a \cdot b \cdot b = c, \quad \text{etc.}$$

The homomorphism is surjective, as can easily be seen, because a, b generate M.   △

EXERCISES

1 Prove that the extension of φ to a monoid homomorphism is unique (in the lemma preceding Theorem 3.2). [*Hint*: Suppose that $\varphi_1, \varphi_2 \colon \Sigma^* \to M$ each have the extension property

$$\varphi_1(\sigma) = \varphi(\sigma) \qquad \varphi_2(\sigma) = \varphi(\sigma)$$

for all $\sigma \in \Sigma$. Show that $\varphi_1 = \varphi_2$.]

2 Indicate explicitly how the monoid of the following are homomorphic images of free monoids:
   (a) Exercise 6(a) of Section 3.1   (b) Exercise 1 of Section 3.2
   (c) Exercise 4(a) of Section 3.2   (d) Example 3.9

3 Show that every free monoid Σ^* is a *cancellation monoid*, in that

$$xy = xz \Rightarrow y = z$$
$$yx = zx \Rightarrow y = z$$

for all x, y, z in Σ^*.

4 Show that the monoid B^e consisting of strings of zeros and ones each having an even number of ones is finitely generated (with the concatenation operation). Indicate explicitly how B^e is a homomorphic image of a free monoid.

*5 Prove or disprove: every submonoid of a free monoid is free.

*6 Prove or disprove: every submonoid of $N = (N, +, 0)$ is isomorphic to N itself.

7 Describe the Morse code as an (injective) homomorphism of free monoids.

3.5   LABELED TREES

We can represent the elements of any free monoid Σ^* as the set of vertices in an (infinite) digraph $G = (\Sigma^*, E)$ by means of the edge relation

$$x \, E \, y \quad \text{iff} \quad y = x\sigma \quad (\sigma \in \Sigma),$$

Figure 3.11 shows a portion of such a digraph, actually a directed tree, for the alphabet $\Sigma = \{a, b, c\}$. Similarly, an infinite tree for B^* is represented in

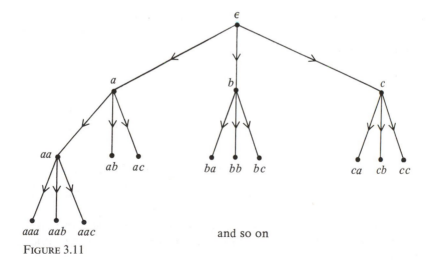

FIGURE 3.11

Figure 3.12; here B is the binary alphabet $B = \{0, 1\}$. As indicated in the figure, we could label the edges in such a way as to suggest an orientation of branches from left to right, but this will hardly be necessary if we simply agree to observe

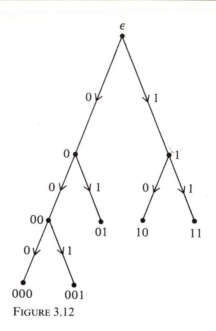

FIGURE 3.12

this ordering in all our pictorial representations of B^*. Similarly, the arrows on the edges can be omitted if we consistently direct all the edges downward, beginning at the *root* of the tree, that is the vertex corresponding to the null word.

In the preceding chapter we encountered several important applications of the concept of a labeled digraph. Most common among all the applications are those in which the underlying digraph is a *binary tree* $T = (V, E)$, which means that V is a finite subset of B^* and that

(i) $x\sigma \in V \Rightarrow x \in V$ $(\sigma \in B)$
(ii) $x E y$ iff $y = x\sigma$ for some $\sigma \in B$

Condition (ii) simply repeats the edge relation mentioned earlier. The first condition assures us that every vertex is joined by a path to the root. Now, at the same time we let the subset

$$L = L(T) = \{x \in V : x\sigma \notin V \text{ for all } \sigma \in B\} \subseteq V$$

denote the set of all terminal nodes or *leaves* of the tree T. Beginning with our use of the word "tree" in describing a connected acyclic graph, a wealth of botanical terms has found its way into graph theory. We have tried not to be excessive in this regard, but it seems quite appropriate that a few concessions be made.

EXAMPLE 3.27   Figure 3.13 shows a binary tree. To illustrate condition (i) in the definition of binary tree, we note that

$$0101 \in V \Rightarrow 010 \in V \Rightarrow 01 \in V \Rightarrow 0 \in V \Rightarrow \epsilon \in V$$

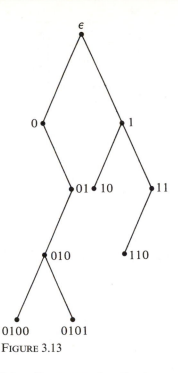

FIGURE 3.13

We might say that condition (i) ensures that the tree contain with any word x, all of its "prefixes." This particular tree T has the following set of leaves:

$$L = L(T) = \{10, 110, 0100, 0101\} \qquad \triangle$$

We will find several applications for *labeled* binary trees, both in this chapter and in Chapter 6. In most cases the binary trees $T = (V, E)$ will be used in connection with certain algorithmic processes, and the labeling of the vertices $V \subseteq B^*$ will take place only during the execution of the algorithm. Thus it is more appropriate to say that T, as a data type, has the array structure

$$T: \textbf{array } B^* \textbf{ of } A$$

where A is the algebra used in labeling the vertices. Since B^* is an infinite set, however, we are certainly not going to label all of these vertices in the course of executing the algorithm. Only a finite subset $V \subseteq B^*$ will actually become labeled. Because of the order in which the vertices are labeled, from the root downward, we are assured that this subset V satisfies condition (i), that is, we will have obtained a binary labeled tree by the time the process terminates.

As our first example in the use of binary labeled trees, we return to the problem of determining the independence number β and the clique number β' of a graph $G = (V, E)$. You will recall that β and β' were two of the important isomorphism invariants introduced in Section 2.6. Interestingly enough, the procedure (Algorithm IX) for determining these invariants first arose in connection with

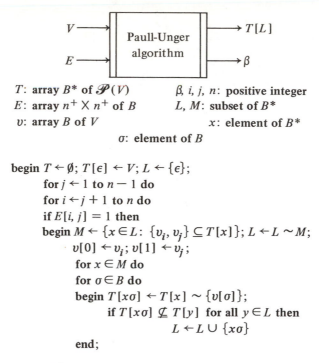

$$T: \textbf{array } B^* \textbf{ of } \mathscr{P}(V) \qquad \beta, i, j, n: \textbf{ positive integer}$$
$$E: \textbf{array } n^+ \times n^+ \textbf{ of } B \qquad L, M: \textbf{ subset of } B^*$$
$$v: \textbf{array } B \textbf{ of } V \qquad\qquad x: \textbf{ element of } B^*$$
$$\sigma: \textbf{ element of } B$$

begin $T \leftarrow \emptyset;\ T[\epsilon] \leftarrow V;\ L \leftarrow \{\epsilon\};$
       for $j \leftarrow 1$ **to** $n-1$ **do**
       for $i \leftarrow j+1$ **to** n **do**
       if $E[i, j] = 1$ **then**
       begin $M \leftarrow \{x \in L: \{v_i, v_j\} \subseteq T[x]\};\ L \leftarrow L \sim M;$
          $v[0] \leftarrow v_i;\ v[1] \leftarrow v_j;$
          for $x \in M$ **do**
          for $\sigma \in B$ **do**
          begin $T[x\sigma] \leftarrow T[x] \sim \{v[\sigma]\};$
             if $T[x\sigma] \not\subseteq T[y]$ **for all** $y \in L$ **then**
                   $L \leftarrow L \cup \{x\sigma\}$
        end;

       end;
       $\beta \leftarrow \max_{x \in L} \{|T[x]|\}$

end
ALGORITHM IX

a problem in the theory of finite state machines. The solution was given by M. C. Paull and S. H. Unger[*], and so we may attribute this particular algorithm to their efforts. As it stands, the algorithm determines the independence number β for any graph $G = (V, E)$, and at the same time lists all of the maximal independent sets—those that are not contained in a larger independent set. But a simple modification,

$$\textbf{if } E[i, j] = 0 \textbf{ then}$$

rather than $E[i, j] = 1$, will produce instead the clique number β' and a list of all of the maximal cliques.

As with most formal descriptions of algorithms, the procedure is nowhere near as complicated as it might appear. The binary tree becomes labeled with subsets of V, the set of vertices of the graph $G = (V, E)$. Initially the tree T

---

[*] M. C. Paull and S. H. Unger, "Minimizing the Number of States in Incompletely Specified Sequential Switching Functions," *IRE Trans. Elec. Computers*, **EC**–8, 356–367 (1959).

is unlabeled; or, if you prefer, you may consider all of its nodes to be labeled initially with the empty subset, as we have indicated in our description of the algorithm. Then the root of the tree is labeled with the entire set V. The subset $L \subseteq B^*$ keeps track of the leaves of the trees as the labeling proceeds. The two **for** statements then search the lower-triangular matrix representation of the graph, looking for pairs of vertices with $v_i E v_j$ ($E[i, j] = 1$). Then M marks all the leaves of the tree whose labels contain this pair as a subset. Subsequently, M is deleted from L, because a subset which contains this pair cannot be independent. Suppose that one of these marked nodes of the tree is $x \in B^*$. Then the most interior "begin ... end" sequence labels the nodes below and to the left and right from x with the maximal subsets of $T[x]$ *not containing both* of the vertices v_i and v_j. (See Figure 3.14.) At the same time we check to see whether

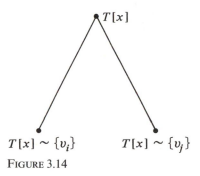

$$T[x] \sim \{v_i\} \qquad T[x] \sim \{v_j\}$$

FIGURE 3.14

the labels of these new leaves are not themselves contained in other labeled leaves of the tree. After all, we are interested only in the maximal independent sets. Depending on the outcome of this inquiry, L may be modified accordingly. Then we are ready to examine a new entry $E[i, j] = 1$. All in all, one should think of the binary tree as consisting of those nodes which at one time or another have been elements of L, the leaves of the developing tree.

EXAMPLE 3.28 In order to illustrate the various steps in the Paull-Unger algorithm, let us consider the graph of Figure 3.15. Initially only the root of our eventual tree is labeled, as shown in Figure 3.16(a). At this point, $L = \{\epsilon\}$. Then we shall begin to examine the entries in the lower-triangular matrix:

b	1					
c	1	0				
d	0	1	0			
e	0	1	1	1		
f	0	1	0	0	0	
g	0	0	0	0	1	1
	a	b	c	d	e	f

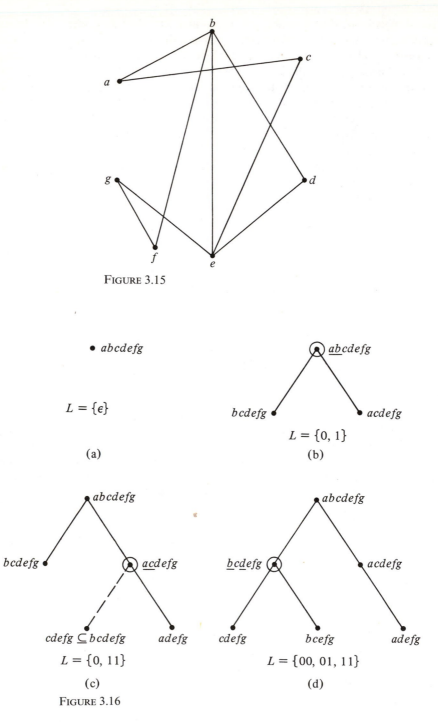

FIGURE 3.15

FIGURE 3.16

looking for 1's. The arrangement of the **for** statements is such that we scan the columns from left to right, top to bottom.

The first 1 we encounter is that which shows that $a \ E \ b$. We look at the leaves (leaf) of the tree to see if any are labeled with the subsets containing $\{a, b\}$. In Figure 3.16(b) we perform the required splitting operation referred to in Figure 3.14. Now our set of leaves is $L = \{0, 1\}$. The next 1 we encounter is that which corresponds to the relation $a \ E \ c$. There is one leaf $(x = 1)$ for which $\{a, c\} \subseteq T[x] = \{a, c, d, e, f, g\}$. So $M = \{1\}$, and this node is deleted from L. The algorithm then sets

$$T[10] = T[x0] \leftarrow T[x] \sim \{a\} = \{c, d, e, f, g\}$$

$$T[11] = T[x1] \leftarrow T[x] \sim \{c\} = \{a, d, e, f, g\}$$

In the next line it checks to see whether the new labels would be contained in those of other leaves of the tree. Since

$$T[10] \subseteq T[0] \quad \text{with} \quad 0 \in L$$

the node 10 does not become a new leaf of the tree. [See Figure 3.16(c).] Now we have $L = \{0, 11\}$. The result of the splitting due to the next 1-entry $(b \ E \ d)$ is shown in Figure 3.16(d), at which point $L = \{00, 01, 11\}$.

In hand computation the entire tree can be generated as one figure. In this case the computation would be summarized as in Figure 3.17. The circled nodes

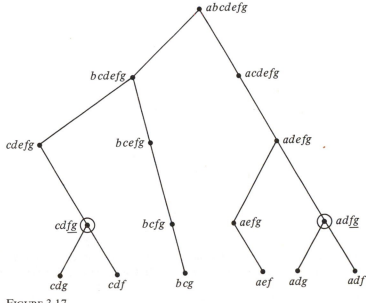

FIGURE 3.17

in the figure correspond to the marked leaves (elements of M) at the time of recognition of the last 1-entry ($f \, E \, g$) found in the matrix. Therefore, at the conclusion we have

$$L = \{0010, 0011, 0111, 1101, 1110, 1111\}$$

and finally

$$\beta = \max_{x \in L} \{|T[x]|\} = 3$$

In fact, the labels of all the leaves found at the conclusion of executing the algorithm will correspond to the maximal independent subsets. Here they are the subsets

$$\{c, d, g\} \quad \{c, d, f\} \quad \{b, c, g\} \quad \{a, e, f\} \quad \{a, d, g\} \quad \{a, d, f\}$$

respectively.   △

EXERCISES

1 Use Algorithm IX to compute the independence number for the following graphs G:
   (a) Exercise 1(b) of Section 2.1   (b) Exercise 3(a) of Section 2.1
   (c) Exercise 5(h) of Section 2.3   (d) Exercise 1 of Section 2.5

2 Use the appropriate variation of Algorithm IX to compute the clique number of the graphs in Exercise 1.

3 List the set of leaves $L = L(T)$ for the binary trees T shown in Figure 3.18:

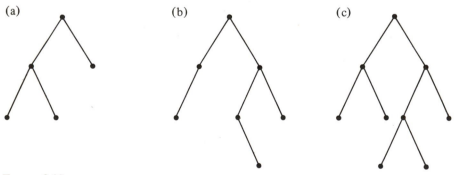

(a)   (b)   (c)

FIGURE 3.18

   (d) List the set of leaves $L = L(T)$ for the tree found at the conclusion of Exercise 1(a).
   (e) List the set of leaves $L = L(T)$ for the tree found at the conclusion of Exercise 2(a).

4 Prove that a binary tree $T = (V, E)$ is a tree.

5 Prove that

$$|V| < 2^a$$

where $a = 1 + \max_{x \in V} |x|$, in any binary tree $T = (V, E)$.

3.6 SEARCHING AND SORTING

Suppose that we want to search a collection of documents or *records* to determine whether a given record is among them. You will immediately recognize this as a fundamental problem in the field of information retrieval. The collection of records is called a *file*. The individual records may contain information on company employees; they may be governmental documents or the listings of a library catalog. Certainly we could select a record more or less at random, check whether it satisfies the search criteria, and continue the process until the desired record is found. That would not be a very efficient way of approaching the problem, however. We must "structure" the information in the file in order to obtain improved search techniques.

The most common file structures associate a *key* with each record. If we think of the individual record as an element in a product set (Section 0.3), for instance a name, its social security number, and job classification in a file of employee records, then we usually use one of these *fields* as the key to the record. Thus the file can be appropriately classified, alphabetically or numerically, according to the keys of the various records. We are all familiar with this technique as it is used in telephone directories, dictionaries, and library classification schemes. For the purpose of our subsequent analysis, it does not matter whether or not these keys are numerical. What does matter is that they be totally ordered. In fact, there is no loss in generality in assuming that the keys are words in a free monoid Σ^*, because we know (Example 3.23) that the natural numbers are thereby included. We can simply use the alphabetical order of the words in Σ^* to sort the records according to increasing order of their keys.

It is not our purpose to examine and compare all of the many sorting techniques in use. That would be a formidable undertaking and we shall instead refer the reader to Reference 3.3. Here we merely wish to contrast certain characteristics of a conventional linear storage technique with those deriving from the use of an underlying tree structure. Suppose that we are reading words from a text and we wish to generate a list of the distinct words occurring in the text. We could make use of a simple linear array, adjoining every new word as it is read, unless it already appears in the array. But there is the hitch. We have to compare the new word with everthing in the array in order to know whether we would be duplicating an entry. Again we can try to cut down the number of comparisons by keeping our array in alphabetical order. Even then, however, to add a new word, we must insert it so that all those words below it must be "pushed down." We will find that in such sorting situations, a tree storage structure can be quite an attractive alternative. Thus we have a very useful application of the preceding material.

Among all the sorting methods, perhaps none is easier to understand than the so-called *selection sort*. If

$$y = (y_1, y_2, \ldots, y_n)$$

is a sequence of keys representing the file to be sorted, we may temporarily think of the y_i as integers. Then, as in Example 1.55, we can find the smallest of the integers $y_1, y_2, \ldots, y_n$ and interchange $\min_i y_i$ with y_1. Next, we can select the smallest of $y_2, \ldots, y_n$ and interchange it with y_2, etc. The complete procedure is described in Algorithm X. The algorithm is so phrased that the keys y_i appear

$$i, j, k, n: \textbf{positive integer}$$
$$y: \textbf{array } n^+ \textbf{ of } \Sigma*$$
$$\text{temp: } \textbf{element of } \Sigma*$$

for $k \leftarrow 1$ **to** n **do**
  begin $i \leftarrow k$;
       for $j \leftarrow k + 1$ **to** n **do**
       if $y[j] < y[i]$ **then** $i \leftarrow j$;
       temp $\leftarrow y[k]$;
       $y[k] \leftarrow y[i]$;
       $y[i] \leftarrow$ temp
  end
ALGORITHM X

to be in an arbitrary free monoid $\Sigma*$. It is quite common to think of the number of comparisons (as occur here in the **if** statement) as an estimate of the "running time" of a sorting technique. The selection sort evidently requires

$$(n - 1) + (n - 2) + \cdots + 2 + 1 = \frac{n(n - 1)}{2} \approx \frac{n^2}{2}$$

comparisons. This figure, derived from Example 1.2, is typical of many sorting techniques. Yet we will see that there is room for improvement.

EXAMPLE 3.29   Suppose that we are given the sequence of integers

$$y = (20, 36, 6, 37, 50, 5, 31, 12, 8, 16, 42, 40, 11, 3, 27, 24)$$

and we desire that they be sorted in increasing order. A trace of the performance of Algorithm X begins as shown in Table 3.3.

TABLE 3.3

n	i	j	k	temp	y
16	1		1		
		2			
	3	3			
		4			
		5			
	6	6			
		7			
		$\vdots$			
	14	14			
		15			
		16		20	$y[1] = 3$ $y[14] = 20$
	2		2		
	3	3			
		4			
		5			
	6	6			
		7			
		$\vdots$			
		16		36	$y[2] = 5$ $y[6] = 36$
				$\cdots$	

At the conclusion of the execution of the algorithm, the sequence will read

$$y = (3, 5, 6, 8, 11, 12, 16, 20, 24, 27, 31, 36, 37, 40, 42, 50)$$

as indeed, it should.   △

Instead of the sorting techniques that use a linear storage arrage, a binary labeled tree can be employed. Suppose again that

$$y = (y_1, y_2, \ldots, y_n)$$

is the sequence of keys to be sorted. Alternatively, they may be the distinct (non-empty) words of a text being read, and without loss of generality, we may suppose that the y_i are words $y_i \neq \epsilon$ in an alphabet Σ. The tree sorting procedure is shown in Algorithm XI. At each stage of the process, all the words alphabetically preceding the node labeled $T[x]$ are on the left branch from x, and those following $T[x]$ are on the right. With each new word considered, we access the tree at $x = \epsilon$ and make alphabetical comparisons all the way down

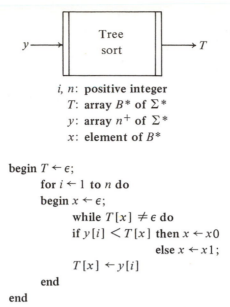

i, n: **positive integer**
T: **array** B^* **of** Σ^*
y: **array** n^+ **of** Σ^*
x: **element of** B^*

begin $T \leftarrow \epsilon$;
    for $i \leftarrow 1$ **to** n **do**
    begin $x \leftarrow \epsilon$;
        while $T[x] \neq \epsilon$ **do**
        if $y[i] < T[x]$ **then** $x \leftarrow x0$
                        else $x \leftarrow x1$;
      $T[x] \leftarrow y[i]$
    end
end
Algorithm XI

the tree, thus determining where to place the new word. Initially the tree is unlabeled or, as we indicated earlier, each of its nodes may be labeled with the null word. The latter alternative is of some advantage in formulating the procedure, especially in connection with the **while** statement of Algorithm XI.

We must be a bit careful in reading this algorithm because there are two alphabets (B and Σ) appearing in it. The words in B^* indicate positions in the tree, and those in Σ^* are used as labels, representing either the keys we are sorting or the words being read from a text. An example will clarify the roles of these two alphabets.

EXAMPLE 3.30   Consider the following famous text:

Four score and seven years ago, our fathers brought forth upon this continent a new nation, . . .

Here we have the initial sequence

$$y_1 = \text{four}, \quad y_2 = \text{score}, \quad \ldots, \quad y_{16} = \text{nation}$$

as input to Algorithm XI. A trace of the algorithm begins as shown in Table 3.4 In this way the algorithm will finally produce the binary labeled tree shown in Figure 3.19. $\triangle$

How do we interpret the resulting binary labeled tree as a sorted sequence of keys? To answer this question, it would have been better to have approached

TABLE 3.4

i	x	T
1	ϵ	$T[\epsilon]$ = four
2	ϵ	
	1	$T[1]$ = score
3	ϵ	
	0	$T[0]$ = and
4	ϵ	
	1	
	11	$T[11]$ = seven
5	ϵ	
	1	
	11	
	111	$T[111]$ = years
6	ϵ	
	0	
	00	$T[00]$ = ago
7	ϵ	
	1	
	10	$T[10]$ = our

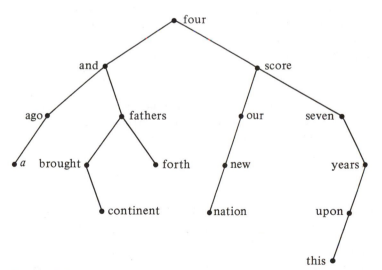

FIGURE 3.19

the whole subject with a *recursive* definition: a *binary labeled tree* is either empty or the triple

$$\text{tree} = (\text{tree, label, tree})$$

More specifically, we could write

$$\text{tree } x = (\text{tree } x0, \ x\text{-label, tree } x1)$$

for each $x \in B^*$, allowing any "tree at x" to be empty. Then in Example 3.30, we would have had

$$\text{tree} = \text{tree } \epsilon = (\text{tree } 0, \ \epsilon\text{-label, tree } 1)$$
$$= (\text{tree } 0, \text{ four, tree } 1)$$
$$= ((\text{tree } 00, \ 0\text{-label, tree } 01), \text{ four, } (\text{tree } 10, \ 1\text{-label, tree } 11))$$
$$= ((\text{tree } 00, \text{ and, tree } 01), \text{ four, } (\text{tree } 10, \text{ score, tree } 11))$$
$$\vdots$$

Finally, after disregarding the punctuation, we would have obtained the sorted sequence

a, ago, and, brought, continent, fathers, forth, four, nation,
new, our, score, seven, this, upon, years

At a later point (see Exercise 8 of Section 6.2) we will develop techniques that allow a nonrecursive sequential examination of a tree storage structure.

For the time being, we can appreciate the ease with which new items can be added to the tree-structured file without upsetting the order. Unlike the situation of simple linear storage we don't have to dislodge any of the existing items. Moreover, it can be shown that on the average only $n \log_2 n$ comparisons are required in performing the tree sort—a figure that certainly compares favorably with that for the selection sort. In fact, it can be shown that this average figure is essentially the best that can be achieved, regardless of the sorting method! But we will have to refer the reader to other sources if he wants to be thoroughly convinced of this fact.

EXERCISES

1 The algorithm shown in Figure 3.20 performs an *exchange sort* on the array y. Give a description and analysis of this sorting procedure.

2 In the *bubble sort* of an array y, we examine $y[j]$ and $y[j + 1]$ (for $1 \le j \le n - 1$). Whenever the entries are out of order, the smaller or "lighter" one is allowed to move upward (that is, "bubble up") until it is in the correct position relative to the others. Devise a formal algorithm for implementing this sorting procedure.

3 Explain why it is that the bubble sort (Exercise 2) requires approximately $\frac{1}{4}n^2$ comparisons on the average.

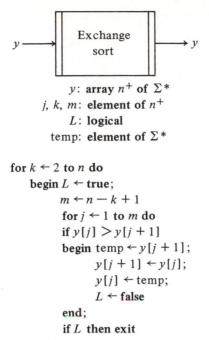

y: **array** n^+ **of** Σ^*
j, k, m: **element of** n^+
L: **logical**
temp: **element of** Σ^*

for $k \leftarrow 2$ **to** n **do**
    begin $L \leftarrow$ **true**;
        $m \leftarrow n - k + 1$
        for $j \leftarrow 1$ **to** m **do**
        if $y[j] > y[j + 1]$
        begin temp $\leftarrow y[j + 1]$;
            $y[j + 1] \leftarrow y[j]$;
            $y[j] \leftarrow$ temp;
            $L \leftarrow$ **false**
        end;
        if L **then exit**
    end

FIGURE 3.20

4 Run a trace on sorting the following data with the exchange sort (Exercise 1):
  (a) 3, 8, 2, 6, 5   (b) 103, 81, 94, 115, 32, 70
  (c) 19, 21, 6, 20, 25   (d) 2, 3, 5, 6, 8

5 Repeat Exercise 4 with the bubble sort (Exercise 2).

6 Repeat Exercise 4 with the selection sort (Algorithm X).

7 Repeat Exercise 4 with the tree sort (Algorithm XI).

8 Continue the trace of the performance of Algorithm X that was begun in Example 3.29, so as to obtain the final values of $y[3]$ and $y[4]$.

9 Use Algorithm XI to sort the following texts into alphabetical order:
  (a) Our father, who art in heaven; hallowed be thy name.
  (b) Yesterday, love was such an easy game to play.
  (c) O say can you see by the dawn's early light, what so proudly we hailed . . .
  (d) Warning: The Surgeon General has determined that cigarette smoking is dangerous to your health.
  (e) When you walk through a storm, keep your head up high; and don't be afraid of the dark

10 Repeat Exercise 9 with Algorithm X.

11 Repeat Exercise 9 with the exchange sort (Exercise 1).

12 Repeat Exercise 9 with the bubble sort (Exercise 2).

3.7 THE HOMOMORPHISM THEOREM

After the digression of the two preceding sections, let us now return to the general theory of monoids. The homomorphism theorem is perhaps the most fundamental result in algebra. With this theorem, we are able to characterize the homomorphic images of an algebra *internally*, with reference only to the existence of certain relations called congruences. Though we shall carry through the construction in its detail only for the monoids, the idea is so fundamental that the reader should be able to generalize it to other algebraic structures (finite state machines, groups, etc.) should the need arise.

Let $M = (M, \cdot, e)$ be any monoid, and suppose that R is an equivalence relation (Section 0.5) on the set M which happens to satisfy the additional property

$$x \ R \ u \text{ and } y \ R \ v \Rightarrow x \cdot y \ R \ u \cdot v$$

Then R is called a *congruence relation* or simply a *congruence* on the monoid M. We may substitute related elements, u for x and v for y, in the product $x \cdot y$, and we are assured of obtaining a related product $u \cdot v$. This gives the clue as to how to proceed in some other algebraic structure: simply insist that the "substitution property" be permissible for every operation that the algebra may possess.

Whenever a congruence relation R is given on a monoid $M = (M, \cdot, e)$, we may construct a corresponding *quotient monoid* M/R whose elements are the equivalence (congruence) classes $[x]$. We multiply in M/R according to the definition

$$[x] \cdot [y] = [x \cdot y]$$

That is, we multiply x and y in M, then take the equivalence class to which this product belongs. Now, were it not for the defining substitution property of congruences, this definition might be ambiguous and thus not really define an operation in M/R at all, for if $x \ R \ u$ and $y \ R \ v$, our "definition" would allow

$$[x] \cdot [y] = [u] \cdot [v] = [u \cdot v] \qquad (= [x \cdot y]?)$$

and $[u \cdot v]$ could be a different equivalence class than $[x \cdot y]$ but for the fact that $x \cdot y \ R \ u \cdot v$. Whenever we define a function on equivalence classes, we have thus to be careful to see that such a function is *well defined*, that its value does not depend on the representatives we might choose from the various equivalence classes.

Having checked that our binary operation on M/R is well defined, we can proceed to show that

$$M/R = (M/R, \cdot, [e])$$

is a monoid. The proof requires that we verify the axioms of associativity and identity:

(1) $[x] \cdot ([y] \cdot [z]) = [x] \cdot [yz] = [x(yz)] = [(xy)z]$
$$= [xy] \cdot [z] = ([x] \cdot [y]) \cdot [z]$$

(2) $[x] \cdot [e] = [xe] = [x] = [ex] = [e] \cdot [x]$

We see that the axioms for M/R follow rather easily from those in M itself.

EXAMPLE 3.31   Let us repeat the multiplication table of the monoid of Example 3.7:

·	e	a	b	c
e	e	a	b	c
a	a	e	c	b
b	b	b	b	b
c	c	c	c	c

According to the discussion of Section 0.5, the partition $\{\overline{e, a}; \overline{b, c}\}$ defines an equivalence relation R on the set $M = \{e, a, b, c\}$. Thus

$$e \, R \, e, \quad e \, R \, a, \quad b \, R \, b, \quad b \, R \, c, \quad \text{etc.}$$

But this relation R is in fact a congruence on the monoid M, for by direct computation, we can verify that the substitution property:

$$x \, R \, u \text{ and } y \, R \, v \Rightarrow x \cdot y \, R \, u \cdot v$$

holds for all x, y, u, v in M. Thus

$$e \cdot b = b \, R \, b = a \cdot c$$
$$e \cdot b = b \, R \, c = a \cdot b, \quad \text{etc.}$$

The corresponding quotient monoid M/R has two elements (equivalence classes), $[e]$ and $[b]$, with the multiplication table:

·	$[e]$	$[b]$
$[e]$	$[e]$	$[b]$
$[b]$	$[b]$	$[b]$

That is,

$$[e] \cdot [e] = [ee] = [e]$$
$$[e] \cdot [b] = [eb] = [b]$$
$$[b] \cdot [e] = [be] = [b]$$
$$[b] \cdot [b] = [bb] = [b] \qquad \triangle$$

EXAMPLE 3.32   In Example 0.37, we saw that the relation

$$a \sim b \Leftrightarrow a - b \text{ is a multiple of } n$$

is an equivalence relation on the set Z of all integers. In fact, this relation is a congruence relation on the monoid $Z = (Z, +, 0)$, for we have

$$a \sim b \text{ and } c \sim d \Rightarrow a - b = pn \quad \text{and} \quad c - d = qn$$
$$\Rightarrow (a + c) - (b + d) = (a - b) + (c - d)$$
$$= pn + qn$$
$$= (p + q)n$$
$$\Rightarrow a + c \sim b + d$$

which is the required substitution property in additive notation. We recall that this particular relation was called *congruence* (*modulo n*) in Example 0.37.

Apparently there is a quotient monoid $Z/\sim$ (mod n) for every integer n. As in Example 0.37, let us look at the case $n = 4$. There are four equivalence classes, and we may compute the following addition table:

+	[0]	[1]	[2]	[3]
[0]	[0]	[1]	[2]	[3]
[1]	[1]	[2]	[3]	[0]
[2]	[2]	[3]	[0]	[1]
[3]	[3]	[0]	[1]	[2]

For example, we have the following sample computations:

$$[0] + [2] = [0 + 2] = [2]$$
$$[1] + [3] = [1 + 3] = [4] = [0]$$
$$[2] + [3] = [2 + 3] = [5] = [1]$$
$$[3] + [3] = [3 + 3] = [6] = [2]$$

It is customary to denote these quotient monoids $Z/\sim$ (mod n) by Z_n and refer to the (additive) monoid of integers mod n.   △

EXAMPLE 3.33   Consider the free monoid Σ^* on the alphabet Σ. Let us define a relation R on Σ^* by writing

$$x \, R \, y \Leftrightarrow |x| = |y|$$

Surely this is an equivalence relation. Moreover,

$$x \, R \, u \text{ and } y \, R \, v \Rightarrow |x| = |u| \quad \text{and} \quad |y| = |v|$$
$$\Rightarrow |xy| = |x| + |y| = |u| + |v| = |uv|$$
$$\Rightarrow xy \, R \, uv$$

showing that R is a congruence. In the corresponding quotient monoid Σ^*/R words of the same length will form an equivalence class, and the definition

$$[x] \cdot [y] = [xy]$$

suggests (since $|xy| = |x| + |y|$) that the quotient is isomorphic to the (additive) monoid $N = (N, +, 0)$. The reader may try to show that the correspondence

$$|x| \leftrightarrow [x]$$

is the required isomorphism from N to Σ^*/R. $\triangle$

EXAMPLE 3.34   Again suppose that we are given a free monoid $\Sigma^* = (\Sigma^*, \cdot, \epsilon)$. Let us consider for a moment the surjective mapping $\varphi \colon \Sigma^* \to N$ as given by

$$\varphi(x) = |x|$$

from Σ^* to the monoid $N = (N, +, 0)$. Since

(i) $\varphi(\epsilon) = |\epsilon| = 0$
(ii) $\varphi(xy) = |xy| = |x| + |y| = \varphi(x) + \varphi(y)$

the mapping is a monoid homomorphism. Now observe that the kernel (Section 0.7) of the mapping φ is precisely the relation R of the preceding example. That is,

$$\varphi(x) = \varphi(y) \Leftrightarrow |x| = |y| \Leftrightarrow x \, R \, y$$

Further, we have just observed that the quotient Σ^*/R is isomorphic to N. Thus the homomorphic image N (of Σ^*) is abstractly equivalent to some quotient of Σ^*. $\triangle$

The circle of ideas as typified by Examples 3.33 and 3.34 is the heart of the homomorphism theorem of algebra: Up to isomorphism, we can obtain all the homomorphic images of a given algebra by considering its various quotient algebras, and each of these quotients is itself a homomorphic image of the given algebra. Returning to the general setting of Section 3.3, the reader should see how we can introduce appropriate definitions for congruences and quotients in an arbitrary algebra and thus be led to a general homomorphism theorem, of which the following is but a special case.

Theorem 3.3   (Homomorphism theorem for monoids)

Let $\varphi \colon M \to M'$ be a surjective monoid homomorphism. Then the kernel of φ is a congruence R on M, and $M' \simeq M/R$. Conversely, if R is a congruence on M, then there exists a surjective homomorphism $\eta \colon M \to M/R$ with kernel R.

Proof   Referring once more to Section 0.7, let R denote the kernel of the mapping $\varphi \colon M \to M'$ and suppose that $x \, R \, u$ and $y \, R \, v$. By the definition of kernel, we have

$$\varphi(xy) = \varphi(x)\varphi(y) = \varphi(u)\varphi(v) = \varphi(uv)$$

showing that $xy \; R \; uv$. So R is a congruence on M.

Next, we show that the mapping

$$\varphi(x) \rightarrow [x]$$

is an isomorphism of M' onto M/R. Since

$$\varphi(x) = \varphi(y) \Leftrightarrow x \; R \; y \Leftrightarrow [x] = [y]$$

the mapping is well defined and injective. To see that it is surjective, we simply observe that we can always find an element that maps onto a given $[x]$ in M/R, namely $\varphi(x)$. Also we have

$$e = \varphi(e) \rightarrow [e]$$

where the latter is the identity element in M/R. Finally, to see that our mapping preserves products, suppose that $\varphi(y) \rightarrow [y]$. Then we have

$$\varphi(x) \cdot \varphi(y) = \varphi(xy) \rightarrow [xy] = [x] \cdot [y]$$

as required.

Conversely, if R is a congruence on the monoid M, then the mapping $\eta(x) = [x]$ is easily seen to be a surjective homomorphism $\eta : M \rightarrow M/R$. And since

$$\eta(x) = \eta(y) \Leftrightarrow [x] = [y] \Leftrightarrow x \; R \; y$$

the kernel of η is the given congruence relation R.   □

EXAMPLE 3.35   In Example 3.19 we established a surjective monoid homomorphism $\varphi : M \rightarrow M'$ for the monoids

$M:$ ·	e	a	b	c		$M':$ ·	E	B
e	e	a	b	c		E	E	B
a	a	e	c	b		B	B	B
b	b	b	b	b				
c	c	c	c	c				

namely the mapping:

$$e \overset{\varphi}{\rightarrow} E$$
$$a \overset{\varphi}{\rightarrow} E$$
$$b \overset{\varphi}{\rightarrow} B$$
$$c \overset{\varphi}{\rightarrow} B$$

At the same time we observed that the kernel of φ is represented by the partition $\{\overline{e, a}; \overline{b, c}\}$ on the set M. In Example 3.31 we established that this kernel is a congruence R on the monoid M, and the quotient M/R obtained in Example 3.31 is easily seen to be isomorphic to M', as predicted by the theorem. $\triangle$

EXERCISES

1 In Example 3.33, prove in detail that $\Sigma^*/R \simeq N$.

2 Find all the congruences on the following monoids M:
   (a) Example 3.7   (b) Example 3.9
   (c) Example 3.14   (d) Exercise 6(a) of Section 3.1
   (e) the multiplication table

·	a	b	c	d	e	f
a	b	e	d	f	a	c
b	e	a	f	c	b	d
c	d	f	c	d	c	f
d	f	c	d	f	d	c
e	a	b	c	d	e	f
f	c	d	f	c	f	d

[*Hint*: See Exercise 5. In each case, a partition of M is the most appropriate representation.]

*3 Characterize all the congruences on the monoid $Z = (Z, +, 0)$.

*4 Devise a formal algorithm for finding all the congruences in any finite monoid.

5 Show that every monoid $M \neq \{e\}$ has at least two congruences, one in which the congruence classes are singleton subsets of M and the other in which there is a single congruence class, M itself. We call these *trival* congruences.

6 For each nontrivial congruence (Exercise 5) on the monoids of Exercise 2, describe the corresponding surjective homomorphism (as given by Theorem 3.3).

7 For the monoid of Exercise 2(e) exhibit a surjective homomorphism $\varphi: M \to M'$, where M' is the monoid of Example 3.19. Show that the kernel of φ is a congruence on M.

8 On the monoid $N = (N, +, 0)$ show that the partition into odds and evens is a congruence R. Describe the quotient monoid N/R.

*9 Given that R_1 and R_2 are congruences on M such that $R_1 \leq R_2$, prove that there is a unique surjective monoid homomorphism

$$\varphi: M/R_1 \to M/R_2$$

10 Show that the meet of two congruence relations is again a congruence relation.

3.8   FINITE STATE MACHINES

A natural language, such as English or French, develops and in fact grows through everyday usage. In contrast, the computer programming languages are more static. Therefore, it is possible to envision a fixed set of grammatical rules that could be used for the generation of proper statements, in ALGOL say. This approach is discussed in Chapter 6. Alternatively, we may imagine an automaton that functions as a recognizer, deciding which strings of symbols correspond to proper statements in the language. In either case, it is convenient to define a *language* λ as a subset $\lambda \subseteq \Sigma^*$, where Σ is an appropriate alphabet. Whether it is English, French, ALGOL, or FORTRAN, only very few words (or "sentences"), considered as strings of symbols in the alphabet, will actually be recognized as legitimate statements in the language. The others are to be rejected. So it is only a subset of the free monoid on which our attention should be focused.

The behavior of our automaton is much akin to a compiler's performing a syntax check. Consequently, the model we present in this section is relevant to either a software or a hardware implementation of a language recognition capability. At its outer reaches, however, the theory of automata is less relevant to such design considerations and rather more concerned with such questions as relate to an overall theory of computation. Nevertheless, such questions as the comparative speed of various computations and the memory space required ought not to be disregarded. In fact, they are at the heart of some of the most important theoretical studies in computer science. Certainly, it is desirable that the student have some awareness of the presuppositions on which the theory is founded. Still, we cannot afford to go into the details here. So we compromise by presenting a bare introduction to the ideas and methods of the theory of automata here and in Chapter 6. The ideas will bear a close relationship to those of the discrete structures presented thus far. At the same time they should provide an adequate foundation for further study.

EXAMPLE 3.36   FORTRAN and ALGOL are languages in the technical sense just introduced. Consider the FORTRAN character set (alphabet) of Example 0.27, that is,

$$\Sigma = \{A, B, \ldots, Z, 0, 1, \ldots, 9, +, -, *, /, =, ', \cdot, \$, (,), \Box\}$$

Not all the strings in Σ^* are FORTRAN language statements. Thus FORTRAN $\subset \Sigma^*$ in that

IF (N.GT.2) GOTO5 $\in$ FORTRAN $X + Y = Z \notin$ FORTRAN
$X = Y + 7.2 \in$ FORTRAN $2XN* - \$ \notin$ FORTRAN

and so on. $\triangle$

EXAMPLE 3.37   French and English are languages. Using the alphabet

$$\Sigma = \{a, b, \ldots, z, A, B, \ldots, Z, ', ", ?, \ldots, !, \Box\}$$

we may say that certain strings represent English sentences while others do
not. Thus

See John run! ∈ English	.eulb si noom ehT ∉ English
The moon is blue. ∈ English	Bonjour Ms. Whatsername. ∉ English

so that English is only a proper subset of Σ^*; English $\subset \Sigma^*$.   △

   The situations of a robot working on an arithmetic problem, a combination
lock, a vending machine, have at least one characteristic in common—the
ability to recall events that took place in the immediate past. Mechanically or
otherwise, the system is in a different "configuration" from one moment to
the next, and that is what allows the system's behavior to be modified as time
goes on, indeed, to react differently to the same input stimulus at different times.
These configurations or situations of the system are referred to, rather abstractly,
as the *states* of the system. The simplest systems that one might design or at
least imagine have rather small, certainly finite, total numbers of configurations.
The resulting capabilities are quite limited. Yet, if we place no a priori bound
on the number of these configurations or states, we might expect to delegate
larger and larger tasks to such systems or machines, and we might well ask,
"Where will this all end?" We will not tackle this interesting philosophical
question here, but related questions will be discussed in Chapter 6. At this
stage we are mainly interested in the formulation and the understanding of a
now standard mathematical model for the study of the *finite state machines*.
Within almost any digital computer there can be found various self-contained
systems or component modules whose performance is not unlike that just
outlined, but which might best be described as in Figure 3.21. The module is

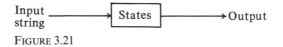

FIGURE 3.21

presented with data in the form of a string of symbols in some alphabet.
Internally, the module possesses a finite memory, which can be implemented
in several ways, depending on the technology employed. You may think of
the individual elements of the memory as consisting of bistable electronic
devices, for instance. In any case, the total configuration of all the memory
elements is collectively referred to once again as the *state* of the module or
machine at that particular time. The purpose of the memory is to recall certain
prior characteristics of the input string, in order to generate an output consistent
with the intended behavior. In accordance with sequential or *on line* mode

of computation, the prior input symbols are no longer available for inspection, and a memory is therefore essential.

EXAMPLE 3.38   Suppose that our module is to examine strings or words $x \in \Sigma^*$, where $\Sigma = \{0, 1\}$, and that the module is to decide whether x has a number (k) of ones for which $k \sim 2$ (mod 4), in the sense of Example 0.37. Then if $k = 2, 6, 10, \ldots$, we want this machine to indicate the fact, perhaps by lighting a bulb, ringing a bell, or some less disruptive action. Obviously, our memory need do nothing more than count (modulo 4). See Figure 3.22, where the states

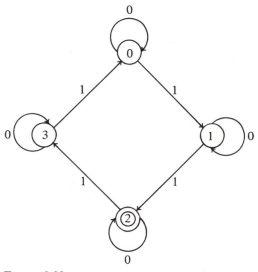

FIGURE 3.22

are represented by circular nodes, labeled 0, 1, 2, 3, and arrows are drawn to indicate the desired state transitions. It is presumed that the experiment begins in state 0; state 2 is the one (doubly circled) that "sounds the alarm."

In case $x = 10111011$, the machine will pass through the states $0 \rightarrow 1 \rightarrow 1 \rightarrow 2 \rightarrow 3 \rightarrow 0 \rightarrow 0 \rightarrow 1 \rightarrow 2$ and the word x will be recognized. On the other hand, if $x = 101110111$, we end up in state 3 so that x is rejected.   △

The *state diagram* of a machine (as in Figure 3.22) is really only a labeled digraph whose vertices are labeled by the states of the machine and whose directed edges indicate the various state transitions. We label these edges with symbols from an alphabet Σ, the *input alphabet*, so as to fully describe the stimuli that give rise to the respective state transitions. We also "distinguish" an *initial state*, ordinarily labeled 0, and single out a collection I of *final states* (in Example 3.38, $I = \{2\}$) to signify the successful termination of the experiment or the recognition of the desired input pattern. It is of course inevitable

that certain terminology and techniques from graph theory be directly trans-
ferable to a discussion of these machines. Nevertheless, instead of the graphical
state diagrams, we prefer a tabular representation, the so-called *state table*,
because the latter comes closer to our eventual definition of a finite state
machine.

EXAMPLE 3.39   For the machine in Example 3.38 we would have the state table

| | Input symbol | |
State	$\sigma = 0$	$\sigma = 1$
0	0	1
1	1	2
2	2	3
3	3	0

$I = \{2\}$

With the understanding that 0 is the initial state, this tabular representation
retains all of the information given in the state diagram of Figure 3.22.   △

EXAMPLE 3.40   Consider a computer language in which floating point numbers
(scientific notation) must be presented in the form

$$\pm dd*.d*E \pm dd$$

Each d is a "digit," an element of the subalphabet

$$D = \{0, 1, 2, 3, 4, 5, 6, 7, 8, 9\}$$

and $d*$ indicates an arbitrary number (perhaps none) of occurrences of digits.
As in FORTRAN, E stands for "times 10 to the," to be followed by an appropriate power. Thus we would have the following representations:

$$6.03 \times 10^{23} \quad \text{as} \quad +6.03E + 23$$

$$0.25 \times 10^{-6} \quad \text{as} \quad +0.25E - 06$$

and so on.

Figure 3.23 gives an abbreviated description of the state diagram of a finite
state machine that would recognize properly formed floating point numbers
in this particular language. Again we have only one final state, $I = \{7\}$. State
8 indicates an unrecognized string; perhaps it will generate an "error flag."
The arrows to state 8 are too numerous to describe in detail. They are all to
be labeled with symbols that do not represent proper exits from the source
of the arrows. Thus we have the arrow $2 \xrightarrow{E} 8$ because a decimal point has not
yet been received.

A compiler for this programming language must examine a rather long
sequence of characters (a program) and recognize appropriate grammatical

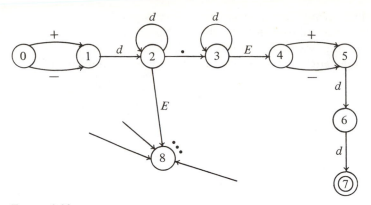

FIGURE 3.23

constituents, among them the floating point numbers of the language. In this line of activity the machine of Figure 3.23 can play an important role. It is not likely to be realized in hardware, however. Instead, the compiler will probably incorporate a simulation of this machine.   △

In light of these preliminary examples, we are ready to define the simple computing or recognition model as proposed, or at least popularized, by Rabin and Scott.* A *finite state machine* (or *automaton*) is an algebraic structure $A = (S, \Sigma, M)$ in which

 (i) S is a finite set of states, with a distinguished element $0 \in S$ and a distinguished subset $I \subseteq S$;
(ii) Σ is an alphabet;
(iii) M is the transition function $M: S \times \Sigma \to S$.

Especially in (iii) we see that the definition is practically synonymous with the idea of a state table. We note that in strict accordance with the definition of an algebraic structure (Section 1.4), we should write $\mathscr{A} = (\mathscr{A}; \mathscr{F}; \mathscr{P})$, where

$$\mathscr{A} = \{S, \Sigma\} \qquad \mathscr{F} = \{M, 0\} \qquad \mathscr{P} = \{I\}$$

In this case, our slight abuse of notation should add to the clarity of the presentation.

The transition function admits an inductive extension to a mapping

$$M: S \times \Sigma^* \to S$$

simply by writing

$$M(s, \epsilon) = s \qquad M(s, x\sigma) = M(M(s, x), \sigma)$$

for x in Σ^* and σ in Σ. Then in using these machines to recognize a language

---

* M. O. Rabin and D. Scott, "Finite Automata and their Decision Problems," *IBM J. Research Develop.*, **3**, 114–125 (1959).

$\lambda \subseteq \Sigma^*$, we say that A *accepts* (or recognizes) the word $x \in \Sigma^*$ just in case $M(0, x) \in I$. So we start the machine in its initial state and present it with the word x. The machine examines the symbols of x sequentially from left to right, making state transitions according to M. As the last symbol is read, A either enters or does not enter one of its distinguished final states and accordingly accepts or rejects x. In this way, we may define the *language λ_A accepted by A*:

$$\lambda_A = \{x \in \Sigma^*: M(0, x) \in I\}$$

Finally, a language $\lambda \subseteq \Sigma^*$ is said to be *regular* if it is accepted by some finite state machine, that is, if $\lambda = \lambda_A$ for some finite state machine A. In Chapter 6 we will find that nonregular languages do, in fact, exist.

EXAMPLE 3.41   If A is the machine of Example 3.38, then

$$S = \{0, 1, 2, 3\} \qquad \Sigma = \{0, 1\}$$

and M is the function given by the table of Example 3.39:

	0	1
0	0	1
1	1	2
2	2	3
3	3	0

$I = \{2\}$

By construction we have

$$\lambda_A = \{x \in \Sigma^*: x \text{ has } k \text{ ones and } k \sim 2 \ (\text{mod } 4)\}$$

In order to demonstrate the use of the inductive extension of M, we compute

$$
\begin{aligned}
M(0, 01010) &= M(M(0, 0101), 0) = M(M(M(0, 010), 1), 0) \\
&= M(M(M(M(0, 01), 0), 1), 0) \\
&= M(M(M(M(M(0, 0), 1), 0), 1), 0) \\
&= M(M(M(M(0, 1), 0), 1), 0) \\
&= M(M(M(1, 0), 1), 0) = M(M(1, 1), 0) = M(2, 0) = 2
\end{aligned}
$$

thus showing that $01010 \in \lambda_A$. $\qquad\qquad\triangle$

In extending the definitions of homomorphism and congruence to automata, we are guided by the general ideas expressed in Sections 3.3 and 3.7. The main thing to bear in mind is that the automata, as algebraic structures, have two sets: the state set and the alphabet. Accordingly, we should say that an *automata homomorphism* $\varphi: A \to B$ is a pair of mappings

$$\varphi: S_A \to S_B \qquad \varphi: \Sigma_A \to \Sigma_B$$

such that

$$\varphi(0_A) = 0_B \qquad \varphi(M_A(s, \sigma)) = M_B(\varphi s, \varphi \sigma)$$

for all s in S_A and σ in Σ_A. Of course, if we were to follow the suggestions of Section 3.3 to the letter, we would also have to insist that $s \in I_A \Leftrightarrow \varphi s \in I_B$. Some authors will do just that. We take the view, instead, that the final states are only of secondary concern, and not so much a part of the basic structure of the machine. In fact, we shall act as though they could be redesignated at the whim of the designer. This attitude will be a distinct advantage in our discussion of the applications of the homomorphism theory. But before proceeding in this direction, let us establish the extension of the preservation of the transition function to words rather than symbols.

Theorem 3.4

Let $\varphi \colon A \to B$ be an automata homomorphism. Then for all s, x we have

$$\varphi(M_A(s, x)) = M_B(\varphi s, \varphi x).$$

Proof   We use induction on the length of words $x \in \Sigma_A^*$ in conjunction with our inductive extension of M. Thus for $x = \epsilon$ we have

$$\varphi(M_A(s, \epsilon)) = \varphi(s) = M_B(\varphi s, \epsilon) = M_B(\varphi s, \varphi \epsilon)$$

Then assuming the validity of our conclusion for some word x in Σ_A^*, we also obtain, with $\sigma \in \Sigma_A$,

$$\begin{aligned}
\varphi(M_A(s, x\sigma)) &= \varphi(M_A(M_A(s, x), \sigma)) \\
&= M_B(\varphi(M_A(s, x)), \varphi\sigma) \\
&= M_B(M_B(\varphi s, \varphi x), \varphi\sigma) \\
&= M_B(\varphi s, \varphi x \cdot \varphi\sigma) = M_B(\varphi s, \varphi(x\sigma)) \qquad \square
\end{aligned}$$

In order to establish the basis for the applications of the homomorphism theory, we must first make a connection to the language recognition capabilities of our machines. At the same time, our motivation for deemphasizing the final states will come into sharper focus. Suppose that A and B are automata. We say that *A has the capability of B* if there exist

(a) A redesignation of I_A
(b) A monoid homomorphism $\Phi \colon \Sigma_B^* \to \Sigma_A^*$ such that
(c) $x \in \lambda_B \Leftrightarrow \Phi(x) \in \lambda_A$

Pictorially, we have the situation of Figure 3.24. Using only a simple transliteration of words, we can cause A to make the decisions that are normally made by B. Note that this definition makes no reference to the internal structure of the two machines. This is the job of the homomorphisms. But our next

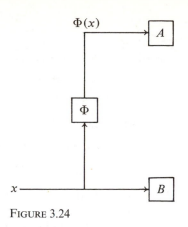

FIGURE 3.24

theorem establishes important connections between structural preservation (homomorphism) and input-output behavior (capability). The result makes special reference to those situations in which

A realizes B   (symbolized: $B \rightarrowtail A$)
B is a *homomorphic image* of *A*   (symbolized: $A \twoheadrightarrow B$)

The terms here are used in the sense of Section 3.3; that is, they refer to the injectivity or surjectivity of the homomorphism. As usual, if the homomorphism $\varphi: A \to B$ is injective and surjective, we say that A and B are *isomorphic* machines ($A \simeq B$).

EXAMPLE 3.42   Suppose A and B are the machines whose state tables are as follows:

A	0	1	2
a	c	d	d
b	b	c	c
c	a	a	d
d	c	d	a

$0_A = a$
$I_A = \{d\}$

B	σ	τ
α	γ	α
β	β	γ
γ	α	α

$0_B = \alpha$
$I_B = \{\alpha\}$

Consider the pair of mappings φ from S_A to S_B and Σ_A to Σ_B (Figure 3.25). We see that

$$\varphi(0_A) = \varphi(a) = \alpha = 0_B$$

and by an exhaustive analysis, we can verify that

$$\varphi(M_A(s, \sigma)) = M_B(\varphi s, \varphi \sigma)$$

FIGURE 3.25

for all $s \in S_A$ and $\sigma \in \Sigma_A$. For example,

$$\varphi(M_A(a, 0)) = \varphi(c) = \gamma = M_B(\alpha, \sigma) = M_B(\varphi a, \varphi 0)$$

$$\varphi(M_A(a, 1)) = \varphi(d) = \alpha = M_B(\alpha, \tau) = M_B(\varphi a, \varphi 1), \quad \text{etc.}$$

It follows that $\varphi: A \to B$ is a surjective automata homomorphism; that is, B is a homomorphic image of A.

For purposes of later analysis, we tabulate the kernel (Section 0.7) of these mappings φ in the form of partitions

$$\{\overline{a, d}; \overline{b}; \overline{c}\} \text{ on } S_A \qquad \{\overline{0}; \overline{1, 2}\} \text{ on } \Sigma_A$$

The reader should not be surprised to learn that this pair of partitions turns out to be an *automata congruence* on A.

Now let us indicate how A can have the capability of B. The final state in B is α, but both of the states a, d are playing the role of α in A. So for the purpose of imitating B, the subset I_A was not defined properly. We should redesignate $I_A = \{a, d\}$. Then if we set

$$\Phi(\sigma) = 0$$
$$\Phi(\tau) = 2 \qquad \text{(or 1, it makes no difference)}$$

according to the lemma preceding Theorem 3.2, Φ extends to a monoid homomorphism $\Phi: \Sigma_B^* \to \Sigma_A^*$. Our next theorem will say that

$$x \in \lambda_B \Leftrightarrow \Phi(x) \in \lambda_A$$

So let us see if this seems to be the case here. Since

$$M_B(\alpha, \sigma\tau) = M_B(M_B(\alpha, \sigma), \tau) = M_B(\gamma, \tau) = \alpha \in I_B$$

we have $\sigma\tau \in \lambda_B$. Now $\sigma\tau$ translates into 02 and

$$M_A(a, 02) = M_A(M_A(a, 0), 2) = M_A(c, 2) = d \in I_A$$

so that $02 = \Phi(\sigma\tau) \in \lambda_A$. The reader may try other examples, but should not be thoroughly convinced until he or she has constructed a proof.   △

Theorem 3.5

$$\begin{array}{c} A \twoheadrightarrow B \\ \text{or} \qquad \Rightarrow A \text{ has the capability of } B \\ B \rightarrowtail A \end{array}$$

Proof   Having just given an illustration of the first situation, suppose that we concentrate instead on the case where A realizes B. If $\varphi: B \to A$ is the injective homomorphism, we redesignate $I_A = \varphi I_B = \{\varphi s: s \in I_B\}$ and take

$$\Phi = \varphi: \Sigma_B^* \to \Sigma_A^*$$

Then we have

$$x \in \lambda_B \Leftrightarrow M_B(0, x) \in I_B \Leftrightarrow \varphi(M_B(0, x)) = M_A(\varphi 0, \varphi x)$$
$$= M_A(0, \Phi x) \in I_A$$
$$\Leftrightarrow \Phi(x) \in \lambda_A$$

which shows that, indeed, A has the capability of B.   □

Corollary

$B \leftarrow C \rightarrowtail A$ (A *covers* B) $\Rightarrow A$ has the capability of B.   □

Given $A = (S, \Sigma, M)$, we should expect an *automata congruence* R to be a pair of equivalence relations (one on S, the other on Σ) such that

$$s \, R \, t \quad \text{and} \quad \sigma \, R \, \tau \Rightarrow M(s, \sigma) \, R \, M(t, \tau)$$

Then the corresponding *quotient machine* A/R is defined, in the usual way, on equivalence classes; that is,

$$A/R = (S/R, \Sigma/R, M/R)$$

where

$$M/R([s], [\sigma]) = [M(s, \sigma)].$$

Moreover, we take $0_{A/R} = [0]$. So closely do the arguments parallel those for monoids that we can omit the proof of the following theorem, trusting that the reader could provide it if necessary after reviewing Theorem 3.3.

Theorem 3.6   (Homomorphism theorem for automata)

Let $\varphi: A \to B$ be a surjective automata homomorphism. Then the kernel R is an automata congruence on A, and $B \simeq A/R$. Conversely, if R is a congruence on A, then there exists a surjective automata homomorphism $\eta: A \to A/R$ with kernel R.   □

EXAMPLE 3.43   Consider the partitions

$$\{\overline{a, d}; \overline{b}; \overline{c}\} \quad \text{and} \quad \{\overline{0}; \overline{1, 2}\}$$

representing the kernel R for the surjective homomorphism $\varphi: A \to B$ of Example 3.42.   Sample calculations in verifying that R is an automata congruence on A are as follows:

$$M(a, 1) = d \, R \, a = M(d, 2)$$
$$M(a, 0) = c \, R \, c = M(d, 0), \quad \text{etc.}$$

The quotient machine A/R has the initial state $[a] = \{a, d\}$ and the transition function

$$M_{A/R}([a], [0]) = [M_A(a, 0)] = [c]$$

$$M_{A/R}([a], [1]) = [M_A(a, 1)] = [d] = [a], \quad \text{etc.}$$

so that A/R has the state table

A/R	$[0]$	$[1]$
$[a]$	$[c]$	$[a]$
$[b]$	$[b]$	$[c]$
$[c]$	$[a]$	$[a]$

We can see immediately, upon establishing the correspondences

$$\begin{matrix} [a] \leftrightarrow \alpha \\ [b] \leftrightarrow \beta \\ [c] \leftrightarrow \gamma \end{matrix} \qquad \begin{matrix} [0] \leftrightarrow \sigma \\ [1] \leftrightarrow \tau \end{matrix}$$

that $B \simeq A/R$ as claimed in the theorem.   △

EXERCISES

1 Obtain the state diagram of a machine that accepts just those words $x \in \Sigma^*$ (where $\Sigma = \{0, 1\}$)
   (a) Having k ones, with $k \sim 3 \pmod 5$
   (b) Ending in a zero
   (c) Having the sequence 011 as a terminal subword
   (d) Having the sequence 011 as a subword (terminal or not)

2 Having the machines $A = (S, \Sigma, M)$ of Exercise 1, compute
   (a) $M(0, 0101111)$ and $M(0, 10011)$
   (b) $M(0, 01011)$ and $M(0, 0110)$
   (c) $M(0, 10010)$ and $M(0, 0011)$
   (d) $M(0, 1010)$ and $M(0, 10110)$
   Show in each case which of the given pair of words is accepted and which is rejected by the respective machines.

3 Show that for every pair x, y in Σ^* and every state s,

$$M(s, xy) = M(M(s, x), y)$$

   in any machine $A = (S, \Sigma, M)$. [*Hint*: Use induction on the length of words $y \in \Sigma^*$.]

*4 Prove that each of the following languages is regular (on the alphabet $\Sigma = \{0, 1\}$):
   (a) $\lambda = \{x : x$ has k ones, with $k \sim r \pmod n\}$ for fixed r, n ($0 \leqslant r < n$)

(b) $\lambda = \{x : x$ has the sequence y as a terminal subword$\}$ for fixed $y \neq \epsilon$ in Σ^*

(c) $\lambda = \{x : x$ has the sequence y as an initial subword$\}$ for fixed $y \neq \epsilon$ in Σ^*

(d) $\lambda = \{x : x$ has the sequence y as a subword$\}$ for fixed $y \neq \epsilon$ in Σ^*

*5 Prove that the union, the intersection, and the complement of regular languages (on the alphabet Σ) are regular.

6 Let A, B be the machines shown below:

A	0	1	2	3
a	c	b	d	b
b	d	b	c	a
c	a	c	a	d
d	a	c	b	c

$0_A = a$
$I_A = \{d\}$

B	σ	τ
0	1	0
1	0	1

$0_B = 0$
$I_B = \{1\}$

Show that the pair of mappings in Figure 3.26 constitute a (surjective) automata homomorphism $\varphi : A \to B$. At the same time show that the kernel is an automata congruence on A.

FIGURE 3.26

7 Illustrate the fact that A has the capability of B for the machines of Exercise 6.

8 Show that A has the capability of B for the following pairs of machines:

(a)

A	0	1	2
a	b	c	b
b	c	d	a
c	b	a	b
d	a	e	c
e	a	d	a

$0_A = a$
$I_A = \{c, d, e\}$

B	σ	τ
0	1	0
1	0	2
2	0	1

$0_B = 0$
$I_B = \{1\}$

(b)

A	α	β	γ
a	a	b	c
b	b	d	e
c	e	c	a
d	a	c	b
e	c	b	a

$0_A = a$
$I_A = \{c, d, e\}$

B	σ	τ
0	2	0
1	0	2
2	0	1

$0_B = 0$
$I_B = \{2\}$

9 Prove that A has the capability of B if $A \twoheadrightarrow B$.

10 Prove the corollary to Theorem 3.5.

11 Show that the pair of partitions

$$\{\overline{a, c}\,;\overline{b, e}\,;\overline{d}\} \text{ and } \{\overline{0, 2}\,;\overline{1}\}$$

represents an automata congruence on A, the machine of Exercise 8(a).

*3.9   MINIMIZATION OF MACHINES

In this section we will consider a question having to do with the economics of machine design. As before, we interpret the performance of a machine $A = (S, \Sigma, M)$ as the language recognized by that machine. Then it is an easy matter to convince ourselves that there are any number of machines capable of the same performance. How, then, do we select an economical representative? First of all, we take the figure $|S|$ as a measure of relative economy, simply because the cost of an implementation of a machine grows with the number of states. Then if we write $B \sim A$ whenever $\lambda_B = \lambda_A$, our question can be phrased as a *minimization problem*: given the machine $A = (S, \Sigma, M)$, find a machine B with the properties

(a) $B \sim A$
(b) $C \sim A \Rightarrow |S_B| \le |S_C|$

Borrowing a bit of terminology from graph theory, we can at least limit our attention to *connected* automata B. Thus we suppose that in each of the machines that we might consider, every state s is reachable from the initial state, in that $s = M(0, x)$ for some $x \in \Sigma^*$. Otherwise, we could delete any unreachable states from the state table without any effect on the performance.

EXAMPLE 3.44   The machine A of Example 3.42 is not connected, because state b is not reachable (from the initial state a). Note that reachability of states is a path problem in a labeled digraph. $\triangle$

EXAMPLE 3.45   Suppose that we let S^c denote the subset of states that are reachable from the initial state and A be the machine of Example 3.42. Then $S^c = \{a, c, d\}$ and the behavior of the connected "subautomaton":

	0	1	2	
a	c	d	d	$A^c = (S^c, \Sigma, M)$
c	a	a	d	$0_{A^c} = 0_A = a$
d	c	d	a	$I_{A^c} = I_A \cap S^c = \{d\}$

is indistinguishable from that of A. $\triangle$

Taking a slightly different point of view, suppose that A is any connected automaton accepting the language λ. Then we can define an obvious equivalence relation on its states by setting

$$s \sim s' \Leftrightarrow M(s, x) \in I \quad \text{iff} \quad M(s', x) \in I$$

that is, if one state is transformed into a final state by x, then so is the other. Because of the implication

$$s \sim s' \Rightarrow \begin{array}{c} M(M(s, \sigma), x) = M(s, \sigma x) \in I \\ \text{iff} \\ M(M(s', \sigma), x) = M(s', \sigma x) \in I \end{array}$$
$$\Rightarrow M(s, \sigma) \sim M(s', \sigma)$$

$\sim$ is an automata congruence. Of course, we should be referring to an equivalence relation on symbols as well as states, but we have in mind the identity relation in which two symbols are regarded as equivalent only if they are equal. Then the reader should anticipate that the quotient

$$\tilde{A} = A/\sim$$

is the solution to the minimization problem. We intend to reinforce this suspicion as we proceed. At the same time we will be looking at an algorithmic process for computing the partition of states (call it $\tilde{\pi}$) that corresponds to this congruence relation. According to the methods of the previous section, that is all we need to derive our "minimal form" $\tilde{A}$ from the given A. Then $\tilde{A}$ is the most economical machine accepting the language λ.

EXAMPLE 3.46   Consider the machine A shown below.

	σ	τ
0	1	2
1	5	6
2	2	0
3	1	4
4	4	3
5	2	4
6	4	1

$$I = \{2, 4, 6\}$$

We never have $s \sim s'$ when s is final but s' is not. Thus $2 \nsim 5$ because in taking $x = \epsilon$ in our definition, we have

$$M(2, \epsilon) = 2 \in I \qquad M(5, \epsilon) = 5 \notin I$$

Similarly, $1 \nsim 5$ because

$$M(1, \sigma) = 5 \notin I \qquad M(5, \sigma) = 2 \in I$$

With other pairs of states, it may take still longer sequences to decide that they are not equivalent. Consider states 2 and 6. They are both final states.

Furthermore,

$$M(2, \sigma) = 2 \in I \qquad M(6, \sigma) = 4 \in I$$
$$M(2, \tau) = 0 \notin I \qquad M(6, \tau) = 1 \notin I$$

and also

$$M(2, \sigma\sigma) = 2 \in I \qquad M(6, \sigma\sigma) = 4 \in I$$
$$M(2, \sigma\tau) = 0 \notin I \qquad M(6, \sigma\tau) = 3 \notin I$$
$$M(2, \tau\sigma) = 1 \notin I \qquad M(6, \tau\sigma) = 5 \notin I$$
$$M(2, \tau\tau) = 2 \in I \qquad M(6, \tau\tau) = 6 \in I$$

Yet, $2 \nsim 6$, as we might anticipate by observing that

$$2 \xrightarrow{\tau} 0 \xrightarrow{\sigma} 1 \qquad 6 \xrightarrow{\tau} 1 \xrightarrow{\sigma} 5$$

whereas we know that $1 \nsim 5$. Thus

$$M(2, \tau\sigma\sigma) = 5 \notin I \qquad M(6, \tau\sigma\sigma) = 2 \in I$$

showing that $2 \nsim 6$. In terms of the definition we are about to present, we would say that $2 \underset{2}{\sim} 6$ but $2 \underset{3}{\nsim} 6$.   △

The algorithm we shall present utilizes a sequence of *nested* partitions, each a refinement of the preceding one. They are the partitions π_i corresponding to the equivalence relations $\underset{i}{\sim}$ defined for $i = 0, 1, 2, \ldots$, as follows:

$$s \underset{i}{\sim} s' \Leftrightarrow M(s, x) \in I \quad \text{iff} \quad M(s', x) \in I \qquad (|x| \le i)$$

Evidently π_0 is the two-block partition $\pi_0 = \{\overline{I}; \overline{S \sim I}\}$. Given this partition, a procedure for determining the desired partition $\tilde{\pi}$ is implicit in the statement of the two lemmas that follow. The first lemma, telling us how to obtain π_{i+1} from π_i, makes use of the functions

$$M_s^{(i)}: \Sigma \to \pi_i \qquad M_s^{(i)}(\sigma) = [M(s, \sigma)]_i$$

That is, $M_s^{(i)}(\sigma)$ is the block of π_i to which $M(s, \sigma)$ belongs.

Lemma 1

$$s \underset{i+1}{\sim} s' \quad \text{iff} \quad s \underset{i}{\sim} s' \quad \text{and} \quad M_s^{(i)} = M_{s'}^{(i)} \qquad \square$$

Lemma 2

$$\pi_{i+1} = \pi_i \quad \text{implies} \quad \tilde{\pi} = \pi_i = \pi_{i+1} = \cdots \qquad \square$$

The second lemma shows that we have reached $\tilde{\pi}$ as soon as the iteration repeats a partition. We leave the proof of these statements as exercises and concentrate instead on understanding the process itself.

EXAMPLE 3.47   For the machine in Example 3.46, we begin with the partition

$$\pi_0 = \{\overline{I}; \overline{S \sim I}\} = \{\overline{2, 4, 6}; \overline{0, 1, 3, 5}\} = \{B_{01}, B_{02}\}$$

We name the blocks of our partitions in order that we may more effectively carry out the comparison required on Lemma 1. Thus in consulting the original state table to determine the functions

$$M_s^{(0)}: \Sigma \to \pi_0$$

we have

	σ	τ	
0	B_{02}	B_{01}	$(1 \in B_{02}, 2 \in B_{01})$
1	B_{02}	B_{01}	$(5 \in B_{02}, 6 \in B_{01})$
2	B_{01}	B_{02}	$(2 \in B_{01}, 0 \in B_{02})$ etc.
3	B_{02}	B_{01}	
4	B_{01}	B_{02}	
5	B_{01}	B_{01}	
6	B_{01}	B_{02}	

According to Lemma 1, we have

$$\pi_1 = \{\overline{2,4,6}; \overline{0,1,3}; \overline{5}\} = \{B_{11}, B_{12}, B_{13}\}$$

Continuing in this way, we can compute

	σ	τ	
0	B_{12}	B_{11}	$(1 \in B_{12}, 2 \in B_{11})$
1	B_{13}	B_{11}	$(5 \in B_{13}, 6 \in B_{11})$
2	B_{11}	B_{12}	$(2 \in B_{11}, 0 \in B_{12})$ etc.
3	B_{12}	B_{11}	
4	B_{11}	B_{12}	
5	B_{11}	B_{11}	$\pi_2 = \{\overline{2,4,6}; \overline{0,3}; \overline{1}; \overline{5}\} = \{B_{21}, B_{22}, B_{23}, B_{24}\}$
6	B_{11}	B_{12}	

	σ	τ	
0	B_{23}	B_{21}	
1	B_{24}	B_{21}	
2	B_{21}	B_{22}	
3	B_{23}	B_{21}	$\pi_3 = \{\overline{2,4}; \overline{6}; \overline{0,3}; \overline{1}; \overline{5}\}$
4	B_{21}	B_{22}	
5	B_{21}	B_{21}	
6	B_{21}	B_{23}	

The reader can verify that $\pi_4 = \pi_3$, so that according to Lemma 2,

$$\tilde{\pi} = \pi_3 = \pi_4 = \cdots$$
$$= \{\overline{2,4}; \overline{6}; \overline{0,3}; \overline{1}; \overline{5}\} = \{B_1, B_2, B_3, B_4, B_5\}. \qquad \triangle$$

EXAMPLE 3.48 With the definition of the preceding section, we can use the results of Example 3.47 to compute the quotient machine $\tilde{A}$:

	σ	τ
B_1	B_1	B_3
B_2	B_1	B_4
B_3	B_4	B_1
B_4	B_5	B_2
B_5	B_1	B_1

$$0_A = B_3$$
$$I_A = \{B_1, B_2\}$$

and this is the machine with the fewest states having the same behavior as the given machine A. △

Having presented an example in some detail, we can now offer the formal procedure with a minimum of comment. In Algorithm XII we begin once

γ, π: partition of S k: positive integer
s: element of S r: nonnegative integer
B: subset of S τ: array π of $\Pi(S)$
A: array r^+ of $\mathscr{P}(s)$

begin $\pi \leftarrow \{\overline{S}\}$;
    while $\pi \neq \gamma$ **do**
    begin $\pi \leftarrow \gamma$;
        for $B \in \pi$ **do**
        begin $r \leftarrow 0$;
            for $s \in B$ **do**
            if $(r = 0) \vee (M_s \neq$ all $M_{A[k]})$ **then**
                begin $r \leftarrow r + 1; A[r] \leftarrow \{s\}$ **end**
            else for $k \leftarrow 1$ **to** r **do**
                if $M_s = M_{A[k]}$ **then**
                    $A[k] \leftarrow A[k] \cup \{s\}$;
                $\tau[B] \leftarrow \{A[k]: k = 1, ..., r\} \cup \{S \sim B\}$
        end;
        $\gamma \leftarrow \prod_{B \in \pi} \tau[B]$
    end
  end
ALGORITHM XII

again with the partition π_0, and the successive iterations of the **while** statement will generate π_{i+1} from π_i. The main decision takes place in the exterior **if** statement, where it is determined whether the state $s \in B$ gives rise to a new block in the next partition being formed. Our presentation of the minimization problem follows along the lines of Moore,* to whom the algorithm might also be credited. By way of comparison, we should also call attention to a more efficient procedure due to Hopcroft.†

EXERCISES

1 Prove Lemma 1.

2 Prove Lemma 2.

3 Show that the equivalence relation $\widetilde{\gamma}$ (together with the identity relation on symbols) is an automata congruence iff $\pi_I = \widetilde{\pi}$.

4 Use Algorithm XII to compute the minimal form $\widetilde{A}$ for the following machines A:

(a)

	0	1
a	c	c
b	d	b
c	d	a
d	b	d

$0 = a$

$I = \{b, d\}$

(b)

	σ	τ
0	1	2
1	3	3
2	4	4
3	3	3
4	2	4

$I = \{2, 4\}$

(c)

	σ	τ
0	1	2
1	1	3
2	2	4
3	1	5
4	6	2
5	5	3
6	4	2

$I = \{2, 4, 6\}$

(d)

	σ	τ
0	0	5
1	0	3
2	1	4
3	4	7
4	0	2
5	7	4
6	5	2
7	1	4

$I = \{3, 4, 5, 6\}$

(e)

	0	1
a	b	c
b	b	e
c	a	d
d	d	e
e	d	c

$0 = a$

$I = \{b, c, d, e\}$

5 Find two nonisomorphic equivalent ($A \sim B$) machines.

*6 Prove that (up to isomorphism) $\widetilde{A}$ is the unique solution to the minimization problem.

* E. F. Moore, "Gedanken Experiments on Sequential Machines," *Automata Studies*, C. E. Shannon and J. McCarthy (ed.), Princeton University Press, Princeton, N.J., 1956, 129–153.
† J. E. Hopcroft, "An *n* log *n* Algorithm for Minimizing States in a Finite Automaton," *Tech. Rept.* CS–190, Computer Science Department, Stanford University, Palo Alto, Cal., 1970.

*7 We say that an automaton is *reduced* if

$$s \sim s' \Rightarrow s = s'$$

Show that (up to isomorphism) $\tilde{A}$ is the unique reduced machine equivalent to A.

*8 Show that

$$\pi_{|S|-1} = \tilde{\pi}$$

for any finite state machine $A = (S, \Sigma, M)$.

*3.10 DECOMPOSITION OF MACHINES

The central idea that we pursue briefly here concerns the possibility of decomposing automata into simpler interconnected ones. The interconnections may be of parallel or serial form. We find that the existence of such decompositions will depend on the nature of the congruences on the machine at hand, but the machine that is actually constructed (decomposed) may have only the capability of the given machine. On the other hand, we learned in Section 3.8 that this much is already sufficient, from the point of view of the language to be recognized.

It is well that we begin by describing just how one might link two given machines, say in a parallel interconnection. So let $A_1 = (S_1, \Sigma_1, M_1)$ and $A_2 = (S_2, \Sigma_2, M_2)$. We should consult Figure 3.27 as we proceed to define the *parallel connection* $A_1 \times A_2$ to be the machine

$$A_1 \times A_2 = (S_1 \times S_2, \Sigma_1 \times \Sigma_2, M_{A_1 \times A_2})$$

where

$$M_{A_1 \times A_2}((s_1, s_2), (\sigma_1, \sigma_2)) = (M_1(s_1, \sigma_1), M_2(s_2, \sigma_2))$$

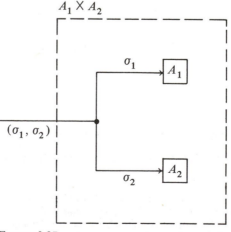

FIGURE 3.27

As indicated in the figure and in the definition of $M_{A_1 \times A_2}$, we simply allow the machines A_1 and A_2 to act independently. Apparently to know the state $s = (s_1, s_2)$ of $A_1 \times A_2$ requires that we know the state of each component machine. Accordingly, it is reasonable to take as our initial state the pair $0_{A_1 \times A_2} = (0_{A_1}, 0_{A_2})$.

EXAMPLE 3.49   Given the two machines $A_1, A_2,$

A_1	a	b	A_2	σ	τ
α	β	γ	0	1	0
β	γ	α	1	0	1
γ	α	γ			

we form the parallel connection $A_1 \times A_2$ as follows:

$A_1 \times A_2$	(a, σ)	(a, τ)	(b, σ)	(b, τ)
$(\alpha, 0)$	$(\beta, 1)$	$(\beta, 0)$	$(\gamma, 1)$	$(\gamma, 0)$
$(\alpha, 1)$	$(\beta, 0)$	$(\beta, 1)$	$(\gamma, 0)$	$(\gamma, 1)$
$(\beta, 0)$	$(\gamma, 1)$	$(\gamma, 0)$	$(\alpha, 1)$	$(\alpha, 0)$
$(\beta, 1)$	$(\gamma, 0)$	$(\gamma, 1)$	$(\alpha, 0)$	$(\alpha, 1)$
$(\gamma, 0)$	$(\alpha, 1)$	$(\alpha, 0)$	$(\gamma, 1)$	$(\gamma, 0)$
$(\gamma, 1)$	$(\alpha, 0)$	$(\alpha, 1)$	$(\gamma, 0)$	$(\gamma, 1)$

Thus

$$M_{A_1 \times A_2}((\alpha, 0), (a, \sigma)) = (M_1(\alpha, a), M_2(0, \sigma)) = (\beta, 1)$$
$$M_{A_1 \times A_2}((\alpha, 0), (a, \tau)) = (M_1(\alpha, a), M_2(0, \tau)) = (\beta, 0), \quad \text{etc.}$$

Needless to say, we will be most interested in the reverse procedure. So how do we ascertain whether a given machine might be decomposable into two machines operating in parallel? We can obtain a clue by partitioning the states (and symbols) of $A_1 \times A_2$ according as the first (or second) coordinates agree:

$$\pi_1 = \{\overline{(\alpha, 0), (\alpha, 1)}; \overline{(\beta, 0), (\beta, 1)}; \overline{(\gamma, 0), (\gamma, 1)}\}$$
$$\pi_2 = \{\overline{(\alpha, 0), (\beta, 0), (\gamma, 0)}; \overline{(\alpha, 1), (\beta, 1), (\gamma, 1)}\}$$

for the states and

$$\pi_1 = \{\overline{(a, \sigma), (a, \tau)}; \overline{(b, \sigma), (b, \tau)}\}$$
$$\pi_2 = \{\overline{(a, \sigma), (b, \sigma)}; \overline{(a, \tau), (b, \tau)}\}$$

for the symbols.   △

What can we learn from the pair of partitions π_i in the last example? For one thing, they represent automata congruences (R_1 and R_2) on the parallel machine.

Moreover, we have $\pi_1 \cdot \pi_2 = 0$ in the algebra of partitions (Section 0.6). This means that

$$
\begin{aligned}
\text{on states:} \ &s \, R_1 \, t \quad \text{and} \quad s \, R_2 \, t \Rightarrow s = t \\
\text{on symbols:} \ &\sigma \, R_1 \, \tau \quad \text{and} \quad \sigma \, R_2 \, \tau \Rightarrow \sigma = \tau
\end{aligned}
$$

or, as we say, the congruences R_1 and R_2 are *orthogonal*. Finally, the corresponding quotient machines $A_1 \times A_2/R_i$ are isomorphic to the original components A_i. All of these observations serve to motivate the statement characterizing parallel realizations:

Theorem 3.7

Let A be a parallel connection $A = A_1 \times A_2$. Then there exist orthogonal congruences R_1, R_2 on A such that $A_i \simeq A/R_i$. Conversely, if there are two orthogonal congruences on a given machine A, then A can be realized by a parallel connection of the corresponding quotient machines.

Proof Let $A = A_1 \times A_2$. We define (surjective) automata homomorphisms $\varphi_i : A \to A_i$ as follows:

$$
\begin{aligned}
&\text{on states} &&\text{on symbols} \\
&\varphi_i(s_1, s_2) = s_i &&\varphi_i(\sigma_1, \sigma_2) = \sigma_i
\end{aligned}
$$

That there are indeed homomorphisms is evident because

$$
\begin{aligned}
\varphi_i M_{A_1 \times A_2}((s_1, s_2), (\sigma_1, \sigma_2)) &= \varphi_i(M_1(s_1, \sigma_1), M_2(s_2, \sigma_2)) \\
&= M_i(s_i, \sigma_i) \\
&= M_i(\varphi_i(s_1, s_2), \varphi_i(\sigma_1, \sigma_2))
\end{aligned}
$$

So we are able to use the homomorphism theorem for automata (Theorem 3.6) to conclude that there are congruences $R_i = \ker \varphi_i$ such that $A_i \simeq A/R_i$. Now, because

$$
\begin{aligned}
(s_1, s_2) \, R_i \, (t_1, t_2) &\Rightarrow \varphi_i(s_1, s_2) = \varphi_i(t_1, t_2) \\
(i = 1, 2) &\qquad\quad (i = 1, 2) \\
&\Rightarrow s_i = t_i \qquad (i = 1, 2) \\
&\Rightarrow (s_1, t_1) = (s_2, t_2)
\end{aligned}
$$

and similarly for symbols, the congruences R_i are orthogonal.

Conversely, if R_1 and R_2 are orthogonal congruences on a machine A, then the mappings

$$
\begin{aligned}
&\text{on states} &&\text{on symbols} \\
&\varphi(s) = ([s]_1, [s]_2) &&\varphi(\sigma) = ([\sigma]_1, [\sigma]_2)
\end{aligned}
$$

constitute an (injective) automata homomorphism

$$
\varphi : A \to A/R_1 \times A/R_2
$$

(Verify) The injectivity follows from the orthogonality; that is,

$$\varphi s = \varphi t \Rightarrow ([s]_1, [s]_2) = ([t]_1, [t]_2)$$
$$\Rightarrow s\, R_1\, t \quad \text{and} \quad s\, R_2\, t$$
$$\Rightarrow s = t$$

and similarly for symbols. □

EXAMPLE 3.50   For the machine A

A	a	b	c
0	6	5	3
1	5	6	4
2	1	0	0
3	3	4	6
4	4	3	5
5	1	0	4
6	0	1	3

we have the pair of orthogonal congruences, as represented by corresponding partitions:

$$R_1: \{\overline{0, 4, 6};\ \overline{1, 2, 3, 5}\} \quad \text{and} \quad \{\overline{a};\ \overline{b, c}\}$$
$$R_2: \{\overline{0, 1};\ \overline{2};\ \overline{3, 4};\ \overline{5, 6}\} \quad \text{and} \quad \{\overline{a, b};\ \overline{c}\}$$

Using the method of Section 3.8, we can obtain quotient machines

A/R_1	$[a]$	$[b]$
$[0]$	$[0]$	$[1]$
$[1]$	$[1]$	$[0]$

A/R_2	$[a]$	$[c]$
$[0]$	$[5]$	$[3]$
$[2]$	$[0]$	$[0]$
$[3]$	$[3]$	$[5]$
$[5]$	$[0]$	$[3]$

The (injective) homomorphism or realization $\varphi: A \to A/R_1 \times A/R_2$ is provided by the mappings described in the proof of Theorem 3.7; that is,

$$0 \xrightarrow{\varphi} ([0], [0])$$
$$1 \xrightarrow{\varphi} ([1], [0])$$
$$2 \xrightarrow{\varphi} ([1], [2]) \qquad a \xrightarrow{\varphi} ([a], [a])$$
$$3 \xrightarrow{\varphi} ([1], [3]) \qquad b \xrightarrow{\varphi} ([b], [a])$$
$$4 \xrightarrow{\varphi} ([0], [3]) \qquad c \xrightarrow{\varphi} ([b], [c])$$
$$5 \xrightarrow{\varphi} ([1], [5])$$
$$6 \xrightarrow{\varphi} ([0], [5])$$

Thus we have realized A by a parallel interconnection of two smaller machines.   .   △

A serial decomposition turns out to require only a single congruence but imposes a restriction on the alphabet of the "tail" machine. This will seem quite natural once we realize that the "head" machine must pass on some information to the tail, as shown in Figure 3.28. Then if

$$A_1 = (S_1, \Sigma, M_1) \qquad A_2 = (S_2, S_1 \times \Sigma, M_2)$$

we may define the *serial connection* $A_1 \circ A_2$ to be the machine

$$A_1 \circ A_2 = (S_1 \times S_2, \Sigma, M_{A_1 \circ A_2})$$

in which

$$M_{A_1 \circ A_2}((s_1, s_2), \sigma) = (M_1(s_1, \sigma), M_2(s_2, (s_1, \sigma)))$$

On any machine there will always be the *trivial* congruences in which we identify all the states or only those that are equal. If we exclude these congruences, we will be able to predict that the component machines are "smaller" in the following characterization of serial decompositions.

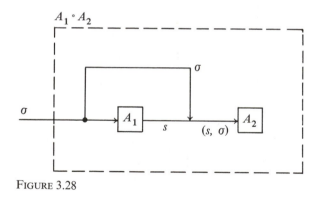

FIGURE 3.28

Theorem 3.8

A finite state machine A can be realized by a serial connection of two machines, each having fewer states than A iff there exists a nontrivial congruence on A.

Proof   Suppose that we have the realization $A \xrightarrow{\varphi} A_1 \circ A_2$, where each A_i has fewer states than A. Then if we let $\varphi(s) = (s_1, s_2)$, we have

$$\varphi(M(s, \sigma)) = M_{A_1 \circ A_2}(\varphi s, \varphi \sigma) = M_{A_1 \circ A_2}((s_1, s_2), \varphi \sigma)$$
$$= (M_1(s_1, \varphi \sigma), M_2(s_2, (s_1, \varphi \sigma)))$$

which shows that the pair of mappings

$$\varphi_1 s = s_1 \qquad \varphi_1 \sigma = \varphi \sigma$$

is an automata homomorphism $\varphi_1 \colon A \to A_1$. (Explain.) Once again we use the homomorphism theorem for automata to postulate the existence of a congruence $R = \ker \varphi_1$ on A with $A/R \simeq \varphi_1 A$ (a subautomaton of A_1, since in general, φ_1 is not surjective). Since A_1 and hence $\varphi_1 A$ have fewer states than A, R cannot be a congruence that identifies states only when they are equal. Neither can R identify all of the states, for we would then have

$$\varphi_1 s = \varphi_1 0 = 0 = s_1$$

for every state s in A. On the other hand, the mapping

$$\varphi s = (s_1, s_2) = (0, s_2)$$

is known to be injective. So we conclude that $|S_2| \geq |S|$, which is a contradiction. It follows that R is nontrivial.

Conversely, if R is a nontrivial congruence on A, then A/R has fewer states than A. Without loss of generality we may assume that R identifies symbols only when they are equal. This way, A/R has the same alphabet as A. Letting π_R denote the partition corresponding to R, we choose another partition π such that $\pi_R \cdot \pi = \theta$ and π has as many blocks as is the size of the largest block in π_R. (Why can this always be done?) Then noting that $|\pi| < |S|$, we consider the machine

$$A_\pi = (\pi, S/R \times \Sigma, M_\pi)$$

in which

$$M_\pi([s]_\pi, ([s]_{\pi_R}, \sigma)) = [M_A(s, \sigma)]_\pi$$

whereas if $[s]_\pi \cap [t]_{\pi_R} = \varnothing$, then we take

$$M_\pi([s]_\pi, ([t]_{\pi_R}, \sigma)) = \text{arbitrary}$$

We leave as an exercise the proof that the machine $A/R \circ A_\pi$ realizes A.   □

EXAMPLE 3.51   Assuming that symbols are identified only when they are equal, we consider the congruence R represented by the partition

$$\pi_R = \{\overline{0, 1, 2, 3}; \overline{4, 5}\}$$

of the states of the machine A shown below.

A	σ	τ
0	3	4
1	1	5
2	2	4
3	0	5
4	5	1
5	4	2

The head machine in the serially connected realization of A is the quotient A/R:

A/R	σ	τ
$[0]$	$[0]$	$[4]$
$[4]$	$[4]$	$[0]$

In order to construct a tail machine A_π, we first look for a partition π with four blocks that is orthogonal to π_R. Evidently

$$\pi = \{\overline{0,4}; \overline{1,5}; \overline{2}; \overline{3}\}$$

will suffice. Then we use the definition of A_π as given in the proof of Theorem 3.8. Following that prescription, we obtain

A_π	$([0], \sigma)$	$([0], \tau)$	$([4], \sigma)$	$([4], \tau)$
$[0]$	$[3]$	$[0]$	$[1]$	$[1]$
$[1]$	$[1]$	$[1]$	$[0]$	$[2]$
$[2]$	$[2]$	$[0]$		
$[3]$	$[0]$	$[1]$	arbitrary	

For instance,

$$M_\pi([0], ([0], \sigma)) = [M(0, \sigma)]_\pi = [3]$$
$$M_\pi([0], ([0], \tau)) = [M(0, \tau)]_\pi = [4] = [0]$$
$$M_\pi([0], ([4], \sigma)) = [M(4, \sigma)]_\pi = [5] = [1], \quad \text{etc.}$$

Four entries are arbitrary because

$$[2]_\pi \cap [4]_{\pi_R} = [3]_\pi \cap [4]_{\pi_R} = \varnothing$$

The realization $A \overset{\varphi}{\rightarrowtail} A/R \circ A_\pi$ is obtained, as in the case of the parallel connections, through the state mapping

$$s \overset{\varphi}{\rightarrow} ([s]_{\pi_R}, [s]_\pi)$$

that is,

$$0 \overset{\varphi}{\rightarrow} ([0], [0])$$
$$1 \overset{\varphi}{\rightarrow} ([0], [1])$$
$$2 \overset{\varphi}{\rightarrow} ([0], [2])$$
$$3 \overset{\varphi}{\rightarrow} ([0], [3])$$
$$4 \overset{\varphi}{\rightarrow} ([4], [0])$$
$$5 \overset{\varphi}{\rightarrow} ([4], [1])$$

$\triangle$

We have presented only the opening themes in an extensive theory of the decomposition of finite state machines. The theory culminates in the Krohn-

Rhodes decomposition theorem* to the effect that

> every finite state machine is covered by a serial-parallel interconnection of "simple" machines.

By a *simple* machine we understand either

(i) a machine with only two states, a "flip flop," to use the technical jargon, or
(ii) a machine A whose semigroup G_A is a "simple" group

The meaning of a covering is explained in the corollary to Theorem 3.5, but the idea of a semigroup G_A of a machine A has yet to be described.

If $A = (S, \Sigma, M)$ is any finite state machine, we define for each $x \in \Sigma^*$ a mapping or transformation

$$M_x: S \to S \qquad M_x(s) = M(s, x)$$

and we impose a multiplication operation in writing

$$M_x \cdot M_y = M_{xy}$$

In fact, with this operation, the semigroup

$$G_A = \{M_x : x \in \Sigma^*\}$$

is actually a finite transformation monoid, as defined in Section 3.2, but it has become quite standard to refer to G_A as the *semigroup of the automaton A*.

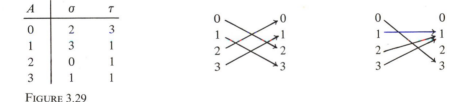

A	σ	τ
0	2	3
1	3	1
2	0	1
3	1	1

FIGURE 3.29

EXAMPLE 3.52   If A is the machine shown at the left in Figure 3.29, then the mappings M_σ and M_τ are just the transformations represented by the respective columns of the state table. Continuing in this vein for longer words x, we have the column representations

M_ϵ	M_σ	M_τ	$M_{\sigma\tau}$	$M_{\tau\sigma}$	$M_{\tau\tau}$	$M_{\sigma\tau\sigma}$	$M_{\tau\tau\sigma}$
0	2	3	1	1	1	3	3
1	3	1	1	3	1	3	3
2	0	1	3	3	1	1	3
3	1	1	1	3	1	3	3

* K. B. Krohn and J. L. Rhodes, "Algebraic Theory of Machines," *Proceedings of the Symposium on the Mathematical Theory of Automata*, Wiley, New York, 1962, pp. 341–384.

For instance,

$$M_{\sigma\tau}(0) = M(0, \sigma\tau) = 1$$
$$M_{\sigma\tau}(1) = M(1, \sigma\tau) = 1$$
$$M_{\sigma\tau}(2) = M(2, \sigma\tau) = 3$$
$$M_{\sigma\tau}(3) = M(3, \sigma\tau) = 1$$

The reader should check that $M_{\sigma\sigma} = M_\epsilon$. As a result, we have

$$M_{\sigma\sigma\sigma} = M_\sigma \qquad M_{\sigma\sigma\tau} = M_{\tau\sigma\sigma} = M_\tau$$

It may also be verified that

$$M_{\sigma\tau\tau} = M_{\tau\sigma\tau} = M_{\tau\tau\tau} = M_{\tau\tau}$$

All the transformations M_x for four-letter words x are found to be equivalent to ones we have already listed. It follows that there are only these eight elements in the monoid G_A (why?). G_A has the multiplication table shown in Table 3.5.

TABLE 3.5

$\cdot$	M_ϵ	M_σ	M_τ	$M_{\sigma\tau}$	$M_{\tau\sigma}$	$M_{\tau\tau}$	$M_{\sigma\tau\sigma}$	$M_{\tau\tau\sigma}$
M_ϵ	M_ϵ	M_σ	M_τ	$M_{\sigma\tau}$	$M_{\tau\sigma}$	$M_{\tau\tau}$	$M_{\sigma\tau\sigma}$	$M_{\tau\tau\sigma}$
M_σ	M_σ	M_ϵ	$M_{\sigma\tau}$	M_τ	$M_{\sigma\tau\sigma}$	$M_{\tau\tau}$	$M_{\tau\sigma}$	$M_{\tau\tau\sigma}$
M_τ	M_τ	$M_{\tau\sigma}$	$M_{\tau\tau}$	$M_{\tau\tau}$	$M_{\tau\tau\sigma}$	$M_{\tau\tau}$	$M_{\tau\tau\sigma}$	$M_{\tau\tau\sigma}$
$M_{\sigma\tau}$	$M_{\sigma\tau}$	$M_{\sigma\tau\sigma}$	$M_{\tau\tau}$	$M_{\tau\tau}$	$M_{\tau\tau\sigma}$	$M_{\tau\tau}$	$M_{\tau\tau\sigma}$	$M_{\tau\tau\sigma}$
$M_{\tau\sigma}$	$M_{\tau\sigma}$	M_τ	$M_{\tau\tau}$	$M_{\tau\tau}$	$M_{\tau\tau\sigma}$	$M_{\tau\tau}$	$M_{\tau\tau\sigma}$	$M_{\tau\tau\sigma}$
$M_{\tau\tau}$	$M_{\tau\tau}$	$M_{\tau\tau\sigma}$	$M_{\tau\tau}$	$M_{\tau\tau}$	$M_{\tau\tau\sigma}$	$M_{\tau\tau}$	$M_{\tau\tau\sigma}$	$M_{\tau\tau\sigma}$
$M_{\sigma\tau\sigma}$	$M_{\sigma\tau\sigma}$	$M_{\sigma\tau}$	$M_{\tau\tau}$	$M_{\tau\tau}$	$M_{\tau\tau\sigma}$	$M_{\tau\tau}$	$M_{\tau\tau\sigma}$	$M_{\tau\tau\sigma}$
$M_{\tau\tau\sigma}$	$M_{\tau\tau\sigma}$	$M_{\tau\tau}$	$M_{\tau\tau}$	$M_{\tau\tau}$	$M_{\tau\tau\sigma}$	$M_{\tau\tau}$	$M_{\tau\tau\sigma}$	$M_{\tau\tau\sigma}$

For instance,

$$M_{\sigma\tau} \cdot M_\tau = M_{\sigma\tau\tau} = M_{\tau\tau}$$
$$M_{\sigma\tau\sigma} \cdot M_\sigma = M_{\sigma\tau\sigma\sigma} = M_{\sigma\tau}, \quad \text{etc.}$$

In closing, the reader should observe that G_A is isomorphic to the transformation monoid T of Example 3.12. We have finally come full circle. $\triangle$

EXERCISES

1 Verify that the pair of mappings in the proof of Theorem 3.7 is an automata homomorphism

$$\varphi: A \to A/R_1 \times A/R_2$$

2 Construct the parallel connection $A/R_1 \times A/R_2$ for the quotient machines of Example 3.50 Then provide a few sample calculations to illustrate the realization

$$\varphi: A \longmapsto A/R_1 \times A/R_2$$

3 In completing the proof of Theorem 3.8, verify that $A/R \circ A_\pi$ realizes A.

4 Construct the serial connection $A/R \circ A_\pi$ for the machines of Example 3.51. Then provide a few sample calculations to illustrate the realization

$$\varphi: A \longmapsto A/R \circ A_\pi$$

5 Compute the multiplication table of the semigroup G_A of the machines A in
  (a) Example 3.39
  (b) Exercise 6 (machine B) of Section 3.8
  (c) Example 3.49 (machine A_1)

6 Obtain a parallel interconnection realizing the following machines:

(a)

	a	b	c
0	4	6	0
1	5	4	1
2	4	5	0
3	6	6	2
4	0	2	4
5	1	0	5
6	0	1	4

(b)

	α	β	γ	δ
0	5	3	2	4
1	4	2	3	5
2	1	3	2	0
3	0	2	3	1
4	3	1	0	2
5	2	0	1	3

(c)

	a	b	c	d	e
0	4	1	2	3	0
1	0	2	1	0	3
2	4	1	2	3	0
3	3	2	1	0	3
4	3	2	2	4	0

Then give an explicit description of the realization in each case.

7 Obtain a serial interconnection realizing each of the following machines:

(a)

	a	b	c
0	4	5	1
1	6	4	2
2	5	5	3
3	5	6	0
4	1	2	5
5	0	1	6
6	3	0	4

(b)

	α	β	γ
0	2	3	4
1	3	2	0
2	4	0	3
3	0	1	2
4	2	2	0

(c) The machine of Exercise 6(a)

Then given an explicit description of the realization in each case.

*8 Let $A = (S, \Sigma, M)$ be a finite state machine. Define the relation R_A on Σ^* by writing

$$x \, R_A \, y \Leftrightarrow M_x = M_y$$

that is, iff the transformations (of S) associated with x and y are equal as functions. Show that R_A is a congruence relation on Σ^*, and infer that $G_A \simeq \Sigma^*/R_A$.

*9 Prove that every finite monoid is (isomorphic to) the semigroup G_A of some finite state automaton A. [*Hint*: Given the monoid G, define an automaton $A = (S, \Sigma, M)$ for which $S = \Sigma = G$ and let M imitate the multiplication in G.]

SUGGESTED REFERENCES

3.1 Ljapin, E. S. *Semigroups.* (English translation from the Russian.) American Mathematical Society, Providence, R.I., 1964.

3.2 Nievergelt, J., J. C. Farrar, and E. M. Reingold. *Computer Approaches to Mathematical Problems.* Prentice-Hall, Englewood Cliffs, N.J., 1974.

3.3 Knuth, D. E. *The Art of Computer Programming, Vol. 3, Sorting and Searching.* Addison-Wesley, Reading, Mass., 1973.

3.4 Booth, T. L. *Sequential Machines and Automata Theory.* Wiley, New York, 1967

3.5 Kohavi, Z. *Switching and Finite Automata Theory.* McGraw-Hill, New York, 1970.

3.6 Ginzburg, A. *Algebraic Theory of Automata.* Academic Press, New York, 1968.

3.7 Arbib, M. A. (ed.). *Algebraic Theory of Machines, Languages, and Semigroups.* Academic Press, New York, 1969.

Monoids as such have not received a great deal of attention in mathematical literature. Except for the book by Ljapin, most texts on the subject would not be appropriate to the level of discussion in this chapter. So our time is perhaps better spent by consulting instead one of the references on group theory (Chapter 5) or in the theory of finite state automata (say Reference 3.7). Two suggested readings relate to sorting algorithms. The first (Reference 3.2) is of broader scope, but does have a nice section which compares various sorting tecnhiques. On the other hand, the book by Knuth contains just about all that one would ever need to know on the subject.

As for finite state machines, two references are included which have somewhat of an engineering flavor. They are followed by two others in a more mathematical vein. The last two books are geared to presenting the Krohn-Rhodes decomposition theory and its ramifications. Before consulting any of these books, however, it is well worth our time to first read the classic article by Rabin and Scott [M. O. Rabin and D. Scott, "Finite Automata and their Decision Problems," *IBM J. Research Develop.* **3**, 114–125 (1959).]

4

Lattices and Boolean Algebras

Boolean algebras are named after the nineteenth-century English mathematician, George Boole, for his pioneering work in symbolic reasoning. Though developed with a different purpose in mind, these algebras represent one of the more important classes of mathematical structures as applied to the computer sciences. In a straightforward manner one could quickly define such a structure by listing the operations and postulates which characterize these algebras. Alternatively, one can first introduce the more general concepts of partially ordered set and lattice. In the latter approach many of the axioms or postulates that one would have stated for Boolean algebras will appear as theorems about lattices. Since partially ordered sets and lattices themselves occur quite frequently in computer applications, we have chosen the second alternative.

The most direct application for all of these principles occurs in switching theory and in the logical design of digital computers. For this reason, we shall turn our attention toward the end of the chapter to an interpretation of the theory in an appropriate language of logic circuits. A study of the behavior of these circuits will help to crystallize the earlier theory in more concrete terms.

4.1 POSETS

The idea of a partial order relation on a set is appropriate for studying or modeling any hierarchical structure. Family trees and the executive organizational chart of a large corporation are examples of everyday situations in which such partial orders arise. It happens, however, that these kinds of structures occur so frequently in mathematics and computer science that the partially ordered sets play a role of central importance in the development of every theory, virtually without exception. At the outset, it would be helpful in understanding the following discussion if the student would briefly review Section 0.3, paying particular attention to the characterizations of the various types of relations.

We will find that in changing only one of the three axioms required of an equivalence relation, we will effect a considerable change in the character of the resulting relations. Thus we say that a *partially ordered set* (or *poset*, as it is usually abbreviated) is an algebraic structure $X = (X, \leq)$ consisting of a set X and a relation $\leq$ on X which satisfies the three conditions:

(i) *reflexivity* $x \leq x$
(ii') *antisymmetry* $x \leq y$ and $y \leq x \Rightarrow x = y$
(iii) *transitivity* $x \leq y,$ $y \leq z \Rightarrow x \leq z$

for all x, y, z in X.

EXAMPLE 4.1   The system $Z = (Z, \leq)$, with the usual interpretation of $\leq$ (less than or equal to), is a poset. $\triangle$

EXAMPLE 4.2   In Example 0.22 we showed that the set of all subsets of a given set S is a partially ordered set $\mathscr{P}(S) = (\mathscr{P}(S), \subseteq)$. $\triangle$

EXAMPLE 4.3   The system $Z^+ = (Z^+, |)$ is a partially ordered set, where "$|$" means "divides" (Section 1.1). To see this, we simply verify the three axioms

(i) $n|n$ $(n = 1 \cdot n)$
(ii') $n|m$ and $m|n \Rightarrow m = qn$ and $n = pm$
                                    $\Rightarrow m = (qp)m$ so $p = q = 1$
                                    $\Rightarrow n = m$
(iii) $n|m, m|k \Rightarrow m = qn, k = pm$
                                    $\Rightarrow k = (pq)n$
                                    $\Rightarrow n|k$ $\triangle$

In comparing our last two examples with Example 4.1, the reader should be cautioned that $\leq$ is simply a generic symbol (as is $\sim$ for equivalence relations) used for any partial order relation whatsoever. It will almost never have its usual meaning of "less than or equal to." In fact, the usual interpretation is not even typical. To explain this, let us temporarily use the symbol $\leq$ in place of $\subseteq$ and $|$ in Examples 4.2 and 4.3. Suppose, in the first case, that $S = \{a, b, c\}$. Then we have

$$\{a, b\} \nleq \{b, c\}   \text{ and }   \{b, c\} \nleq \{a, b\}$$

Similarly, in Example 4.3, we have

$$5 \nleq 6   \text{ and }   6 \nleq 5$$

So in a general poset $X = (X, \leq)$ there may be *incomparable* pairs of elements x, y in that

$$x \leq y   \text{ and }   y \leq x$$

are both false. Such situations do not arise in Example 4.1 because $Z = (Z, \leq)$ with the usual interpretation of $\leq$ is *totally* (or linearly) *ordered* in that for each pair of elements x and y, either $x \leq y$ or $y \leq x$. This explains the use of the word "partial" in reference to the complex nonlinear orderings that can arise in a general poset $X = (X, \leq)$.

EXAMPLE 4.4 In Section 0.6 we learned that the *covering* relation

$$R_1 \leq R_2 \quad \Leftrightarrow \quad (x \; R_1 \; y \Rightarrow x \; R_2 \; y)$$

is reflexive, antisymmetric, and transitive on the set of all relations on a fixed set S. It follows that $\mathscr{R}(S) = (\mathscr{R}(S), \leq)$ is a poset. △

EXAMPLE 4.5 Again in Section 0.6 we defined a relation called *refinement* among the partitions of

$$\Pi(S) = \{\pi : \pi \text{ is a partition of } S\}$$

In effect, we wrote

$$\pi_1 \leq \pi_2 \quad \Leftrightarrow \quad \begin{array}{l} \text{every block of } \pi_1 \text{ is contained} \\ \text{in some block of } \pi_2 \end{array}$$

As is quite commonly the case, the reflexivity and transitivity are fairly obvious; it is only the antisymmetry that needs to be checked in order to verify that $\Pi(S) = (\Pi(S), \leq)$ is a partially ordered set.

If $\pi_1 \leq \pi_2$ and $\pi_2 \leq \pi_1$, let us consider any block $A \in \pi_1$. We have $A \subseteq B \subseteq A'$, where B is a block of π_2 and A' is a block of π_1. But in a partition, we can have $A \subseteq A'$ only if $A = A'$. It follows that $A = B = A'$, which shows that every block of π_1 is a block of π_2. The converse is established in exactly the same way. This means that $\pi_1 = \pi_2$ as required. △

EXAMPLE 4.6 If $X = \{x_1, x_2, x_3, x_4\}$ and R is the relation described by the following matrix:

R	x_1	x_2	x_3	x_4
x_1	1	0	1	0
x_2	0	1	1	0
x_3	0	0	1	0
x_4	1	1	1	1

then the reader can use the characterizations of Section 0.3 to verify that $X = (X, R)$ is a poset. △

The last example suggests that we can decide algorithmically whether a given finite system $X = (X, R)$ is a poset, R being a relation on the set $X = \{x_1, x_2, \ldots, x_n\}$. Indeed, we can. Algorithm XIII provides for an exhaustive

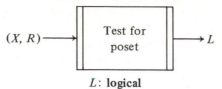

$$L: \text{logical}$$
$$i, j, k: \text{ element of } n^+$$

```
begin L ← false;
        for i ← 1 to n do
        if xᵢ Ɽ xᵢ then exit;
        for i ← 2 to n do
        for j ← 1 to i − 1 do
        begin if (xᵢ R xⱼ) ∧ (xⱼ R xᵢ) then exit;
                if xᵢ R xⱼ then
                    for k ← 1 to n do
                    if (xⱼ R xₖ) ∧ (xₖ R xᵢ) then exit
                else for k ← 1 to n do
                    if (xᵢ R xₖ) ∧ (xₖ R xⱼ) then exit
        end;
            L ← true
end
```

ALGORITHM XIII

check of the axioms of reflexivity, antisymmetry, and transitivity. In case any of the axioms fail, we "exit" with $L =$ **false**. Only when an exhaustive verification reveals that the three axioms hold do we set $L =$ **true**. Also note that if we replaced the antisymmetry test with a test for symmetry, the same procedure would be suitable for deciding whether R was an equivalence relation on X. Of course, we do not mean to imply that there are not more efficient algorithms for the latter purpose.

Assuming that we have verified that a given system $X = (X, \leq)$ is indeed a poset, then its hierarchical structure is best revealed through the construction of the *digraph of the poset*. Let each element of the set X be represented by a vertex or node, so placed on an oriented sheet of paper that the vertex for x is drawn below that for y just in case $x < y$ ($x \leq y$ but $x \neq y$). Then we draw a directed edge from vertex x to vertex y whenever y *covers* x, that is, $x < y$ and there are no elements "in between". Obviously, we would say that z is *between* x and y if $x < z < y$.

That all of this information can be extracted from a given finite poset X is best illustrated by Algorithm XIV. If we take as input for the algorithm the relation matrix E for the poset, then at the time of termination, E will have been transformed into the relation matrix for the digraph of the poset. First, we set all the diagonal elements of E to zero, thus obtaining the relation $<$ from

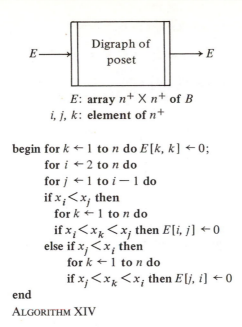

$$E: \textbf{array } n^+ \times n^+ \textbf{ of } B$$
$$i, j, k: \textbf{ element of } n^+$$

begin for $k \leftarrow 1$ **to** n **do** $E[k, k] \leftarrow 0$;
    for $i \leftarrow 2$ **to** n **do**
    for $j \leftarrow 1$ **to** $i - 1$ **do**
    if $x_i < x_j$ **then**
      for $k \leftarrow 1$ **to** n **do**
      if $x_i < x_k < x_j$ **then** $E[i, j] \leftarrow 0$
    else if $x_j < x_i$ **then**
      for $k \leftarrow 1$ **to** n **do**
      if $x_j < x_k < x_i$ **then** $E[j, i] \leftarrow 0$
end
ALGORITHM XIV

$\leq$. Then thinking of E as the relation matrix for a digraph, we systematically exclude edges corresponding to $x_i < x_j$ when an element is found in between. These are the entries $E[i, j]$ (or $E[j, i]$) set to zero by the algorithm.

EXAMPLE 4.7   If we use the array of Example 4.6 as input to Algorithm XIV, then a trace of the algorithm will proceed as shown in Table 4.1.

TABLE 4.1

i	j	k	E
		1	$E[1, 1] = 0$
		2	$E[2, 2] = 0$
		3	$E[3, 3] = 0$
		4	$E[4, 4] = 0$
2	1		
		$\vdots$	
4	3	1	$E[4, 3] = 0$

Thus with $i = 4$ and $j = 3$, we find that

$$x_4 = x_i < x_j = x_3$$

and with $k = 1$, we find an element in between:

$$x_4 = x_i < x_k < x_j = x_3$$

This causes the assignment $E[4, 3] \leftarrow 0$.

For this example, we obtain the output matrix E:

E	x_1	x_2	x_3	x_4
x_1	0	0	1	0
x_2	0	0	1	0
x_3	0	0	0	0
x_4	1	1	0	0

Now the digraph of the poset can be obtained immediately. We arrive at the picture of Figure 4.1(a). Since all the arrows will be directed upward, they are

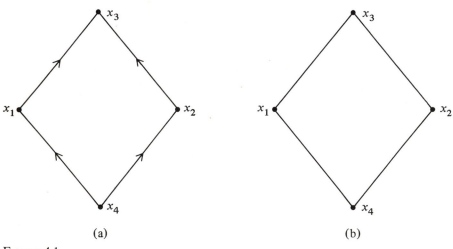

(a)   (b)

FIGURE 4.1

really superfluous. If we omit the arrows, however, as in Figure 4.1(b), then we will refer to the *diagram* of the poset. Note particularly the relationship between the incomparable elements x_1 and x_2 in this figure. △

EXAMPLE 4.8   In Example 0.22 we depicted a relation matrix for the containment of one subset in another, in the case of the set of all subsets of $S = \{a, b, c\}$. If Algorithm XIV were applied to this matrix, we would be led to the diagram of Figure 4.2. △

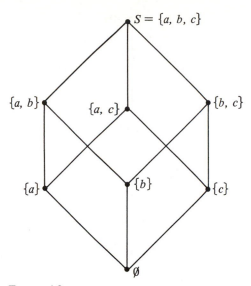

FIGURE 4.2

Evidently the passage from posets to diagrams can be reversed. If we are given the diagram of some hierarchy, we can construct a poset on the nodes $x, y, z, \ldots$, by agreeing to write $x \le y$ iff it is possible to climb from x to y along ascending line segments of the diagram. In case the diagram was already constructed by the method of Algorithm XIV, beginning with some partially ordered set $X = (X, \le)$, then the reverse procedure will return us to the original poset X. It follows that hierarchical diagrams and posets are synonymous. Each is but an alternative representation of the other.

EXAMPLE 4.9   In Figure 4.2 we can read off the diagram the fact that $\varnothing \subseteq \{a, c\}$ by virtue of the ascending path from $\varnothing$ through $\{a\}$, say, to $\{a, c\}$. In fact, the entire containment relation can be reconstructed in this way.   △

EXAMPLE 4.10   Consider the set

$$X = \{1, 2, 3, 5, 6, 10, 15, 30\}$$

with the relation of divisibility (Example 4.3). After a little experience, we should be able to draw the diagram of a poset without direct reference to Algorithm XIV. In the present example we are lead to the diagram of Figure 4.3. Again we note that the elements 5 and 6 are incomparable. It is not possible to climb from 5 to 6 (nor from 6 to 5) along ascending line segments of the diagram. The path from 5 through 10, then through 2 to 6, does not consist entirely of ascending segments. The edge from 10 to 2 is descending.   △

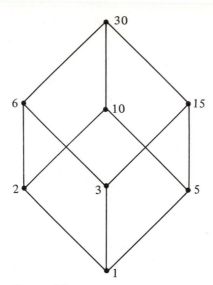

FIGURE 4.3

EXAMPLE 4.11   Given the hierarchy of Figure 4.4, we may use our reverse procedure to construct the following relation matrix:

$\leq$	I	0	1
I	1	0	0
0	1	1	0
1	1	0	1

We are assured that this relation on the set

$$C = \{0, 1, I\}$$

is a partial order.   △

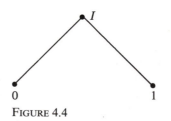

FIGURE 4.4

EXERCISES

1 Derive the digraph (or diagram) of each of the posets:
   (a) $X = \{n \in Z^+ : n|105\}$ (b) $X = \{n \in Z^+ : n|180\}$
   (c) $X = \{n \in Z^+ : n|192\}$ (d) $X = \{n \in Z^+ : n|120\}$

   where the relation is divisibility (Example 4.3).

2 Obtain relation matrices for the diagrams shown in Figure 4.5.

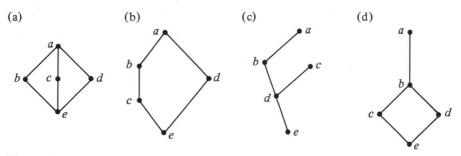

(a)   (b)   (c)   (d)

FIGURE 4.5

3 Use Algorithm XIII to test whether or not the following relation matrices represent posets:

(a)

	x_1	x_2	x_3	x_4	x_5
x_1	1	1	0	1	1
x_2	0	1	0	1	1
x_3	0	0	1	1	1
x_4	0	0	0	1	1
x_5	0	0	0	0	1

(b)

	x_1	x_2	x_3	x_4	x_5
x_1	1	0	0	1	0
x_2	1	1	0	0	1
x_3	0	0	1	0	0
x_4	0	1	1	1	0
x_5	0	0	0	1	1

(c)

	x_1	x_2	x_3	x_4	x_5
x_1	1	0	0	1	0
x_2	1	1	0	1	0
x_3	1	0	1	1	0
x_4	0	0	0	1	0
x_5	1	1	1	1	1

(d)

	x_1	x_2	x_3	x_4	x_5
x_1	1	0	0	0	1
x_2	1	1	1	0	1
x_3	1	0	0	0	1
x_4	1	1	1	1	1
x_5	0	0	0	0	1

4 List all the pairs of incomparable elements in the posets of Exercise 2.

5 Show that there cannot exist n distinct elements $x_1, x_2, \ldots, x_n$ in a poset $X = (X, \leq)$, where $n > 1$ and such that $x_1 \leq x_2 \leq \cdots \leq x_n \leq x_1$.

6 Define the concept of isomorphism for posets. Draw diagrams of all the nonisomorphic posets of sizes:

(a) 3   (b) 4   (c) 5

7 Show that if $a < b$ in a finite poset, then there exists a sequence

$$a = x_0 < x_1 < \cdots < x_n = b$$

such that x_i covers x_{i-1} $(1 \leq i < n)$. [*Hint*: Use induction on the number of elements between a and b.]

8 What can you say about a partially ordered equivalence relation?

*9 Let $X = (X, \leq)$ be a totally ordered set. In X^n we introduce the *lexicographical order* relation:

$$(x_1, \ldots, x_n) \leq (y_1, \ldots, y_n)$$

iff

$$x_1 < y_1 \text{ or for some maximal } k \begin{cases} x_i = y_i & (1 \leq i \leq k \leq n) \\ \text{and} \\ x_{k+1} < y_{k+1} & (\text{if } k < n) \end{cases}$$

Show that this is a partial order relation on X^n. Is it total?

*10 How would you define lexicographical (dictionary) order in

(a) $\bigcup_{n=1}^{\infty} X^n$ [for a totally ordered set $X = (X, \leq)$]

(b) Σ^* (for a totally ordered alphabet Σ)

[*Hint*: See Exercise 9 and Figure 3.11.]

4.2 LATTICES

Let $X = (X, \leq)$ be a partially ordered set and suppose that x and y are elements of X. Then an element u is called a *least upper bound* (abbreviated lub) of x and y if

(a) $u \geq x$ and $u \geq y$ (u is an *upper bound* of x and y)
(b) $w \geq x$ and $w \geq y \Rightarrow w \geq u$

Condition (b) asserts that among all the upper bounds of x and y, u is a "smallest." Of course, the notation $a \geq b$ is equivalent to $b \leq a$.

EXAMPLE 4.12   In Example 4.10, 6 is a lub of 2 and 3. Note that 30 is an upper bound for 2 and 3, since

$$30 \geq 2 \quad \text{and} \quad 30 \geq 3$$

but 30 is not a least upper bound for 2 and 3 because condition (b) fails. The statement

$$w \geq 2 \quad \text{and} \quad w \geq 3 \Rightarrow w \geq 30$$

is blatantly false (let $w = 6$ to see this). △

EXAMPLE 4.13   In Example 4.6 x_3 is an upper bound of x_2 and x_4, but again x_3 is not a least upper bound of x_2 and x_4, because with $w = x_2$, the statement (b)

$$w \geq x_2 \quad \text{and} \quad w \geq x_4 \Rightarrow w \geq x_3$$

is false. Evidently x_2 is the lub of x_2 and x_4. △

EXAMPLE 4.14   In Figure 4.2 the subset $\{a, b\}$ is the lub of the subsets $\{a\}$ and $\{b\}$. In fact, there is every reason to suspect that for arbitrary subsets A, B in a poset $\mathscr{P}(S) = (\mathscr{P}(S), \subseteq)$, the lub of A and B will coincide with the set union $A \cup B$. Far from being just a suspicion, however, we have proved the properties [(a), (b) in Example 0.10] attesting to this fact. △

In a dual fashion, an element v is called a *greatest lower bound* (glb) of the elements x and y provided that

(a′) $v \leq x$ and $v \leq y$ (v is a *lower bound* of x and y)
(b′) $w \leq x$ and $w \leq y \Rightarrow w \leq v$

Again condition (b′) says that among all the lower bounds of x and y, v is the "largest." The reader should appreciate the fact that each partially ordered set $X = (X, \leq)$ has an associated dual poset $(X, \geq)$ in which the elements are related as $x \geq y$ just as $y \leq x$ in the original poset. The diagram of the dual poset is obtained by turning the diagram of the original poset upside down. Surely it is apparent that lub's are transformed into glb's (and vice versa) as we pass from X to its dual.

EXAMPLE 4.15   For the poset of Example 4.10, the glb of 6 and 10 is 2, and 3 is the glb of 6 and 15. On the other hand, the glb of 3 and 6 is 3 itself. (See the statement of our forthcoming lemma.) The reader is correct to guess that glb's coincide with greatest common divisors in the case of a poset of integers organized around the relation of divisibility. In fact, that is exactly the sense of the properties (a), (b) in the definition of gcd's (Section 1.1). △

EXAMPLE 4.16   In Figure 4.4, the elements 0 and 1 have no glb. In fact, they do not even have a lower bound. There is not an element v such that $v \leq 0$ and $v \leq 1$. On the other hand, the element I is the lub of 0 and 1. △

If we have occasionally used the definite article "the" in reference to glb's and lub's, it is because of the following proposition.

Theorem 4.1

Let $X = (X, \leq)$ be a partially ordered set and let x and y be elements of X. Then if x and y have a lub u, they have only this one lub. Similarly, glb's are unique whenever they exist.

   Proof Suppose that u_1 and u_2 are both lub's for x and y. Then $u_1 \geq x$ and $u_1 \geq y$, and also $u_2 \geq x$ and $u_2 \geq y$. According to condition (b) of the definition of lub, these would imply that $u_1 \geq u_2$ and $u_2 \geq u_1$ (taking w first as u_1, then as u_2). Then we are able to conclude that $u_1 = u_2$ because of the antisymmetry property (ii') of a poset. Obviously the proof for uniqueness of glb's will be entirely dual to that just given.   □

   We know from Example 4.16 that glb's (or lub's) need not exist for particular pairs of elements in a poset. On the other hand, our theorem assures that they are always uniquely determined whenever they do exist for a given pair of elements. On the strength of this result, we can obtain an interesting (and as it turns out, very useful) class of posets simply by postulating the existence of lub's and glb's for all pairs of elements. Thus we say that a *lattice L* is a partially ordered set $L = (L, \leq)$ in which every pair of elements, x and y has a lub (which we denote by $x + y$) and a glb (denoted by $x \cdot y$). The uniqueness theorem guarantees that $+$ and $\cdot$ are well-defined binary operations $L \times L \to L$, and we write

$$u = \text{lub } (x, y) = x + y$$

$$v = \text{glb } (x, y) = x \cdot y$$

to emphasize the interpretation of lub and glb as bona fide binary operations on L. Some authors will use $\vee$ and $\wedge$ where we have used $+$ and $\cdot$. It is also quite common to refer to the "join" and "meet" of pairs of elements in L. We find it more convenient to use the standard algebraic operational symbols and to speak of "sums" and "products" of the elements.

   It will be useful to reformulate conditions (a) and (b) and their duals in our new algebraic notation. In any lattice $L = (L, \leq) = (L, +, \cdot)$, we necessarily have

(a) $x + y \geq x$ and $x + y \geq y$ (a') $x \cdot y \leq x$ and $x \cdot y \leq y$
(b) $w \geq x$ and $w \geq y \Rightarrow w \geq x + y$ (b') $w \leq x$ and $w \leq y \Rightarrow w \leq x \cdot y$

When we wish to emphasize the binary operations of the lattice, we use the notation $L = (L, +, \cdot)$ rather than $(L, \leq)$. Both of these interpretations are implicit in the four important properties and also in the lemma that follows.

Lemma

In any lattice L, the following three conditions are equivalent:

$$x + y = y \qquad x \cdot y = x \qquad x \leq y$$

Proof Obviously we intend to prove a cycle of three implications. So suppose $x + y = y$. Then we have $y \geq x$ and $y \geq y$ on account of condition (a). But $x \geq x$ because of reflexivity, and

$$y \geq x \text{ and } x \geq x \Rightarrow x \cdot y \geq x$$

according to (b'). Since we already have $x \geq x \cdot y$ from (a'), we obtain $x \cdot y = x$ by antisymmetry. Altogether we have shown

$$x + y = y \Rightarrow x \cdot y = x$$

The two remaining implications are left as an exercise. □

EXAMPLE 4.17   Evidently the poset of Figure 4.1 (Examples 4.6 and 4.7) is a lattice, for we are able to find sums (lub's) and products (glb's) for each pair of elements, in accordance with the following operation tables:

$+$	x_1	x_2	x_3	x_4			x_1	x_2	x_3	x_4
x_1	x_1	x_3	x_3	x_1		x_1	x_1	x_4	x_1	x_4
x_2	x_3	x_2	x_3	x_2		x_2	x_4	x_2	x_2	x_4
x_3	x_3	x_3	x_3	x_3		x_3	x_1	x_2	x_3	x_4
x_4	x_1	x_2	x_3	x_4		x_4	x_4	x_4	x_4	x_4

Presumably the reader is developing a facility in looking for lub's and glb's on the diagram of a poset. Thus the sum $a + b$ is the node at which ascending paths from a and b first come together. Similarly, look for descending paths for $a \cdot b$. With these observations and the special situation described in the lemma, the entries in our tables should be clear.

As an example of the situation described in the lemma, we cite the following:

$$x_2 = \text{lub} \, (x_4, x_2) = x_4 + x_2$$
$$x_4 = \text{glb} \, (x_4, x_2) = x_4 \cdot x_2$$
$$x_4 \leq x_2$$

Thus all three of the statements of the lemma will hold simultaneously (or not at all). △

EXAMPLE 4.18   Consider the poset $Z^+ = (Z^+, |)$ of positive integers relative to divisibility. The discussion of Example 4.15 shows that every pair of elements n and m will have the glb

$$n \cdot m = \text{glb} \, (n, m) = \text{gcd} \, (n, m)$$

In complete duality with these greatest common divisors, we have

$$n + m = \text{lub} \, (n, m) = \text{lcm} \, (n, m)$$

for the *least common multiple* (lcm) of n and m. This particular lattice forms the framework in the study of number theory, a rich and most fascinating branch of mathematics.

A portion of the (infinite) diagram of this lattice appears in Figure 4.3. That we have

$$2 + 3 = 6 \qquad 2 \cdot 3 = 1$$
$$10 + 15 = 30 \qquad 10 \cdot 15 = 5$$
$$3 + 6 = 6 \qquad 3 \cdot 6 = 3 \qquad (3 \leq 6, \text{ that is, } 3|6)$$

only serves to emphasize that we are not dealing with ordinary addition and multiplication here. The last of these calculations again illustrates the meaning of our lemma. $\triangle$

EXAMPLE 4.19   Since 0 and 1 have no glb, the poset $C = (C, \leq)$ of Example 4.11 is not a lattice. $\triangle$

Our next theorem gives a characterization of lattices solely in terms of their two binary operations. A list of eight properties in dual pairs turns out to be necessary and sufficient conditions for an algebra with two binary operations to be a lattice. These properties would have been among the postulates for a Boolean algebra if we had desired a more direct exposition of that theory, but they seem somewhat more clearly motivated when approached from a study of posets and lattices.

Theorem 4.2

Let $L = (L, +, \cdot)$ be a lattice. Then

(P1a) $x + x = x$ (P1b) $x \cdot x = x$
(P2a) $x + (y + z) = (x + y) + z$ (P2b) $x \cdot (y \cdot z) = (x \cdot y) \cdot z$
(P3a) $x + y = y + x$ (P3b) $x \cdot y = y \cdot x$
(P4a) $x + (x \cdot y) = x$ (P4b) $x \cdot (x + y) = x$

for all x, y, and z in L. Conversely, given a set L with two binary operations satisfying these eight conditions (known in pairs as the *idempotent*, *associative*, *commutative*, and *absorption* laws), there exists a partial order relation on L, relative to which L is a lattice with the given $+$, $\cdot$ as lub and glb operations.

Proof   To illustrate the style of argument used in the proof, we shall choose to verify property (P2a). Let $w = x + (y + z)$. Then by condition (a) in the definition of lub, we have $w \geq x$ and $w \geq y + z$. Using (a) again together with transitivity, we have $w \geq y$ and $w \geq z$. Recombining, we obtain $w \geq x + y$, according to (b). And since $w \geq z$ as well, we have $w \geq (x + y) + z$, according to (b) again. Similarly, if $v = (x + y) + z$, we can show that $v \geq x + (y + z) = w$. Having $v \geq w$ and $w \geq v$, we obtain $x + (y + z) = (x + y) + z$ by antisymmetry.

Let us now verify property (P4b). We have $x \leq x + y$ because of (a) in the definition of lub. Then according to the lemma, $x \cdot (x + y) = x$ is an immediate consequence.

Since the other verifications employ similar arguments, we shall content ourselves now with a short discussion of the converse. Suppose that the eight properties are satisfied by $L = (L, +, \cdot)$. Then we take our cue from the lemma and define a relation on L by writing:

$$x \leq y \quad \Leftrightarrow \quad x + y = y$$

Then property (P1a) implies reflexivity of the relation. As for the antisymmetry, we have

$$x \leq y \text{ and } y \leq x \Rightarrow x + y = y \quad \text{and} \quad y + x = x$$
$$\Rightarrow x = y + x = x + y = y$$

according to (P3a). Finally, we use (P2a) to obtain the transitivity:

$$x \leq y \text{ and } y \leq z \Rightarrow x + y = y \quad \text{and} \quad y + z = z$$
$$\Rightarrow x + z = x + (y + z) = (x + y) + z$$
$$= y + z = z$$
$$\Rightarrow x \leq z$$

Altogether we have shown that $L = (L, \leq)$ is a poset. The reader may go on to verify that $+, \cdot$ serve as lub and glb operations for this partial order. $\square$

EXAMPLE 4.20   If the properties (P1a) through (P4b) seem familiar, it is because you have seen them before. In Section 0.2, the same properties were established for the set algebras

$$\mathscr{P}(S) = (\mathscr{P}(S), \subseteq) = (\mathscr{P}(S), \cup, \cap)$$

Of course, we used the symbols $\cup$ and $\cap$ instead of $+$ and $\cdot$, but abstractly the operations are the same. It follows that $\mathscr{P}(S) = (\mathscr{P}(S), \cup, \cap)$ is a lattice for any set S. $\triangle$

EXAMPLE 4.21   If we are given only the two tabulations of $+$ and $\cdot$ in Example 4.17, an exhaustive check of the eight properties will reveal, according to the converse statement in Theorem 4.2, that we indeed have a lattice. Thus, with respect to (P4a), we will have

$$x_1 + (x_1 \cdot x_1) = x_1 + x_1 = x_1$$
$$x_1 + (x_1 \cdot x_2) = x_1 + x_4 = x_1$$
$$x_1 + (x_1 \cdot x_3) = x_1 + x_1 = x_1$$
$$x_1 + (x_1 \cdot x_4) = x_1 + x_4 = x_1 \quad \text{etc.}$$

According to our proof of the converse portion of the theorem, we can obtain the original poset (going back to Example 4.6) by defining

$$x \leq y \Leftrightarrow x + y = y$$

Thus we will obtain

$$x_1 \leq x_1 \quad \text{since} \quad x_1 + x_1 = x_1$$
$$x_4 \leq x_1' \quad \text{since} \quad x_4 + x_1 = x_1, \quad \text{etc.} \qquad \triangle$$

In connection with Example 4.20, you will recall that the set algebras of Section 0.2 satisfied a fifth pair of properties: distributivity. It happens that some lattices will satisfy these additional properties while others will not. Therefore, it is necessary to give a separate postulate to distinguish the important class of lattices with the distributive properties. We say that a lattice is *distributive* if the following laws hold:

(P5a) $x \cdot (y + z) = x \cdot y + x \cdot z$ (P5b) $x + (y \cdot z) = (x + y) \cdot (x + z)$

We note that only the distributive law (P5a) is encountered in ordinary arithmetic. Property (P5b) is actually a consequence of (P5a) and is further evidence of the duality which pervades the theory of lattices.

EXAMPLE 4.22   Evidently the lattice of Figure 4.6 is not distributive. If we consider the elements x, y, z as shown, we can compute

$$x \cdot (y + z) = x \cdot u = x \neq v = v + v = x \cdot y + x \cdot z$$

Here again we are using the suggestions outlined in Example 4.17 to compute sums (lub's) and products (glb's) in reference to the diagram. Ascending paths from y and z (first) come together at the node labeled u, and so $y + z = u$. Then $x \cdot u = x$ because of the lemma preceding Theorem 4.2; that is, $x \leq u$

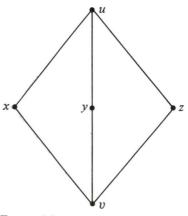

FIGURE 4.6

by virtue of the path along ascending line segments (just one here) from x up to u. Again $x \cdot y = x \cdot z = v$ because descending paths from x and y or x and z (first) come together at the node labeled v. △

EXERCISES

1 Of the diagrams in Figure 4.5 which are lattices? Of those which are lattices, which are distributive? Explain your answers.

2 Let X be the set of all continuous real-valued functions on the interval $0 \le x \le 1$. Define

$$f \le g \Leftrightarrow f(x) \le g(x) \text{ for all } x \text{ in the interval}$$

(a) Show that $(X, \le)$ is a poset.
(b) Show that it is not totally ordered.
(c) Define $f + g$ and $f \cdot g$ appropriately, and prove that $(X, +, \cdot)$ is a lattice.

3 Show that in any lattice

$$x \ge y \Rightarrow x \cdot (y + z) \ge y + (x \cdot z)$$

4 In any lattice, verify:
(a) Property (P1a) (b) Property (P3a) (c) Property (P4a)
as a part of the proof of Theorem 4.2.

5 Complete the proof of Theorem 4.2 as suggested, showing that $x + y$ and $x \cdot y$ are lub's and glb's respectively, for x and y.

6 Complete the proof of the lemma.

7 Devise a formal algorithm for testing
(a) whether a given finite poset represents a lattice
(b) whether a given finite lattice is distributive

8 In the poset shown in Figure 4.7 determine (if indeed they exist):
(a) glb (o, p) (b) lub (u, q) (c) lub (q, r) (d) glb (o, q)
(e) glb (t, v) (f) lub (p, z) (g) lub (p, y) (h) lub (u, v)

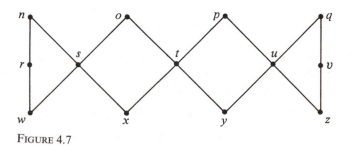

FIGURE 4.7

9 Define the concept of isomorphism for lattices. Draw diagrams of all the nonisomorphic lattices of sizes
   (a) 3   (b) 4   (c)5

10 Show that the following properties hold in any lattice:
   (a) $x \cdot y + x \cdot z \leq x \cdot (y + z)$
   (b) $x + (y \cdot z) \leq (x + y) \cdot (x + z)$

11 Show that Property (P5a) implies Property (P5b).

4.3 FREE DISTRIBUTIVE LATTICES

Suppose that we are going to order dinner in a restaurant from an a la carte menu. Certainly we should be concerned with the nutritional aspect and we would want to make selections that add up to a well-balanced meal. To simplify the problem, suppose that the menu consists of the following items:

A. Spanish omelet
B. chicken enchilada
C. Waldorf salad
D. steak
E. potato
F. liver with onions

Assume that these foods offer the nutritional benefits described in Table 4.2 where the entry 1 or 0 indicates that

TABLE 4.2

	Protein	Carbohydrates	Vitamins	Minerals
A	1	0	1	1
B	1	1	0	0
C	0	0	1	1
D	1	0	0	0
E	0	1	1	0
F	1	0	0	1

the nutritional benefit in question is or is not provided.

In order to ensure that our meal has sufficient protein content, we see from the first column that we must select A *or* B *or* D *or* F. Suppose that we symbolize this requirement by the sum A + B + D + F. Altogether, we can express the four nutritional requirements as a product:

$$(A + B + D + F)(B + E)(A + C + E)(A + C + F)$$

with each multiplication interpreted as the word *and*. Then if we simply "multiply out" in an appropriate lattice algebra, we will find that we obtain the desired list of well-balanced meals!

Of course we will assume the use of the distributive laws in multiplying. More importantly, however, we should point out the significance of the properties.

(P1b) $x \cdot x = x$
(P4a) $x + x \cdot y = x$

The idempotent law (P1b) symbolizes the fact that we need not order more than one serving of a particular dish, assuming that one serving will sufficiently provide the stated nutritional benefits. The absorption law (P4a) is addressed to the economics of the situation. If "apple pie" *or* "apple pie *and* berry pie" meet a certain set of nutritional requirements, then an economical solution would not include both. Needless to say, our model does not allow the diner to indulge, but he is permitted some latitude in choosing from the final list of well-balanced meals. Even here, however, we might insist that economics enter the picture. Thus we might be interested in ordering a well-balanced meal involving the fewest number of items or one which minimizes the total cost. These additional requirements can be imposed easily enough, once the "sum" of all the well-balanced meals is obtained.

EXAMPLE 4.23   The lattice polynomial f of the diner's problem is transformed from a product of sums to a sum of products as follows:

$$\begin{aligned}
f &= (A + B + D + F)(B + E)(A + C + E)(A + C + F) \\
&= (B + AE + DE + EF)(A + C + E)(A + C + F) \\
&= (AB + BC + BE + AE + DE + EF)(A + C + F) \\
&= AB + AE + BC + EF + CDE
\end{aligned}$$

The last line, which is to be read "A *and* B, *or* A *and* E, *or*, ...," gives a list of all the well-balanced meals.

Let us give a detailed description of the first multiplication, just in case the mechanics of the process are not yet clear. By the distributive law (P5a), we have

$$\begin{aligned}
(A + B + D + F)(B + E) &= (A + B + D + F)B + (A + B + D + F)E \\
&= AB + BB + DB + FB + AE + BE + DE + FE \\
&= AB + B + BD + BF + AE + BE + DE + EF \\
&= B + AE + DE + EF
\end{aligned}$$

Note particularly how B "absorbs" the products AB, BD, BF, and BE.   △

The diner will choose from the meals:

A and B = Spanish omelet and chicken enchilada
A and E = Spanish omelet and potato
B and C = chicken enchilada and Waldorf salad

E and F = potato and liver with onions
C and D and E = Waldorf salad and steak and potato

You will note that in each case, the selected meals include a 1 from every column. This is the nature of the general covering problems to which the so-called free distributive lattices are applicable. We will find that these covering problems arise in so many different settings that it is best to cast the whole discussion in a general framework. As we have found on a number of occasions, it is an advantage to view the various combinations of items as being chosen from an arbitrary alphabet of symbols. Whether these symbols represent items on an a la carte menu, vertices of a graph, or whatever, the algebraic structure is always the same.

Let $\Sigma = \{\sigma_1, \sigma_2, \ldots, \sigma_n\}$ be an alphabet of symbols. We recall (Section 0.4) that a combination from Σ is a word:

$$x = \sigma_{i_1}\sigma_{i_2} \cdots \sigma_{i_k} \qquad (i_1 < i_2 < \cdots < i_k)$$

that is, a selection without replications. Such selections are appropriate, as in the diner's problem, whenever duplications ought not to be recognized. Also it is convenient to interpret the concatenation of symbols linguistically as "and." Thus x is the selection of σ_{i_1} *and* σ_{i_2} *and* ... *and* σ_{i_k}, there being no repetitions. Now if we also have the combination $y = \sigma_{j_1} \sigma_{j_2} \cdots \sigma_{j_l}$, then we will say that x *absorbs* y (written $x \succ y$) if every symbol appearing in x occurs also in y. The terminology here is suggested by the absorption law (P4a).

EXAMPLE 4.24   For the combinations $\sigma_1\sigma_3$ and $\sigma_1\sigma_2\sigma_3\sigma_5$ we have

$$\sigma_1\sigma_3 \succ \sigma_1\sigma_2\sigma_3\sigma_5$$

in agreement with the absorption law, as if we were allowed to write

$$\sigma_1\sigma_3 + \sigma_1\sigma_2\sigma_3\sigma_5 = \sigma_1\sigma_3 + \sigma_1\sigma_3(\sigma_2\sigma_5)$$
$$= \sigma_1\sigma_3 \qquad\qquad \triangle$$

In examining Examples 4.23 and 4.24, it appears that we should be considering formal sums of combinations:

$$f = \sum_i x_i \qquad \begin{array}{l} x_i = \sigma_{i_1}\sigma_{i_2} \cdots \sigma_{i_k} \\ (i_1 < i_2 < \cdots < i_k) \end{array}$$

formal in that two sums are not to be regarded as equal unless they are identical—except, perhaps, for the arrangement of the summands. Then we will say that f is *irredundant* if $|f| = f$. Here the *absolute-value* sign refers not to an absolute value at all, but to an exhaustive use of the *absorption* law in reducing f to the point where the law can no longer be used to shorten the sum. We will use such irredundant expressions in the formation of our free distributive lattices.

EXAMPLE 4.25   If

$$f = \sigma_1\sigma_3\sigma_5 + \sigma_2\sigma_4 + \sigma_2\sigma_4\sigma_5 + \sigma_3 + \sigma_3$$

then

$$|f| = \sigma_2\sigma_4 + \sigma_3$$

Thus one of the copies of σ_3 absorbs the other as well as the combination $\sigma_1\sigma_3\sigma_5$, and $\sigma_2\sigma_4$ absorbs $\sigma_2\sigma_4\sigma_5$. At this point we have arrived at $|f|$ since no further absorption is possible.   △

   In making the set

$$L(n) = \{f : f = \sum_i x_i \quad \text{and} \quad |f| = f\}$$

into a lattice, we will be creating a kind of "combinational calculus," useful in any number of combinatorial situations, particularly those involved in covering problems. The first step in the introduction of a lattice structure is the definition of a partial order relation among the irredundant expressions f. We write

$$g \leq f \Leftrightarrow \begin{array}{c} \text{every combination } y \text{ in } g \text{ is absorbed} \\ x \succ y \\ \text{by some combination } x \text{ in } f \end{array}$$

The reader can verify (the proof is much like that of Example 4.5) that indeed we have defined a reflexive, antisymmetric, and transitive relation on $L(n)$. More importantly, however, it can be shown that lub's and glb's exist for each pair of elements, namely

$$f + g = \left| \sum_i x_i + \sum_j x_j \right|$$

$$f \cdot g = \left| \sum_{i,j} x_i \cdot x_j \right|$$

whenever $f = \sum_i x_i$ and $g = \sum_j x_j$. Of course, in the glb $f \cdot g$ the multiplication of combinations must be performed in accordance with the properties (P1b) and (P3b), so that there be no resulting duplication of symbols. Then in summary we are able to conclude that

$$L(n) = (L(n), \leq) = (L(n), +, \cdot)$$

is a lattice.

EXAMPLE 4.26   The following are sample calculations in $L(5)$:

$$(\sigma_1\sigma_2 + \sigma_3 + \sigma_2\sigma_5) + (\sigma_2 + \sigma_3\sigma_4) = |\sigma_1\sigma_2 + \sigma_3 + \sigma_2\sigma_5 + \sigma_2 + \sigma_3\sigma_4|$$
$$= \sigma_2 + \sigma_3$$

$(\sigma_1\sigma_2 + \sigma_3 + \sigma_2\sigma_5) \cdot (\sigma_2 + \sigma_3\sigma_4)$

$= |\sigma_1\sigma_2 \cdot \sigma_2 + \sigma_3 \cdot \sigma_2 + \sigma_2\sigma_5 \cdot \sigma_2 + \sigma_1\sigma_2 \cdot \sigma_3\sigma_4 + \sigma_3 \cdot \sigma_3\sigma_4 + \sigma_2\sigma_5 \cdot \sigma_3\sigma_4|$

$= |\sigma_1\sigma_2 + \sigma_2\sigma_3 + \sigma_2\sigma_5 + \sigma_1\sigma_2\sigma_3\sigma_4 + \sigma_3\sigma_4 + \sigma_2\sigma_3\sigma_4\sigma_5|$

$= \sigma_1\sigma_2 + \sigma_2\sigma_3 + \sigma_2\sigma_5 + \sigma_3\sigma_4$ △

EXAMPLE 4.27   All the elements of $L(3)$ are listed below:

0	$\sigma_1 + \sigma_2$	$\sigma_1\sigma_2 + \sigma_3$
1	$\sigma_1 + \sigma_3$	$\sigma_1\sigma_3 + \sigma_2$
σ_1	$\sigma_2 + \sigma_3$	$\sigma_2\sigma_3 + \sigma_1$
σ_2	$\sigma_1\sigma_2$	$\sigma_1\sigma_2 + \sigma_1\sigma_3$
σ_3	$\sigma_1\sigma_3$	$\sigma_1\sigma_2 + \sigma_2\sigma_3$
$\sigma_1\sigma_2\sigma_3$	$\sigma_2\sigma_3$	$\sigma_1\sigma_3 + \sigma_2\sigma_3$
$\sigma_1 + \sigma_2 + \sigma_3$	$\sigma_1\sigma_3 + \sigma_2\sigma_3 + \sigma_1\sigma_2$	

Here 0 corresponds to an empty summation, and 1 represents a summation with only one summand—an empty product—but it is perhaps just as well to think of these special elements only in terms of the relationship:

$$0 \le f \le 1$$

for all f. How are we then to compute $0 + f$, $0 \cdot f$, $1 + f$, and $1 \cdot f$? [*Hint*: Use the lemma preceding Theorem 4.2.]

How do we know that our list of the elements of $L(3)$ is complete? A direct verification that the sum or product of any two members results in another member of the list should be convincing enough, especially if no other irredundant sums of combinations come to mind. But the reader may still be surprised to learn that

$$|L(3)| = 20$$
$$|L(4)| = 168$$
$$|L(5)| = 7581$$
$$|L(6)| = 7,828,354$$

whereas a general formula for $|L(n)|$ is not known!   △

We call $L(n)$ the *free distributive lattice* on n generators: $\sigma_1, \sigma_2, \ldots, \sigma_n$. The reason for this terminology can be found in our discussion (Section 3.3) of homomorphisms of algebras and freely generated algebras. It happens that we will be rather less interested in the homomorphism theory for lattices than we were in the case of monoids. Nevertheless, it does seem appropriate that we call $L(n)$ by its proper name. Having done this, however, we now move on to some rather interesting applications having to do with the computation of the important isomorphism invariants for undirected graphs.

You will recall (Section 2.6) the many situations in which the solution to a particular combinatorial problem regarding a graph $G = (V, E)$ reduces to that of determining one or another of the invariants:

$$\alpha = \min_{\alpha} \{v_1, v_2, \ldots, v_\alpha : \text{covering set}\}$$
$$\beta = \max_{\beta} \{v_1, v_2, \ldots, v_\beta : \text{independent set}\}$$
$$\delta = \min_{\delta} \{v_1, v_2, \ldots, v_\delta : \text{dominating set}\}$$
$$\kappa = \min_{\kappa} \{V_1, V_2, \ldots, V_\kappa : \text{proper coloring}\}$$

These invariants were respectively called the covering, independence, domination, and chromatic numbers of the graph G. The independence number β (and also the clique number β') can be determined by Algorithm IX (Section 3.5), but we have yet to describe procedures for computing the covering, domination, and chromatic numbers. We will find, however, that the free distributive lattices are exactly what we need.

In the case of the covering number α we are interested in the fewest number of vertices that will cover all the edges. So in Algorithm XV we merely form the

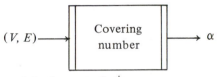

$(V, E) \longrightarrow$ Covering number $\longrightarrow \alpha$

i, j: **element of** n^+
α: **nonnegative integer**
f: **element of** $L(n) = (L(n), +, \cdot)$

> **begin** $f \leftarrow 1$;
>   **for** $j \leftarrow 1$ **to** $n - 1$ **do**
>   **for** $i \leftarrow j + 1$ **to** n **do**
>   **if** $v_i \ E \ v_j$ **then**
> $f \leftarrow f \cdot (v_i + v_j)$;
> $\alpha \leftarrow \min_{x \text{ in } f} \{|x|\}$
> **end**

ALGORITHM XV

product of sums $v_i + v_j$, one for each edge $v_i \ E \ v_j$. In contrast, the final expression f in Algorithm XVI will be a product of sums g_j which represent all the vertices that dominate the vertex v_j. In either case, however, the computations take place in the lattice $L(n)$ with the vertices themselves as generators (as symbols of the underlying alphabet), and the resulting expression f will represent, as a sum, all the covering sets (dominating sets). All that remains is to ascertain the length of a smallest word (or combination) in the expression f, and that will be the covering number (or domination number).

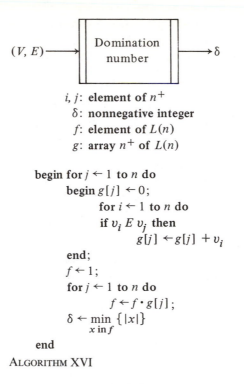

i, j: **element of** n^+

δ: **nonnegative integer**

f: **element of** $L(n)$

g: **array** n^+ **of** $L(n)$

begin for $j \leftarrow 1$ **to** n **do**
    begin $g[j] \leftarrow 0$;
        for $i \leftarrow 1$ **to** n **do**
        if $v_i \, E \, v_j$ **then**
                $g[j] \leftarrow g[j] + v_i$
    end;
    $f \leftarrow 1$;
    for $j \leftarrow 1$ **to** n **do**
        $f \leftarrow f \cdot g[j]$;
    $\delta \leftarrow \min_{x \,\text{in}\, f} \{|x|\}$

end

ALGORITHM XVI

EXAMPLE 4.28   Let us use the graph of Figure 4.8 to illustrate Algorithm XV. Essentially, the algorithm scans the lower-triangular matrix representation of the graph, "cascading" the sums $v_i + v_j$ corresponding to edges of the graph by multiplication in the appropriate free distributive lattice. Thus for Figure 4.8 we obtain

$$
\begin{aligned}
f &= (a + b)(a + c)(b + d)(b + e)(b + f)(c + e)(d + e)(e + g)(f + g) \\
&= (a + bc)(b + d) \cdots \\
&= (ab + bc + ad)(b + e) \cdots \\
&= (ab + bc + ade)(b + f) \cdots \\
&= (ab + bc + adef)(c + e) \cdots \\
&= (bc + abe + adef)(d + e) \cdots \\
&= (bcd + adef + bce + abe)(e + g) \cdots \\
&= (adef + bce + abe + bcdg)(f + g) \\
&= adef + bcef + abef + bceg + abeg + bcdg
\end{aligned}
$$

Accordingly, $\alpha = 4$ is found to be the covering number.   △

EXAMPLE 4.29   If we use the same graph to illustrate Algorithm XVI, we will first determine the sums g_j as follows:

        $(a + b + c)$ for vertex a
        $(a + b + d + e + f)$ for vertex b,   etc.

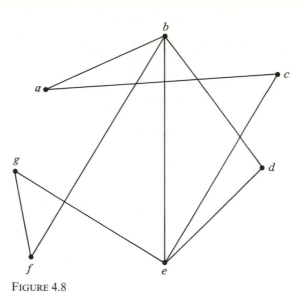

FIGURE 4.8

indicating the collection of vertices that dominate vertex a, b, etc. Then the expression f generated by Algorithm XVI will be the product of these sums in the appropriate free distributive lattice; that is,

$$
\begin{aligned}
f &= (a + b + c)(a + b + d + e + f)(a + c + e)(b + d + e) \\
&\qquad (b + c + d + e + g)(b + f + g)(e + f + g) \\
&= (a + b + cd + ce + cf)(a + c + e) \cdots \\
&= (a + bc + cd + ce + cf + be)(b + d + e) \cdots \\
&= (ab + bc + be + ad + cd + ae + ce)(b + c + d + e + g) \cdots \\
&= (ab + bc + be + ad + cd + ae + ce)(b + f + g) \cdots \\
&= (ab + bc + be + adf + cdf + aef + cef + adg + cdg \\
&\qquad\qquad\qquad\qquad + aeg + ceg)(e + f + g) \\
&= be + aef + cef + aeg + ceg + abf + bcf + adf + cdf \\
&\qquad\qquad\qquad\qquad + abg + bcg + adg + cdg
\end{aligned}
$$

Accordingly, $\delta = 2$ because $\{b, e\}$ is a smallest dominating set.   △

If we want to use the free distributive lattices for determining the chromatic number of a graph, we must first apply Algorithm IX (Section 3.5). That is, we must assume that all the maximal independent sets have been found and that they are named $B_1, B_2, \ldots, B_m$. Then we may speak of the *covering matrix* (b_{ij}), whose row headings are the maximal independent sets B_i, and whose column headings are the vertices v_j of the graph $G = (V, E)$. As usual, we let $|V| = n$, so our matrix has m rows and n columns. As entries in the matrix, we let

$$
b_{ij} = \begin{cases} 1 & \text{if } v_j \in B_i \\ 0 & \text{otherwise} \end{cases}
$$

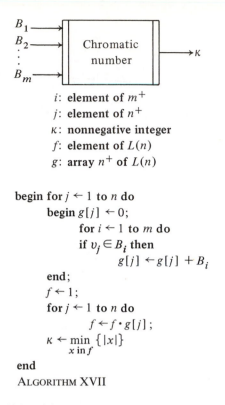

i: **element of** m^+
j: **element of** n^+
κ: **nonnegative integer**
f: **element of** $L(n)$
g: **array** n^+ **of** $L(n)$

begin for $j \leftarrow 1$ **to** n **do**
　　　begin $g[j] \leftarrow 0$;
　　　　　　for $i \leftarrow 1$ **to** m **do**
　　　　　　if $v_j \in B_i$ **then**
　　　　　　　　　　$g[j] \leftarrow g[j] + B_i$
　　　end;
　　　$f \leftarrow 1$;
　　　for $j \leftarrow 1$ **to** n **do**
　　　　　　$f \leftarrow f \cdot g[j]$;
　　　$\kappa \leftarrow \min_{x\ \text{in}\ f} \{|x|\}$

end

ALGORITHM XVII

Then the procedure (Algorithm XVII) for computing the chromatic number is much like that for the domination number. We simply multiply the sums g_j that account for the maximal independent sets that include the vertex v_j. And if we wish, we may think of this information as being summarized in column j of the covering matrix. In any case, the size of a smallest combination in the resulting product f will be the chromatic number of the given graph G. (Why?)

EXAMPLE 4.30　The graph of Example 3.28 is the same as that in Figure 4.8. In that example we obtained the maximal independent sets:

$$A = cdg \quad B = cdf \quad C = adg \quad D = adf \quad E = aef \quad F = bcg$$

So the appropriate covering matrix is the following:

	a	b	c	d	e	f	g
A	0	0	1	1	0	0	1
B	0	0	1	1	0	1	0
C	1	0	0	1	0	0	1
D	1	0	0	1	0	1	0
E	1	0	0	0	1	1	0
F	0	1	1	0	0	0	1

The expression f generated by Algorithm XVII will be

$$f = (C + D + E)F(A + B + F)(A + B + C + D)E(B + D + E)(A + C + F)$$
$$= (CF + DF + EF)(A + B + F) \cdots$$
$$= (CF + DF + EF)(A + B + C + D) \cdots$$
$$= (CF + DF + AEF + BEF)E \cdots$$
$$= (CEF + DEF + AEF + BEF)(B + D + E) \cdots$$
$$= (CEF + DEF + AEF + BEF)(A + C + F)$$
$$= CEF + DEF + AEF + BEF$$

It follows that the chromatic number of the graph of Figure 4.8 is $\kappa = 3$.

If we wish to obtain a proper coloring of the graph with only three colors, we will choose any combination in f of size three, say

$$CEF: \quad C = \{a, d, g\} \qquad E = \{a, e, f\} \qquad F = \{b, c, g\}$$

These three blocks will cover V, but we have to "back off" to a partition:

$$C' = \{d, g\} \qquad E' = \{a, e, f\} \qquad F' = \{b, c\}$$

so as not to paint the same vertex twice. △

EXERCISES

1 Perform the following calculations in $L(5)$:
(a) $(\sigma_1\sigma_2 + \sigma_3 + \sigma_2\sigma_5) + (\sigma_5 + \sigma_3)$
(b) $(\sigma_1\sigma_2 + \sigma_3 + \sigma_2\sigma_5) \cdot (\sigma_5 + \sigma_3)$
(c) $(\sigma_2 + \sigma_3) + (\sigma_2\sigma_3 + \sigma_2\sigma_4 + \sigma_3\sigma_4)$
(d) $(\sigma_1\sigma_3 + \sigma_1\sigma_5 + \sigma_3\sigma_5) \cdot (\sigma_2 + \sigma_3 + \sigma_4)$

2 Compute $|f|$ for the following sums:
(a) $f = \sigma_1\sigma_3\sigma_4 + \sigma_2\sigma_5 + \sigma_3\sigma_4 + \sigma_1\sigma_2\sigma_3\sigma_4 + \sigma_2\sigma_3\sigma_5$
(b) $f = \sigma_1\sigma_5 + \sigma_2\sigma_3 + \sigma_5 + \sigma_2\sigma_3\sigma_5 + \sigma_1\sigma_2\sigma_3$
(c) $f = \sigma_2\sigma_4 + \sigma_3 + \sigma_2\sigma_4 + \sigma_3\sigma_5 + \sigma_1\sigma_3 + \sigma_2\sigma_4\sigma_5$

3 Show that Example 4.27 lists all the elements of $L(3)$.

4 Use Algorithm XV to compute the covering number α for each of the following graphs. At the same time list all the minimal covering sets.
(a) Exercise 1(a) of Section 2.1   (b) Exercise 3(b) of Section 2.1
(c) Exercise 3(c) of Section 2.1   (d) Exercise 1(b) of Section 2.1
(e) Exercise 1(c) of Section 2.1   (f) Exercise 3(a) of Section 2.1
(g) Figure 2.15a   (h) Exercise 1 of Section 2.5
(i) Exercise 5(h) of Section 2.3

5 Use Algorithm XVI to compute the domination number δ for each of the graphs of Exercise 4. At the same time list all the minimal dominating sets.

6 Use Algorithm XVII to compute the chromatic number κ for each of the graphs of Exercise 4. Provide in each case a minimal proper coloring of the vertices.

4.4 LATTICE OF THE n-CUBE

The unit interval on the real line

$$I = \{x: x \in R \text{ and } 0 \leq x \leq 1\}$$

is studied from various points of view in mathematics. In point set topology, the theory of real variables, and calculus, one ordinarily studies the structure of I relative to its collection $\mathscr{U}$ of "open" sets. Without getting into all the details, we simply say that $(\mathscr{U}, \subseteq)$ is a partially ordered set with respect to the ordinary inclusion relation of subsets of I. It is an infinite (in fact, an uncountably infinite) poset, whose structure is by no means simple to describe. All the more complicated is the topological description of the n-*cube*

$$I^n = \{(x_1, x_2, \ldots, x_n): x_i \in I \text{ for each } i\}$$

after a suitable generalization of the one-dimensional open sets has been given.

In contrast, the replacement of $\mathscr{U}$ by the finite poset $C = \{0, 1, I\}$ of Figure 4.4 (Example 4.11) effects a discretization which emphasizes the "cellular" structure of I. That is to say, considering Figure 4.9, we see that the three-element

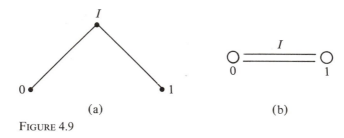

(a)   (b)

FIGURE 4.9

poset $C = (C, \leq)$ represents only the composition of I as a line segment with its two included endpoints or vertices. Everything else is considered irrelevant. Accordingly, the set

$$C^n = \{(c_1, c_2, \ldots, c_n): c_i \in C \text{ for each } i\}$$

together with the extended order relation

$$(c_1, c_2, \ldots, c_n) \subseteq (c'_1, c'_2, \ldots, c'_n) \Leftrightarrow c_i \leq c'_i \qquad \text{(all } i)$$

is sometimes called the *cellular n-cube* $C^n = (C^n, \subseteq)$.

EXAMPLE 4.31   We have labeled various elements of C^3 in Figure 4.10, but the reader should be aware that this figure simply depicts the geometric realiza-

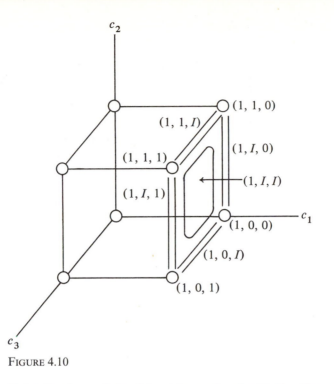

FIGURE 4.10

tion of the cellular 3-cube and should not be confused with the diagram of the poset C^3. The distinction is the same as that between Figure 4.9(b) and 4.9(a). Note, however, how the inclusions

$$(1, 1, 0) \subseteq (1, I, 0) \quad \text{since} \quad 1 \le 1, 1 \le I, 0 \le 0 \text{ in } C$$
$$(1, 0, 0) \subseteq (1, I, 0)$$
$$(1, I, 0) \subseteq (1, I, I), \quad \text{etc.}$$

are consistent with our geometric intuition. Yet, the definition of the order relation is purely algebraic. It follows that we can refer to this inclusion as well as other geometric notions even for n greater than three, where the actual realization of the n-cube is difficult to visualize.   △

EXAMPLE 4.32   Figure 4.11(b) shows a geometric realization of C^2. Ignoring for the moment the dashed lines, the student should compare this figure with the diagram of the poset C^2 as shown in Figure 4.11(a). In the latter each vertex is included in exactly two line segments, and each of the line segments in turn includes just two vertices.   △

Of course, we intend that these cellular n-cubes C^n be considered as examples of lattices. Together with the examples in Section 4.2 and the discussion of Section 4.3, the present examples attest to the great diversity of lattices that one encounters. The particular model of the n-cube has numerous applications.

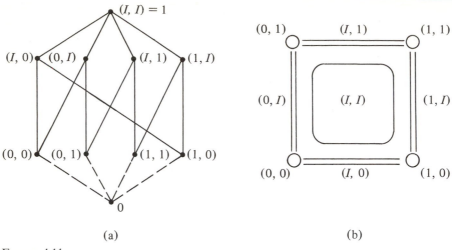

(a)   (b)

FIGURE 4.11

We will be particularly concerned with the applications to switching theory and coding theory. In trying to view C^n as a lattice, however, we are hampered somewhat by the fact that 0 and 1 have no glb in the poset C of Example 4.11. As a direct consequence, there are pairs of elements in C^n which appear not to have glb's.

This defect, however, is easily remedied by simply postulating the existence of a *zero element* 0 to be annexed at the bottom of the diagram of the poset C^n [as shown in Figure 4.11(a) for C^2]. Then in writing $1 = (I, I, \ldots, I)$, we would have the same situation

$$0 \subseteq c \subseteq 1 \qquad (\text{all } c \in C^n)$$

as obtained in the case of the free distributive lattices $L(n)$. Now, it would appear that we might have notational difficulties when $n = 1$ and $C^1 = C = \{0, 1, I\}$. But if we imagine that the elements of C^1 are enclosed in parentheses (as for C^n when $n > 1$), then there can be no confusion between the zero element 0 and (0), or between (1) and $1 = (I)$.

In general, the elements $c = (c_1, c_2, \ldots, c_n)$ of C^n are called *cells* (or k-cells, where k is the number of coordinates $c_i = I$). In particular, a 0-cell is called a *vertex*. It is difficult, perhaps impossible, to visualize a cell 0 (our zero element) that is to be considered as included in every vertex. So if you prefer to think of this special element as entirely fictitious, it is quite all right. But we do need to so augment C^n in order to obtain a lattice structure. Henceforth we will consider this special element 0 to be a cell in good standing.

We will now show that the augmented poset $C^n = (C^n, \subseteq)$ is a lattice. If $c = (c_1, c_2, \ldots, c_n)$ and $d = (d_1, d_2, \ldots, d_n)$ are cells of C^n, then we set

$$c + d = (c_1 + d_1, c_2 + d_2, \ldots, c_n + d_n)$$

where the coordinatewise lub's (they exist!) are taken in the poset C. In view of the lemma preceding Theorem 3.2, we have (since $0 \subseteq c$)

$$0 + c = c + 0 = c$$

for every cell c, and so we do not need to consider the zero element as one of the cells c, d entering into the general formula. As for products, we can write

$$c \cdot d = \sum_{\substack{e \subseteq c \\ e \subseteq d}} e$$

for the sum just defined. Note that this is not an empty summation because $e = 0$ always works. In fact, it is sometimes the only choice for e. The reader should see, however, that this definition gives exactly the same product as would be obtained if we were to set

$$c \cdot d = (c_1 \cdot d_1, c_2 \cdot d_2, \ldots, c_n \cdot d_n)$$

with the exception that $c \cdot d = 0$ whenever one of the coordinatewise glb's fails to exist in the poset C. The latter event will occur only when the corresponding coordinates are 0 and 1 (or vice versa) in c and d.

EXAMPLE 4.33   In order to perform additions and multiplications in C^n, we have only to consult the coordinatewise tables

$+$	0	1	I
0	0	I	I
1	I	1	I
I	I	I	I

$\cdot$	0	1	I
0	0		0
1		1	1
I	0	1	I

obtained by interpreting the poset C of Figure 4.9(a). Then, for instance in C^3, we may compute

$$(1, I, 0) + (1, 0, I) = (1 + 1, I + 0, 0 + I) = (1, I, I)$$
$$(1, I, I) \cdot (I, I, 0) = (1 \cdot I, I \cdot I, I \cdot 0) = (1, I, 0)$$

whereas

$$(1, I, 1) \cdot (I, I, 0) = 0$$

since $1 \cdot 0$ does not exist in C.

The reader should interpret the same results for Figure 4.10. Thus the 2-cell $(1, I, I)$ is the smallest cell containing both of the 1-cells or the lines $(1, I, 0)$ and $(1, 0, I)$; and $(1, I, 0)$ is the largest cell included in both of the 2-cells $(1, I, I)$ and $(I, I, 0)$, the latter being the back face of the cube as pictured in Figure 4.10.   △

We will begin to find applications for the lattice of the n-cube later in this chapter. At this point, however, it is convenient to introduce a few more geometric ideas that will be used in the later developments. Most of these ideas are related to the ordinary concept of distance, so our geometric intuition will come into play. Suppose we first consider the subset $B^n \subseteq C^n$,

$$B^n = \{v = (b_1, b_2, \ldots, b_n): b_i \in B = \{0, 1\}\}$$

that is, the set of vertices of the n-cube. We can introduce various notions of "distance" in B^n. By a *distance function*

$$d: B^n \times B^n \to R$$

we mean a real-valued function satisfying the properties

(1) $d(u, v) \geq 0$ and $d(u, v) = 0 \Rightarrow u = v$
(2) $d(u, v) = d(v, u)$
(3) $d(u, w) \leq d(u, v) + d(v, w)$

These conditions should seem entirely natural to the reader. The third condition is usually called the *triangle inequality*, and all three properties are satisfied by the Euclidean distance, with which the student is most familiar. In the context of the n-cube, it is sometimes more convenient to employ the *Hamming distance*[*] $\Delta(u, v)$, which simply counts the number of coordinates in which u and v differ. Then the whole idea can easily be extended to the cells of C^n by defining

$$\Delta(c, d) = \min_{\substack{u \in B^n \\ v \in B^n}} \{\Delta(u, v): u \subseteq c \text{ and } v \subseteq d\}$$

EXAMPLE 4.34   In Figure 4.10 (Example 4.31), where $u = (1, 1, 0)$ and $v = (1, 0, 1)$, we have

$$\Delta(u, v) = 2$$

because u and v differ in the second and third coordinates. Note that the Euclidean distance between u and v is $\sqrt{2}$. Now if $c = (1, I, I)$ and $d = (0, 0, I)$, then

$$\Delta(c, d) = 1$$

because c includes the vertices $(1, 0, 0)$, $(1, 0, 1)$, $(1, 1, 0)$, $(1, 1, 1)$ and d contains $(0, 0, 0)$ and $(0, 0, 1)$. The minimum among all the distances between the vertices $u \subseteq c$ and the vertices $v \subseteq d$ is that achieved by taking $u = (1, 0, 0)$ and $v = (0, 0, 0)$, or else $u = (1, 0, 1)$ and $v = (0, 0, 1)$. In either case, the distance is one.

Note in particular how the Hamming distance extends beyond our power of visualization, as was the case with the order relation in C^n. Thus if $u = (1, 0, 0, 1, 1)$ and $v = (0, 0, 1, 1, 0)$, we simply count $\Delta(u, v) = 3$ in C^5. The

---

[*] R. W. Hamming, "Error Detecting and Error Correcting Codes," *Bell System Tech. J.*, **29**, 147–160 (1950).

calculation requires no recourse to geometric visualization, any more than we would need a picture to deduce that $(1, I, 0, 1, 0) \subseteq (1, I, I, I, 0)$. $\triangle$

Many other geometric notions can be introduced. Though the idea is most commonly associated with vertices, we may say that two cells c, d are *adjacent* (and we write c adj d) if $\Delta(c, d) = 1$. Of course, in that event there would also exist vertices $u \subseteq c$ and $v \subseteq d$ such that $\Delta(u, v) = 1$. The next definition requires that both cells be of the same *order* (the k in k-cell) but is otherwise consistent with our geometric intuition. We say that two cells $c = (c_1, c_2, \ldots, c_n)$ and $d = (d_1, d_2, \ldots, d_n)$ are *parallel* (and we write $c \parallel d$) if

$$c_i = I \Leftrightarrow d_i = I$$

that is, if they have I's in exactly the same coordinates. Just as before, these questions can be resolved without using a picture.

EXAMPLE 4.35   In C^5, $(1, I, I, 0, 1)$ and $(0, 0, I, 0, I)$ are adjacent, but $(I, I, 0, 1, 0)$ and $(0, I, 1, 0, I)$ are not. The reader can verify these assertions by computing the respective distances.

The cells $(1, I, 0, I, 1)$ and $(0, I, 0, I, 1)$ are parallel because each has I's in the second and fourth coordinates. On the other hand, none of the four preceding cells are parallel, even though they are all of the same order, two. $\triangle$

EXERCISES

1 Obtain a formula for the number of k-cells in C^n.

2 Perform the following calculations in C^5:
   (a) $(1, 0, I, 0, 1) + (I, 1, I, 0, 0)$ (b) $(1, 0, I, 0, 1) \cdot (I, 0, I, 0, 1)$
   (c) $(1, 1, I, I, 0) + (I, 1, 0, I, 1)$ (d) $(1, 1, I, I, 0) \cdot (1, I, 0, 1, 1)$

3 Show that C^n is not a distributive lattice when $n \geq 2$.

4 Obtain the digraph (diagram) of the lattice C^3.

5 Compute the following distances in C^4:
   (a) $\Delta((1, 0, I, 1), (1, 1, I, I))$ (b) $\Delta((0, 1, 1, 0), (1, 0, 0, 0))$
   (c) $\Delta((I, I, 0, 1), (I, 1, 1, I))$ (d) $\Delta((1, 1, 0, 1), (0, 0, 1, 0))$

6 Decide whether the following pairs of cells are adjacent.
   (a) $(1, 0, I, 1)$ and $(1, 1, I, I)$ (b) $(0, 1, 1, 0)$ and $(1, 0, 0, 0)$
   (c) $(I, 0, 1, 1)$ and $(1, 1, 0, I)$ (d) $(I, 1, 0, 0)$ and $(1, 1, 0, 1)$

7 Decide whether the following pairs of cells are parallel.
   (a) $(1, 0, I, 1)$ and $(1, 1, I, I)$ (b) $(0, 1, 1, 0)$ and $(1, 0, 0, 0)$
   (c) $(I, I, 0, I)$ and $(I, I, 1, I)$ (d) $(1, 0, I, 0)$ and $(1, 1, I, 1)$

8 Describe the order of a cell as a transfer function (Section 1.4).

*9 Devise formal algorithms for
   (a) Determining whether two given cells are parallel
   (b) Computing the Hamming distance between two cells
   (c) Determining whether two given cells are adjacent

10 Show that the graph (of the vertices and edges) of C^n is nonplanar for $n > 3$.

4.5 BOOLEAN ALGEBRAS

It should be clear by now that every finite lattice

$$L = (L, +, \cdot) = (L, \leq)$$

has a *zero element* 0 and a *one element* 1, as defined by the conditions

$$0 \leq x \leq 1$$

for all x in L. And according to the lemma preceding Theorem 4.2, these elements enjoy rather special properties with regard to the addition and multiplication operations, namely:

(P6a) $x + 0 = x$ (P6b) $x \cdot 0 = 0$
(P7a) $x + 1 = 1$ (P7b) $x \cdot 1 = x$

for every x in L. Whether finite or not, a lattice L with zero and one is said to be *complemented* if corresponding to each element x in L there is an element x' (a *complement* of x) such that

(P8a) $x + x' = 1$ (P8b) $x \cdot x' = 0$

Finally, a *Boolean algebra* is a distributive, complemented lattice (with zero and one). It follows that all the laws (P1) through (P8) hold in a Boolean algebra.

EXAMPLE 4.36   The most obvious examples of Boolean algebra are the algebras of subsets of a given set S, for we obtained precisely these eight pairs of identities in Section 0.2 with union and intersection as the addition and multiplication operations. In particular, we found that the zero and one elements were given by $0_{\mathscr{P}(S)} = \varnothing$ and $1_{\mathscr{P}(S)} = S$. The complementary subsets

$$\sim A = \{x \colon x \in S \text{ and } x \notin A\}$$

satisfied the conditions

(P8a) $A \cup (\sim A) = S$ (P8b) $A \cap (\sim A) = \varnothing$

required of a Boolean algebra.   △

Not only is it true that every $\mathscr{P}(S) = (\mathscr{P}(S), \cup, \cap, \sim)$ is a Boolean algebra, but there is a converse. It goes by the name of *Stone's representation theorem*, to the effect that every Boolean algebra is (isomorphic to) an algebra of sets.

The reader will have an opportunity to discover the essence of this important result in Exercises 8 through 10 of Section 4.6.

EXAMPLE 4.37   An elementary Boolean algebra was introduced in Example 1.15. Changing notations, we now write

$$B = \{0, 1\}$$

with the operations of addition, multiplication, and complementation as follows:

+	0	1
0	0	1
1	1	1

·	0	1
0	0	0
1	0	1

$$0' = 1$$
$$1' = 0$$

By exhaustive checking we can see that all eight pairs of identities (P1) through (P8) are satisfied by this algebra.   △

EXAMPLE 4.38   Since the lattice of Figure 4.6 (Example 4.22) is not distributive, it is not a Boolean algebra. But it does have a zero element ($0 = v$) and a one element ($1 = u$), and we are even able to find complements for each of the five elements. Note, however, that the complements are not uniquely determined. Thus for the element x, we find two elements ($x' = y$ or $x' = z$) with the properties: $x + x' = u$ and $x \cdot x' = v$. Compare this fact with the statement of Theorem 4.3.   △

The following theorem asserts the uniqueness of complements in a Boolean algebra and states two important laws relating complementation to the binary operations of addition and multiplication. They are often called *DeMorgan's laws*, after the nineteenth-century English mathematician and logician Augustus deMorgan.

Theorem 4.3

In a Boolean algebra the complement x' of any element x is uniquely determined. Furthermore $(x')' = x$ and

$$(x + y)' = x' \cdot y' \qquad (x \cdot y)' = x' + y'$$

Proof   Suppose x has two complements, x' and x'', so that

$$x + x' = 1 \qquad x + x'' = 1$$
$$x \cdot x' = 0 \qquad x \cdot x'' = 0$$

Then we have

$$x' = x' \cdot 1 = x' \cdot (x + x'') = x'x + x'x'' = 0 + x'x''$$
$$= xx'' + x'x'' = (x + x') \cdot x'' = 1 \cdot x'' = x''$$

showing the uniqueness of the complement.

By what we have just proved, there is a unique complement $(x')'$ for the element x':

$$x' + (x')' = 1 \qquad x' \cdot (x')' = 0$$

Since x itself meets these requirements, we must have $(x')' = x$, as claimed. As for DeMorgan's laws, we write

$$(x + y) + x' \cdot y' = (x + y + x') \cdot (x + y + y') = 1 \cdot 1 = 1$$
$$(x + y) \cdot (x' \cdot y') = xx'y' + yx'y' = 0 + 0 = 0$$

using the two distributive laws. This shows that $x' \cdot y'$ is a complement of $x + y$. By the uniqueness of complements, we may in fact write $(x + y)' = x' \cdot y'$. In a similar way, we see that $(x \cdot y)' = x' + y'$. $\qquad\square$

There is a general method for constructing a Cartesian product of posets by coordinatewise use of the order relations in the various component posets. In fact, this is what was done in our extension of the order relation in C to the construction of the poset C^n in Section 4.4. In general, if we are given partially ordered sets $X_i = (X_i, \le)$ for every $i = 1, 2, \ldots, n$, then we may define

$$(x_1, x_2, \ldots, x_n) \le (y_1, y_2, \ldots, y_n) \quad \Leftrightarrow \quad x_i \le y_i \qquad (\text{all } i)$$

The reader should have no difficulty in seeing that the product set $\bigtimes_{i=1}^{n} X_i$ becomes a poset relative to this *coordinatewise order relation*. When combining posets in this fashion, it is of interest to ask whether certain algebraic properties of the component posets are inherited by the product. Certainly it is clear that if each component poset has a zero element, then the same will be true of the product poset. In fact,

$$0_{\times X_i} = (0, 0, \ldots, 0)$$

as is easily understood. Similarly, the possession of a one element by each X_i will ensure the existence of a one element for their Cartesian product. These simple observations are needed in deducing the following general result.

Theorem 4.4

Let $X_i = (X_i, \le)$ be posets for $i = 1, 2, \ldots, n$. Then $\bigtimes_{i=1}^{n} X_i$ with the coordinatewise order relation is

(I) A lattice if each X_i is a lattice.
(II) A Boolean algebra if each X_i is a Boolean algebra.

Furthermore,

$$(x_1, x_2, \ldots, x_n) + (y_1, y_2, \ldots, y_n) = (x_1 + y_1, x_2 + y_2, \ldots, x_n + y_n)$$

$$(x_1, x_2, \ldots, x_n) \cdot (y_1, y_2, \ldots, y_n) = (x_1 \cdot y_1, x_2 \cdot y_2, \ldots, x_n \cdot y_n)$$

and in case (II),

$$(x_1, x_2, \ldots, x_n)' = (x'_1, x'_2, \ldots, x'_n) \qquad \square$$

We leave the details of the proof as an exercise for the reader. Here we shall concentrate on an illustration of some importance. We consider the product set

$$B^n = \{(b_1, b_2, \ldots, b_n): b_i \in B \text{ for each } i = 1, 2, \ldots, n\}$$

Now, we learned in Example 4.37 that B is itself a Boolean algebra. So according to Theorem 4.4, B^n is also a Boolean algebra with the operations

$$(a_1, a_2, \ldots, a_n) + (b_1, b_2, \ldots, b_n) = (a_1 + b_1, a_2 + b_2, \ldots, a_n + b_n)$$
$$(a_1, a_2, \ldots, a_n) \cdot (b_1, b_2, \ldots, b_n) = (a_1 \cdot b_1, a_2 \cdot b_2, \ldots, a_n \cdot b_n)$$
$$(b_1, b_2, \ldots, b_n)' = (b'_1, b'_2, \ldots, b'_n)$$

and the zero and one elements

$$0_{B^n} = (0, 0, \ldots, 0) \qquad 1_{B^n} = (1, 1, \ldots, 1)$$

Thus the addition, multiplication, and complementation in B^n are performed coordinatewise in B, and B^n is a Boolean algebra with 2^n elements.

EXAMPLE 4.39   If $n = 4$, then $B^n = B^4$ is a 16-element Boolean algebra. Some sample calculations in B^4 are

$$(1, 0, 1, 0) + (0, 1, 1, 0) = (1, 1, 1, 0)$$
$$(1, 0, 1, 0) \cdot (0, 1, 1, 0) = (0, 0, 1, 0)$$
$$(1, 0, 1, 1)' = (0, 1, 0, 0)$$

Can you draw the diagram for B^4? $\triangle$

EXERCISES

1 Show that every finite lattice has a zero and a one element.

2 Show that for the elements a, b, c in any Boolean algebra:
   (a) $(ac + bc')' = a'c + b'c'$
   (b) $((a + c)(b + c'))' = (a' + c)(b' + c')$
   (c) $(a + b)(a' + c) = ac + a'b$

3 Prove the following identities:
   (a) $(a + b)(a' + b) = b$ (b) $ab' + c(a' + b + d) = c + ab'$
   (c) $ab'd + a'bc = (ad + bc)(a' + b')$ (d) $ad + a'c + bc = ab + a'c$

4 Given that $ab = ac$ and $a + b = a + c$, show that $b = c$.

[*Note.* In a Boolean algebra, division and subtraction have no meaning whatever, so the conclusion $b = c$ would not follow from just one of the conditions alone.]

5 Give an example of a Boolean algebra in which there exist three elements a, b, c satisfying (see the note for Exercise 4) the conditions:

(a) $ab = ac$ but $b \neq c$ (b) $a + b = a + c$ but $b \neq c$

*6 In any Boolean algebra, define the *Peirce function* (after the American logician C. S. Peirce) by

$$f \downarrow g = f' \cdot g'$$

Show that every Boolean expression (see Section 4.7) $f(x_1, x_2, \ldots, x_n)$ can be expressed in terms of the x_i by means of only the Peirce arrows. [*Hint:* Use the inductive definition of Boolean expression.]

7 Express $ab' + a'bc$ by means of only the Peirce arrows (see Exercise 6).

8 Draw the digraph (or diagram) of the lattice B^4.

9 Show that there is no three-element Boolean algebra.

10 Perform the following calculations in the Boolean algebra B^5:

(a) $(0, 1, 0, 1, 1) + (1, 1, 0, 1, 0)$ (b) $(0, 1, 0, 1, 1) \cdot (1, 1, 0, 1, 0)$

(c) $(0, 1, 0, 1, 1)'$ (d) $(1, 1, 0, 0, 1) + (0, 1, 0, 1, 0)$

(e) $(1, 1, 0, 0, 1) \cdot (0, 1, 0, 1, 0)$ (f) $(1, 1, 0, 0, 1)'$

4.6   FREE BOOLEAN ALGEBRAS

In the theory of monoids an important role is played by the free monoids (Section 3.4), those generated by concatenation of the symbols of an alphabet. In Section 4.3, we introduced the free distributive lattices $L(n)$, which were found to have many interesting applications. Now we are ready to study the free Boolean algebras. In all these instances, the situation is the same as that described in Section 3.3 for an algebra that is freely generated by the elements of a certain finite subset. With the free Boolean algebras, however, the generating subset will not be immediately apparent; it will emerge only in the course of the development.

We begin with the set of all functions from the product set B^n into B. These functions are called *truth tables*. We will describe a means for adding, multiplying, and complementing these truth tables, so that

$$B(n) = \{F: F: B^n \to B\}$$

becomes a Boolean algebra. The trick here is to add, multiply, and complement the functional values. Thus if we have the truth tables

$$F: B^n \to B \qquad G: B^n \to B$$

we set

$$(F + G)(b_1, b_2, \ldots, b_n) = F(b_1, b_2, \ldots, b_n) + G(b_1, b_2, \ldots, b_n)$$
$$(F \cdot G)(b_1, b_2, \ldots, b_n) = F(b_1, b_2, \ldots, b_n) \cdot G(b_1, b_2, \ldots, b_n)$$
$$F'(b_1, b_2, \ldots, b_n) = (F(b_1, b_2, \ldots, b_n))'$$

Then the easiest way to see that $B(n) = (B(n), +, \cdot, ')$ is a Boolean algebra is by direct verification of the properties (P1) through (P8). The results are an immediate consequence of the corresponding properties in the Boolean algebra B. We call this new algebra $B(n)$ the *free Boolean algebra* on n generators. Again the reason for this terminology will become apparent as we proceed.

EXAMPLE 4.40   Let us verify property (P4a) for the algebra $B(n)$. We have

$$(F + (F \cdot G))(b_1, b_2, \ldots, b_n)$$
$$= F(b_1, b_2, \ldots, b_n) + (F \cdot G)(b_1, b_2, \ldots, b_n)$$
$$= F(b_1, b_2, \ldots, b_n) + F(b_1, b_2, \ldots, b_n) \cdot G(b_1, b_2, \ldots, b_n)$$
$$= F(b_1, b_2, \ldots, b_n)$$

by the property (P4a) in the Boolean algebra B. According to the definition of equality of functions (Section 0.7), we have shown that $F + F \cdot G = F$.   △

EXAMPLE 4.41 $B(2)$ has 16 elements (functions), which we may collect in a table (Table 4.3),

TABLE 4.3

x_1	x_2	F_0	F_1	F_2	F_3	F_4	F_5	F_6	F_7	F_8	F_9	F_{10}	F_{11}	F_{12}	F_{13}	F_{14}	F_{15}
0	0	0	1	0	1	0	1	0	1	0	1	0	1	0	1	0	1
0	1	0	0	1	1	0	0	1	1	0	0	1	1	0	0	1	1
1	0	0	0	0	0	1	1	1	1	0	0	0	0	1	1	1	1
1	1	0	0	0	0	0	0	0	0	1	1	1	1	1	1	1	1

and the Boolean algebra $B(2)$ has the zero and one elements

$$0_{B(2)} = F_0 \qquad 1_{B(2)} = F_{15}$$

respectively.   △

EXAMPLE 4.42   We shall present a few computational illustrations for $B(3)$. Suppose that F and G are the truth tables in Table 4.4. Then $F + G$, $F \cdot G$, and F' are as indicated in Table 4.5.   △

We claim that the Boolean algebra $B(n)$ is freely generated by the *coordinate truth tables* $X_1, X_2, \ldots, X_n$, those having the definition

$$X_i(b_1, b_2, \ldots, b_n) = 1 \quad \Leftrightarrow \quad b_i = 1$$

TABLE 4.4

x_1	x_2	x_3	F	G
0	0	0	0	1
0	0	1	1	1
0	1	0	1	0
0	1	1	0	0
1	0	0	1	1
1	0	1	0	1
1	1	0	0	0
1	1	1	1	0

TABLE 4.5

x_1	x_2	x_3	$F + G$	$F \cdot G$	F'
0	0	0	1	0	1
0	0	1	1	1	0
0	1	0	1	0	0
0	1	1	0	0	1
1	0	0	1	1	0
1	0	1	1	0	1
1	1	0	0	0	1
1	1	1	1	0	0

In any case, we use the latter to define the *atomic truth tables* (or *atoms*) M_j ($0 \le j \le 2^n - 1$) as products:

$$M_j = X_1{}^{b_1} X_2{}^{b_2} \cdots X_n{}^{b_n} \qquad (b_1 b_2 \cdots b_n \to j)$$

the factors being the coordinates X_i or their complements in the algebra $B(n)$. That is to say, we use the notational convenience

$$X_i{}^0 = X_i' \qquad X_i{}^1 = X_i$$

and let the corresponding index j be determined by a *binary-to-decimal conversion*, symbolized $b_1 b_2 \cdots b_n \to j$. All of these ideas are really quite elementary, as the following example will show.

EXAMPLE 4.43   If $n = 3$, then X_2 and M_6 are the truth tables depicted in Table 4.6.

TABLE 4.6

x_1	x_2	x_3	X_2	M_6	
0	0	0	0	0	
0	0	1	0	0	
0	1	0	1	0	
0	1	1	1	0	
1	0	0	0	0	
1	0	1	0	0	
1	1	0	1	1	$(110 \to 6)$
1	1	1	1	0	

The reader should check that

$$M_6 = X_1{}^1 \cdot X_2{}^1 \cdot X_3{}^0 = X_1 \cdot X_2 \cdot X_3' \qquad (110 \to 6)$$

in accordance with the notation just introduced, keeping in mind that the multiplications are to be carried out in the Boolean algebra $B(3)$. $\triangle$

The actual verification of the $B(n)$ being free Boolean algebras requires two lemmas. These lemmas are important in their own right, because they summarize the most useful properties of the atomic truth tables. On the other hand, the arguments presented here can be imitated (see Exercises 8 through 10 at the end of this section) in devising a proof of the important Stone representation theorem for finite Boolean algebras alluded to earlier in the discussion following Example 4.36. If the ideas of Example 4.43 have been clearly understood, then our argument will seem to be quite straightforward.

Lemma 1

The atoms of $B(n)$ satisfy the properties

$$\text{(I)} \quad M_j \cdot M_k = \begin{cases} M_j & (j = k) \\ 0 & (j \neq k) \end{cases}$$

$$\text{(II)} \quad \sum_{j=0}^{2^n - 1} M_j = 1$$

Proof   Every M_j has exactly one 1 in its truth table, namely at position j $(0 \leq j \leq 2^n - 1)$. $\square$

Lemma 2

Every truth table F has a unique representation

$$F = \sum_{j \in J} M_j$$

as a sum of atoms.

Proof   Using property (II) of Lemma 1, we obtain

$$F = 1 \cdot F = \left(\sum_{j=0}^{2^n - 1} M_j \right) \cdot F = \sum_{j=0}^{2^n - 1} M_j \cdot F = \sum_{j \in J} M_j$$

where the last sum is taken over only those indices or positions j at which F takes on the value 1. As for uniqueness, suppose that we have

$$\sum_{j \in J} M_j = \sum_{k \in K} M_k$$

Now, if $j_1 \in J$, we use property (I) of Lemma 1 to write

$$M_{j_1} = \sum_{j \in J} M_j \cdot M_{j_1} = \left(\sum_{j \in J} M_j \right) \cdot M_{j_1}$$

$$= \left(\sum_{k \in K} M_k \right) \cdot M_{j_1} = \sum_{k \in K} M_k \cdot M_{j_1}$$

then the last expression shows that $j_1 \in K$. Similarly it is shown that $K \subseteq J$, so the two summations are really the same.   □

Theorem 4.5

The Boolean algebra $B(n)$ is freely generated by its coordinate truth tables $X_1, X_2, \ldots, X_n$.

Proof According to the definition in Section 3.3, we must first show that every truth table F has a representation as an algebraic expression in the X_i. To do this, we use Lemma 2 in obtaining

$$F = \sum_{j \in J} M_j = \sum_{\substack{j \in J \\ b_1 b_2 \cdots b_n \to j}} X_1^{b_1} X_2^{b_2} \cdots X_n^{b_n}$$

Certainly the last expression is an algebraic expression in the coordinate truth tables. Then if A is any Boolean algebra, and we have an assignment $\varphi(X_i) = a_i$, we can use the above representation to extend φ to a homomorphism $\varphi: B(n) \to A$ of Boolean algebras. We shall leave the details to the reader.   □

EXAMPLE 4.44 Once more in $B(3)$ we consider the truth table F in Table 4.7. Then we may represent F as an algebraic expression in X_1, X_2, X_3 as follows:

$$\begin{aligned}
F = M_2 + M_5 + M_6 &= X_1{}^0 X_2{}^1 X_3{}^0 + X_1{}^1 X_2{}^0 X_3{}^1 + X_1{}^1 X_2{}^1 X_3{}^0 \\
&= X_1' X_2 X_3' + X_1 X_2' X_3 + X_1 X_2 X_3' \qquad \triangle
\end{aligned}$$

TABLE 4.7

x_1	x_2	x_3	F	
0	0	0	0	
0	0	1	0	
0	1	0	1	$(010 \to 2)$
0	1	1	0	
1	0	0	0	
1	0	1	1	$(101 \to 5)$
1	1	0	1	$(110 \to 6)$
1	1	1	0	

EXERCISES

1 Perform the following calculations in the Boolean algebra $B(3)$:
  (a) $F + G$ (b) $F \cdot G$ (c) F'
  (d) $G \cdot H$ (e) $G + H$ (f) G'

x_1	x_2	x_3	F	G	H
0	0	0	0	1	1
0	0	1	1	1	0
0	1	0	1	1	0
0	1	1	0	0	0
1	0	0	1	0	1
1	0	1	1	1	1
1	1	0	1	1	0
1	1	1	0	0	0

2 Obtain a formula for $|B(n)|$. Explain your procedure.

3 Write the following truth tables as sums of atoms and as algebraic expressions in the coordinate truth tables X_1, X_2, X_3:
   (a) F in Exercise 1   (b) G in Exercise 1   (c) H in Exercise 1

4 Repeat Exercise 3 for the following truth tables:

x_1	x_2	x_3	F	G	H
0	0	0	0	1	0
0	0	1	1	1	0
0	1	0	1	0	1
0	1	1	0	0	1
1	0	0	1	1	0
1	0	1	0	1	0
1	1	0	0	0	1
1	1	1	1	1	0

5 Complete the details of the proof of Theorem 4.5.

*6 Define *coatoms*

$$L_i = X_1^{b_1} + X_2^{b_2} + \cdots + X_n^{b_n} \qquad (b_1 b_2 \cdots b_n \to i)$$

and prove the statements dual to those of Lemma 1. Then state and prove a result that is dual to Lemma 2 regarding products of coatoms.

7 Prove that coatoms are covered by the one element in the digraph or diagram of the Boolean algebra $B(n)$. [*Hint*: See Exercise 6.]

8 In any Boolean algebra an element a is called an *atom* if it covers the zero element. Show that in this sense, the atomic truth tables are the atoms of $B(n)$.

*9 In any finite Boolean algebra show that each element x has a unique representation, namely

$$x = \sum_{a \le x} a$$

as a sum of atoms. (See the definition in Exercise 8.) [*Hint*: Proceed along the lines of Lemma 1 and Lemma 2.]

*10 Show that every finite Boolean algebra is isomorphic to a set algebra (a power set with the operations $\cup$, $\cap$, $\sim$). This is Stone's representation theorem for finite Boolean algebras. [*Hint*: Use the representation of Exercise 9 and define a mapping

$$\Phi(x) = \{a : a \leq x\}$$

from the given Boolean algebra to the power set of its atoms. Then show that this mapping is an isomorphism of Boolean algebras.]

11 Show that every finite Boolean algebra has an order that is a power of 2. [*Hint*: See Exercise 10.]

4.7   INTERPRETATION AND CANONICAL FORMS

In this section we shall begin an interpretation of the theory of lattices and Boolean algebras for the purpose of treating certain problems arising in switching theory and the logical design of digital circuitry. Our point of departure is related to the algebraic expressions of Section 1.3. Accordingly, we introduce the class $\mathscr{B}(n)$ of *Boolean expressions* $f(x_1, x_2, \ldots, x_n)$ in the symbols $x_1, x_2, \ldots, x_n$ as follows:

(i) every x_i is in $\mathscr{B}(n)$, as is every constant element of $B = \{0, 1\}$;
(ii) $f \in \mathscr{B}(n)$ and $g \in \mathscr{B}(n)$ $\Rightarrow$ $f + g$, $f \cdot g$, and f' are in $\mathscr{B}(n)$

The direction of our applications will depend on our making an appropriate electronic interpretation of the meaning of the symbols x_i.

EXAMPLE 4.45 $f(x_1, x_2, x_3) = ((x_1 \cdot x_2' + x_3) \cdot x_1)' + x_1 \cdot x_2$ is a Boolean expression, that is, a member of the class $\mathscr{B}(3)$.   △

Given the finite set

$$X = \{x_1, x_2, \ldots, x_n\}$$

we now suppose that the elements x_i are wires, each of which is assumed to be at either a high or a low voltage at the discrete points in time, $t = 1, 2, \ldots$. If x and y are two such wires, then we let $x + y$ denote the output wire of an electronic circuit [Figure 4.12(a)] which behaves in such a way that

$x + y$ is a high-voltage wire $\Leftrightarrow$ x is high or y is high (or both)

Similarly, we let $x \cdot y$ denote the output wire of an electronic circuit [Figure 4.12(b)] with the property that

$x \cdot y$ is high $\Leftrightarrow$ x is high and y is high

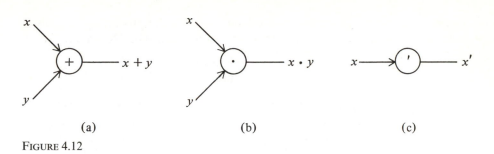

FIGURE 4.12

Electrical engineers can describe several different realizations of these particular specifications. Regardless of the implementation, they would call them *or* and *and* circuits, respectively. But to complete the picture here, we should also include the idea of *inverting* or *complementing* circuits [Figure 4.12(c)] according to which the output is low when the input is high, and vice versa. Of course, these are only *symbolic circuits*, representing an abstraction of the actual electronic configurations. The latter may involve transistor or vacuum tube realizations, but such details are not essential to our discussion. It is necessary only to understand how in principle one can construct an infinite collection $\mathcal{C}(n)$ of *logic circuits* in one-to-one correspondence with the members of the class $\mathcal{B}(n)$ of Boolean expressions.

EXAMPLE 4.46 Corresponding to the expression of Example 4.45 is the logic circuit of Figure 4.13. The logical designer, however, is generally concerned only with the input-output behavior of these circuits. Thus, as indicated in

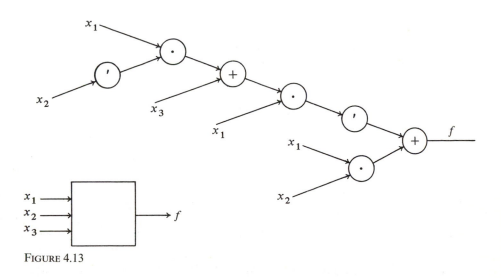

FIGURE 4.13

Figure 4.13, we may think of the circuit as a 3-input, 1-output box whose behavior is specified by the truth table shown in Table 4.8. Here we make the replacements

$$\text{low} = 0 \qquad \text{high} = 1$$

and the overall behavior of the box is governed by the nature and interconnection of the individual symbolic circuits. For instance, the second line of our truth table is determined according to the analysis of Figure 4.14. We simply set $x_1 = 0$, $x_2 = 0$, $x_3 = 1$, and use the defining specifications of the symbolic circuits to obtain $f = 1$ for this set of inputs. Note that the symbolic circuits simply implement the operations of the Boolean algebra B of Example 4.37.

TABLE 4.8

x_1	x_2	x_3	F
0	0	0	1
0	0	1	1
0	1	0	1
0	1	1	1
1	0	0	0
1	0	1	0
1	1	0	1
1	1	1	1

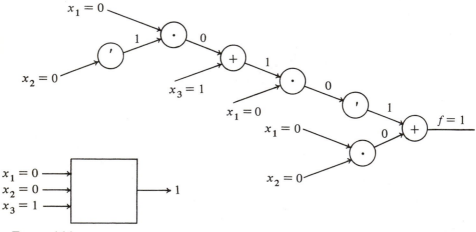

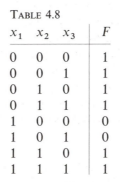

FIGURE 4.14

In order to fully appreciate the usefulness of Boolean algebra to the designer of logical circuitry, we present the following calculation:

$$f(x_1, x_2, x_3) = ((x_1 \cdot x_2') + x_3) \cdot x_1)' + x_1 \cdot x_2$$
$$= (x_1 x_2' + x_1 x_3)' + x_1 x_2$$
$$= (x_1 x_2')'(x_1 x_3)' + x_1 x_2$$
$$= (x_1' + x_2)(x_1' + x_3') + x_1 x_2$$
$$= x_1' + x_1' x_3' + x_1' x_2 + x_2 x_3' + x_1 x_2$$
$$= x_1' + x_2 x_3' + x_1 x_2 + x_1' x_2$$
$$= x_1' + x_2 x_3' + (x_1 + x_1') x_2$$
$$= x_1' + x_2 x_3' + x_2$$
$$= x_1' + x_2$$

The student should note the use of DeMorgan's laws (Theorem 4.3) and the other laws [(P1) through (P8)] of Boolean algebra.

Now we observe that the (much more economical) circuit of Figure 4.15 corresponds to the expression in last line; and most importantly, we find that

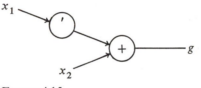

FIGURE 4.15

the behavior of this circuit is identical to that of Figure 4.13! That is, because this new circuit has the truth table of Table 4.9, there is no way to distinguish the boxes shown in Figure 4.16 by means of external analysis. If you were paying the bill, which one would you prefer?   △

TABLE 4.9

x_1	x_2	x_3	G
0	0	0	1
0	0	1	1
0	1	0	1
0	1	1	1
1	0	0	0
1	0	1	0
1	1	0	1
1	1	1	1

The preceding discussion should serve as adequate motivation for the following idea. We will say that an *interpretation I* of the Boolean expressions

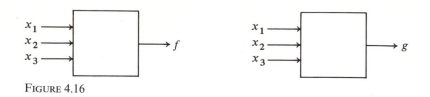

FIGURE 4.16

is a mapping

$$I: \mathscr{B}(n) \to B(n)$$

into the free Boolean algebra [we write $I(f) = F$] such that

(i) $I(x_i) = X_i$
(ii) $I(f + g) = I(f) + I(g)$ $I(f \cdot g) = I(f) \cdot I(g)$ $I(f') = (I(f))'$

The interpretation will ensure that $I(f) = F$ is the truth table representing the input-output behavior of the logic circuit corresponding to the expression f.

We saw in Example 4.46 that I is not an injective mapping. Two (formally) different Boolean expressions may yet have the same truth table. Now, the kernel (Section 0.7) of the interpretation I is an equivalence relation on $\mathscr{B}(n)$ that groups all those expressions having a given truth table into the same equivalence class

$$[f] = \{g \in \mathscr{B}(n): I(g) = I(f) = F\}$$

With a "canonical form" in $\mathscr{B}(n)$, we can single out a unique representative from each equivalence class. Two Boolean expressions are equivalent if and only if they have the same canonical form. The latter fact allows for the algorithmic determination of whether the logic circuits corresponding to two given expressions have the same behavior.

The clue to the canonical representation is contained in the proof of the second lemma preceding Theorem 4.5. Thus we write

$$F = \sum_{j \in J} M_j   J \subseteq \{0, 1, \ldots, 2^n - 1\}$$

where J is the set of indices for which the truth table F takes the value 1. Then, according as

$$M_j = X_1^{b_1} X_2^{b_2} \cdots X_n^{b_n}$$

we set

$$m_j = x_1^{b_1} x_2^{b_2} \cdots x_n^{b_n}$$

Such products of the x_i or their complements are called *minterms*. The terminology derives from the position of the M_j in a diagram of the poset $B(n)$; the M_j cover the zero element. In any case, we have

$$I(\sum m_j) = \sum I(m_j) = \sum M_j = F$$

showing that $\sum m_j \in [f]$. The resulting sum of minterms $\sum_{j \in J} m_j$ is called the *minterm canonical form* of the Boolean expression f.

EXAMPLE 4.47   Let f be the Boolean expression of Example 4.45, that is,

$$f(x_1, x_2, x_3) = ((x_1 \cdot x_2' + x_3) \cdot x_1)' + x_1 \cdot x_2$$

Without any reference to the corresponding logic circuit, we may compute the truth table of f by substitution and evaluation in the Boolean algebra B. Thus we obtain

$$f(0, 0, 0) = ((0 \cdot 0' + 0) \cdot 0)' + 0 \cdot 0 = ((0 + 0) \cdot 0)' + 0 = 0' + 0 = 1$$
$$f(0, 0, 1) = ((0 \cdot 0' + 1) \cdot 0)' + 0 \cdot 0 = ((0 + 1) \cdot 0)' + 0 = 0' + 0 = 1$$

etc., with the same result as before (see Table 4.10).

TABLE 4.10

x_1	x_2	x_3	F	
0	0	0	1	$(000 \to 0)$
0	0	1	1	$(001 \to 1)$
0	1	0	1	$(010 \to 2)$
0	1	1	1	$(011 \to 3)$
1	0	0	0	
1	0	1	0	
1	1	0	1	$(110 \to 6)$
1	1	1	1	$(111 \to 7)$

Then the minterm canonical form of f is the Boolean expression

$$m_0 + m_1 + m_2 + m_3 + m_6 + m_7$$

or, in the less-compressed notation,

$$x_1'x_2'x_3' + x_1'x_2'x_3 + x_1'x_2x_3' + x_1'x_2x_3 + x_1x_2x_3' + x_1x_2x_3$$

Alternatively, we may use the laws of Boolean algebra to compute

$$
\begin{aligned}
f(x_1, x_2, x_3) &= ((x_1 \cdot x_2' + x_3) \cdot x_1)' + x_1 \cdot x_2 \\
&= (x_1x_2' + x_1x_3)' + x_1x_2 \\
&= (x_1x_2')'(x_1x_3)' + x_1x_2 \\
&= (x_1' + x_2)(x_1' + x_3') + x_1x_2 \\
&= x_1' + x_1'x_3' + x_1'x_2 + x_2x_3' + x_1x_2 \\
&= x_1' + x_2x_3' + x_1x_2 \\
&= x_1'(x_2' + x_2)(x_3' + x_3) + x_2x_3'(x_1' + x_1) + x_1x_2(x_3' + x_3) \\
&= x_1'x_2'x_3' + x_1'x_2'x_3 + x_1'x_2x_3' + x_1'x_2x_3 + x_1x_2x_3' + x_1x_2x_3 \quad \triangle
\end{aligned}
$$

EXERCISES

1 Draw logic circuits to correspond to the following Boolean expressions:
(a) $((x_1' \cdot x_2)' + x_3) \cdot x_1'$
(b) $((x_1 + x_3') \cdot x_2') \cdot x_3 + x_2$
(c) $((x_1 \cdot x_3)' \cdot x_2) + x_4$

2 Obtain truth tables for the logic circuits of Exercise 1.

3 Show that the following pairs of expressions have the same interpretations.
(a) $((x_1' \cdot x_2)' + x_3) \cdot x_1'$ and $(x_1 + x_2 \cdot x_3')'$
(b) $((x_1 + x_3') \cdot x_2') \cdot x_3 + x_2$ and $x_2 \cdot x_3' + (x_1 + x_2) \cdot x_3$

4 Show that the atoms M_j cover the zero element in the digraph or diagram of the poset $B(n)$.

5 Prove that $I(f) = I(g)$ iff f and g have the same minterm canonical form.

6 Obtain the minterm canonical form of the following expressions:
(a) Exercise 1(a)   (b) Exercise 1(b)   (c) Exercise 1(c)
(d) $x_2'x_3' + x_1x_2 + x_2x_3 + x_1'x_3'$
(e) $(x_1x_2' + x_3'x_4)(x_2 + x_4)(x_1 + x_3'x_4')$
(f) $x_1x_2'x_4 + x_1x_3x_4' + x_1'x_2'x_4 + x_1x_2x_3$.

*7 Define *maxterms*:

$$l_i = x_1^{b_1} + x_2^{b_2} + \cdots + x_n^{b_n}$$

and prove the existence of a maxterm canonical form

$$f = \underset{i \in I}{\mathsf{X}} l_i$$

for each Boolean expression f. [*Hint*: See Exercise 6 of Section 4.6. Whereas

$$j \in J \Leftrightarrow f(b_1, b_2, \ldots, b_n) = 1 \qquad (b_1 b_2 \cdots b_n \to j)$$

we have

$$i \in I \Leftrightarrow f(b_1', b_2', \ldots, b_n') = 0 \qquad (b_1 b_2 \cdots b_n \to i)]$$

8 Obtain the maxterm canonical form of the expressions in Exercise 6. [*Hint*: See Exercise 7.]

*4.8   THE MINIMIZATION PROBLEM

We have already hinted at the existence of a minimization problem in the design of logic circuits or switching circuits, as they are sometimes called. Consider all the Boolean expressions in one of our equivalence classes

$$[f] = \{g \in \mathscr{B}(n): I(g) = I(f) = F\}.$$

They all have the same truth table F, but each equivalence class contains infinitely many expressions! So how do we go about selecting an "economical" representative from a given class? The precise statement and the different attempts at solution of problems of this type are quite obviously of great practical importance to the logical designer.

Various measures of cost or relative economy have been proposed. Typical of these is the *number* #(f) *of arrows* found in the logic circuit corresponding to the expression f, disregarding complementation (inversion) arrows. It is known that this figure gives a close approximation to the number of electronic components (transistors, vacuum tubes, or whatever) used in an actual implementation of the logic circuit. It also happens that the cost of complementation may often be negligible. Other assumptions could also be made, but the variations from one technology to another do not materially influence the main thrust of our argument.

EXAMPLE 4.48 Consider the running example of the previous section. We discussed three different Boolean expressions

$$f = ((x_1 x_2' + x_3)x_1)' + x_1 x_2$$

$$g = x_1' + x_2$$

$$h = x_1' x_2' x_3' + x_1' x_2' x_3 + x_1' x_2 x_3' + x_1' x_2 x_3 + x_1 x_2 x_2' + x_1 x_2 x_3$$

with

$$I(f) = I(g) = I(h) = F$$

So the three corresponding logic circuits will have the same input-output behavior. Nevertheless, in consulting Figures 4.13 and 4.15, we find that

$$\#(f) = 10$$
$$\#(g) = 2$$
$$\#(h) = (3 + 3 + 3 + 3 + 3 + 3) + 6 = 24$$

In the case of h, we assumed the existence of m-input *or* and m-input *and* circuits, for $m \geq 2$. This much is consistent with existing technology, so that we will take advantage of this fact in our subsequent development. In any event, it is certainly clear that g is the most economical among the three representations. △

To the logical designer, a problem will usually begin as a truth table. That is to say, he is usually faced with the problem of designing a "box," as in Figure 4.17, with a prescribed input-output behavior. The *minimization problem*

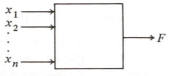

FIGURE 4.17

of switching theory may then be stated: to find a Boolean expression $g \in \mathscr{B}(n)$ with the properties

(a) $I(g) = F$
(b) $I(h) = F \Rightarrow \#(g) \leq \#(h)$

so that g is of minimum cost among all expressions h having the given truth table F. (Compare this problem with that of Section 3.9.) This minimization problem is quite difficult to solve, mainly because there are infinitely many h for which (b) must be tested. In fact, it is an unsolved problem! So why have we brought it up? Primarily because an appreciation of this ideal problem is quite helpful in gaining an understanding of a more realistic problem, one whose solution *is* known.

We can simplify the problem at hand if we look only at those expressions which are sums of (Boolean) products. A *Boolean product* in $x_1, x_2, \ldots, x_n$ is an expression

$$p = x_{i_1}^{b_{i_1}} x_{i_2}^{b_{i_2}} \cdots x_{i_k}^{b_{i_k}} \qquad (i_1 < i_2 < \cdots < i_k)$$

with the usual convention on 0 or 1 exponents; that is, $x_i^0 = x_i'$ and $x_i^1 = x_i$. Then a minterm (Section 4.7) is simply a Boolean product in which every variable appears. Of course, the variables are among the members of the set

$$P_n = \{p : p \text{ is a Boolean product in } x_1, x_2, \ldots, x_n\}$$

and each expression

$$g = \sum p_j \qquad (p_j \text{ in } P_n)$$

is called a *sum of* (Boolean) *products*.

EXAMPLE 4.49   The following are Boolean products from P_5:

$$x_1 x_3' x_4 \qquad x_2 x_4' x_5 \qquad x_3 x_5$$

Consequently,

$$g = x_1 x_3' x_4 + x_2 x_4' x_5 + x_3 x_5$$

is a sum of products. Also $\#(g) = (3 + 3 + 2) + 3 = 11$, in accordance with our earlier remark (Example 4.48). See Figure 4.18.   △

Given a truth table $f \in \mathscr{B}(n)$, our *modified minimization problem* becomes: find a sum g of products such that

(a) $I(g) = f$
(b) $I(h) = f$, where h is a sum of products $\Rightarrow \#(g) \leq \#(h)$

We will present the classical algorithmic solution to the modified problem in our next section. To prepare for this presentation, we shall first establish a geometric framework for visualizing the techniques involved.

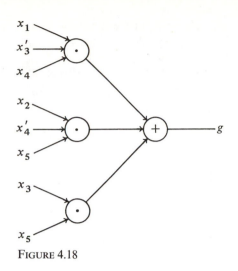

FIGURE 4.18

It happens that the Boolean products of P_n are in one-to-one correspondence with the cells of the n-cube C^n (Section 4.4). This feature is what makes possible a geometric interpretation of the modified minimization problem. The correspondence is as follows. To the product

$$p = x_{i_1}^{b_{i_1}} x_{i_2}^{b_{i_2}} \cdots x_{i_k}^{b_{i_k}} \qquad (i_1 < i_2 < \cdots < i_k)$$

in P_n, we associate the cell c

$$p \leftrightarrow c = (c_1, c_2, \ldots, c_n)$$

such that the coordinates

$$c_i = \begin{cases} 0 & \text{if } x_i \text{ appears in } p \text{ as } x_i^0 = x_i' \\ 1 & \text{if } x_i \text{ appears in } p \text{ as } x_i^1 = x_i \\ I & \text{if } x_i \text{ does not appear in } p \end{cases}$$

This association establishes a bijection $P_n \leftrightarrow C^n$. Now suppose that $K^0 = K^0(f) \subseteq B^n \subseteq C^n$ simply denotes the collection of vertices $(b_1, b_2, \ldots, b_n)$ of the n-cube for which the given truth table f takes on the value 1. Then if $h = \sum p_j$ is a sum of products and $p_j \leftrightarrow c_j$, we must have

$$I(h) = f \Leftrightarrow \{c_j\} \text{ is a } \textit{covering} \text{ of } K^0(f)$$

by which we mean that

(i) All the vertices in each c_j fall within K^0
(ii) Every vertex of K^0 is included in some c_j

EXAMPLE 4.50   The last idea—that of a covering of the "true" vertices of our truth table by cells of the n-cube—is crucial to the geometric interpretation of the modified minimization problem. Once this idea has been grasped, our geometric intuition can take over, and everything should become abundantly clear. So we should make every effort to get the most out of this particular example.

Suppose that we are given the truth table shown at the left in Table 4.11.

TABLE 4.11

x_1	x_2	x_3	x_4	$f\ (= H_3)$	H_1	H_2
0	0	0	0	0	0	0
0	0	0	1	1	1	1
0	0	1	0	1	1	0
0	0	1	1	0	0	0
0	1	0	0	1	1	1
0	1	0	1	0	1	0
0	1	1	0	1	1	1
0	1	1	1	0	0	0
1	0	0	0	0	0	0
1	0	0	1	1	1	1
1	0	1	0	0	0	0
1	0	1	1	1	1	1
1	1	0	0	1	1	1
1	1	0	1	1	1	1
1	1	1	0	0	0	0
1	1	1	1	0	0	0

Then

$$K^0(f) = \{(0, 0, 0, 1), (0, 0, 1, 0), (0, 1, 0, 0), (0, 1, 1, 0),$$
$$(1, 0, 0, 1), (1, 0, 1, 1), (1, 1, 0, 0), (1, 1, 0, 1)\}$$

and we can consider three different sums of products:

$$h_1 = \quad x_2'x_3'x_4 \quad + \quad x_1'x_3x_4' \quad + \quad x_2x_3' \quad + \quad x_1x_2'x_4$$
$$\updownarrow \qquad\qquad \updownarrow \qquad\qquad \updownarrow \qquad\qquad \updownarrow$$
$$\{(I, 0, 0, 1), \quad (0, I, 1, 0), \quad (I, 1, 0, I), \quad (1, 0, I, 1)\}$$

$$h_2 = \quad x_2'x_3'x_4 \quad + \quad x_1'x_2x_4' \quad + \quad x_1x_2'x_3x_4 \quad + \quad x_1x_2x_3'$$
$$\updownarrow \qquad\qquad \updownarrow \qquad\qquad \updownarrow \qquad\qquad \updownarrow$$
$$\{(I, 0, 0, 1), \quad (0, 1, I, 0), \quad (1, 0, 1, 1), \quad (1, 1, 0, I)\}$$

$$h_3 = \quad x_1'x_2'x_3'x_4 \quad + \quad x_1'x_3x_4' \quad + \quad x_2x_3'x_4' \quad + \quad x_1x_2'x_4 \quad + \quad x_1x_2x_3'$$
$$\updownarrow \qquad\qquad\quad \updownarrow \qquad\qquad\quad \updownarrow \qquad\qquad\quad \updownarrow \qquad\qquad\quad \updownarrow$$
$$\{(0, 0, 0, 1), \quad (0, I, 1, 0), \quad (I, 1, 0, 0), \quad (1, 0, I, 1), \quad (1, 1, 0, I)\}$$

Note the cells corresponding to the various products, as determined by our bijection $P_n \rightarrow C^n$.

Now $I(h_1) = H_1 \neq f$, and accordingly, the corresponding collection of cells is not a covering of $K^0(f)$ because condition (i) fails:

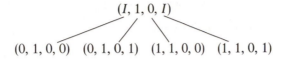

$(I, 1, 0, I)$

$(0, 1, 0, 0) \quad (0, 1, 0, 1) \quad (1, 1, 0, 0) \quad (1, 1, 0, 1)$

The vertex $(0, 1, 0, 1) \subseteq (I, 1, 0, I)$ is not in $K^0(f)$. On the other hand, condition (ii) fails for h_2:

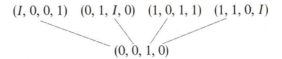

$(I, 0, 0, 1) \quad (0, 1, I, 0) \quad (1, 0, 1, 1) \quad (1, 1, 0, I)$

$(0, 0, 1, 0)$

The vertex $(0, 0, 1, 0) \in K^0$ is in none of the cells corresponding to products of h_2.

The sum of products h_3 satisfies both conditions. For condition (ii), we observe the inclusions

$(0, 0, 0, 1) \quad (0, I, 1, 0) \quad (I, 1, 0, 0) \quad (1, 0, I, 1) \quad (1, 1, 0, I)$

$(0,0,0,1) \ (0,0,1,0) \ (0,1,0,0) \ (0,1,1,0) \ (1,0,0,1) \ (1,0,1,1) \ (1,1,0,0) \ (1,1,0,1)$

For condition (i), we obtain

$(0, 0, 0, 1) \qquad (0, I, 1, 0) \qquad\qquad (I, 1, 0, 0)$

$\qquad\qquad | \qquad\qquad \diagup \diagdown \qquad\qquad \diagup \diagdown$

$(0, 0, 0, 1) \quad (0, 0, 1, 0) \quad (0, 1, 1, 0) \quad (0, 1, 0, 0) \quad (1, 1, 0, 0)$

$(1, 0, I, 1) \qquad\qquad (1, 1, 0, I)$

$(1, 0, 0, 1) \quad (1, 0, 1, 1) \quad (1, 1, 0, 0) \quad (1, 1, 0, 1)$

and all of these vertices are in $K^0(f)$. Accordingly,

$$\{(0, 0, 0, 1), (0, I, 1, 0), (I, 1, 0, 0), (1, 0, I, 1), (1, 1, 0, I)\}$$

is a covering of $K^0(f)$. At the same time, we have

$$I(h_3) = H_3 = f$$

so that the circuit corresponding to the expression h_3 will have the input-output behavior of the given truth table f.

If we wish to evaluate the "cost" of a covering, we have only to remember (considering the correspondence $p \leftrightarrow c$) that "the larger the cell, the smaller the cost." More precisely, we should take $\#(c) = n -$ order of (c). Then, after accounting for the arrows required by an *or* circuit, we obtain, for example,

$$\#(h_3) = \#\{(0, 0, 0, 1), (0, I, 1, 0), (I, 1, 0, 0), (1, 0, I, 1), (1, 1, 0, I)\}$$
$$= (4 + 3 + 3 + 3 + 3) + 5 = 21$$

in agreement with the cost of the expression itself. $\triangle$

The geometric rephrasing of our modified minimization problem thus becomes: find a *minimal covering* of $K^0(f)$. Here the minimality must be measured according to the cost criteria described at the close of Example 4.50, and it is apparent from condition (i) that we must choose the cells of the coverings from the *cell complex*

$$K(f) = \{c \in C^n : f(v) = 1 \text{ for every vertex } v \subseteq c\}$$

As in the notation K^0, we will let K^q denote the set of all cells of order q in the complex K. Then keeping in mind our cost criteria, we can expect the maximal cells of the cell complex to play an important role. These cells are not necessarily the largest cells, for we use the term "maximal" in the sense of not being contained in a larger cell of the complex. Thus $b \in K(f)$ is said to be a *basic cell* if

$$b \subseteq c \quad \text{and} \quad c \in K(f) \Rightarrow b = c$$

(The corresponding Boolean products $p \to b$ are called *prime implicants*.)

We hope to have provided sufficient insight with Example 4.50 for the following theorem to be more or less apparent, even at first glance.

Theorem 4.6   (Quine*)

Every minimal covering consists of basic cells.

Proof   Rather than presenting a detailed proof, we shall simply remark that the argument depends on the fact that

$$b \subseteq c \Rightarrow \#(c) \leq \#(b)$$

as dictated by our cost criteria. $\square$

EXAMPLE 4.51   Suppose that we are given the truth table f shown in Table 4.12. Then the complex $K(f)$ is the following collection of cells:

$$K(f) = \{(0, 0, 0), (0, 1, 0), (0, 1, 1), (1, 1, 0), (1, 1, 1), (0, I, 0),$$
$$(0, 1, I), (1, 1, I), (I, 1, 0), (I, 1, 1), (I, 1, I)\}$$

(See Figure 4.19.) These are all of the cells of C^3 with the property that f takes the value "1" at each vertex. A glance at the figure shows that $(0, I, 0)$ and $(I, 1, I)$ are basic cells, but $(0, 1, I)$, for instance, is not a basic cell since we can

* W. V. Quine, "The Problem of Simplifying Truth Functions," *Am. Math. Monthly*, **59**, 521–531 (1952).

TABLE 4.12

x_1	x_2	x_3	f
0	0	0	1
0	0	1	0
0	1	0	1
0	1	1	1
1	0	0	0
1	0	1	0
1	1	0	1
1	1	1	1

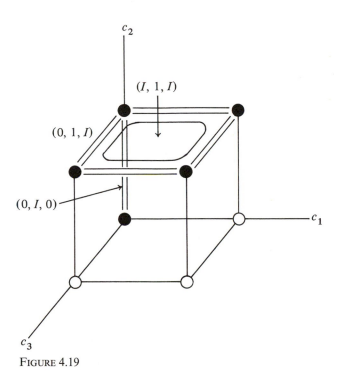

FIGURE 4.19

find a cell $c \in K(f)$, namely $c = (I, 1, I)$, such that

$$(0, 1, I) \subseteq c \quad \text{and} \quad (0, 1, I) \neq c$$

In the actual solution of the modified minimization problem (Section 4.9), all of the relevant information will be obtained without reference to pictures. At this stage, however, we are simply interested in developing our geometric intuition. $\triangle$

EXAMPLE 4.52   For the truth table f in Example 4.50, the following collection of cells is a covering (of $K^0(f)$):

$$\{(0, 0, 0, 1), (0, I, 1, 0), (0, 1, 0, 0), ((1, 0, I, 1), (1, 1, 0, I)\}$$

But $(0, 1, 0, 0)$ is not a basic cell, since $(0, 1, 0, 0) \subseteq (I, 1, 0, 0)$ in $K(f)$. According to Quine's theorem, the covering is not a minimal one. The reader should compare the cost of this covering with that corresponding to h_3 of Example 4.50.   △

EXERCISES

1 Compute the cost $\#(f)$ for the following expressions f:
  (a) Exercise 1(a) of Section 4.7   (b) Exercise 1(b) of Section 4.7
  (c) Exercise 1(c) of Section 4.7   (d) Exercise 6(d) of Section 4.7
  (e) Exercise 6(e) of Section 4.7   (f) Exercise 6(f) of Section 4.7

2 Associate cells of C^5 with the following Boolean products:
  (a) $x_2'x_3x_5'$ (b) x_2
  (c) $x_4'x_5$ (d) $x_1'x_2x_3'x_4x_5'$
  (e) x_1x_5' (f) x_3x_5

3 Which of the following collections of cells are coverings of $K^0(f)$, f being the truth table shown below?

x_1	x_2	x_3	x_4	f
0	0	0	0	1
0	0	0	1	1
0	0	1	0	1
0	0	1	1	1
0	1	0	0	1
0	1	0	1	1
0	1	1	0	0
0	1	1	1	0
1	0	0	0	0
1	0	0	1	0
1	0	1	0	1
1	0	1	1	1
1	1	0	0	1
1	1	0	1	1
1	1	1	0	0
1	1	1	1	1

(a) $\{(1, 1, I, 1), (1, 0, 1, I), (I, 1, 0, 0), (0, 0, I, I)\}$
(b) $\{(1, 0, 1, 1), (I, 0, 1, I), (1, 1, 0, I), (0, I, 0, I), (1, 1, 1, 1)\}$
(c) $\{(1, 1, I, 1), (1, 0, 1, I), (I, 1, 0, 0), (0, 0, I, I), (0, 1, 0, I)\}$
(d) $\{(1, 0, 1, 1), (I, 0, 1, I), (1, 1, I, I), (0, I, 0, I)\}$

What is the "cost" of each collection of cells?

4 Repeat Exercise 3 for the following:

x_1	x_2	x_3	x_4	f
0	0	0	0	1
0	0	0	1	0
0	0	1	0	1
0	0	1	1	0
0	1	0	0	0
0	1	0	1	0
0	1	1	0	1
0	1	1	1	1
1	0	0	0	1
1	0	0	1	1
1	0	1	0	1
1	0	1	1	1
1	1	0	0	1
1	1	0	1	0
1	1	1	0	1
1	1	1	1	1

(a) $\{(0, 0, I, 0), (0, 1, 1, I), (1, 0, I, I), (1, 1, 1, I)\}$
(b) $\{(I, 0, I, 0), (0, I, 1, 0), (0, 1, I, 0), (1, 0, I, I), (1, 1, I, 0)\}$
(c) $\{(0, 0, I, 0), (1, 0, I, I), (1, I, 1, I), (1, 1, I, 0), (0, 1, 1, I)\}$
(d) $\{(I, 0, I, 0), (1, 0, I, I), (I, 1, 1, I), (1, I, I, 0)\}$

5 For the following truth tables f, list the elements of the cell complex $K(f)$ and identify all the basic cells, drawing figures as in Example 4.51 (Figure 4.19).

(a) x_1	x_2	x_3	f	(b) x_1	x_2	x_3	f	(c) x_1	x_2	x_3	f
0	0	0	1	0	0	0	1	0	0	0	0
0	0	1	1	0	0	1	1	0	0	1	1
0	1	0	1	0	1	0	0	0	1	0	0
0	1	1	1	0	1	1	1	0	1	1	1
1	0	0	0	1	0	0	1	1	0	0	1
1	0	1	1	1	0	1	0	1	0	1	1
1	1	0	0	1	1	0	1	1	1	0	1
1	1	1	0	1	1	1	1	1	1	1	1

6 A *complex K* in a lattice L is a subset with the property

$$x \in K \quad \text{and} \quad y \le x \Rightarrow y \in K$$

Show that the cell complexes $K(f)$ are complexes in C^n.

*4.9   THE QUINE-McCLUSKEY PROCEDURE

Some aspects of the solution to the sum-of-products minimization problem will appear almost exhaustive in nature. Certainly they are exhausting if performed by hand for a reasonably large problem. Nevertheless, the procedure we shall present, due to Quine and McCluskey,* is completely systematic and not particularly difficult to follow.

We learned from Quine's theorem that in seeking a minimal covering of $K^0(f)$, we do not have to consider all the cells in C^n, indeed, not even all of the cells in the complex $K(f)$, but only the basic ones. There are two phases to the Quine-McCluskey procedure as represented by Algorithm XVIII and

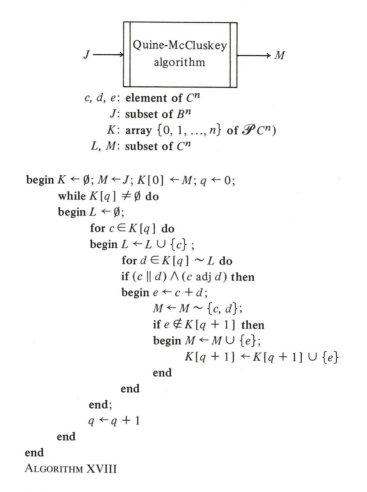

c, d, e: **element of** C^n
J: **subset of** B^n
K: **array** $\{0, 1, \ldots, n\}$ **of** $\mathscr{P}C^n$)
L, M: **subset of** C^n

```
begin K ← ∅; M ← J; K[0] ← M; q ← 0;
    while K[q] ≠ ∅ do
    begin L ← ∅;
        for c ∈ K[q] do
        begin L ← L ∪ {c} ;
            for d ∈ K[q] ~ L do
            if (c ∥ d) ∧ (c adj d) then
            begin e ← c + d;
                M ← M ~ {c, d};
                if e ∉ K[q + 1] then
                begin M ← M ∪ {e};
                    K[q + 1] ← K[q + 1] ∪ {e}
                end
            end
        end;
        q ← q + 1
    end
end
```
ALGORITHM XVIII

Algorithm XIX. The first one simply consists in finding all the basic cells, for it is from these that the selection will be made. Fortunately, the second phase represents nothing new. The selection from the basic cells is a covering problem,

---

* E. J. McCluskey, "Minimization of Boolean Functions," *Bell System Tech. J.*, **35**, 1417–1444 (1956).

not unlike those discussed earlier (Section 4.3) in connection with the determination of the various graphical invariants.

Though we are interested only in the basic cells, we find that Algorithm XVIII actually determines the entire complex $K(f)$ in the process; and since the complex is generated out of K^0, moving from K^q to K^{q+1}, the nonbasic cells are easily recognized. The initial data J is taken to be the list K^0 of "true" vertices for the function f. After a little thought, the reader should be convinced that e is a cell of K^{q+1} if and only if $e = c + d$ for cells c, d in K^q that are parallel and adjacent. This is the key to the Quine-McCluskey procedure for deriving K^{q+1} from K^q. We look at all pairs c and d in K^q, to determine whether the conditions

$$c \parallel d \quad \text{and} \quad c \text{ adj } d$$

are satisfied. Whenever such a pair is found, the addition $e = c + d$ takes place in the lattice $C^n = (C^n, +, \cdot)$, as in

$$e = (I, 0, 1, I, 0) + (I, 1, 1, I, 0) = (I, I, 1, I, 0)$$

At the same time, both c and d are deleted from the tentative list (the eventual output M) of basic cells, while e is adjoined to M and to K^{q+1} if it was not already there. The whole process terminates when we arrive at q for which $K^q = \varnothing$.

EXAMPLE 4.53   It is well to begin with a very elementary problem, that of Example 4.51. The tabulations of the K^q as would be derived in Algorithm XVIII are:

K^0			K^1			K^2		
$\checkmark$0	0	0	0	I	0	I	1	I
$\checkmark$0	1	0	$\checkmark$0	1	I			
$\checkmark$0	1	1	$\checkmark I$	1	0			
$\checkmark$1	1	0	$\checkmark I$	1	1			
$\checkmark$1	1	1	$\checkmark$1	1	I			

Only the list of K^0 is available initially, however. We note that vertices are always parallel (they have I's in the same positions, namely, nowhere) so that in the case of K^0, it is only the adjacency condition that needs to be tested. Thus we make the following comparisons.

000 with 010, 011, 110, 111:
   form $0I0 = 000 + 010$ and check 000, 010 (nonbasic)

010 with 011, 110, 111:
  form $01I = 010 + 011$ and check 010, 011
  form $I10 = 010 + 110$ and check 010, 110
011 with 110, 111:
  form $I11 = 011 + 111$ and check 011, 111
110 with 111:
  form $11I = 110 + 111$ and check 110, 111

Here we are using check marks to record the nonbasic cells, whereas in the statement of Algorithm XVIII, these cells are deleted from the subset M of maximal or basic cells.

At this stage, we have determined the subset K^1. We then proceed in the same way to generate K^2 from K^1. Now, however, we must remember that cells c and d must be parallel as well as adjacent when we make our comparisons.

$0I0$ with $01I$, $I10$, $I11$, $11I$:
  no action here
$01I$ with $I10$, $I11$, $11I$:
  form $I1I = 01I + 11I$ and check $01I$, $11I$
$I10$ with $I11$, $11I$:
  form $I1I = I10 + I11$ and check $I10$, $I11$
$I11$ with $11I$:
  no action here

Note that the cell $(I, 1, I)$ was generated twice. Of course, we do not duplicate the entry in K^2. That is, in Algorithm XVIII you will notice that e is adjoined to K^{q+1} only if it does not already appear there. The algorithm terminates with the basic (unchecked) cells $(0, I, 0)$ and $(I, 1, I)$.   △

One important time-saving device was not incorporated into the statement of Algorithm XVIII, for the sake of clarity. The device concerns the partitioning of each K^q into blocks of cells (cell blocks?) each having the same number of coordinates equal to one. It is then necessary to compare the cells of a given block only with those in the following block, if we assume that the blocks are arranged according to increasing number of ones, for otherwise, we can never have the distance 1 that is required by the adjacency condition. The reader is advised to use such a partitioning in any of the problems he or she tries.

EXAMPLE 4.54   For purposes of illustrating the partitioning of the K^q, we consider the somewhat larger problem ($n = 4$) represented by the truth table (Table 4.13):

TABLE 4.13

x_1	x_2	x_3	x_4	f	K^0	K^1	K^2
0	0	0	0	1	√0 0 0 0	0 *I* 0 0	0 1 *I* *I*
0	0	0	1	0		*I* 0 0 0	
0	0	1	0	0	√0 1 0 0		
0	0	1	1	1	√1 0 0 0	√0 1 0 *I*	
0	1	0	0	1		√0 1 *I* 0	
0	1	0	1	1	√0 0 1 1	1 0 *I* 0	
0	1	1	0	1	√0 1 0 1		
0	1	1	1	1	√0 1 1 0	0 *I* 1 1	
1	0	0	0	1	√1 0 1 0	*I* 0 1 1	
1	0	0	1	0		√0 1 *I* 1	
1	0	1	0	1	√0 1 1 1	√0 1 1 *I*	
1	0	1	1	1	√1 0 1 1	1 0 1 *I*	
1	1	0	0	0			
1	1	0	1	0			
1	1	1	0	0			
1	1	1	1	0			

It is fortunate that having once arranged the vertices of K^0 according to an increasing number of ones in the coordinates, the same ordering is automatically preserved in the subsequent lists K^q. In any case, the advantage of the partitioning is evident from the fact that we need only compare

0000 with 0100, 1000
0100 with 0011, 0101, 0110, 1010
1000 with 0011, 0101, 0110, 1010

$\vdots$

0*I*00 with 010*I*, 01*I*0, 10*I*0

etc. Again the basic cells are those that remain unchecked at the end of the procedure. △

So much for the first phase of the procedure. The second is a covering problem. Accordingly, the algebra of the free distributive lattices (Section 4.3) will come into play. We use the information found in the first phase to construct a *basic cell matrix* $M = (m_{ij})$. It has row headings corresponding to the various basic cells b_i and column headings labeled by the vertices v_j of $K^0(f)$. We set

$$m_{ij} = \begin{cases} 1 & \text{if } v_j \subseteq b_i \\ 0 & \text{otherwise} \end{cases}$$

This array indicates the choices that can be made in covering K^0 by basic cells. If a column v_j has only one 1, say $m_{ij} = 1$, then b_i is said to be an *essential cell*. The reader should have no difficulty in seeing that essential cells necessarily belong to every minimal covering. Accordingly, *we select* the essential cell b_i *and delete* from the matrix:

(i) *row i*
(ii) and *all columns k with* $m_{ik} = 1$

the latter because such vertices v_k (including v_j, of course) are now covered (by b_i).

EXAMPLE 4.55   Continuing the problem of Example 4.54, we construct the following basic cell matrix.

	0	3	4	5	6	7	8	10	11
$(0, 1, I, I)$			1	1	1	1			
$(0, I, 0, 0)$	1	1							
$(I, 0, 0, 0)$	1						1		
$(1, 0, I, 0)$							1	1	
$(0, I, 1, 1)$		1				1			
$(I, 0, 1, 1)$		1							1
$(1, 0, 1, I)$								1	1

For the sake of clarity, we have omitted the zeros of the matrix. Moreover, we have converted the vertices v_j of K^0 from binary to decimal to make for more convenient column headings. Thus we determine that

$(0, 1, I, I)$ contains vertices

$$0100 \rightarrow 4$$
$$0101 \rightarrow 5$$
$$0110 \rightarrow 6$$
$$0111 \rightarrow 7$$

$(0, I, 0, 0)$ contains vertices

$$0000 \rightarrow 0$$
$$0100 \rightarrow 4$$

etc., which is all the information that we need for constructing the matrix.

Now we see that $(0, 1, I, I)$ is essential to vertex 5. So *we select* $(0, 1, I, I)$ and delete this row, as well as the columns 4, 5, 6, 7. The resulting matrix no longer has any essential cells:

	0	3	8	10	11
$A = (0, I, 0, 0)$	1				
$B = (I, 0, 0, 0)$	1		1		
$C = (1, 0, I, 0)$			1	1	
$D = (0, I, 1, 1)$		1			
$E = (I, 0, 1, 1)$		1			1
$F = (1, 0, 1, I)$				1	1

We always continue the selection and deletion process as long as essential cells can still be found.   △

The "selection and deletion" process associated with the essential cells is not crucial to the solution of the covering problem. It could be by-passed, but then we would have missed the opportunity to reduce the matrix acting as input to Algorithm XIX. As in previous covering problems, the algorithm uses the free

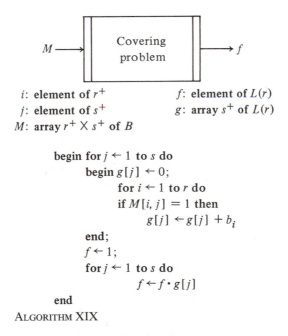

i: **element of** r^+ f: **element of** $L(r)$

j: **element of** s^+ g: **array** s^+ **of** $L(r)$

M: **array** $r^+ \times s^+$ **of** B

> **begin for** $j \leftarrow 1$ **to** s **do**
> > **begin** $g[j] \leftarrow 0$;
> > > **for** $i \leftarrow 1$ **to** r **do**
> > > **if** $M[i, j] = 1$ **then**
> > > > $g[j] \leftarrow g[j] + b_i$
>
> **end;**
> > $f \leftarrow 1$;
> > **for** $j \leftarrow 1$ **to** s **do**
> > > $f \leftarrow f \cdot g[j]$
>
> **end**

ALGORITHM XIX

distributive lattice (generated by the basic cells of the matrix) to convert a product of sums to a sum of products. The sums correspond to the rows (cells) covering the various vertices, and the resulting sum of products gives all the irredundant coverings of K^0 (or what is left of it after the deletions have been made).

EXAMPLE 4.56   Continuing with Examples 4.54 and 4.55, we see that Algorithm XIX will perform the computation:

$$f = (A + B)(D + E)(B + C)(C + F)(E + F)$$
$$= (AD + AE + BD + BE)(B + C)(C + F)(E + F)$$
$$= (ACD + ACE + BD + BE)(C + F)(E + F)$$
$$= (ACD + ACE + BCD + BCE + BDF + BEF)(E + F)$$
$$= ACE + BCE + BEF + ACDF + BDF$$

indicating that we may choose to cover our five remaining vertices with

> *A* and *C* and *E*, or
> *B* and *C* and *E*, or
> *B* and *E* and *F*, or
> *A* and *C* and *D* and *F*, or
> *B* and *D* and *F*

Rejecting *ACDF* as costing more than the others, *we choose ACE*, say. This selection [together with the $(0, 1, I, I)$ already selected] gives us the minimal covering:

$$\{(0, 1, I, I)\} \cup \{(0, I, 0, 0), (1, 0, I, 0), (I, 0, 1, 1)\}$$

Accordingly, a minimal-cost sum-of-products expression for the original truth table (Example 4.54) is as follows:

$$x_1' x_2 + x_1' x_3' x_4' + x_1 x_2' x_4' + x_2' x_3 x_4$$

If we had chosen to by-pass the selection and deletion of the essential cell, say $G = (0, 1, I, I)$, the result would have been the same. We would have found G appearing in every product of our sum-of-products expression f, again showing that it is essential.   △

EXERCISES

1 Use Algorithm XVIII to find all the basic cells for each of the following truth tables.

(a) x_1	x_2	x_3	f
0	0	0	0
0	0	1	0
0	1	0	1
0	1	1	0
1	0	0	1
1	0	1	1
1	1	0	1
1	1	1	1

(b) x_1	x_2	x_3	f
0	0	0	1
0	0	1	0
0	1	0	1
0	1	1	1
1	0	0	1
1	0	1	1
1	1	0	0
1	1	1	1

(c) x_1	x_2	x_3	x_4	f
0	0	0	0	1
0	0	0	1	0
0	0	1	0	1
0	0	1	1	1
0	1	0	0	1
0	1	0	1	1
0	1	1	0	0
0	1	1	1	0
1	0	0	0	0
1	0	0	1	0
1	0	1	0	1
1	0	1	1	1
1	1	0	0	1
1	1	0	1	1
1	1	1	0	1
1	1	1	1	1

(d) x_1	x_2	x_3	x_4	f
0	0	0	0	1
0	0	0	1	1
0	0	1	0	0
0	0	1	1	1
0	1	0	0	0
0	1	0	1	1
0	1	1	0	1
0	1	1	1	1
1	0	0	0	1
1	0	0	1	1
1	0	1	0	0
1	0	1	1	1
1	1	0	0	0
1	1	0	1	1
1	1	1	0	0
1	1	1	1	1

(e) Exercise 3 of Section 4.8   (f) Exercise 4 of Section 4.8

2 For each truth table of Exercise 1 construct the basic-cell matrix and identify all the essential cells.

3 Continuing from Exercise 2, extract the essential cells and use Algorithm XIX to find minimal coverings for the truth tables of Exercise 1. Transform your results into minimal sums of products.

*4 Rewrite Algorithm XVIII with a partitioning of the K^q into blocks of cells having the same number of coordinates equal to one.

*5 Justify the following *column domination rule* (in solving basic-cell matrices): If

$$m_{ij} = 1 \quad \Rightarrow \quad m_{ik} = 1$$

then column k can be deleted.

6 Find a minimal sum-of-products representations for each of the following:
(a) $x_1' x_2' x_3' x_4 + x_1' x_2' x_3 x_4 + x_1' x_2 x_3 x_4' + x_1 x_2 x_3' x_4'$
$\qquad + x_1' x_2 x_3 x_4 + x_1 x_2 x_3 x_4' + x_1 x_2 x_3 x_4$
(b) $x_1 x_3 (x_2 + x_4 + x_2 x_5') + x_1' x_2' x_5 (x_3 x_4' + x_2'(x_5' + x_1' x_4))$
(c) $x_1 x_2 x_3 x_4 x_5 + x_1 x_2 x_3 x_4 x_5' + x_1 x_2 x_3 x_4' x_5 + x_1 x_2 x_3' x_4 x_5'$
$\qquad + x_1 x_2 x_3' x_4' x_5 + x_1 x_2 x_3' x_4' x_5' + x_1 x_2' x_3 x_4 x_5$
$\qquad + x_1 x_2' x_3 x_4 x_5' + x_1 x_2' x_3 x_4' x_5 + x_1 x_2' x_3' x_4 x_5'$
$\qquad + x_1' x_2 x_3 x_4 x_5 + x_1' x_2 x_3 x_4' x_5' + x_1' x_2 x_3' x_4 x_5$
$\qquad + x_1' x_2' x_3 x_4' x_5 + x_1' x_2' x_3 x_4 x_5' + x_1' x_2' x_3' x_4 x_5'$
$\qquad + x_1' x_2' x_3' x_4' x_5'$
(d) $(x_2' + x_4)(x_1 + x_2 + x_4)(x_1' + x_2 + x_3 + x_4')(x_1' + x_2 + x_3' + x_4)$

*7 Describe a Quine-McCluskey procedure for finding a minimal product-of-sums representation for Boolean expressions.

SUGGESTED REFERENCES

4.1 Rutherford, D. E. *Introduction to Lattice Theory.* Hafner, New York, 1965.

4.2 Gratzer, G. *Lattice Theory.* Freeman, San Francisco, 1971.

4.3 Birkhoff, G. *Lattice Theory*, 3rd ed. A.M.S. Colloquium Publications, 1967.

4.4 Birkhoff, G., and T. C. Bartee. *Modern Applied Algebra.* McGraw-Hill, New York, 1970.

4.5 Hohn, F. E. *Applied Boolean Algebra.* Macmillan, New York, 1966.

4.6 Halmos, P. *Lectures on Boolean Algebra.* Van Nostrand, Princeton, N.J., 1963.

4.7 McCluskey, E. J. *Introduction to the Theory of Switching Circuits.* McGraw-Hill, New York, 1965.

4.8 Prather, R. E. *Introduction to Switching Theory.* Allyn and Bacon, Boston, 1967.

Lattice theory has not quite found the range of application enjoyed by Boolean algebra, but as a mathematical discipline, it has seen a tremendous growth over the last century, and especially in recent years. This situation is reflected in the references, which are dominated by purely theoretical treatments of lattice theory and texts slanted toward applications in the case of Boolean algebra.

Either of the first two references will provide a readable account of elementary lattice theory. The book by Birkhoff is the classical treatment of lattices, fascinating in its development of the many facets of the theory. It is probably too concise for most undergraduate students, however. The same author has collaborated with Bartee on a text in applied algebra, and a sizeable portion of this book is devoted to lattices and Boolean algebras; but here again some parts of the text are quite compact.

The applications of Boolean algebra to logic and switching circuits is nicely presented in the book by Hohn. In contrast, the book by Halmos gives an excellent treatment of Boolean algebra from a purely mathematical point of view. Almost any book on switching theory will begin with an elementary introduction to Boolean algebra. So perhaps they should be consulted. On the other hand, they are quite numerous. So without any intention to slight the others, we simply cite two examples which the reader may find useful, particularly in respect to the minimization problems.

5

Groups and Combinatorics

To most well-educated persons, whether specializing in the arts or in the sciences, the concept of symmetry is commonplace. On the other hand, only someone familiar with mathematics will know that there is a whole algebra of symmetry lying behind this universal concept. This algebraic theory—the theory of groups—forms a cornerstone in any modern program of mathematical studies, and though its applications to computer science are widely scattered, they are steadily growing and their importance should not be underestimated.

In order to focus our attention on a concrete situation, let us briefly consider the rigid motions of a geometric figure. For definiteness, suppose we are considering a subset of the plane: a circle, a square, or some less regular figure. Already we have hinted at a distinction concerning symmetry in our use of the word "regular," as if irregular figures were less appealing esthetically. A re-arrangement of the points of a figure by means of a transformation T is said to be a *rigid motion* if the distance between any pair of points remains the same after the transformation as before. Thus in a rigid motion we are dealing with a cardboard figure, as it were, where the relative position of the points in the figure cannot be altered. Loosely speaking, the degree of symmetry of a figure is reflected in the complexity of its "group of rigid motions." For this reason, the latter is often called the *symmetry group* of the given figure.

To be more specific, let us now assume that our figure is a regular polygon of n sides, where $n \geq 3$. [See Figure 5.1(a).] It is clear that a rigid motion must carry vertices into vertices. For the vertex v_1 at position 1 we have n choices of image $T(v_1)$. With each such choice there are only two alternatives for $T(v_2)$, because the two vertices v_1 and v_2 must remain adjacent. It is evident that the two pieces of information,

(1) vertex v_1 to have its position rotated clockwise by $k \cdot 2\pi/n$ ($k = 0, 1, 2, \ldots,$
$n - 1$),

(2) vertex v_2 to be relocated, adjacent to $T(v_1)$, clockwise or counterclockwise, will specify T completely. So we conclude that there are $2n$ symmetries for the n-sided polygon (n-gon). Note that these symmetries will by no means account for all of the permutations of the n vertices, there being $n!$ of them.

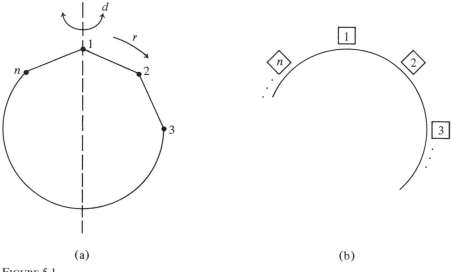

(a)   (b)

FIGURE 5.1

How are such considerations connected with combinatorics? Let us discuss a problem of no particular significance in itself, but one that can be related rather easily to the foregoing presentation. Suppose that we want to seat n persons at a round table [Figure 5.1(b)], and we wish to count the number of possible seating arrangements. You might say that there are $n!$ arrangements, because we can seat the first person in any one of the n seats, and that will leave $n - 1$ choices for the second person, and so on. On the other hand, we would not ordinarily distinguish between two seating arrangements where one is simply a rotation of the other. What we really want is the number of non-equivalent arrangements after allowing for the symmetries of the round table. Actually, the problem is once more that of a regular polygon of n sides, since the seats are equivalent to polygonal sides; but because we would probably want to distinguish between the clockwise and counterclockwise arrangements in the same order, there would be only n symmetries this time. Thus you would be right if you guessed that there are $n!/n = (n - 1)!$ distinct seating arrangements. On the other hand, you are oversimplifying the problem if you think that such questions are always so easy to decide.

Later in the chapter we will arrive at a general theory for counting configurations in the face of symmetry considerations. This theory is applicable to a number of different situations arising in computer science and related fields. First, however, we need to become familiar with the algebra of symmetry itself. In presenting this subject, we will not delve very deeply into the theory of groups but will be satisfied with only the fundamentals. For most of the elementary computer science applications, the fundamentals will suffice.

5.1 GROUPS AND SUBGROUPS

The student is already familiar with the ordinary algorithmic methods for solving linear, quadratic, and perhaps cubic equations. Interestingly enough, it was the search for methods of solving general polynomial equations that led to the first significant research into group theory. Two tragic figures of the nineteenth century pioneered these studies: Evariste Galois of France and Niels Abel of Norway. Both men died in their twenties, but the circumstances surrounding Galois' death were particularly intriguing. (The student should take the time to read a brief account of the lives of these two famous mathematicians.) Insurmountable difficulties—the misfortune of poverty in the case of Abel and the arrogance shown by the academies toward Galois—could not entirely hold back the originality of these two brilliant young mathematicians. Among other things, their works showed that algorithmic methods for solving even the general fifth-degree equation cannot exist.

It is indeed a very special class of monoids that allows for unique solutions of linear equations $ax = b$, where a and b are arbitrary members. The existence of unique solutions of such equations is seen to depend on the existence of (unique) reciprocals a^{-1} for each element a in the monoid. Thus

$$x = ex = (a^{-1}a)x = a^{-1}(ax) = a^{-1}b$$

in all such cases. Note in particular the use of associativity and the property of the identity element—the two characteristic features of monoids. If the third property, the existence of reciprocals, holds as well, then the monoid is said to be a *group*.

Axiomatically, then, a *group* $G = (G, \cdot, e)$ is an algebraic structure satisfying the monoid axioms:

(1) *associativity* $\qquad x \cdot (y \cdot z) = (x \cdot y) \cdot z$
(2) *identity* $\qquad\qquad x \cdot e = e \cdot x = x$

and have the further property that for each element $x \in G$ there is an element x^{-1} (*inverse* of x), also in G, such that

(3) *inverse* $\qquad\qquad x^{-1} \cdot x = x \cdot x^{-1} = e$

The uniqueness of inverses is easily established, for if x^* also satisfies (3), then we have

$$x^* = x^*e = x^*(xx^{-1}) = (x^*x)x^{-1} = ex^{-1} = x^{-1}$$

A group G is said to be *commutative* (or *abelian*, after Abel) if in addition to properties (1) through (3) we have

(4) *commutativity* $x \cdot y = y \cdot x$

for all x and y in G. We will find that the elementary theory of groups parallels the corresponding monoid theory rather closely. But owing to property (3), our results will be much "sharper" in the case of groups.

EXAMPLE 5.1   As with monoids, many of the most familiar groups are groups of numbers, the binary operation being taken as the addition or multiplication of the corresponding arithmetic. Thus we may take $G = Z$ with the operation of addition. Then the identity element is the number 0, and inverses are "negatives." Writing $-x$ rather than x^{-1}, we see that property (3) is the familiar law:

$$(-x) + x = x + (-x) = 0 \qquad \triangle$$

EXAMPLE 5.2   Let G be the set of positive real numbers with ordinary multiplication. The identity element is 1 and the inverse of x is usually denoted by $1/x$. $\triangle$

EXAMPLE 5.3   The set of all real numbers $R = (R, +, 0)$ is an (additive) group. Inverses are characterized in the same way as for the group $Z = (Z, +, 0)$ of Example 5.1. In the sense of a definition that should be anticipated, we say that Z is a subgroup of R. $\triangle$

EXAMPLE 5.4   The set of all complex numbers $C = (C, +, 0)$ is another additive group. One might think of R as a subgroup of C. At the very least, R is isomorphic to a subgroup of C. $\triangle$

EXAMPLE 5.5   The set of all nonzero complex numbers is a group with respect to complex multiplication. The number $1 = 1 + 0i$ is the identity element. $\triangle$

EXAMPLE 5.6   The set $\{1, i, -1, -i\}$ is a group if we take as our operation the following:

$\cdot$	1	i	-1	$-i$
1	1	i	-1	$-i$
i	i	-1	$-i$	1
-1	-1	$-i$	1	i
$-i$	$-i$	1	i	-1

This group is a subgroup of the group of Example 5.5. All of our examples thus far have been commutative groups. In the case of a finite group whose operation is given by a multiplication table (often called the *Cayley table* of the group), we have commutativity if and only if the table is symmetric with respect to the diagonal. △

Given a multiplication table, the suggestions for locating the identity element of the group (if, indeed, we have a group) and for verifying the associativity property, are the same as those given in Example 3.7 for monoids; but with groups, the *cancellation laws*

$$xy = xz \Rightarrow y = z \qquad \text{(multiply by } x^{-1} \text{ on the left)}$$
$$yx = zx \Rightarrow y = z \qquad \text{(multiply by } x^{-1} \text{ on the right)}$$

show that in the multiplication table of any finite group, each row and column must be a permutation of all the group elements. Furthermore, each row (column) must represent a different permutation. (Why?) These characteristics will help simplify the task of showing that a given table cannot represent a group. With them in mind, the reader should try to construct a catalog of "small" groups, say of *order* $|G| = 1, 2, 3, 4, 5, 6, 7, 8$. Again, the general definition of isomorphism should be kept in mind so that abstractly equivalent groups can be detected. Homomorphism or isomorphism is defined in the same way for groups as that for monoids (Example 3.17), except that we do not need to insist that $\varphi(e) = e$. (Why?)

EXAMPLE 5.7   Table 5.1, reproduced from Example 1.12, represents a group

TABLE 5.1

·	a	b	c	d	e	f
a	e	d	f	b	a	c
b	c	e	a	f	b	d
c	b	f	d	e	c	a
d	f	a	e	c	d	b
e	a	b	c	d	e	f
f	d	c	b	a	f	e

with e as the identity element. Note that each row and column of the table represents a different permutation of the six elements. As for inverses, we have

$$a^{-1} = a \qquad \text{(since } aa = e)$$
$$b^{-1} = b \qquad \text{(since } bb = e)$$
$$c^{-1} = d \qquad \text{(since } cd = dc = e) \quad \text{etc.}$$
$$d^{-1} = c$$
$$e^{-1} = e$$
$$f^{-1} = f$$

Evidently this group is noncommutative, since

$$cb = f \neq a = bc$$

for instance.　　　　　　　　　　　　　　　　　　　　　　　　　　　△

EXAMPLE 5.8　One monoid of Example 3.11 had the following table:

	w	x	y	z
w	z	z	w	w
x	z	y	x	w
y	w	x	y	z
z	w	w	z	z

with y as the identity element. This monoid cannot be a group, because the first row, for instance, is not a permutation of the elements w, x, y, z.　　△

EXAMPLE 5.9　The free monoids Σ^* (Section 3.4) cannot be groups because $|xy| \geq |x|$. It follows that whenever $x \neq \epsilon$, there will not be a word y such that $xy = yx = \epsilon$.　　　　　　　　　　　　　　　　　　　　　△

EXAMPLE 5.10　With Examples 1.16 and 1.17, we obtain a whole class of finite groups $B^n = (B^n, \oplus, 0)$ where 0 is the vector

$$0 = (0, 0, \ldots, 0)$$

You may recall that the operation of B^n is as follows:

$$(a_1, \ldots, a_n) \oplus (b_1, \ldots, b_n) = (a_1 \oplus b_1, \ldots, a_n \oplus b_n)$$

with the coordinatewise sum given as follows:

$\oplus$	0	1
0	0	1
1	1	0

In the additive notation, we write $-x$ for the inverse of x, but in this case we have $-x = x$ for every x in B^n.　　　　　　　　　　　　　　　　△

EXAMPLE 5.11　The group $B^3 = (B^3, \oplus, 0)$ has Table 5.2 as its addition table. In general we will refer to these groups $B^n = (B^n, \oplus, 0)$ as *binary groups*.　△

　　As with monoids, a *subgroup* H of a group $G = (G, \cdot, e)$ is a subset $H \subseteq G$ which itself forms a group relative to the same group operation as in G and having the same identity. Thus a subgroup H must satisfy the *closure conditions*:

TABLE 5.2

$\oplus$	$(0,0,0)$	$(0,0,1)$	$(0,1,0)$	$(0,1,1)$	$(1,0,0)$	$(1,0,1)$	$(1,1,0)$	$(1,1,1)$
$(0,0,0)$	$(0,0,0)$	$(0,0,1)$	$(0,1,0)$	$(0,1,1)$	$(1,0,0)$	$(1,0,1)$	$(1,1,0)$	$(1,1,1)$
$(0,0,1)$	$(0,0,1)$	$(0,0,0)$	$(0,1,1)$	$(0,1,0)$	$(1,0,1)$	$(1,0,0)$	$(1,1,1)$	$(1,1,0)$
$(0,1,0)$	$(0,1,0)$	$(0,1,1)$	$(0,0,0)$	$(0,0,1)$	$(1,1,0)$	$(1,1,1)$	$(1,0,0)$	$(1,0,1)$
$(0,1,1)$	$(0,1,1)$	$(0,1,0)$	$(0,0,1)$	$(0,0,0)$	$(1,1,1)$	$(1,1,0)$	$(1,0,1)$	$(1,0,0)$
$(1,0,0)$	$(1,0,0)$	$(1,0,1)$	$(1,1,0)$	$(1,1,1)$	$(0,0,0)$	$(0,0,1)$	$(0,1,0)$	$(0,1,1)$
$(1,0,1)$	$(1,0,1)$	$(1,0,0)$	$(1,1,1)$	$(1,1,0)$	$(0,0,1)$	$(0,0,0)$	$(0,1,1)$	$(0,1,0)$
$(1,1,0)$	$(1,1,0)$	$(1,1,1)$	$(1,0,0)$	$(1,0,1)$	$(0,1,0)$	$(0,1,1)$	$(0,0,0)$	$(0,0,1)$
$(1,1,1)$	$(1,1,1)$	$(1,1,0)$	$(1,0,1)$	$(1,0,0)$	$(0,1,1)$	$(0,1,0)$	$(0,0,1)$	$(0,0,0)$

(i) $x \in H$ and $y \in H \Rightarrow x \cdot y \in H$

(ii) $x \in H \Rightarrow x^{-1} \in H$

If H is any subgroup, then the (left) *cosets* of G modulo H are the subsets

$$Hx = \{hx : h \in H\}$$

for fixed x in G. The idea is simple enough. We just multiply x on the left by each of the elements in H. Of course, it is conceivable that $Hx_1 = Hx_2$ even though $x_1 \neq x_2$, so that, a priori, we cannot say how many cosets will result from a given subgroup H. Nevertheless, we call this number the *index* $[G : H]$ of H in G. The precise number will be provided by the important theorem named after Lagrange, the foremost French algebraist in the period immediately preceding the life of Galois. It goes without saying that we are mainly thinking of finite groups throughout this discussion, even though some of the ideas make perfectly good sense in the infinite case.

EXAMPLE 5.12   Checking the closure conditions, we find that $aa = e \in H$ and $a^{-1} = a \in H$ in the case of the subset $H = \{e, a\}$, showing that it is a subgroup of the group G of Example 5.7. Evidently H itself is a coset, for with $x = e$ we have

$$He = \{ee, ae\} = \{e, a\} = H$$

If we let $x = a$, we get the same coset all over again, that is,

$$Ha = \{ea, aa\} = \{a, e\} = He = H$$

Next, suppose that $x = b$ or $x = c$. Then

$$Hb = \{eb, ab\} = \{b, d\} \qquad Hc = \{ec, ac\} = \{c, f\}$$

The reader should check that $Hd = Hb$ and $Hf = Hc$. Thus there are exactly three cosets altogether. As Lagrange's theorem will show, it is no coincidence that

$$[G : H] = 3 = 6/2 = |G|/|H| \qquad \triangle$$

EXAMPLE 5.13   If we take the subgroup $H = \{e, c, d\}$ in this same group G, then we have

$$He = Hc = Hd = \{e, c, d\} = H$$
$$Ha = Hb = Hf = \{a, b, f\}$$

and again,

$$[G : H] = 2 = 6/3 = |G|/|H| \qquad \triangle$$

Lemma

The cosets of G modulo a subgroup H form a partition of G, each coset having the same size as H itself.

Proof   Let x be any element of G. Since

$$x = ex \in Hx$$

(because e is in H), we see that the cosets form a covering of G. So in order to prove that $\{Hx : x \in G\}$ is a partition of G, we need only show that distinct cosets do not intersect.

If $Hx \cap Hy \neq \varnothing$, then there is an element z such that $z \in Hx$ and $z \in Hy$, in which case we can write

$$z = h_1 x = h_2 y$$

where h_1 and h_2 are in H. Now, for any $h \in H$ we have

$$hx = h(h_1^{-1}h_2 y) = (hh_1^{-1}h_2)y$$

which shows that $Hx \subseteq Hy$. Similarly,

$$hy = (hh_2^{-1}h_1)x$$

showing that $Hy \subseteq Hx$ as well. Altogether we have proved that

$$Hx \cap Hy \neq \varnothing \Rightarrow Hx = Hy$$

that is, different cosets are disjoint.

It only remains to show that each coset Hx is of the same size as the subgroup H. For this purpose we have a bijection $H \to Hx$ provided by the mapping $h \to hx$. Obviously the mapping is surjective. The injectivity follows from the cancellation law:

$$hx = hy \Rightarrow x = h^{-1}(hy) = (h^{-1}h)y = ey = y$$

This completes the proof of the lemma. $\square$

Theorem 5.1   (Lagrange)

$$[G : H] = |G|/|H|$$

Proof   If we multiply the number $[G : H]$ of cosets by the size $|H|$ of each coset, we will obtain $|G|$ because the cosets partition G. $\square$

Let us close this section with a remark concerning the rewriting of the statement of Lagrange's theorem in the form

$$|G| = |H| \cdot [G : H]$$

Evidently this equation shows that a group G cannot have subgroups of arbitrary order—the order or size of each subgroup must divide the order of G itself! Note at the same time what this statement does not say. It does not say that there is necessarily a subgroup of size corresponding to every divisor of the group order. This is an important distinction to keep in mind.

EXAMPLE 5.14   The group G of Example 5.7 has the order $|G| = 6$. Of course G has the "trivial" subgroups G and $\{e\}$, of sizes 6 and 1, respectively. Besides these subgroups, each subgroup of G must be of order 2 or 3, for these numbers are the only other divisors of 6.   △

EXAMPLE 5.15   The binary groups of Example 5.10 are of the order 2^n. According to Lagrange's theorem, each subgroup of B^n must have order 2^k for some $k, 0 \leq k \leq n$. Does B^n have subgroups of every order 2^k?   △

EXERCISES

1 Show that $\varphi(e) = e$ in any group homomorphism φ.

2 Show that there are exactly two nonisomorphic groups of order four.

*3 Find five nonisomorphic groups of order eight. [*Hint*: Three are abelian and two are not.]

4 Explain why each row (column) in the multiplication table of a group must represent a *distinct* permutation of the group elements.

*5 Show that there is at least one group of every finite order. [*Hint*: See Section 5.3.]

6 Prove that a subset $H \subseteq G$ of a group G is a subgroup if and only if

$$x \text{ and } y \text{ in } H \Rightarrow xy^{-1} \in H$$

7 Find all the subgroups of the group of Example 5.7. Show that they are subgroups and that there cannot be any more.

8 Describe the coset decompositions corresponding to the given subgroups of the following groups G:
   (a) Example 5.7 with $H = \{e, b\}$
   (b) Example 5.7 with $H = \{e\}$
   (c) Example 5.3 with $H = Z$
   (d) Example 5.2 with $H = Q$ (rational numbers)
   (e) Example 5.5 with the subgroup H of Example 5.6
   (f) Example 5.11 with $H = \{(0, 0, 0), (0, 1, 0)\}$
   (g) Example 5.11 with $H = \{(0, 0, 0), (0, 0, 1), (0, 1, 0), (0, 1, 1)\}$
   (h) Example 5.16 with $H = \{1, r, r^2, r^3\}$

9 For groups of each given order, use Lagrange's theorem to list the orders of possible subgroups:

(a) 24   (b) 50   (c) 60   (d) 120   (e) 17

*10 A subgroup N of a group G is said to be *normal* if $x^{-1}nx \in N$ for each $x \in G$ and $n \in N$. Show that corresponding to each congruence relation R on G (considered as a monoid) there exists a normal subgroup $N_R = [e]_R$ and show that this correspondence is one to one among all the congruences and normal subgroups. [*Hint*: Show that with each normal subgroup $N \subseteq G$ one can associate the congruence relation R_N:

$$x\, R_N\, y \Leftrightarrow xy^{-1} \in N$$

and that

$$R_{N_R} = R \quad \text{and} \quad N_{R_N} = N]$$

*11 Show that if N is a normal subgroup of G, then the set of all cosets Nx forms a group (we call it the *quotient group* G/N) with respect to the operation

$$Nx \cdot Ny = N(x \cdot y)$$

[*Hint*: See Exercise 10 and Section 3.7.]

12 Define the *kernel K* of a group homomorphism $\varphi: G \to H$ to be the subset

$$K = \{x \in G : \varphi(x) = e\}$$

Show that K is always a normal subgroup (see the definition of "normal" in Exercise 10).

*13 Prove the homomorphism theorem for groups: Let $\varphi: G \to H$ be a surjective group homomorphism. Then the kernel is a normal subgroup N and $G/N \simeq H$. Conversely, if N is a normal subgroup of G, then there is a surjective group homomorphism $\varphi: G \to G/N$ with kernel N. [*Hint*: See Exercise 10 through 12 and Theorem 3.3.]

14 Find all the normal subgroups of the group of Example 5.7 (see the definition of "normal" in Exercise 10).

15 Find all the subgroups of the group of symmetries of the square (Example 5.16). Which of these subgroups are normal? (See the definition of "normal" in Exercise 10.)

*16 Prove that a subgroup of index two is always normal. (Again, see the definition of "normal" in Exercise 10.)

5.2 PERMUTATION GROUPS

Let us return briefly to our discussion of the rigid motions of a regular polygon of n sides. Imagine that we have a cardboard figure as depicted in Figure 5.1(a). Then our earlier analysis shows that the symmetry group of the polygon is generated by the following pair of transformations:

$$r = \text{clockwise rotation through an angle } 2\pi/n$$
$$d = \text{reflection through the vertical diameter}$$

If we perform r, n times in succession, or if we perform d twice, we will be back where we started. So if 1 denotes the identity transformation, then at least one of the following relations is clear:

$$r^n = 1 = d^2$$

$$dr = r^{-1}d$$

where the multiplication refers to composition of transformations. Evidently r^{-1} is a counterclockwise rotation through the same angle $2\pi/n$. The validity of the second identity is best seen by considering Figure 5.2. Owing to these two relationships, we are able to express every composition of r and d in the

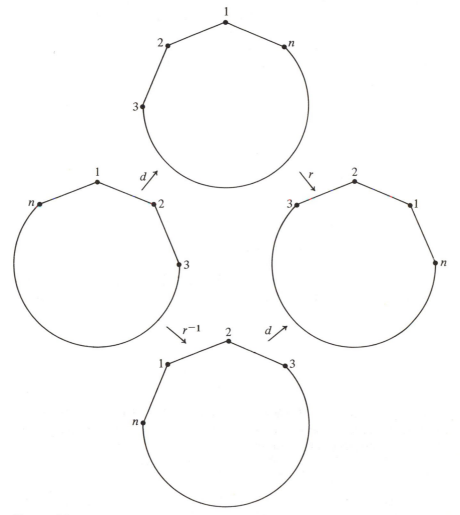

FIGURE 5.2

form $r^k d^j$ (for $k = 0, 1, \ldots, n - 1$ and $j = 0, 1$). Again we see that there are $2n$ symmetries of the polygon.

EXAMPLE 5.16   There are eight symmetries of the square, which may be represented as in the multiplication table (Table 5.3).

TABLE 5.3

$\cdot$	1	r	r^2	r^3	d	rd	r^2d	r^3d
1	1	r	r^2	r^3	d	rd	r^2d	r^3d
r	r	r^2	r^3	1	rd	r^2d	r^3d	d
r^2	r^2	r^3	1	r	r^2d	r^3d	d	rd
r^3	r^3	1	r	r^2	r^3d	d	rd	r^2d
d	d	r^3d	r^2d	rd	1	r^3	r^2	r
rd	rd	d	r^3d	r^2d	r	1	r^3	r^2
r^2d	r^2d	rd	d	r^3d	r^2	r	1	r^3
r^3d	r^3d	r^2d	rd	d	r^3	r^2	r	1

Thus

$$r^2d \cdot r = r^2(dr) = r^2(r^{-1}d) = (r^2r^{-1})d = rd$$

$$r^2d \cdot rd = r^2(dr)d = r^2(r^{-1}d)d = (r^2r^{-1})(dd) = r$$

$$r^2 \cdot r^3d = (r^2r^3)d = r^5d = rd$$

where we used the relationships:

$$r^4 = 1 = d^2 \qquad dr = r^{-1}d \qquad \triangle$$

EXAMPLE 5.17   As generators of the group of symmetries of the square, r and d may be represented as transformations of the set of vertices as follows (see Figure 5.3):

$$
\begin{array}{ll}
1 \xrightarrow{r} 2 & 1 \xrightarrow{d} 1 \\
2 \xrightarrow{r} 3 & 2 \xrightarrow{d} 4 \\
3 \xrightarrow{r} 4 & 3 \xrightarrow{d} 3 \\
4 \xrightarrow{r} 1 & 4 \xrightarrow{d} 2
\end{array}
$$

In this way any member of the multiplication table of Example 5.16 may be interpreted in concrete terms. Thus in order to visualize the symmetry rd, we have only to compose r and d to obtain

$$
r \circ d: \quad
\begin{array}{l}
1 \rightarrow 4 \\
2 \rightarrow 3 \\
3 \rightarrow 2 \\
4 \rightarrow 1
\end{array}
$$

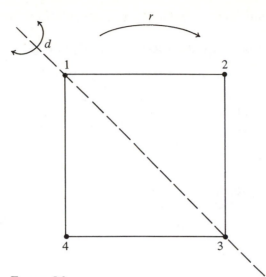

FIGURE 5.3

Then it is clear that *rd* is a reflection through a line parallel to two opposite sides, as shown in Figure 5.4. Other members of the group of symmetries will have similar interpretations. △

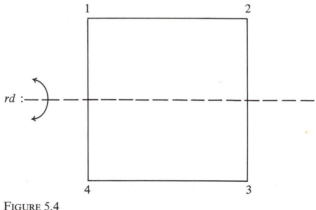

FIGURE 5.4

Quite apart from geometrical considerations, we may seek to form groups of transformations of an arbitrary set, but in order that such transformations be invertible, as we know from our characterization of Section 0.7, they must be bijective. We recall that the bijective transformations of the set

$$n^+ = \{1, 2, \ldots, n\}$$

(or any set, for that matter) are called permutations. So we may expect to find a theory of permutation groups analogous to that of the transformation monoids of Section 3.2.

Let us first consider the full *symmetric group* S_n of degree n. As a set,

$$S_n = \{\sigma: \sigma \text{ is a permutation of } n^+\}$$

Among the permutations is the identity permutation 1_{n^+}. Also the compositional product $\sigma \circ \tau$ of two permutations is surely another permutation. So we definitely have a transformation monoid. Being bijections, however, each permutation σ will have an inverse σ^{-1}, so that as functions,

$$\sigma^{-1} \circ \sigma = \sigma \circ \sigma^{-1} = 1$$

and this is all we need to conclude that $S_n = (S_n, \circ, 1)$ is a group. Clearly $|S_n| = n!$, as we remarked on several occasions.

Instead of using the usual functional descriptions, it is customary to denote the permutation $\sigma: n^+ \to n^+$ by

$$\sigma = \begin{pmatrix} 1 & 2 & \cdots & n \\ \sigma(1) & \sigma(2) & \cdots & \sigma(n) \end{pmatrix}$$

where the value of the function is written immediately below its argument. Now suppose we have the situation described in Figure 5.5, where certain members of n^+ are transformed in a circle (while the others remain *fixed*, that is, $\sigma(i) = i$).

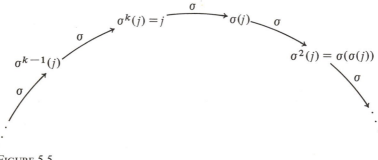

FIGURE 5.5

Then σ is said to be a *cycle* of length k. For such permutations we will ordinarily use the more concise notation:

$$\sigma = (j \ \sigma(j) \ \sigma^2(j) \cdots \sigma^{k-1}(j))$$

to symbolize the situation in Figure 5.5. In particular, we assume that

$$\sigma^{k-1}(j) \overset{\sigma}{\to} j$$

and that members of n^+ that do not appear in the presentation are fixed (by σ). We trust that the next several examples will clarify the notation in all of its aspects. Of course, the reader should be aware that an arbitrary permutation will contain more than one cycle (see τ in Example 5.20). On the other hand, a representation theorem (Theorem 5.2) will show that even such a permutation may be expressed as a product of cycles.

EXAMPLE 5.18   As in Example 0.47, with $n = 3$, the symmetric group $S_n = S_3$ consists of the $6 = 3!$ permutations:

$$1 = \begin{pmatrix} 1 & 2 & 3 \\ 1 & 2 & 3 \end{pmatrix} \begin{pmatrix} 1 & 2 & 3 \\ 2 & 1 & 3 \end{pmatrix} \begin{pmatrix} 1 & 2 & 3 \\ 3 & 2 & 1 \end{pmatrix} \begin{pmatrix} 1 & 2 & 3 \\ 1 & 3 & 2 \end{pmatrix} \begin{pmatrix} 1 & 2 & 3 \\ 2 & 3 & 1 \end{pmatrix} \begin{pmatrix} 1 & 2 & 3 \\ 3 & 1 & 2 \end{pmatrix}$$

Thus the permutation

$$\sigma = \begin{pmatrix} 1 & 2 & 3 \\ 2 & 1 & 3 \end{pmatrix}$$

is the same function as that which we wrote in the form of Figure 5.6 before.

FIGURE 5.6

Other than the identity, we have cycles of lengths 2, 2, 2, 3, 3. In cycle notation, the six permutations are represented as

$$1 \quad (12) \quad (13) \quad (23) \quad (123) \quad (132)$$

respectively, where a special notation is reserved for 1. Thus in the case of

$$\sigma = \begin{pmatrix} 1 & 2 & 3 \\ 2 & 1 & 3 \end{pmatrix}$$

we have

$$\overset{\sigma}{\underset{\sigma}{1 \rightleftarrows 2}} \qquad 3 \overset{\sigma}{\to} 3$$

so that $\sigma = (12)$ in cycle notation. △

EXAMPLE 5.19   In order to obtain the multiplication table of the group S_3, we have only to perform the various functional compositions. Thus

$$(12) \circ (132) = (1 \to 2) \circ (1 \to 3 \to 2)$$

$$= \begin{pmatrix} 1 \to 2 \to 1 \\ 2 \to 1 \to 3 \\ 3 \to 3 \to 2 \end{pmatrix} = (23)$$

$$(132) \circ (132) = (1 \to 3 \to 2) \circ (1 \to 3 \to 2)$$

$$= \begin{pmatrix} 1 \to 3 \to 2 \\ 2 \to 1 \to 3 \\ 3 \to 2 \to 1 \end{pmatrix} = (123)$$

etc. In this way, we obtain Table 5.4.

TABLE 5.4

∘	1	(12)	(13)	(23)	(123)	(132)
1	1	(12)	(13)	(23)	(123)	(132)
(12)	(12)	1	(123)	(132)	(13)	(23)
(13)	(13)	(132)	1	(123)	(23)	(12)
(23)	(23)	(123)	(132)	1	(12)	(13)
(123)	(123)	(23)	(12)	(13)	(132)	1
(132)	(132)	(13)	(23)	(12)	1	(123)

△

EXAMPLE 5.20   In the symmetric group S_7 the permutation

$$\sigma = \begin{pmatrix} 1 & 2 & 3 & 4 & 5 & 6 & 7 \\ 1 & 2 & 5 & 4 & 7 & 3 & 6 \end{pmatrix}$$

is a cycle of length four:

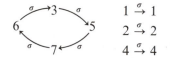

Accordingly, we write $\sigma = (3576)$. In the same group the permutation

$$\tau = \begin{pmatrix} 1 & 2 & 3 & 4 & 5 & 6 & 7 \\ 2 & 4 & 5 & 1 & 6 & 7 & 3 \end{pmatrix}$$

is not a cycle, there being two "orbits":

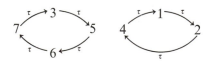

However, it appears that these two orbital rotations could be performed in either order, just because they do not overlap. We might say that

$$\tau = (3567)(124)$$
$$= \begin{pmatrix} 1 & 2 & 3 & 4 & 5 & 6 & 7 \\ 1 & 2 & 5 & 4 & 6 & 7 & 3 \end{pmatrix} \circ \begin{pmatrix} 1 & 2 & 3 & 4 & 5 & 6 & 7 \\ 2 & 4 & 3 & 1 & 5 & 6 & 7 \end{pmatrix}$$

and refer to τ as a product of disjoint cycles. There is, to be sure, a certain nonuniqueness about such a product representation, because

$$(3567)(124) = (124)(3567) = (5673)(412) = \cdots$$

Nevertheless, such distinctions are not essential. Note the parenthetical qualification in the following statement.   △

Theorem 5.2

Every permutation has an (essentially) unique representation as a product of disjoint cycles.

Proof If σ is a permutation of $n^+ = \{1, 2, \ldots, n\}$, then the relation

$$x \sim y \qquad x = \sigma^m(y) \qquad \text{(for some integer } m\text{)}$$

is an equivalence relation on n^+. The equivalence classes (*orbits*) partition n^+ into nonoverlapping subsets on which the effect of σ is a cycle. $\qquad\qquad\square$

EXAMPLE 5.21   Consider the permutation

$$\sigma = \begin{pmatrix} 1 & 2 & 3 & 4 & 5 & 6 & 7 & 8 & 9 & 0 \\ 4 & 5 & 1 & 3 & 7 & 2 & 6 & 8 & 0 & 9 \end{pmatrix}$$

in S_{10}. We find the following orbits:

$$
\begin{array}{ll}
1 \overset{\sigma}{\to} 4 \overset{\sigma}{\to} 3 \overset{\sigma}{\to} 1 & (143) \\
2 \overset{\sigma}{\to} 5 \overset{\sigma}{\to} 7 \overset{\sigma}{\to} 6 \overset{\sigma}{\to} 2 & (2576) \\
9 \overset{\sigma}{\to} 0 \overset{\sigma}{\to} 9 & (90)
\end{array}
$$

whereas 8 remains fixed. Thus we may write

$$\sigma = (143)(2576)(90)$$

which is a product of disjoint cycles. $\qquad\qquad\triangle$

There are any number of computerized investigations in which we require the use of the set of all permutations of n^+, for some integer n. In these cases dealing with the sheer magnitude of the collection ($|S_n| = n!$) can be a problem in itself. To read in or to store all these permutations is certainly to be avoided, if that is possible. Fortunately, effective schemes have been devised to meet this problem. One procedure is Algorithm XX, due to Langdon.* Here basically the idea is to generate the successive permutations internally, each one from its predecessor. Then if the intended uses of the permutation are handled as the permutations are generated, it should not be necessary to store them all at once.

We shall revert to the original flowchart language in our description of Algorithm XX. The main reason for doing so concerns the repetitive nature of the output. With most of our previous algorithms, perhaps all of them, the output values are available at the time the procedure is terminated. Therefore, we can get by without a formal output statement. Here, however, it is important to see how the various permutations are generated, one after another, throughout the course of execution of the procedure. Not all of the values of the output variable will belong to this sequence, so it is well that we include an output

* G. G. Langdon, "An Algorithm for Generating Permutations," *Commun. A.C.M.*, **10**, 298–299 (1967).

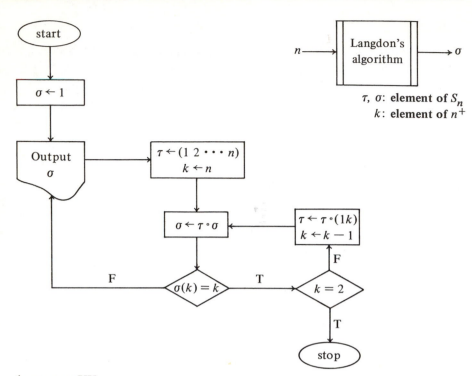

ALGORITHM XX

statement to clarify matters. All of this is perhaps best explained in the original flowchart language.

In Algorithm XX the successive values printed out as σ will constitute a complete listing of S_n, although in actual practice they may only represent an input to some other procedure, as mentioned earlier. At first, we set

$$\tau = (1\ 2 \cdots n)$$

but from time to time this cycle will be reduced, inasmuch as

$$(1\ 2 \cdots k)(1k) = (1\ 2 \cdots k - 1)$$

And the principal use of the cycle $(1\ 2 \cdots k)$ takes place in the assignment $\sigma \leftarrow \tau \circ \sigma$ where σ is the permutation

$$\sigma = \begin{pmatrix} 1 & 2 & \cdots & n \\ \sigma(1) & \sigma(2) & \cdots & \sigma(n) \end{pmatrix}$$

It is most important to note that this assignment statement has the simple effect of rotating the first k images of σ:

$$\tau \circ \sigma = \begin{pmatrix} 1 & 2 & \cdots & k & k+1 & \cdots & n \\ 2 & 3 & \cdots & 1 & k+1 & \cdots & n \end{pmatrix} \circ \begin{pmatrix} 1 & 2 & \cdots & k & \cdots & n \\ \sigma(1) & \sigma(2) & \cdots & \sigma(k) & \cdots & \sigma(n) \end{pmatrix}$$

$$= \begin{pmatrix} 1 & 2 & \cdots & k-1 & k & k+1 & \cdots & n \\ \sigma(2) & \sigma(3) & \cdots & \sigma(k) & \sigma(1) & \sigma(k+1) & \cdots & \sigma(n) \end{pmatrix}$$

This observation is the whole key to understanding the mechanics of Langdon's procedure.

EXAMPLE 5.22   We shall present a trace of the algorithm as it generates the $3! = 6$ permutations of S_3 as follows. The output variable is given in two different forms (see Table 5.5). In the second representation (in the style of Section 0.4) the effect of rotating the images of σ is most easily seen. At the same time we shall include an output column, where a check mark ($\checkmark$) indicates that the corresponding permutation has been printed.

TABLE 5.5

k	τ	σ		output
3	(123)	1	123	$\checkmark$
		(123)	231	$\checkmark$
		(132)	312	$\checkmark$
		1	123	
2	(12)	(12)	213	$\checkmark$
3	(123)	(23)	132	$\checkmark$
		(13)	321	$\checkmark$
		(12)	213	
2	(12)	1	123	

The reader should try to use the algorithm to generate S_4, so as to be sure that he or she understands the procedure.   △

Any subgroup of the group of all permutations of the (perhaps infinite) set S is said to be a *permutation group*. And the universality of this structure is fully expressed in the extension of Theorem 3.1 to groups:

Theorem 5.3   (Cayley's theorem for groups)

Every group is isomorphic to a permutation group.

Proof   In rereading the proof of Theorem 3.1, simply observe that in the case of groups, the transformations

$$T_g : G \to G \qquad T_g(x) = x \cdot g$$

become permutations of the set G. They are injective because of the cancellation law:

$$T_g(x) = T_g(y) \Rightarrow xg = yg \Rightarrow x = y$$

and they are surjective because we can express any z in G in the form xg, namely $z = (zg^{-1})g$, so that

$$T_g(zg^{-1}) = (zg^{-1})g = z(g^{-1}g) = ze = z \qquad \square$$

Corollary

Up to isomorphism, the only finite groups are the subgroups of the symmetric groups S_n. $\qquad \square$

One observation perhaps should be made in connection with Cayley's theorem for groups (as with the case of monoids). The theorem states that each group is isomorphic to a subgroup of some full symmetric group. In no sense does it preclude the possibility that this could happen in several ways. Thus in the case of a finite group G there is not a unique S_n having G as a subgroup. Also in applications, where we are perhaps interested in a representation as a permutation group for computational purposes, we would naturally look for the smallest n for which S_n realizes G. In general, this will not be the representation provided by Cayley's theorem.

EXAMPLE 5.23   Consider the group of symmetries of the square as described in Example 5.16. The representation provided by Cayley's theorem is that of a subgroup of S_8. Yet, we know from the discussion of Example 5.17 that our group is isomorphic to a subgroup of S_4. In fact, rd is represented as the permutation $(14)(23)$, and the other symmetries have similar interpretations (see Table 5.6).

TABLE 5.6

$\circ$	1	(1234)	(13)(24)	(1432)	(24)	(14)(23)	(13)	(12)(34)
$1 \to 1$	1	(1234)	(13)(24)	(1432)	(24)	(14)(23)	(13)	(12)(34)
$r \to (1234)$	(1234)	(13)(24)	(1432)	1	(14)(23)	(13)	(12)(34)	(24)
$r^2 \to (13)(24)$	(13)(24)	(1432)	1	(1234)	(13)	(12)(34)	(24)	(14)(23)
$r^3 \to (1432)$	(1432)	1	(1234)	(13)(24)	(12)(34)	(24)	(14)(23)	(13)
$d \to (24)$	(24)	(12)(34)	(13)	(14)(23)	1	(1432)	(13)(24)	(1234)
$rd \to (14)(23)$	(14)(23)	(24)	(12)(34)	(13)	(1234)	1	(1432)	(13)(24)
$r^2d \to (13)$	(13)	(14)(23)	(24)	(12)(34)	(13)(24)	(1234)	1	(1432)
$r^3d \to (12)(34)$	(12)(34)	(13)	(14)(23)	(24)	(1432)	(13)(24)	(1234)	1

In this multiplication table the group of symmetries of the square is represented as a subgroup of the full symmetric group S_4. $\qquad \triangle$

EXERCISES

1 Show that the groups of Examples 5.7 and 5.19 are isormorphic.

2 Write the following permutations, including those in Figure 5.7, as products of disjoint cycles:

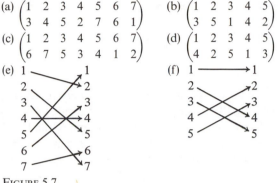

(a) $\begin{pmatrix} 1 & 2 & 3 & 4 & 5 & 6 & 7 \\ 3 & 4 & 5 & 2 & 7 & 6 & 1 \end{pmatrix}$ (b) $\begin{pmatrix} 1 & 2 & 3 & 4 & 5 \\ 3 & 5 & 1 & 4 & 2 \end{pmatrix}$

(c) $\begin{pmatrix} 1 & 2 & 3 & 4 & 5 & 6 & 7 \\ 6 & 7 & 5 & 3 & 4 & 1 & 2 \end{pmatrix}$ (d) $\begin{pmatrix} 1 & 2 & 3 & 4 & 5 \\ 4 & 2 & 5 & 1 & 3 \end{pmatrix}$

FIGURE 5.7

3 Show that in the proof of Theorem 5.2

$$x \sim y \Leftrightarrow x = \sigma^m(y) \qquad \text{(for some integer } m\text{)}$$

is an equivalence relation on n^+.

4 Use Algorithm XX to generate S_4.

5 Give the representation (resulting from Cayley's theorem) of the group of Example 5.7 as a permutation group on the set $S = \{a, b, c, d, e, f\}$, writing each group element as a product of disjoint cycles. Considering Exercise 1, show that this group is realized by a full symmetric group in more than one way.

6 Show that every permutation can be expressed as a product of *transpositions*, that is, cycles of length two.

*7 A permutation is said to be *even* (or *odd*) according as there are an even (or odd) number of transpositions in the representation of Exercise 6. Show that this definition is not ambiguous. Why does it seem to be ambiguous?

8 Show that the even permutations (see Exercise 7) form a normal subgroup A_n of the full symmetric group S_n. (See Exercise 10 of Section 5.1 for the definition of "normal.") We call A_n the *alternating group* of degree n.

9 A group G is said to be *simple* if it has no normal (Exercise 10 of Section 5.1) subgroups except G and $\{e\}$. Show that each group of prime order is simple.

*10 Show that A_n (Exercise 8) is simple for all $n > 4$.

11 List the elements of A_4 (Exercise 8) and show that A_4 is not a simple group. (See Exercise 9 for the definition of "simple.")

12 Express the permutations of Exercise 2 as products of transpositions (see Exercise 6) and use the criteria of Exercise 7 to decide their oddness or evenness.

*13 Devise a formal algorithm for deciding whether a permutation is odd or even.

14 Prove that the group S_n is nonabelian for all $n \geq 3$.

5.3   CYCLIC GROUPS

Perhaps the most elementary of all the groups that one is likely to encounter are those which are each generated by a single element. We call them *cyclic* groups. They happen to be commutative groups; in fact, they are the building blocks in a very comprehensive decomposition theory for general abelian groups. Though we will make contact with this theory only in one very special instance (Theorem 5.4), the nature of such decompositions should perhaps be spelled out ahead of time. All such decompositions take the form of product representations. By analogy with the general outline of a product algebra (Section 1.2), we may construct from the groups $G_1, G_2, \ldots, G_r$ the *product group*

$$G = \overset{r}{\underset{i=1}{\times}} G_i$$

with the operation

$$(g_1, g_2, \ldots, g_r) \cdot (g_1', g_2', \ldots, g_r') = (g_1 \cdot g_1', g_2 \cdot g_2', \ldots, g_r \cdot g_r')$$

It is not difficult to see that G is again a group, with the identity element $e = (e, e, \ldots, e)$ and with inverses given by the formula

$$(g_1, g_2, \ldots, g_r)^{-1} = (g_1^{-1}, g_2^{-1}, \ldots, g_r^{-1})$$

The same definition is used whether the G_i are commutative or not. But in the case of commutative groups, it is quite common to use the additive notation instead of the one used here, and there should be no confusion if we choose to do so.

EXAMPLE 5.24   The product of the groups of Examples 5.6 and 5.7 will result in a group of order $4 \cdot 6 = 24$. Without constructing the complete multiplication table for the product group, we might just list a few of the computations for clarification:

$$(a, i) \cdot (b, 1) = (a \cdot b, i \cdot 1) = (d, i)$$

$$(c, i) \cdot (d, i) = (c \cdot d, i \cdot i) = (e, -1)$$

$$(f, 1) \cdot (b, i) = (f \cdot b, 1 \cdot i) = (c, i)$$

The identity element for the product group is $(e, 1)$, and the following are examples showing the computation of inverses:

$$(c, i)^{-1} = (c^{-1}, i^{-1}) = (d, -i)$$

$$(f, -1)^{-1} = (f^{-1}, -1^{-1}) = (f, -1) \qquad \triangle$$

EXAMPLE 5.25   If R is the group of Example 5.3, then the product group $R \times R$ is isomorphic to C, the additive group of complex numbers (Example 5.4). $\qquad \triangle$

EXAMPLE 5.26   The binary groups of Examples 5.10 and 5.11 are products of the elementary group B with itself. $\qquad \triangle$

There are several ways in which one could introduce the cyclic groups. (See Exercise 2 of Section 5.3.) It seems, however, that the approach through integral congruences may be the most accessible, from the student's point of view. We recall (Example 3.32) that if we define

$$a \sim b \,(\text{mod } n) \Leftrightarrow n|a - b$$

then we have introduced a congruence on the monoid $Z = (Z, +, 0)$. Actually Z is a commutative group, as mentioned in Example 5.1. The quotient monoid is denoted by

$$Z_n = Z/\sim (\text{mod } n) = \{[0], [1], [2], \ldots, [n - 1]\}$$

and its members are the equivalence or congruence classes:

$$[a] = [a]_n = \{x: x \sim a \,(\text{mod } n)\}$$

Now recall that the monoid Z_n has the usual quotient operation

$$[a] + [b] = [a + b]$$

and if we use the fact that $n \sim 0 \,(\text{mod } n)$, we find that inverses exist for each element $[a]$ in Z_n:

$$[a] + [n - a] = [a + (n - a)] = [n] = [0]$$

All of this implies that Z_n is a (commutative) group. We will call it the (additive) *cyclic group* of integers modulo n.

EXAMPLE 5.27   If $n = 6$, then

$$Z_6 = \{[0], [1], [2], [3], [4], [5]\} \qquad (\text{or simply } \{0, 1, 2, 3, 4, 5\})$$

has the addition table (Table 5.7).

Note that when it is convenient to write

$$Z_n = \{0, 1, 2, \ldots, n - 1\}$$

we use the circled group operation symbol $\oplus$ or $\oplus_n$ (as in binary groups) as a

TABLE 5.7

+	[0]	[1]	[2]	[3]	[4]	[5]
[0]	[0]	[1]	[2]	[3]	[4]	[5]
[1]	[1]	[2]	[3]	[4]	[5]	[0]
[2]	[2]	[3]	[4]	[5]	[0]	[1]
[3]	[3]	[4]	[5]	[0]	[1]	[2]
[4]	[4]	[5]	[0]	[1]	[2]	[3]
[5]	[5]	[0]	[1]	[2]	[3]	[4]

or

$\oplus$	0	1	2	3	4	5
0	0	1	2	3	4	5
1	1	2	3	4	5	0
2	2	3	4	5	0	1
3	3	4	5	0	1	2
4	4	5	0	1	2	3
5	5	0	1	2	3	4

reminder that the elements are only representatives of their respective equivalence classes, and that the addition is to be performed modulo n.   △

In the next section we will introduce the residue number systems, whose usefulness derives from the properties of the cyclic groups Z_n. These systems allow for an implementation of computer arithmetic operations with a faster execution time than can be obtained by the more conventional methods. The feasibility of such schemes is found to be dependent on the following classical decomposition theorem.

Theorem 5.4   (Chinese remainder theorem)

If $n = p_1^{\alpha_1} p_2^{\alpha_2} \cdots p_r^{\alpha_r}$ is the decomposition of the integer n into distinct prime powers $p_i^{\alpha_i} = q_i$, then the cyclic group Z_n has the product representation

$$Z_n \simeq Z_{q_1} \times Z_{q_2} \times \cdots \times Z_{q_r}$$

Proof   We consider the mapping $\varphi \colon Z_n \to \bigtimes_{i=1}^{r} Z_{q_i}$ as given by

$$\varphi([a]_n) = ([a]_{q_1}, [a]_{q_2}, \ldots, [a]_{q_r})$$

That is, we simply take the appropriate congruence class in each of the coordinate cyclic groups $Z_{q_i} = Z_{p_i}^{\alpha_i}$. In order to see that φ is injective, suppose that $\varphi([a]_n) = \varphi([b]_n)$. Then we have $[a]_{q_i} = [b]_{q_i}$ for $i = 1, 2, \ldots, r$ and so

$$a \sim b \,(\mathrm{mod}\ q_i) \qquad (i = 1, 2, \ldots, r)$$

This means that every q_i divides the difference $a - b$. By the definition of least common multiples, we have

$$n = \mathrm{lcm}\,(q_1, q_2, \ldots, q_r)|(a - b)$$

which in turn implies that $a \sim b \,(\mathrm{mod}\ n)$ or $[a]_n = [b]_n$, so that φ is indeed injective. Then because

$$|Z_n| = n = q_1 q_2 \cdots q_r = |Z_{q_1}||Z_{q_2}| \cdots |Z_{q_r}| = \left| \bigtimes_{i=1}^{r} Z_{q_i} \right|$$

we may use the pigeonhole principle (Section 0.7) to conclude that φ is surjective as well.

Finally, we compute

$$
\begin{aligned}
\varphi([a]_n + [b]_n) &= \varphi([a + b]_n) = ([a + b]_{q_1}, \ldots, [a + b]_{q_r}) \\
&= ([a]_{q_1} + [b]_{q_1}, \ldots, [a]_{q_r} + [b]_{q_r}) \\
&= ([a]_{q_1}, \ldots, [a]_{q_r}) + ([b]_{q_1}, \ldots, [b]_{q_r}) \\
&= \varphi([a]_n) + \varphi([b]_n)
\end{aligned}
$$

showing that φ is a group isomorphism.   □

Corollary

The cyclic groups Z_n have subgroups of order k for each divisor k of n.   □

EXAMPLE 5.28   Consider the cyclic group Z_{24}. Since 24 factors into prime powers as

$$24 = 3 \cdot 2^3$$

we learn from the Chinese remainder theorem that

$$Z_{24} \simeq Z_3 \times Z_8$$

A few correspondences in the isomorphism $\varphi: Z_{24} \to Z_3 \times Z_8$ are as follows:

$$17 \overset{\varphi}{\to} (2, 1) \qquad \text{since } 17 \sim 2 \ (\text{mod } 3) \text{ and } 17 \sim 1 \ (\text{mod } 8)$$
$$13 \overset{\varphi}{\to} (1, 5) \qquad \text{since } 13 \sim 1 \ (\text{mod } 3) \text{ and } 13 \sim 5 \ (\text{mod } 8)$$

$$6 \overset{\varphi}{\to} (0, 6)$$

Note in particular that the additions in Z_{24} and in $Z_3 \times Z_8$ result in corresponding sums, that is,

$$17 \oplus 13 = 6 \qquad (2, 1) \oplus (1, 5) = (2 \oplus 1, 1 \oplus 5) = (0, 6)$$

whereas

$$6 \overset{\varphi}{\to} (0, 6)$$

This is the property that has proved useful in the design of fast arithmetic units for certain applications of digital computers.   △

EXERCISES

1 Show that, as additive groups, $R \times R \simeq C$.

*2 Consider a multiplicative group G generated by a single element x (so that every group element has a representation x^r for some integer r) with the group operation

$$x^r \cdot x^s = x^{r+s}$$

It is natural to write $x^0 = e$ (the identity element of the group) and to distinguish the two mutually exclusive cases:

I. All the x^r are distinct.

II. There are integers $r \neq s$ such that $x^r = x^s$.

(a) Show that in Case I, $G \simeq Z$.

(b) Show that in Case II, $G \simeq Z_n$ for some integer n. [*Hint*: In Case II we have $x^{r-s} = x^r \cdot x^{-s} = x^0 = e$ so we may let n be the smallest positive integer for which $x^n = e$.]

3 Construct the addition table for the group Z_6.

4 Express each of the following as a product of cyclic groups of prime-power order:

(a) Z_{60} (b) Z_{240} (c) Z_{225} (d) Z_{450}

5 Corresponding to the groups Z_n of Exercise 4, determine respectively the images of each of the following pairs of elements and their sums, according to the isomorphism of Theorem 5.4.

(a) 14, 57   (b) 92, 193   (c) 105, 122   (d) 303, 222

*6 Show that every subgroup of a finite cyclic group (as characterized in Exercise 2) is itself a cyclic group.

*7 Prove that every finite abelian group is isomorphic to a direct product of cyclic groups.

8 The group of symmetries of a regular n-gon is isomorphic to $Z_2 \times Z_n$. True or false? Explain.

5.4   COMPUTER ARITHMETIC

Certainly it is because of the easy availability of binary logical circuitry that computer arithmetic has largely been based on the binary number system. The discussion of Example 1.16 may serve as a convenient point of departure. We recall that although we use an algebra in which $1 + 1 = 0$, we must remember that in fact we want $1 + 1 = 10$, which "carries" a one into the next position whenever two (or more) ones are added. Therefore, at any position in a pair of binary numbers a, b we may describe the addition according to the pair of truth tables:

a	b	carry	sum
0	0	0	0
0	1	0	1
1	0	0	1
1	1	1	0

where the last line shows the situation just described.

Using the methods of Chapter 4, we may obtain the following expressions for the carry and the sum:

$$\text{carry} = ab$$
$$\text{sum} = a'b + ab' \qquad [= (a + b)(ab)']$$

The equivalent expression for the sum shown in brackets might be preferable. All other things being equal, it has two features in its favor. For one thing, the subexpression ab coincides with the required carry expression. Secondly, one fewer inverter is needed compared with $a'b + ab'$. Sometimes this second feature is not all that important. As Figure 5.8 indicates (with the small circle), logic circuits often have a complemented output available "free of charge." Of course, if there is not one, then the small circle can be taken as an inverter circuit, so that the original argument will stand. In any case, a two-input two-output logic circuit (such as shown in Figure 5.8) having the behavior of the preceding truth tables is called a *half-adder*.

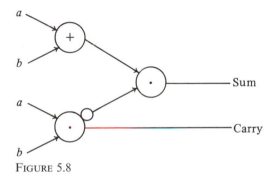

FIGURE 5.8

The reason for this name is obvious. We have forgotten that there might have been a previous carry from the positions to the right. So the real sum S is obtained only after another addition with c, the previous carry. If there is a carry in either half-adder (or in both), then we are to send a carry C to the next position, to the left. Accordingly, Figure 5.9 represents the appropriate logic to be implemented at each position of the binary addition.

Why have we gone to the trouble of describing the details of a conventional binary adder? Mainly to show the reasons for the time involved in performing an addition. Consider the sum of the following pair of binary numbers:

$$0\ 0\ 1\ 1\ 0\ 1\ 1\ 1$$
$$+\ 0\ 1\ 0\ 0\ 1\ 0\ 0\ 1$$

The sum in the left-most position would be zero but for the pair of ones clear over at the right. Thus the sum at any given position may depend on the entries at each and every previous position. In a conventional adder, the speed of addition is held up by this *carry propagation time*. One more glance at

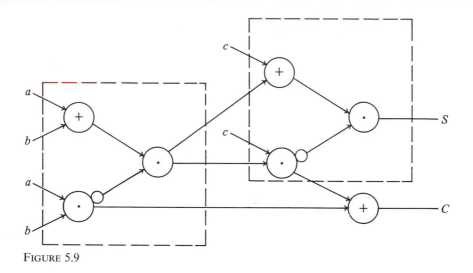

FIGURE 5.9

Figure 5.9 will show that the carry signal must pass through something like $2n$ levels of logical circuitry, n being the overall word length. Needless to say, this requirement places a severe limitation on the speed at which addition can be accomplished. One can try to design faster logic circuits—and there are certain sophisticated "speed-up" techniques that can be employed—but perhaps none of them can effect such a dramatic improvement as the change in philosophy we are about to present.

Computer operations are inherently confined to an arithmetic modulo n for some (large) n, but the Chinese remainder theorem suggests that we might profit by seeking a decomposition of n into distinct prime powers

$$n = q_1 q_2 \cdots q_r$$

for we then have

$$Z_n \simeq Z_{q_1} \times Z_{q_2} \times \cdots \times Z_{q_r}$$

and can perform our arithmetic in the separate Z_{q_i}, with a faster addition time. Of course, all of this would require our having a means for converting from a conventional binary representation to the factored form, and vice versa, as indicated in Figure 5.10. Neglecting this problem for the moment, we see that because the component additions are carried out simultaneously, the internal addition time is dependent only on that for an addition in the largest Z_{q_i}; and we could choose n and the prime powers q_i in such a way that the saving is substantial.

Now let us see whether the required encoding and decoding processes are feasible. If so, let us try to determine just what kind of computations would need to be carried out. In the case of the conversion to the factored form, the situation

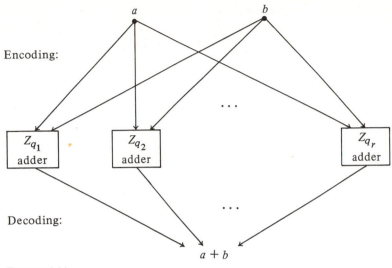

Figure 5.10

<small>FIGURE 5.10</small>

is quite clear. We need to be able to calculate remainders upon division by the q_i. Surely this is feasible. In fact, our first algorithm (Algorithm I, Section 1.7) shows how this can be done. Of course, we must not forget that the time required for these computations must be balanced against the prospective savings. At this stage it is easy to come away with the impression that we are almost certainly fighting a losing battle. But the reader should try to reserve his judgment until such time as we are able to put everything into perspective.

Next, let us consider the problem of decoding the factored form of a number and then converting it back into the number system Z_n. Since we have encoded n by means of a bijective mapping φ (the isomorphism of the Chinese remainder theorem), an inverse mapping necessarily exists. We are interested, however, in the nature of the computations involved. The answer is provided in the summation formula of the following theorem.

Theorem 5.5

Let

$$x_1 = \varphi^{-1}(1, 0, 0, \ldots, 0)$$

$$x_2 = \varphi^{-1}(0, 1, 0, \ldots, 0), \quad \text{etc.}$$

where φ is the isomorphism of the Chinese remainder theorem. Then for each a in Z_n we have

$$a = \varphi^{-1}(a \bmod q_1, a \bmod q_2, \ldots, a \bmod q_r) = \oplus \sum_{i=1}^{r} x_i(a \bmod q_i)$$

Proof   Since φ is an isomorphism, we may write

$$\varphi\left(\sum_{i=1}^{r} x_i(a \bmod q_i)\right) = \sum_{i=1}^{r} \varphi(x_i(a \bmod q_i))$$

$$= \sum_{i=1}^{r} (0, 0, \ldots, 0, a \bmod q_i, 0, \ldots, 0)$$

$$= (a \bmod q_1, a \bmod q_2, \ldots, a \bmod q_r)$$

which is equivalent to the statement in the theorem.   □

EXAMPLE 5.29   Suppose that $n = 42 = 2 \cdot 3 \cdot 7$. Then the result of encoding

TABLE 5.8

0—000	7—110	14—020	21—100	28—010	35—120
1—111	8—021	15—101	22—011	29—121	36—001
2—022	9—102	16—012	23—122	30—002	37—112
3—103	10—013	17—123	24—003	31—113	38—023
4—014	11—124	18—004	25—114	32—024	39—104
5—125	12—005	19—115	26—025	33—105	40—015
6—006	13—116	20—026	27—106	34—016	41—126

each a, $0 \le a \le 41$, is shown in Table 5.8. We note that

$$x_1 = \varphi^{-1}(1, 0, 0) = 21$$
$$x_2 = \varphi^{-1}(0, 1, 0) = 28$$
$$x_3 = \varphi^{-1}(0, 0, 1) = 36$$

According to Theorem 5.5, these are all we need for decoding. For instance,

$$\varphi^{-1}(1, 2, 4) = x_1 \cdot 1 + x_2 \cdot 2 + x_3 \cdot 4$$
$$= 21 \cdot 1 + 28 \cdot 2 + 36 \cdot 4$$
$$= 21 + 56 + 144$$
$$= 11 \ (\bmod 42)$$

Similarly,

$$\varphi^{-1}(1, 0, 2) = x_1 \cdot 1 + x_2 \cdot 0 + x_3 \cdot 2$$
$$= 21 \cdot 1 + 28 \cdot 0 + 36 \cdot 2$$
$$= 21 + 72$$
$$= 9 \ (\bmod 42)$$

Both calculations agree with the table.   △

Our theory shows that using only the standard arithmetic techniques, we can with no particular difficulty convert numbers back and forth between a standard number system Z_n and a *residue number system*, as these product representations are called. For some time now, computer designers have shown great interest in these residue systems, for obvious reasons. There are serious drawbacks, however, in using these systems. One cannot expect the full range of algorithmic processes to lend themselves to implementation with as much ease as addition; and in fact, even a simple magnitude comparison will cause great difficulty, as will the problem of overflow detection. Of course, this does not mean that the residue systems do not have their place. For certain kinds of calculations, where comparisons, conversions, etc., are infrequent, the residue representation can offer great advantages.

EXERCISES

1 Use the isomorphism of the Chinese remainder theorem to encode each element from:

(a) Z_{35} into $Z_5 \times Z_7$ (b) Z_{18} into $Z_2 \times Z_9$

(c) Z_{60} into $Z_3 \times Z_4 \times Z_5$ (d) Z_{50} into $Z_2 \times Z_{25}$

(e) Z_{90} into $Z_2 \times Z_5 \times Z_9$ (f) Z_{21} into $Z_3 \times Z_7$

2 Corresponding to the respective isomorphisms of Exercise 1, determine the "weights" x_i to be used in Theorem 5.5.

3 Corresponding to the weights of Exercise 2 and the isomorphisms of Exercise 1, decode the following into the respective cyclic groups of Exercise 1.

(a) $(2, 5), (3, 1),$ and $(4, 4)$ (b) $(1, 7), (1, 5),$ and $(0, 4)$

(c) $(2, 2, 2), (2, 1, 4),$ and $(1, 0, 3)$ (d) $(1, 17), (0, 18),$ and $(1, 22)$

(e) $(1, 4, 6), (1, 3, 7),$ and $(0, 2, 8)$ (f) $(1, 6), (2, 5),$ and $(2, 3)$

*4 Show that the isomorphism of the Chinese remainder theorem preserves multiplication (as well as addition).

*5 Explain the difficulties of magnitude comparison with the residue number systems.

5.5   BINARY CODES

In transmitting digital data through various channels (telephone lines, space communications, etc.) there is always a probability that errors will be introduced. Consider the case of two computer systems communicating with each other by bouncing signals off an orbiting satellite. The atmospheric conditions alone could distort the signals to such an extent that the receiving system would not be able to make correct interpretations. In order to offset such difficulties, certain "redundancy" coding systems are used which allow for the detection or even correction of the most likely errors.

Suppose that, instead of transmitting the messages directly, we intervene at both the sending and receiving ends of the system with encoding and decoding mechanisms, such as shown in the diagram of Figure 5.11. The idea is to augment the original message, by supplying additional but redundant information in the encoding process so that the added information can be used by the decoder to recognize the occurrence of errors and thereby restore the

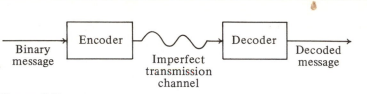

FIGURE 5.11

message to its original form. Within limits, such a procedure can indeed be successful, given that an appropriate coding system has been devised.

With Figure 5.11 we have enlarged the overall communication system to the scheme

$$B^m \xrightarrow{\epsilon} B^n \xrightarrow{\tau} B^n \xrightarrow{\delta} B^m$$

where

ϵ is the coding function
τ is the transmission
δ is the decoding function

for binary messages $a \in B^m$. Actually, we would probably want to send messages $x \in B^*$, that is, words of arbitrary length from the alphabet $B = \{0, 1\}$, but if we can think of x as being split up into message units $a \in B^m$ for some convenient integer m, then our model is still appropriate. Together with the binary groups of Examples 5.10 and 5.11, the geometric model of the n-cube (Section 4.4) is useful in discussing the resulting coding theory.

EXAMPLE 5.30   The Hamming distance of the n-cube can be used to establish a "magnitude" or *norm*

$$|a| = \Delta(a, 0)$$

for each $a \in B^n$. Evidently $|a|$ simply counts the number of ones among the coordinates of $a = (a_1, a_2, \ldots, a_n)$. However, it does allow for a sensible definition of the *sphere* $S(a, r)$ with center a and radius r; that is,

$$S(a, r) = \{x \in B^n : |x - a| \le r\}$$

It follows that

$$x \in S(a, r) \Leftrightarrow x = a \oplus e, \qquad |e| \le r$$

We now have a convenient means for discussing neighboring "code words."   △

EXAMPLE 5.31   In B^4 let $a = (1, 0, 1, 1)$ and $r = 2$. Then

$$
\begin{aligned}
S(a, r) = \{&(1, 0, 1, 1), (0, 0, 1, 1), (1, 1, 1, 1), (1, 0, 0, 1), (1, 0, 1, 0) \\
&(0, 1, 1, 1), (0, 0, 0, 1), (0, 0, 1, 0), (1, 1, 0, 1), (1, 1, 1, 0) \\
&(1, 0, 0, 0)\}
\end{aligned}
$$

We only have to change coordinates in a, zero, one, and two at a time, in all possible ways. Note that every $x \in S(a, r)$ has the form $x = a \oplus e$ with $|e| \leq 2$. Thus

$$(1, 0, 0, 1) = (1, 0, 1, 1) \oplus (0, 0, 1, 0) \qquad |(0, 0, 1, 0)| = 1 \leq 2 \qquad \triangle$$

One obvious difficulty with our communication model is that $\tau : B^n \to B^n$ is unknown and probably not a function at all, since there are generally spurious or random influences which account for the "noise" in the transmission environment. So we speak with some imprecision if we say that ideally

$$\epsilon \circ \tau \circ \delta = 1_{B^m}$$

But at least this is a description of the behavior of the system when all goes well. Now suppose that we had a perfect transmission channel so that $\tau = 1_{B^n}$. Then we would surely want the preceding identity to hold. This consideration thus motivates the following definition. We say that the pair (ϵ, δ) is an *admissible* (binary) *code* provided that $\epsilon \circ \delta = 1_{B^m}$, so that with perfect transmission no errors will occur, and any errors that do occur can be attributed to a faulty transmission.

EXAMPLE 5.32   Consider the so-called *parity check* codes (ϵ, δ) in which the functions

$$\epsilon : B^m \to B^{m+1} \qquad \delta : B^{m+1} \to B^m$$

are determined as follows. For $a \in B^m$ and $b \in B^{m+1}$ we set

$$\epsilon(a_1, a_2, \ldots, a_m) = (a_1, a_2, \ldots, a_m, a_{m+1}) \quad \text{where} \quad a_{m+1} = \begin{cases} 0 & \text{if } |a| \text{ even} \\ 1 & \text{if } |a| \text{ odd} \end{cases}$$

$$\delta(b_1, b_2, \ldots, b_m, b_{m+1}) = (b_1, b_2, \ldots, b_m)$$

Then we compute

$$\begin{aligned}(\epsilon \circ \delta)(a) = \delta(\epsilon(a)) &= \delta(a_1, a_2, \ldots, a_{m+1}) \\ &= (a_1, a_2, \ldots, a_m) = a = 1_{B^m}(a)\end{aligned}$$

showing that (ϵ, δ) is an admissible code. $\qquad \triangle$

The word "parity" is used to distinguish oddness from evenness. With the parity check codes, the redundant a_{m+1} is chosen so that $|\epsilon(a)|$ is even for each and every *code word* $\epsilon(a)$. Then, should an error occur in just one coordinate during transmission (so that $|\tau(\epsilon(a))| = $ odd), it would surely be detectable. On the other hand, we can show by an example that such an error may go uncorrected. In preparation for this example and a forthcoming lemma, we observe that in the relation $R = \text{kernel}(\delta)$, as defined in Section 0.7, words are related in pairs, in the case of the parity check codes. That is

$$(a_1, a_2, \ldots, a_m, 0) \, R \, (a_1, a_2, \ldots, a_m, 1)$$

because

$$(a_1, a_2, \ldots, a_m, 0) \overset{\delta}{\searrow}$$
$$\overset{\delta}{\to}(a_1, a_2, \ldots, a_m) = a$$
$$(a_1, a_2, \ldots, a_m, 1) \overset{\delta}{\nearrow}$$

In the absence of transmission errors, one of these two words is the code word $\epsilon(a)$.

EXAMPLE 5.33   If $m = 2$, then an explicit description of the parity check code is as shown in Figure 5.12. Suppose that we send the message $(1, 0)$, encoded as the code word $(1, 0, 1)$. If an error should occur in just one position during the transmission, then $(0, 0, 1)$, $(1, 1, 1)$, or $(1, 0, 0)$ would be received. Since $|\tau(\epsilon(1, 0))| = |\tau(1, 0, 1)| = $ odd in all three cases, the error could be detected. But

$$\delta(0, 0, 1) = (0, 0)$$

$$\delta(1, 1, 1) = (1, 1)$$

$$\delta(1, 0, 0) = (1, 0)$$

so it would be only a coincidence if the received message was properly interpreted. The code words are "too close together" in B^3 to permit a proper correction.   △

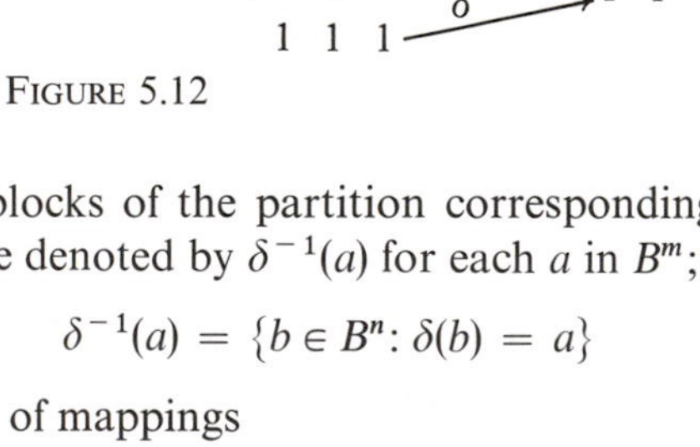

FIGURE 5.12

Suppose that the blocks of the partition corresponding to the kernel of δ (see Example 0.54) are denoted by $\delta^{-1}(a)$ for each a in B^m; that is,

$$\delta^{-1}(a) = \{b \in B^n : \delta(b) = a\}$$

Here (ϵ, δ) is any pair of mappings

$$B^m \overset{\epsilon}{\to} B^n \overset{\delta}{\to} B^m$$

We find that the admissible codes (or simply *codes*, as we will now call them) may be characterized as follows.

Lemma

(ϵ, δ) is a code iff $\epsilon(a) \in \delta^{-1}(a)$ for all $a \in B^m$

Proof First, suppose that the condition fails to hold. Then for some $a \in B^m$ we would have $\epsilon(a) \notin \delta^{-1}(a)$. Therefore, $\epsilon \circ \delta \neq 1$ because $\delta(\epsilon(a)) \neq a$. On the other hand, if the condition does hold, then for every $a \in B^m$,

$$(\epsilon \circ \delta)(a) = \delta(\epsilon(a)) = a = 1_{B^m}(a)$$

and (ϵ, δ) is a code. ☐

Theorem 5.6

If (ϵ, δ) is a code, then ϵ is injective and δ is surjective. ☐

EXAMPLE 5.34 Immediately we see that ϵ is injective and δ is surjective in the case of the parity check code of Example 5.33. △

EXAMPLE 5.35 The pair of mappings shown in Figure 5.13, which are injective and surjective, respectively, show that the conditions of Theorem 5.6 are necessary but not sufficient for (ϵ, δ) to be a code. △

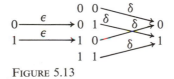

FIGURE 5.13

We have established that any errors in a coded communication system [one where (ϵ, δ) is a code, as defined here] are due solely to imperfections in the transmission channel. So if we trace a message unit through the system:

$$a \xrightarrow{\epsilon} b \xrightarrow{\tau} b' \xrightarrow{\delta} a'$$

we may write

$$\tau(b) = b' = b \oplus e$$

and it is appropriate to call $e = (e_1, e_2, \ldots, e_n)$ the *error*. Note that $e \in B^n$ and we are using the addition of the binary group $B^n = (B^n, \oplus, 0)$ of Examples 5.10 and 5.11. Of course, the random nature of our nonfunction τ would necessitate a similar comment regarding e. Nonetheless, what is known in probability theory as the maximum-likelihood estimation assures us that we could minimize the probability of misjudgment in the design of our decoding scheme δ by always assuming that the most likely error has occurred. We are thus referred back to the norm of Example 5.30. That is, we must assume that $|e| = 0$ is most likely, that $|e| = 1$ is more likely than $|e| = 2$, etc.

Now let us consider the spheres of Example 5.30 as an aid in discussing the proximity of code words. We learned in Theorem 5.5 that if (ϵ, δ) is to be a code, then the mapping

$$\epsilon: B^m \to B^n$$

must be injective. Accordingly, there are $|B^m| = 2^m$ distinct code words

$$\epsilon(a) = b^0, b^1, b^2, \ldots, b^{2^m - 1}$$

in B^n. We will say that $\epsilon: B^m \to B^n$ is *k-detectable* (allows for the detection of all combinations of k or fewer errors) if for all $i \neq j$,

$$b^i \notin S(b^j, k)$$

and is *k-correctable* if

$$S(b^i, k) \cap S(b^j, k) = \varnothing$$

whenever $i \neq j$.

So much for definitions. How about the motivation behind them? First, we note that if $b^i \in S(b^j, k)$ with $i \neq j$, then according to Example 5.30,

$$b^i = b^j \oplus e \qquad |e| \leq k$$

Therefore, if such an error were to occur in the transmission, then

$$\tau(b^j) = b^j \oplus e = b^i$$

and the error would go undetected because b^i is another code word. Conversely, if ϵ is k-detectable, then with each $i \neq j$ and $|e| \leq k$, we have

$$\tau(b^j) = b^j \oplus e \neq b^i$$

Thus we can detect not having received a code word—evidence that some kind of error has occurred. As for the other definition, we shall state our motivation somewhat more formally, as follows.

Theorem 5.7

If $\epsilon: B^m \to B^n$ is k-correctable, then there exists a decoding $\delta: B^m \to B^n$ with the property that for all errors e in the transmission

$$\tau(b^j) = b^j \oplus e \quad \text{where} \quad |e| \leq k$$

we have

$$\epsilon \circ \tau \circ \delta = 1$$

Proof If ϵ is k-correctable, then all of the spheres $S(b^i, k)$ are pairwise disjoint. So if we define

$$\delta(S(b^j, k)) = a^j$$

[where $\epsilon(a^j) = b^j$] and otherwise extend it arbitrarily to a mapping $\delta: B^n \to B^m$, then

$$a^j \overset{\epsilon}{\to} b^j \overset{\tau}{\to} b^j \oplus e \overset{\delta}{\to} a^j$$

as required.   □

EXAMPLE 5.36   The encoding $\epsilon: B^2 \to B^3$ of Example 5.33 is 1-detectable because

$$S((0, 0, 0), 1) = \{(0, 0, 0), (0, 0, 1), (0, 1, 0), (1, 0, 0)\}$$

$$S((0, 1, 1), 1) = \{(0, 1, 1), (0, 1, 0), (0, 0, 1), (1, 1, 1)\}$$

$$S((1, 0, 1), 1) = \{(1, 0, 1), (1, 0, 0), (1, 1, 1), (0, 0, 1)\}$$

$$S((1, 1, 0), 1) = \{(1, 1, 0), (1, 1, 1), (1, 0, 0), (0, 1, 0)\}$$

showing that each sphere contains only one code word (at the center). On the other hand, ϵ is not 1-correctable because the spheres overlap.   △

EXAMPLE 5.37   The coding $\epsilon: B^2 \to B^5$ given by

$$(0, 0) \overset{\epsilon}{\to} (0, 0, 0, 0, 0)$$
$$(0, 1) \overset{\epsilon}{\to} (0, 1, 1, 1, 0)$$
$$(1, 0) \overset{\epsilon}{\to} (1, 0, 1, 0, 1)$$
$$(1, 1) \overset{\epsilon}{\to} (1, 1, 0, 1, 1)$$

is 1-correctable, because the four spheres $S(\epsilon(a), 1)$ are nonoverlapping in B^5. The student should compute these spheres, just to make sure.   △

Some very simple necessary and sufficient conditions can be given for testing whether a given encoding function $\epsilon: B^m \to B^n$ is k-detectable or k-correctable. The tests involve the *minimum distance* $\Delta(\epsilon)$ between code words:

$$\Delta(\epsilon) = \min_{i \neq j} \{\Delta(b^i, b^j)\}$$

If we think in terms of the radii of the spheres entering into the definitions of k-detectable and k-correctable, then the following theorem will not require a proof.

Theorem 5.8

Let $\epsilon: B^m \to B^n$ be an encoding function. Then ϵ is k-detectable iff $\Delta(\epsilon) \geq k + 1$; it is k-correctable iff $\Delta(\epsilon) \geq 2k + 1$.   □

EXAMPLE 5.38   For the encoding of Example 5.33 we have the minimum distance $\Delta(\epsilon) = 2$; and it follows from the theorem that ϵ is 1-detectable, in agreement with the analysis of Example 5.36.   △

EXAMPLE 5.39   In Example 5.37 the minimum distance between code words is 3. For instance, $\Delta((0, 1, 1, 1, 0), (1, 1, 0, 1, 1)) = 3$, whereas there are other pairs of code words separated by a larger distance. So, according to Theorem 5.8, ϵ is 1-correctable and 2-detectable. △

Certain codes, called *group codes*, offer several advantages, practical as well as theoretical. The decoding process associated with them makes use of the coset construction of Lagrange's theorem (more precisely, the lemma preceding Lagrange's theorem), at least conceptually. We say that $\epsilon: B^n \to B^m$ is a *group encoding function* if the set of code words $H = \{b^i\} = \{\epsilon(a^i)\}$ is a subgroup of $B^n = (B^n, \oplus, 0)$. The minimum distance of such an encoding function is determined more easily than usual.

Lemma

The minimum distance of a group encoding function is equal to the minimum norm of its nonzero code words.

Proof   First, suppose that b^i and b^j are code words for which $\Delta(\epsilon) = \Delta(b^i, b^j)$. Then $b^i \oplus b^j$ is also a code word, because H is a subgroup. But $|b^i \oplus b^j| = \Delta(b^i, b^j)$, so there is a code word whose norm is equal to the minimum distance. Conversely, if $|b^i|$ is minimum among the norms of the nonzero code words, then we can easily find two code words separated by this distance, namely 0 and b^i. □

Theorem 5.9

A group encoding function is k-correctable iff

$$\min_{H \sim \{0\}} \{|b^i|\} \geq 2k + 1$$

Proof   Use the lemma, together with Theorem 5.8. □

Corollary

Given the condition of the theorem, the cosets $H \oplus e$ $(|e| \leq k)$ are mutually disjoint.

Proof   We know from the lemma preceding Lagrange's theorem that two cosets, $H \oplus e_1$ and $H \oplus e_2$, are equal if they are not disjoint. We are trying to prove that they are disjoint whenever $e_1 \neq e_2$ where each $|e_i| \leq k$. Suppose the contrary:

$$H \oplus e_1 = H \oplus e_2$$

Then $e_1 \oplus e_2 \in H$ (why?), and we would have

$$|e_1 \oplus e_2| \leq |e_1| + |e_2| \leq k + k = 2k$$

contradicting the given condition. □

The corollary allows for a convenient decoding scheme. We prepare a table whose first column is the subgroup H (the set of code words), and whose successive columns are the various cosets $H \oplus e$, where $|e| \leq k$. Since the columns are disjoint, so are the rows. But the latter are the spheres $S(b^i, k)$ required of an optimal decoding scheme! (See the proof of Theorem 5.7.)

EXAMPLE 5.40   Consider the encoding function $\epsilon: B^3 \rightarrow B^6$ as shown below:

$$000 \overset{\epsilon}{\rightarrow} 000000$$
$$001 \overset{\epsilon}{\rightarrow} 001111$$
$$010 \overset{\epsilon}{\rightarrow} 010011$$
$$011 \overset{\epsilon}{\rightarrow} 011100$$
$$100 \overset{\epsilon}{\rightarrow} 100110$$
$$101 \overset{\epsilon}{\rightarrow} 101001$$
$$110 \overset{\epsilon}{\rightarrow} 110101$$
$$111 \overset{\epsilon}{\rightarrow} 111010$$

The sum (in the binary group B^6) of any two code words is again a code word. (Also the inverse of any code word is the same code word.) For instance,

$$(0, 0, 1, 1, 1, 1) \oplus (0, 1, 0, 0, 1, 1) = (0, 1, 1, 1, 0, 0)$$
$$(0, 0, 1, 1, 1, 1) \oplus (0, 1, 1, 1, 0, 0) = (0, 1, 0, 0, 1, 1)$$

Therefore, ϵ is a group encoding function. We have

$$\Delta(\epsilon) = \min_{H \sim \{0\}} \{|b^i|\} = 3$$

since there are several nonzero code words of norm 3, whereas the others are of norm 4. Setting $3 = 2k + 1$, we find that ϵ is 1-correctable.

Following the implications of the corollary, we proceed to compute the cosets $H \oplus e$ where $|e| \leq 1$. When $e = 0$, we obtain $H \oplus e = H \oplus 0 = H$, the original set of code words. When $e = (1, 0, 0, 0, 0, 0)$, we have

$$(0, 0, 0, 0, 0, 0) \oplus (1, 0, 0, 0, 0, 0) = (1, 0, 0, 0, 0, 0)$$
$$(0, 0, 1, 1, 1, 1) \oplus (1, 0, 0, 0, 0, 0) = (1, 0, 1, 1, 1, 1)$$
$$(0, 1, 0, 0, 1, 1) \oplus (1, 0, 0, 0, 0, 0) = (1, 1, 0, 0, 1, 1), \quad \text{etc.}$$

which form the second column of our decoding table. Altogether, we obtain Table 5.9.

TABLE 5.9

$000 \overset{\epsilon}{\rightarrow} 000000$	100000	010000	001000	000100	000010	$000001 \overset{\delta}{\rightarrow} 000$
$001 \overset{\epsilon}{\rightarrow} 001111$	101111	011111	000111	001011	001101	$001110 \overset{\delta}{\rightarrow} 001$
$010 \overset{\epsilon}{\rightarrow} 010011$	110011	000011	011011	010111	010001	$010010 \overset{\delta}{\rightarrow} 010$
$011 \overset{\epsilon}{\rightarrow} 011100$	111100	001100	010100	011000	011110	$011101 \overset{\delta}{\rightarrow} 011$
$100 \overset{\epsilon}{\rightarrow} 100110$	000110	110110	101110	100010	100100	$100111 \overset{\delta}{\rightarrow} 100$
$101 \overset{\epsilon}{\rightarrow} 101001$	001001	111001	100001	101101	101011	$101000 \overset{\delta}{\rightarrow} 101$
$110 \overset{\epsilon}{\rightarrow} 110101$	010101	100101	111101	110001	110111	$110100 \overset{\delta}{\rightarrow} 110$
$111 \overset{\epsilon}{\rightarrow} 111010$	011010	101010	110010	111110	111000	$111011 \overset{\delta}{\rightarrow} 111$

To decode we simply locate the received signals in the table and read the message according to the rows in which the signals are found. What should be done about signals that cannot be found in the table?   △

EXERCISES

1 Determine all the vertices in the sphere $S(a, r)$, given:
(a) $a = (0, 1, 1, 0)$ and $r = 2$ (b) $a = (1, 0, 0, 1, 1)$ and $r = 2$
(c) $a = (0, 1, 0, 1, 0)$ and $r = 1$ (d) $a = (1, 1, 0, 0)$ and $r = 3$

2 Give an explicit description of the parity check codes for
(a) $m = 3$ (b) $m = 4$

3 Prove Theorem 5.6.

4 For each of the following, compute the spheres required to show that the given encoding is 1-correctable. Then define a decoding function $\delta: B^5 \to B^2$ according to the prescription in the proof of Theorem 5.7.
(a) Example 5.37   (b) Exercise 7(b)

5 Compute the spheres which show that the parity check code of Exercise 2(a) is 1-detectable.

6 Prove Theorem 5.8.

7 Use Theorem 5.8 to determine the detectability and correctability of the following encodings:
(a) $000 \to 000000$ (b) $00 \to 00001$
$\quad 001 \to 001110$ $\quad 01 \to 01010$
$\quad 010 \to 010101$ $\quad 10 \to 10100$
$\quad 011 \to 011011$ $\quad 11 \to 11111$
$\quad 100 \to 100011$ (c) $00 \to 000000000$
$\quad 101 \to 101101$ $\quad 01 \to 011111110$
$\quad 110 \to 110110$ $\quad 10 \to 101010101$
$\quad 111 \to 111000$ $\quad 11 \to 110101011$

8 Show that the following are group encoding functions:
(a) Example 5.37   (b) Exercise 7(a)

9 Obtain the group decoding tables for the functions of Exercise 8.

*10 Let $M(n, k)$ denote the maximum number of code words to be found in a k-correcting code $\epsilon: B^m \to B^n$. Show that

$$M(n, k) \leq \frac{2^n}{\sum_{j=0}^{k} \binom{n}{j}}$$

[*Hint*: Count the number of vertices in the various spheres.]

*11 Let $M(k)$ be the number of code words in a k-correcting group code $\epsilon: B^m \to B^n$. Show that

$$M(k) \le \frac{4k + 2}{4k + 2 - n}$$

[*Hint*: First show that among the coordinates in all the code words there are at least $[M(k) - 1](2k + 1)$ ones. Then establish the fact that each column in a list of code words has as many ones as it has zeros (unless it is identically zero). Use this fact to obtain an upper bound on the number of ones among the coordinates in all the code words.]

5.6 COMBINATORICS

The origins of combinatorial mathematics can be traced back for centuries. For a time, the combinatorial problems were closely associated with mysticism, in particular numerology. As if this were not enough of a handicap, the subject later became improperly identified as "recreational" mathematics, since it formed the basis for many of the popular mathematical puzzles of the day.

> A schoolteacher takes her class of fifteen girls on a daily walk. The girls are arranged in five rows of three, so that every girl will always have two companions. In how many ways can we arrange the class, day by day, on a walking schedule, so that for seven consecutive days, no girl will have another as a companion in a triplet more than once?

This famous schoolgirl problem is a combinatorial problem because it concerns the feasibility and the enumeration of arrangements of elements into sets. After showing that a particular arrangement does indeed exist (it is not at all obvious in the case of the schoolgirl problem), the combinatorial mathematician will generally proceed to consider the question, "In how many ways?" The latter question then defines the enumerative phase of the investigation.

In view of our concern with discrete, even finite, mathematics, it was almost inevitable that combinatorial questions would arise. Ever since the brief introduction to the more elementary enumerative techniques in Sections 0.4 and 0.8, such questions have come up at almost every turn. The more sophisticated notions connected with symmetry, however, have been deferred until now. To deal with these ideas requires some familiarity with elementary group theory, which accounts for our resumption of the discussion of combinatorics at this point.

The methods of combinatorial theory often seem unsystematic, especially to the uninitiated. One way to avoid giving this impression is to concentrate on a particular framework that encompasses most of the specific problems one expects to encounter, even though the framework may not be completely general from the point of view of the specialist.

Thus let us begin by supposing that we are dealing simply with a collection

$$F = \{f: P(f)\}$$

of functions $f: X \to A$ satisfying the property (or properties) P. Then X and A are assumed to be finite sets, and we are generally not interested in all of the functions from X to A, only those which satisfy a given (combinatorial) property. Another commonly assumed feature has to do with the "indistinguish-ability" among the various functions in the class F. That is to say, there will often be an equivalence relation ($\sim$) on F, perhaps having to do with the inherent symmetry of the problem, so that equivalent functions are not to be distinguished, but to be counted as one. Then the two phases of the combina-torial investigation may be characterized as follows:

(I) *Existence question*: $F \neq \varnothing$?
(II) *Enumerative aspect*: count the quotient set $F/\sim$.

Note that (II) automatically solves (I). On the other hand, if (I) has a negative answer, then a consideration of phase (II) is unnecessary.

EXAMPLE 5.41   In the schoolgirl problem, we may take X to be a three-dimensional $7 \times 5 \times 3$ schedule of slots or cells, and A may be taken as a set of fifteen girl's names (Figure 5.14). Then every $f: X \to A$ is a labeling of the slots by the names, and the combinatorial properties (P) we require are the following:

(i) The restriction of f to any given day be a one-to-one correspondence of slots and names.
(ii) No 2-element subset of A should be repeated among the labeled rows.

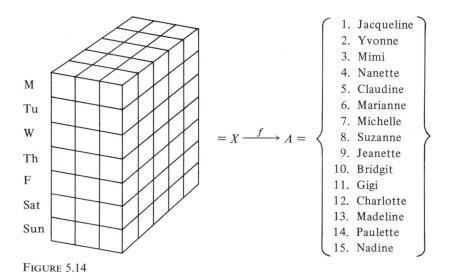

FIGURE 5.14

Now, we would not want to distinguish between two functions f and g that merely

(a) rearrange the daily schedules;
(b) rearrange the rows within a daily schedule;
(c) rearrange the three girls within a given row;

or are combinations of these rearrangements. We have thus established an equivalence relation ($\sim$) on the set $F = \{f : P(f)\}$ of admissible labelings.

The answer to the existence question is yes, by virtue of our being able to exhibit a schedule meeting the properties P:

Monday	Tuesday	Wednesday
$\{1, 2, 5\}$	$\{1, 3, 9\}$	$\{1, 4, 15\}$
$\{3, 14, 15\}$	$\{2, 8, 15\}$	$\{2, 9, 11\}$
$\{4, 6, 12\}$	$\{4, 11, 13\}$	$\{3, 10, 12\}$
$\{7, 8, 11\}$	$\{5, 12, 14\}$	$\{5, 7, 13\}$
$\{9, 10, 13\}$	$\{6, 7, 10\}$	$\{6, 8, 14\}$

Thursday	Friday	Saturday	Sunday
$\{1, 6, 11\}$	$\{1, 8, 10\}$	$\{1, 7, 14\}$	$\{1, 12, 13\}$
$\{2, 7, 12\}$	$\{2, 13, 14\}$	$\{2, 4, 10\}$	$\{2, 3, 6\}$
$\{3, 8, 13\}$	$\{3, 4, 7\}$	$\{3, 5, 11\}$	$\{4, 5, 8\}$
$\{4, 9, 14\}$	$\{5, 6, 9\}$	$\{6, 13, 15\}$	$\{7, 9, 15\}$
$\{5, 10, 15\}$	$\{11, 12, 15\}$	$\{8, 9, 12\}$	$\{10, 11, 14\}$

The reader can verify that no girl has the same companion more then once during the week. Since this particular problem is obviously not of central concern to us but was used only for its vivid illustration of the underlying combinatorial concepts, we simply state the answer ($|F/\sim| = 7$) to the enumeration problem, giving no indication as to the method of solution. $\triangle$

EXAMPLE 5.42 In Section 0.4 we introduced the (combinatorial) concept of a k-permutation of an n-set, say $n^+ = \{1, 2, \ldots, n\}$, which may be viewed simply as one of injective functions

$$f : k^+ \rightarrow n^+$$

In the current framework, we may consider $X = k^+$, $A = n^+$, and the property of being injective, as we form the admissible family

$$F = \{f : f \text{ is injective}\}$$

When it came to counting the number of k-combinations of n^+ (Section 0.8), we observed that a combination is an equivalence class of permutations. In the present context we would say that

$$f \sim g \Leftrightarrow \text{range}(f) = \text{range}(g)$$

Fortunately, each such equivalence class was found to be of the same size, namely $P(k, k)$, so that we were able to compute

$$C(n, k) = |F/{\sim}| = \frac{P(n, k)}{P(k, k)} = \frac{n!}{k!(n - k)!} \qquad \triangle$$

EXAMPLE 5.43   Recall that a partition of the finite set S into m blocks (an m-partition in Section 0.2) is a collection $\pi = \{A_1, A_2, \ldots, A_m\}$ of m mutually disjoint subsets $A_i \neq \varnothing$ whose union is S. At the same time π can also be viewed as a function

$$\pi: S \to m^+ = \{1, 2, \ldots, m\}$$

That is, we can write $\pi(s) = i$ according as $s \in A_i$. Then, we are not interested in all the functions from S to m^+ but only the family

$$F = \{\pi: \pi \text{ is surjective}\}$$

so as to ensure that each $A_i \neq \varnothing$.

If we are to count the m-partitions, we will not want to count F as it is, because two functions π, γ correspond to the same partition whenever they agree after permuting the integers of m^+. Thus the following pair:

$$
\begin{array}{ll}
1 \xrightarrow{\pi} 1 & 1 \xrightarrow{\gamma} 2 \\
2 \xrightarrow{\pi} 2 & 2 \xrightarrow{\gamma} 1 \\
3 \xrightarrow{\pi} 1 & 3 \xrightarrow{\gamma} 2 \\
4 \xrightarrow{\pi} 1 & 4 \xrightarrow{\gamma} 2 \\
5 \xrightarrow{\pi} 3 & 5 \xrightarrow{\gamma} 3 \\
6 \xrightarrow{\pi} 3 & 6 \xrightarrow{\gamma} 3 \\
7 \xrightarrow{\pi} 2 & 7 \xrightarrow{\gamma} 1
\end{array}
$$

though different as functions in F, correspond to the same partition $\{\overline{1, 3, 4};$ $\overline{2, 7}; \overline{5, 6}\}$. This observation gives rise to an equivalence relation $(\sim)$ on F, which must be taken into account before we enumerate. Then if S has n elements, the number $S(n, m)$ of partitions of S into m blocks is given by

$$S(n, m) = |F/{\sim}|$$

which are the Stirling numbers that we determined in Section 0.8 by means of a recursion formula. $\qquad \triangle$

These examples show that even when we concentrate on a class of problems that fit a common framework, the methods of solution may be quite varied indeed. As yet, there is no single unifying approach to combinatorics. Nevertheless, it is possible to understand the nature of combinatorial mathematics and become reasonably adept in the standard methods of enumeration. As a matter of fact, these are among the goals of a course in discrete structures. We have taken the view that the familiarity we seek is most easily achieved through a graduated introduction. That is why topics of a combinatorial nature have been sprinkled throughout the text. Beginning with the most elementary enumerative techniques of the preliminary chapter (some already known to the students beforehand), we have proceeded through an introduction to graph theory, itself a branch of combinatorial mathematics, and then studied the use of free distributive lattices in the exhaustive enumerative technique. In Section 5.8 we will come to a most beautiful theory of enumeration, attributed mainly to the contemporary mathematician George Pólya* (though anticipated to some extent in the earlier work by Redfield†). This theory fits the framework we have outlined here but has an added feature in that the equivalence relation on the functions arises out of symmetry considerations. The latter characteristic accounts for the connection with the theory of groups.

EXERCISES

1 Discuss each of the following enumeration problems in the context of a family $F = \{f: P(f)\}$ of functions and an assoicated equivalence relation.
   (a) Seating arrangements at a round table
   (b) k-selections and k-samples from an n-set
   (c) Graphs
   (d) Monoids
   (e) *Latin squares*: $n \times n$ arrays with entries from n^+ in which each number appears exactly once in every row and every column

*2 Obtain a solution to the schoolgirl problem that is not equivalent to that of Example 5.41.

3 Obtain a solution to the following combinatorial problem: To test six drugs on ten subjects (guinea pigs) over a three-day period, so that with each subject taking only one drug per day,

   (i) All six drugs are equally tested (on $(10 \times 3)/6 = 5$ subjects).
   (ii) Every pair of drugs is tested by just two subjects.

   [*Hint*: You want a 10×3 array with entries from 6^+.]

4 Express the enumeration problem of Exercise 3 in the context of a family of functions and an associated equivalence relation.

*5.7 ENUMERATION BY GROUP ACTION

A typical application of the Pólya theory is found in relation to the logic circuits of Sections 4.7 through 4.9. The free Boolean algebra $B(n)$ represents a complete catalog of all the n-input truth tables. But in the terminology of the preceding

---

* G. Pólya, "Kombinatorische Anzahlbestimmungen für Gruppen, Graphen und Chemische Verbindungen," *Acta Math.*, **68**, 145–254 (1937).
† J. J. Redfield, "The Theory of Group Reduced Distributions," *Am. J. Math.*, **49**, 433–455 (1927).

section (with $X = B^n$ and $A = B$), the switching theorist must deal with an exceedingly large family of functions

$$B(n) = \{f \colon X \to A\}$$

even for moderate values of n. Thus the values of $|B(n)| = 2^{2^n}$ are as tabulated below.

n	2^{2^n}
1	4
2	16
3	256
4	65,536
5	4,294,967,296
$\vdots$	

In order to partially compensate for the sheer magnitude of the numbers involved, one can introduce an equivalence relation on truth tables, writing $f \sim g$ just in case there is a permutation σ in S_n with the identity

$$f(x_1, x_2, \ldots, x_n) = g(x_{\sigma_1}, x_{\sigma_2}, \ldots, x_{\sigma_n})$$

Note that symmetry will follow from the use of the inverse permutation.

The motivation for this definition of functional equivalence is best understood by considering Figure 5.15. The *permutation network* σ simply consists of wires, permuted within so as to arrange the x_i differently from the order in which they

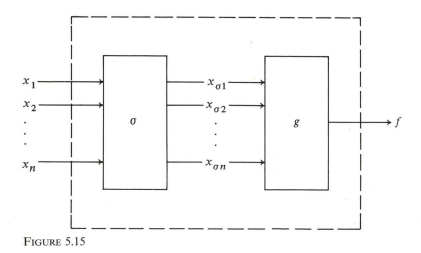

FIGURE 5.15

enter. Inasmuch as σ transforms g into f, we may also view it as a function

$$\sigma \colon B(n) \to B(n)$$

for which

$$(\sigma g)(x_1, x_2, \ldots, x_n) = g(x_{\sigma_1}, x_{\sigma_2}, \ldots, x_{\sigma_n})$$
$$(= f(x_1, x_2, \ldots, x_n))$$

Then it follows that σg is indistinguishable from f, which is the meaning of $f \sim g$: any logic circuit for g will serve equally well for f, provided we are willing to rearrange the input wires.

Accordingly, the same basic logic circuit may be used to implement several truth tables. Then the total number of basic circuits needed in an inventory or stockpile for the n-input truth tables is precisely the number of equivalence classes. We would therefore be interested to learn just how this number compares with $|B(n)| = 2^{2^n}$. For certain values of n it may even be feasible and worthwhile to compile a complete listing of minimal cost circuits, one representative from each equivalence class. In fact, this has been done to some extent, and we will summarize the results later.

EXAMPLE 5.44   Consider the truth tables f and g shown in Table 5.10.

TABLE 5.10

x_1	x_2	x_3	x_4	f	g
0	0	0	0	0	0
0	0	0	1	1	1
0	0	1	0	0	0
0	0	1	1	0	1
0	1	0	0	1	1
0	1	0	1	1	1
0	1	1	0	0	0
0	1	1	1	0	0
1	0	0	0	1	1
1	0	0	1	1	1
1	0	1	0	1	0
1	0	1	1	0	0
1	1	0	0	1	1
1	1	0	1	1	1
1	1	1	0	0	0
1	1	1	1	0	0

Although $f \neq g$ in $B(4)$, we have $f \sim g$ since

$$f(x_1, x_2, x_3, x_4) = g(x_2, x_4, x_3, x_1)$$

That is,

$$f(0, 0, 0, 0) = g(0, 0, 0, 0) = 0$$
$$f(0, 0, 0, 1) = g(0, 1, 0, 0) = 1$$
$$f(0, 0, 1, 0) = g(0, 0, 1, 0) = 0$$
$$f(0, 0, 1, 1) = g(0, 1, 1, 0) = 0$$
$$f(0, 1, 0, 0) = g(1, 0, 0, 0) = 1, \quad \text{etc.}$$

showing that $f = \sigma g$ for the permutation

$$\sigma = \begin{pmatrix} 1 & 2 & 3 & 4 \\ 2 & 4 & 3 & 1 \end{pmatrix} = (124).$$

Using the methods of Sections 4.8 and 4.9, we can compute a minimal sum-of-products expression

$$g = x_1 x_3' + x_1' x_2' x_4 + x_2 x_3'$$

(see Figure 5.16), and we only have to prefix the appropriate permutation network (Figure 5.17) in order to obtain a corresponding logic circuit for f. $\triangle$

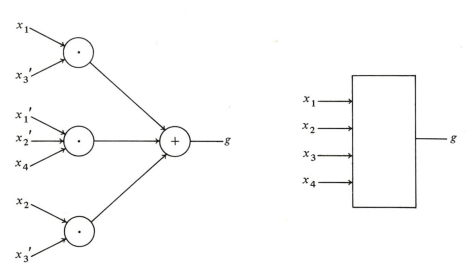

FIGURE 5.16

Using a process of abstraction, let us now see if we can somehow describe the essential features of the preceding situation. As in Section 5.6, we have an equivalence relation on a family of functions

$$F = \{f : X \to A\}$$

But the equivalence relation derives from or is induced by the "action of a group" on the domain of the functions. Let us give this idea a precise interpre-

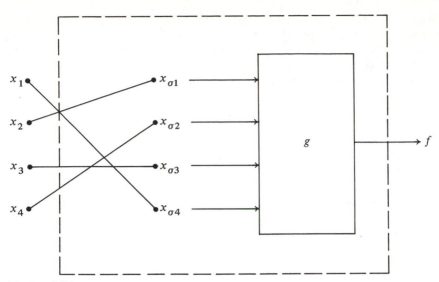

FIGURE 5.17

tation and then go on to discuss methods for computing the number of resulting equivalence classes.

Let G be a group and X a finite set. We say that G *acts on* X (as a group of permutations of X) if there is a function $G \times X \to X$, denoted by

$$(\sigma, x) \to \sigma x$$

such that

(i) $1x = x$ (with 1 the identity of G)
(ii) $(\sigma\tau)x = \tau(\sigma x)$

for all $x \in X$ and $\sigma, \tau \in G$. Under these conditions alone we will see that the elements σx always constitute a permutation of X for the restriction to a fixed group element σ. Thus if

$$\sigma x = \sigma y$$

we must have

$$x = 1x = (\sigma\sigma^{-1})x = \sigma^{-1}(\sigma x) = \sigma^{-1}(\sigma y) = (\sigma\sigma^{-1})y = 1y = y$$

which shows that the restriction $\sigma: X \to X$ is injective (and hence also surjective). So we may think of G as a group of permutations of the set X. Conversely, a group G of permutations of X can easily be identified as an action of G on X as previously defined.

On the functional level the same group elements $\sigma \in G$ will give rise to a group of permutations of F if we define

$$(\sigma g)(x) = g(\sigma x)$$

for all $x \in X$. In fact, we can show that the two groups are isomorphic if $|A| > 1$. (See Exercise 11 of Section 5.7.) Though we are primarily interested in F, the situations are analogous whether we speak of X or F. The problem is to determine the number of equivalence classes and perhaps obtain a representative of each equivalence class given that

$$x \sim y \quad \text{iff} \quad x = \sigma y \quad (\text{respectively, } f \sim g \quad \text{iff} \quad f = \sigma g)$$

for some group element $\sigma \in G$. Here, however, an extra word of caution is in order. We are speaking of two different equivalence relations (one on X, the other on F), so that entirely different partitions are involved. It is merely a matter of convenience to use the same relational symbol in both cases.

Since the problem is similar in structure for the two cases (though the answers may be quite different), let us try to determine the number of equivalence classes when G acts on a set X. If you replace X by F, the analysis will be the same. In any case, a special role is played by the dual subsets:

$$G_x = \{\sigma \in G : \sigma x = x\} \subseteq G \qquad \text{for fixed } x \in X$$
$$X_\sigma = \{x \in X : \sigma x = x\} \subseteq X \qquad \text{for fixed } \sigma \in G$$

We might call these subsets the *subgroup fixing* x (see the statement of the forthcoming lemma), and the *subset fixed by* σ, respectively.

Lemma

The collection

$$G_x = \{\sigma \in G : \sigma x = x\}$$

is a subgroup of G. Moreover

$$|[x]| = [G : G_x] \qquad (= |G|/|G_x|)$$

Proof   If σ and τ are in G_x so that $\sigma x = \tau x = x$, then

$$(\sigma \tau) x = \tau(\sigma x) = \tau x = x$$
$$\sigma^{-1} x = \sigma^{-1}(\sigma x) = (\sigma \sigma^{-1}) x = 1x = x$$

showing that G_x is a subgroup of G. Note in particular the use of the defining properties of a group action.

Now we can assert that

$$\sigma x = \tau x \Leftrightarrow \sigma \tau^{-1} \in G_x \Leftrightarrow G_x \sigma = G_x \tau$$

Thus there are exactly as many elements equivalent to x as there are cosets modulo G_x; that is, $|[x]| = [G : G_x]$ as claimed. We shall prove only the first assertion:

$$\sigma x = \tau x \Rightarrow (\sigma\tau^{-1})x = \tau^{-1}(\sigma x) = \tau^{-1}(\tau x)$$
$$= (\tau\tau^{-1})x = 1x = x$$
$$\sigma\tau^{-1} \in G_x \Rightarrow \tau x = \tau(\sigma\tau^{-1}x) = ((\sigma\tau^{-1})\tau)x$$
$$= (\sigma(\tau^{-1}\tau))x = (\sigma 1)x = \sigma x$$

The second assertion is left as an exercise.   □

EXAMPLE 5.45   The symmetric group $G = S_3$ acts on $X = B^3$ by rearrangement of the coordinates as shown in Table 5.11.

TABLE 5.11

1	(12)	(13)
000 → 000	000 → 000	000 → 000
001 → 001	001 → 001	001 → 100
010 → 010	010 → 100	010 → 010
011 → 011	011 → 101	011 → 110
100 → 100	100 → 010	100 → 001
101 → 101	101 → 011	101 → 101
110 → 110	110 → 110	110 → 011
111 → 111	111 → 111	111 → 111

(23)	(123)	(132)
000 → 000	000 → 000	000 → 000
001 → 010	001 → 100	001 → 010
010 → 001	010 → 001	010 → 100
011 → 011	011 → 101	011 → 110
100 → 100	100 → 010	100 → 001
101 → 110	101 → 110	101 → 011
110 → 101	110 → 011	110 → 101
111 → 111	111 → 111	111 → 111

The subsets fixing particular elements of B^3 are seen to be subgroups of S_3:

$$G_{000} = S_3$$
$$G_{001} = \{1, (12)\}$$
$$G_{010} = \{1, (13)\}$$
$$G_{011} = \{1, (23)\}, \quad \text{etc.}$$

and in every case

$$|[x]| = [G : G_x]$$

as predicted by the lemma. Thus we have

$$|[000]| = |\{000\}|$$
$$= 1 = 6/6$$
$$= |S_3|/|G_{000}|$$
$$= [S_3 : G_{000}]$$
$$|[001]| = |\{001, 010, 100\}|$$
$$= 3 = 6/2$$
$$= |S_3|/|G_{001}|$$
$$= [S_3 : G_{001}]$$

and so on.   △

EXAMPLE 5.46   Let us return to the problem of partitioning $B(n)$ into its equivalence classes for the purpose of obtaining a minimal inventory for the n-input logic circuits. A naive approach would be the following:

1. Choose any function $g \in B(n)$.
2. Apply the permutations σ of S_n in succession (note the usefulness of Algorithm XX here) to obtain all the functions $f = \sigma g$ equivalent to g.

3. Select a new function, one not previously obtained.
4. Continue until all of $B(n)$ have appeared.

In such a program, the result of the lemma,

$$|[f]| = [G : G_f] = [S_n : G_f]$$

together with Lagrange's theorem shows that the size of each equivalence class must divide the group order $|S_n| = n!$.   △

EXAMPLE 5.47   Let $g \in B(4)$ be the truth table of Example 5.44. From its sum-of-products representation

$$g = x_1 x_3' + x_1' x_2' x_4 + x_2 x_3'$$

we recognize a symmetry regarding the interchange of x_1 and x_2. In fact, it appears that $\sigma g = g$ only if $\sigma = 1$ or $\sigma = (12)$. It follows that

$$G_g = \{1, (12)\}$$

and

$$\begin{aligned}
|[g]| &= [S_4 : G_g] \\
&= |S_4|/|G_g| \\
&= 24/2 = 12
\end{aligned}$$

$\triangle$

EXAMPLE 5.48   Let G be the group of symmetries of the square, viewed as a permutation group on its set of vertices (Example 5.23). Then

$$G_{v_1} = G_{v_3} = \{1, (24)\}$$
$$G_{v_2} = G_{v_4} = \{1, (13)\}$$

That is, each vertex is left fixed by the identity permutation and by a reflection through a diagonal. According to the lemma,

$$\begin{aligned}
|[v_i]| &= [G : G_{v_i}] \\
&= |G|/|G_{v_i}| \\
&= 8/2 = 4
\end{aligned}$$

for each vertex. Of course, this is consistent with the fact that there is a rigid motion carrying any one vertex into every other. $\triangle$

Now we shall use the lemma to establish an important result, usually credited to William Burnside (1852–1927), though the main idea was already implicit in an 1896 paper by Frobenius. The counting technique used in Burnside's theorem is typical of many combinatorial arguments. Instead of enumerating the equivalence classes by a summation over the elements of the set, we sum over the group elements. The Burnside technique is most effective when the group is small compared with the size of the set on which it acts.

Theorem 5.10   (Burnside)

$$|X/\sim| = \frac{1}{|G|} \sum_{\sigma \in G} |X_\sigma|$$

Proof   Consider the 0, 1 array (g_{ij}) of size $n \times m$ relative to

$$G = \{\sigma_1, \ldots, \sigma_n\}$$
$$X = \{x_1, \ldots, x_m\}$$

in which

$$g_{ij} = \begin{cases} 1 & \text{if } \sigma_i \text{ fixes } x_j, \text{ that is, if } \sigma_i x_j = x_j \\ 0 & \text{otherwise} \end{cases}$$

Eventually we will obtain the Burnside identity by counting the number of ones in this array in two different ways. We begin by observing that the column corresponding to the element $x \in X$ will have $|G_x|$ ones, where

$$|G_x| = |G|/[G : G_x]$$
$$= |G|/|[x]|$$

according to the lemma. This equation shows that the sizes $|G_x|$ are identical for equivalent elements of X. It follows that the columns pertaining to any one equivalence class will have exactly $|G|$ ones altogether. So if we multiply by the number of equivalence classes, we will obtain $|G| \cdot |X/\sim|$ as the total number of ones in the array.

On the other hand, if we now count by the rows of the array, we obtain a total of $\sum_{\sigma \in G} |X_\sigma|$ ones, so the identity follows immediately.   □

EXAMPLE 5.49   Let us count the number of "patriotic" bracelets with four beads. Using the colors red, white, and blue, we might expect $3^4 = 81$ distinct colorings, but we would probably not want to distinguish between two colorings that differ only by a rotation. So we should count the number of equivalence classes under the action of the cyclic group of rotations (Z_4, essentially). All 81 colorings are fixed by the identity element of the group. Denoting the group elements by $1, r, r^2, r^3$ (as a subgroup of the group of symmetries of the square) the identity, 90° rotation, etc., we have

$$|F_1| = 81$$
$$|F_r| = 3$$
$$|F_{r^2}| = 9$$
$$|F_{r^3}| = 3$$

Under a 90° rotation, the only colorings left fixed are those in which all four beads are of the same color, either red, white, or blue, so $|F_r| = 3$. The same is true for the 270° rotation. These colorings are also fixed by the 180° rotation. But in this case, we have to consider those colorings in which diagonally opposite beads have the same color. There are six ways. Applying Burnside's theorem, we obtain·

$$|F/\sim| = \tfrac{1}{4}(81 + 3 + 9 + 3)$$
$$= \tfrac{96}{4} = 24$$

which is the number of distinct colorings after rotational symmetry has been considered.   △

EXAMPLE 5.50   Again we would use rotational symmetry in counting the number of distinct seating arrangements for n persons at a round table. Since $|Z_n| = n$ and there are no seating arrangements left fixed by a genuine rotation (only the identity element of the group will leave any arrangement fixed), we obtain

$$\frac{1}{n}(n! + 0 + 0 + \cdots + 0) = \frac{n!}{n} = (n-1)!$$

as the number of distinct circular seating arrangements. △

EXAMPLE 5.51 Consider the 16 elements of $B(2)$, as tabulated in Example 4.41. The only truth tables left fixed by an interchange of x_1 and x_2 are the following:

$$F_0 = 0$$
$$F_1 = x_1'x_2'$$
$$F_6 = x_1x_2' + x_1'x_2$$
$$F_7 = x_1' + x_2'$$
$$F_8 = x_1x_2$$
$$F_9 = x_1'x_2' + x_1x_2$$
$$F_{14} = x_1 + x_2$$
$$F_{15} = 1$$

Therefore, using Burnside's theorem, we have

$$|B(2)/\sim| = \tfrac{1}{2}(16 + 8) = \tfrac{24}{2} = 12$$ △

EXAMPLE 5.52 Similarly, one can show, in reference to the group action of Example 5.45, that the three transpositions will leave 64 of the 256 truth tables of $B(3)$ fixed, whereas only 16 functions are fixed by the 3-cycles. It follows from Burnside's theorem that

$$|B(3)/\sim| = \tfrac{1}{6}(256 + 64 + 64 + 64 + 16 + 16)$$
$$= \tfrac{480}{6} = 80$$ △

The problem of determining the number of equivalence classes in $B(n)$ was first undertaken by a group of researchers at Harvard University in the early 1950s. They also obtained a catalog of "best" circuits for the nonequivalent truth tables, representative of the various equivalence classes when $n \le 4$. Table 5.12 shows these first results, as compared with the values of $|B(n)|$.

TABLE 5.12

n	$|B(n)| = 2^{2^n}$	$|B(n)/\sim|$
1	4	4
2	16	12
3	256	80
4	65,536	3,984
5	4,294,967,296	37,333,248
$\vdots$		

Included are the results of Examples 5.51 and 5.52. We can see that already at $n = 5$ the second phase of the overall investigation is no longer feasible.

Yet, with all things considered, the project must be called a success. For one thing, various researchers have carried the analysis still further, using larger symmetry groups (for instance, allowing for complementation of the variables and complementation of the function) with the result that the number of equivalence classes is even smaller than is shown here.

Much of this later work made considerable use of the Pólya theory of enumeration, which we are about to describe. The entire theory of enumeration, with its symmetry considerations, has remarkably extensive applications. The student should not try to make a judgment on the basis of the examples presented here. Particularly in the case of the truth tables of $B(n)$, we were mainly interested in establishing a focal point for the discussion. While the example is certainly typical, it gives no indication as to the variety of instances in which these ideas are so crucial.

EXERCISES

1 Prove the assertion

$$\sigma\tau^{-1} \in G_x \Leftrightarrow G_x\sigma = G_x\tau$$

required in the proof of the lemma.

2 Use the "naive" approach for partitioning $B(n)$ into its equivalence classes for

(a) $n = 2$ (b) $n = 3$

For $n = 3$ a mere indication of the procedure will suffice.

3 Complete the verification that $f = \sigma g$ in Example 5.44.

4 Compute the truth table of the function $f = \sigma g$, given that g is the truth table of Example 5.44 and

(a) $\sigma = (13)(24)$ (b) $\sigma = (1423)$
(c) $\sigma = (13)$ (d) $\sigma = (134)$

5 Compute the truth table of the function $f = \sigma g$, given that g is defined by the truth table and σ is the following permutation:

x_1	x_2	x_3	g	
0	0	0	0	(a) $\sigma = (132)$
0	0	1	1	(b) $\sigma = (12)$
0	1	0	1	(c) $\sigma = (13)$
0	1	1	0	(d) $\sigma = (123)$
1	0	0	0	
1	0	1	1	
1	1	0	1	
1	1	1	1	

6 Show that $f \sim g$ as we defined it is indeed an equivalence relation on $B(n)$.

7 Compute the subsets
  (a) G_{0001} (b) G_{0101} (c) G_{1110} (d) G_{0110}
  and show that in each case we obtain a subgroup of S_4.

8 Verify the equality stated in the lemma for the respective circumstances in Exercise 7.

9 Use the lemma to predict (as in Example 5.47) the size of the equivalence class $[g]$, given that g is the truth table

(a) x_1 x_2 x_3	g	(b) x_1 x_2 x_3	g	(c) x_1 x_2 x_3	g	(d) x_1 x_2 x_3	g
0 0 0	0	0 0 0	0	0 0 0	0	0 0 0	0
0 0 1	1	0 0 1	0	0 0 1	0	0 0 1	1
0 1 0	1	0 1 0	1	0 1 0	0	0 1 0	0
0 1 1	0	0 1 1	0	0 1 1	0	0 1 1	0
1 0 0	0	1 0 0	1	1 0 0	0	1 0 0	1
1 0 1	1	1 0 1	0	1 0 1	1	1 0 1	1
1 1 0	1	1 1 0	1	1 1 0	0	1 1 0	0
1 1 1	0	1 1 1	0	1 1 1	0	1 1 1	0

10 Use Burnside's theorem, choosing an appropriate group of symmetries in each instance, to compute the number of
  (a) "Patriotic" bracelets with six beads
  (b) Seatings of students, computer science faculty, and mathematics faculty at a round table with five seats
  (c) Paintings of the vertices of a square with three colors (using the entire group of symmetries of the square)
  (d) Black-and-red colorings of a 4×4 chess board
  (e) Equivalence classes in $B(2)$ when complementation of the variables is allowed
  (f) Equivalence classes in $B(2)$ when complementation of the variables and complementation of the function are allowed
  (g) Black-and-white colorings of the faces of a regular tetrahedron

11 Let G be a group of permutations of a set X, and let $F = \{f: X \to A\}$. Then define a mapping $\sigma \to \sigma'$ from G into the group of permutations of F by writing

$$(\sigma'g)(x) = g(\sigma x)$$

for all $x \in X$. Show that this mapping is an injective group homomorphism if $|A| > 1$. Conclude that

$$G' = \{\sigma': \sigma \to \sigma'\}$$

is isomorphic to G whenever $|A| > 1$.

*5.8 PÓLYA THEORY OF ENUMERATION

At times Burnside's theorem proves to be somewhat less than adequate to the task of enumerating the equivalence classes, especially when the group order is fairly large. In this section we shall present a more powerful technique, one which will not only carry out the required enumeration but also give substantially more information concerning the equivalence classes, information that could not be learned from a simple Burnside enumeration.

Once again, we denote our family of functions by

$$F = \{f : X \to A\}$$

but now we may think of A as an alphabet of symbols. Each function f is to be given a *weight*

$$w(f) = \bigtimes_{x \in X} f(x)$$

not strictly to be considered as a word on the alphabet A, because we want to think of our multiplication as being commutative. It should be considered merely as a formal "descriptor" tabulating the number of image elements of the various A-values. Thus $w(f) = w(g) = 0^8 1^8$ in the case of the functions of Example 5.44. Again we have an equivalence relation on F induced by a group action on X. It has become standard in describing the Pólya theory to refer to these equivalence classes as *patterns*. First, we will show, as in the example just cited, that equivalent functions always have the same weight: if $f \sim g$, then

$$w(f) = \bigtimes_{x} f(x) = \bigtimes_{\sigma x} g(\sigma x) = \bigtimes_{x} g(x) = w(g)$$

That is, the symbols from A are merely accumulated in a different order in compiling each of the two product expressions. Once we have realized this, we can define the *weight of a pattern*, written $w[f]$, to be the weight of any member in the class. Finally, the *pattern inventory* of F is taken to be the formal sum

$$\text{inv}\,(F) = \sum_{[f]} w[f]$$

over all the equivalence classes (patterns).

EXAMPLE 5.53   Continuing with Example 5.51, let us consider the symmetric group S_2 acting on the 16 truth tables of $B(2)$ as represented in Example 4.41. The function $F_{10} = x_2$ has the weight $w(F_{10}) = 0^2 1^2$. Now, $F_{10} \sim F_{12}$, and you will note that F_{12} has the same weight. On the other hand, $w(F_6) = 0^2 1^2$ as well, but this function is not equivalent to the other two. Thus the weight alone does not completely classify a function; we can have nonequivalent functions of the same weight, as will be reflected in the pattern inventory. Thus we can determine that

$$\text{inv}\,(B(2)) = 0^4 + 3(0^3 1) + 4(0^2 1^2) + 3(01^3) + 1^4$$

This is the kind of information that will become available with Pólya's theorem. Besides showing that there are 12 equivalence classes—information that we could obtain from Burnside's theorem (see Example 5.51)—the Pólya theorem will also tell us that there is just one class with 4 zeros in the truth table, three classes with 3 zeros and 1 one, four classes with 2 zeros and 2 ones, etc.   △

We need only one more definition before stating Pólya's fundamental counting theorem, and it is a pure group-theoretic notion concerning the group G acting as a group of permutations on the set X. For each $\sigma \in G$ we form a product

$$z_1^{n_1} z_2^{n_2} \cdots z_m^{n_m}$$

in which n_i is the number of cycles of length i in the cycle representation of σ as a permutation of the set X. (Note that $n_1 + 2n_2 + 3n_3 + \cdots + mn_m = |X|$, so that we have an effective checking device for our computations.) The formal sum

$$Z_G(z_1, z_2, \ldots, z_m) = \frac{1}{|G|} \sum_{\sigma \in G} z_1^{n_1} z_2^{n_2} \cdots z_m^{n_m}$$

is called the *cycle index polynomial* of G (on X).

EXAMPLE 5.54   Considering Example 5.18, we see that the cycle index polynomial of S_3 acting on 3^+ is

$$Z_{S_3}(z_1, z_2, z_3) = \tfrac{1}{6}(z_1^3 + 3z_1 z_2 + 2z_3)$$

Let us explain. The identity permutation consists of three cycles of length 1 (every element of 3^+ is fixed). On the other hand, the permutation (12) consists of one cycle of length two (and one cycle of length one), so the product $z_1 z_2$ is associated with this group element. The same product is obtained for the other two transpositions. The two 3-cycles of the group will each contribute a product z_3. Summing up and dividing by $|S_3| = 6$, we obtain the given cycle index polynomial.   △

EXAMPLE 5.55   Consulting Example 5.23, we see that the cycle index polynomial for the symmetry group G of the square (G acting on 4^+) is

$$Z_G(z_1, z_2, z_3, z_4) = \tfrac{1}{8}(z_1^4 + 2z_1^2 z_2 + 3z_2^2 + 2z_4)$$ △

Theorem 5.11   (Pólya)

$$\text{inv}\,(F) = Z_G\left(\sum_{a \in A} a, \sum_{a \in A} a^2, \ldots, \sum_{a \in A} a^m \right)$$

Proof   As in the proof of Burnside's theorem, we construct an $n \times p$ array (g_{ij}) relative to

$$G = \{\sigma_1, \sigma_2, \ldots, \sigma_n\} \qquad F = \{f_1, f_2, \ldots, f_p\}$$

This time, however, we set

$$g_{ij} = \begin{cases} w(f_j) & \text{if } \sigma_i \text{ fixes } f_j \\ 0 & \text{otherwise} \end{cases}$$

Then once again we sum the elements of the array in two different ways to obtain the desired result.

The sum of the entries for column f yields

$$|G_f| \, w(f) = \frac{|G|}{|[f]|} \, w(f)$$

according to the lemma preceding Burnside's theorem. Again the column sums pertaining to a given equivalence class $[f]$ are identical, and so the sum of all these columns contributes $|G| \, w[f]$. It follows that the sum of all the entries in the array is $|G|$ times the pattern inventory.

If we next add all the entries by rows, we obtain

$$\sum_{\sigma \in G} \left(\sum_{\sigma f = f} w(f) \right) = \sum_{\sigma \in G} \left(\sum_{a \in A} a \right)^{n_1} \left(\sum_{a \in A} a^2 \right)^{n_2} \cdots \left(\sum_{a \in A} a^m \right)^{n_m}$$

$$= |G| \, Z_G \left(\sum_{a \in A} a, \sum_{a \in A} a^2, \ldots, \sum_{a \in A} a^m \right)$$

where the first equality certainly requires some explanation. The connection between the two expressions,

$$\sum_{\sigma f = f} w(f) = \left(\sum a \right)^{n_1} \left(\sum a^2 \right)^{n_2} \cdots \left(\sum a^m \right)^{n_m}$$

hinges on the fact that

$$\sigma f = f \Leftrightarrow f \text{ is constant on the cycles of } \sigma$$

Thus if

$$f \equiv a_{i_1} \text{ on cycle } C_1 \text{ of length } n_{i_1}$$
$$f \equiv a_{i_2} \text{ on cycle } C_2 \text{ of length } n_{i_2}$$
$$\vdots$$

then the term $w(f)$ will appear in the right-hand expression by virtue of the product $a_{i_1}{}^{n_{i_1}} a_{i_2}{}^{n_{i_2}} \cdots$ out of the corresponding sums, and conversely.

The theorem follows by canceling $|G|$ in the two expressions obtained for the sum of the entries in the array. □

Corollary

$$|F/{\sim}| = Z_G(|A|, |A|, \ldots, |A|)$$

Proof   View the Pólya theorem as a polynomial identity in the indeterminates $a \in A$. Then substitute 1 for all a. The pattern inventory becomes

$\sum_{[f]} w[f] = \sum_{[f]} 1 = |F/\sim|$, and the cycle index polynomial reduces to

$$Z_G(\textstyle\sum a, \sum a^2, \ldots, \sum a^m) = Z_G(|A|, |A|, \ldots, |A|)$$

The "trick" works because the theory would remain unchanged if we had chosen to use integer weights. □

EXAMPLE 5.56 In calculating the pattern inventory for $B(2)$, as in Example 5.53, let us consider the action of the group S_2 on B^2 (the domain of the functions):

$$
\begin{array}{ll}
00 \xrightarrow{1} 00 & 00 \xrightarrow{(12)} 00 \\
01 \to 01 & 01 \to 10 \\
10 \to 10 & 10 \to 01 \\
11 \to 11 & 11 \to 11
\end{array}
$$

so as to obtain the cycle index polynomial

$$Z_{S_2}(z_1, z_2) = \tfrac{1}{2}(z_1{}^4 + z_1{}^2 z_2)$$

(Important: This is not the same as the cycle index polynomial of S_2 as acting on 2^+!) Using Pólya's theorem, we compute

$$
\begin{aligned}
Z_G(0 + 1, 0^2 + 1^2, 0^3 &+ 1^3, 0^4 + 1^4) \\
&= \tfrac{1}{2}[(0+1)^4 + (0+1)^2(0^2 + 1^2)] \\
&= \tfrac{1}{2}[(0^4 + 4(0^3 1) + 6(0^2 1^2) + 4(0 1^3) + 1^4) + (0^2 + 2(01) + 1^2)(0^2 + 1^2)] \\
&= 0^4 + 3(0^3 1) + 4(0^2 1^2) + 3(0 1^3) + 1^4
\end{aligned}
$$

the same result as that described in Example 5.53.

If we had merely wanted the number of equivalence classes, then we could have used the corollary to obtain

$$|B(2)/\sim| = Z_{S_2}(2, 2) = \tfrac{1}{2}(2^4 + 2^2 \cdot 2) = \tfrac{1}{2}(16 + 8) = 12$$

(Compare with Example 5.51.) △

EXAMPLE 5.57 Let us revisit the problem of the "patriotic" beads (Example 5.49). For the cyclic group of rotations of the four positions, we obtain the cycle index polynomial

$$Z_G(z_1, z_2, z_3, z_4) = \tfrac{1}{4}(z_1{}^4 + 2z_4 + z_2{}^2)$$

Using the corollary to Pólya's theorem, we obtain

$$|F/\sim| = \tfrac{1}{4}(3^4 + 2 \cdot 3 + 3^2) = \tfrac{1}{4}(81 + 6 + 9) = 24$$

the same answer as before; but here the advantage of the Pólya enumeration, in comparison with the Burnside technique, is evident.

In order to show the additional information obtainable with the Pólya theory, we compute the pattern inventory:

$$Z_G(r + w + b, r^2 + w^2 + b^2, r^3 + w^3 + b^3, r^4 + w^4 + b^4)$$
$$= \tfrac{1}{4}[(r + w + b)^4 + 2(r^4 + w^4 + b^4) + (r^2 + w^2 + b^2)^2]$$
$$= r^4 + w^4 + b^4 + 2r^2w^2 + 2r^2b^2 + 2w^2b^2 + 3r^2wb + 3rw^2b + 3rwb^2$$
$$+ r^3w + r^3b + rw^3 + w^3b + rb^3 + wb^3$$

Here we can see that there are two patterns or equivalence classes with two red beads and two white beads, three patterns with two red beads, one white bead, and one blue bead, etc. △

EXERCISES

1　Determine the weights of the functions in Exercise 9 of Section 5.7.

2　Verify the pattern inventory inv $(B(2))$ given in Example 5.53.

3　Obtain the cycle index polynomial for the following group actions on 4^+:
  (a) The symmetric group S_4
  (b) The cyclic subgroup generated by (1234)
  (c) The subgroup that leaves 4 fixed

4　Obtain appropriate cycle index polynomials for the circumstances described in Exercise 10 of Section 5.7.

5　Use Pólya's theorem to compute the pattern inventories corresponding to the circumstances described in Exercise 10 of Section 5.7. [*Hint*: See Exercise 4.]

6　Use the corollary to Pólya's theorem to count the number of equivalence classes of functions in the circumstances described in Exercise 10 of Section 5.7. [*Hint*: See Exercise 4.]

SUGGESTED REFERENCES

5.1　Kurosh, A. G. *Theory of Groups, Vol.* 1. Chelsea, New York, 1960.

5.2　Hall, M. *Theory of Groups.* Macmillan, New York, 1959.

5.3　Baumslag, B., and B. Chandler. *Group Theory.* Schaum, New York, 1968.

5.4　Szabo, N. S., and R. I. Tanaka. *Residue Arithmetic and its Applications to Computer Technology.* McGraw-Hill, New York, 1967.

5.5　Peterson, W. W. *Error-Correcting Codes.* MIT Press, Cambridge, Mass. 1961.

5.6　Harrison, M. *Introduction to Switching and Automata Theory.* McGraw-Hill New York, 1965.

5.7　de Bruijn, N. G. "Pólya's Theory of Counting," in *Applied Combinatorial Techniques.* E. F. Beckenbach (ed.), Wiley, New York, 1964.

5.8　Berge, C. *Principles of Combinatorics.* Academic Press, New York, 1971.

There are any number of fine books on group theory. The books by Kurosh and Hall are representative, the latter being somewhat more advanced. In both books however, the development of the theory goes well beyond the introductory material, as presented here, so it is perhaps better to consult a less specialized algebra text, such as References 1.1 and 1.2. The selection (5.3) from the Schaum Outline Series may also be useful, especially its presentation of well-chosen illustrative examples.

One may find entire books devoted to some of the applications we have discussed. Thus Peterson's book concentrates on codes (not specifically group codes), and the book by Szabo and Tanaka on the applications of the residue number systems. For a detailed treatment of the application of Pólya's technique to switching theory, one could not do better than to consult the book by Harrison.

Two references on combinatorics were already listed in Chapter 0. Reference 0.4 is particularly well suited for those who would like to see a more thorough presentation of Pólya's theory of enumeration as well as a more generous collection of examples. Two other excellent treatments of this theory are found in the de Bruijn article and the book by Berge. The latter gives some interesting insights into combinatorics generally, quite apart from the discussion of the Pólya theory.

6

Logic and Languages

In many ways mathematical logic and formal language theory share a common ideological point of view. Each in turn has a direct bearing on the directions taken by computer science. Thus we will find that the parenthesis-free notation, so convenient in formal logic, is equally effective in the computer processing of arithmetic expressions. The axiomatic techniques of logic find their way into computer-aided theorem-proving investigations. And formal language theory's hierarchical classification of abstract machine capabilities has much to offer in helping us understand the outer limits of computability. Although we cannot elaborate on any of these ideas at this time, we nevertheless hope that the intended applications and the subtle inner consistencies among the subjects taken up in this final chapter will become increasingly apparent to the student as our discussions proceed.

The first real mathematical subject to be encountered here will be propositional calculus, itself a branch of mathematical logic. Our discussion will serve several purposes. First of all, it will provide an alternative interpretation of the Boolean expressions of Section 4.7, an interpretation which happens to be required for the logical expressions of the computer programming languages (ALGOL, FORTRAN, etc.). At the same time, by focusing on two new relations—semantic implication and semantic equivalence—we will be able to provide a formal setting for analyzing our logical thought processes. So far we have dealt with these processes quite casually since their introduction in Section 0.9. Moreover, we will obtain an axiomatic scheme for deriving the theorems of propositional calculus, which in turn will establish a style and an ideological framework for introducing the mathematical language theory. The latter is a vast and rapidly developing area of investigation. We will concentrate on the Chomsky grammars and the Turing machines, however, because of their obvious importance to computer science.

6.1   POLISH EXPRESSIONS

When we use arithmetic expressions in computer programming or in ordinary algebra, we find that it is necessary to dictate the order in which the various operations are to be performed. This may be done with parentheses, with a hierarchical convention for the operations, or with some combination of these two schemes. In this way we can outline the schedule of operations and indicate the corresponding operands. The obvious ambiguities notwithstanding, the real subtleties of a proper interpretation of algebraic expressions are not always fully appreciated. It requires considerable thought to design an algorithm that will interpret such expressions correctly. For these and other reasons, the ordinary way of writing algebraic expressions is not at all well suited to computer processing and analysis. In fact, most computer systems will immediately translate such expressions into a more convenient form. Let us consider these matters in some detail.

Example 6.1   The following expression:

$$A \cdot B + C + D / E$$

might be considered ambiguous as it stands. But if we write

$$A \cdot (B + C) + D / E$$

and also understand that multiplication ($\cdot$) and division ($/$) have a higher priority than addition ($+$) in our hierarchical scheme, then the ambiguity will be removed. $\triangle$

We will find that the binary labeled trees (Section 3.5) provide a most convenient means for specifying the order of execution of the various operations in an algebraic expression. Suppose for the moment that all the operations occurring in the expression are binary. Then each execution of an operation may be represented as in Figure 6.1(a). If this idea is extended inductively to the expression as a whole, we will obtain a labeled binary tree, whose leaves identify the distinct occurrences of variables or constants (operands) in the expression and whose nonterminal nodes are labeled by the various binary operations. In the interpretation of such tree structures, each left and right subtree is to be executed before the operation point at which the two subtrees are "grafted." [See Figure 6.1(b).] As a consequence, no parentheses or hierarchical ordering of operations need accompany the tree representation. The structure of the tree tells the whole story.

Example 6.2   Consider once more the expression

$$A \cdot (B + C) + D / E$$

of Example 6.1. With the hierarchical conventions already described, the interpretation of this expression is as clear as that of the *fully parenthesized* version:

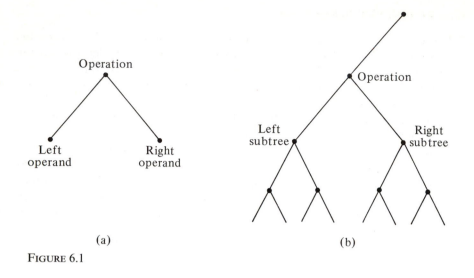

FIGURE 6.1

$$(A \cdot (B + C)) + (D/E)$$

Let us take the second + in the schedule of operations as the "root" of our tree. Grafting the subexpressions $A \cdot (B + C)$ and D/E as left and right subtrees at this root, we obtain the diagram in Figure 6.2(a). Continuing inductively, we will finally obtain the labeled binary tree in Figure 6.2(b). The intended schedule of operations and their respective operands are completely specified by the tree structure itself. △

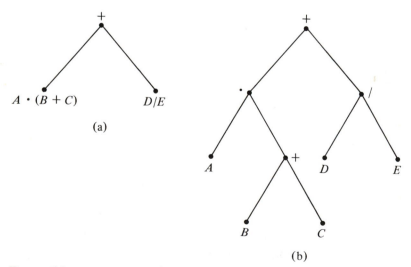

FIGURE 6.2

EXAMPLE 6.3   The trees of Figure 6.3 represent two more possible interpretations of the original unparenthesized expression $A \cdot B + C + D / E$.   △

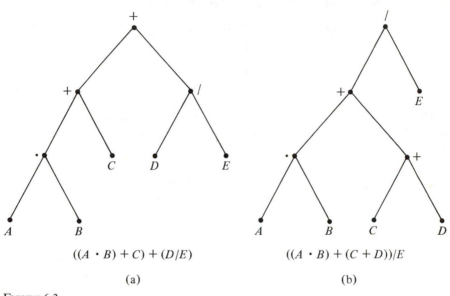

$((A \cdot B) + C) + (D/E)$

(a)

$((A \cdot B) + (C + D))/E$

(b)

FIGURE 6.3

To those having any familiarity with computer data structures, it should come as no surprise that we are less than completely satisfied with these tree representations, helpful though they may be. The problem is that storing the tree-structured information may not be so convenient; certainly the available means are not as concise as we would like. Fortunately, a linear data representation, developed by the Polish logician Jan Lukasiewicz, will retain all the structural information of the tree. Probably because the word "Polish" is easier to pronounce, this representation is called the *Polish form* (or the *postfix form*) of the corresponding fully parenthesized expression. We simply write:

left operand – right operand – operation

instead of the *infix form*

left operand – operation – right operand

that we are most accustomed to seeing. Once again there is no need for parentheses. For this reason, the Polish expressions are often called "parenthesis-free" expressions.

EXAMPLE 6.4   Once again we consider the expression of Example 6.2:

$$((A \cdot (B + C)) + (D/E))$$

Transferring the last scheduled operation ($+$) to the rear, we write

$$(A \cdot (B + C))(D/E) +$$

If the same is done for the subexpressions, we will have

$$(A \cdot (B + C)) \to A\,(B + C) \cdot \to A\,B\,C\,+\,\cdot$$
$$(D\,/\,E) \to D\,E\,/$$

Finally, we will obtain the Polish expression

$$A\,B\,C\,+\,\cdot\,D\,E\,/\,+$$

One might say that this parenthesis-free expression would make sense only if we could reconstruct the tree. But the reconstruction is in fact quite straightforward, if we use the following *underlining procedure*. Read the expression from left to right, looking for an operation symbol. When one is found, we back up to locate its two operands and underline all three symbols. In the present example, we read to the first operator ($+$) and back up to find B and C as operands. Underlining these symbols, we have:

$$A\,\underline{B\,C\,+}\,\cdot\,D\,E\,/\,+$$

Now if this group is treated as an operand in itself, then we can start all over again; that is, draw a line back from the multiplication symbol and under its two operands, A and the group just underlined. Continuing to scan from left to right, looking for new operations, we find the division operator ($/$) and back up to find D and E as its operands. Underlining these, we obtain

$$\underline{A\,\underline{B\,C\,+}\,\cdot}\,\underline{D\,E\,/}\,+$$

Finally, we have

$$\underline{\underline{A\,\underline{B\,C\,+}\,\cdot}\,\underline{D\,E\,/}\,+}$$

From this representation, it is an easy matter to reconstruct the tree of the expression. Thus we obtain Figure 6.2(b) once more.   △

The inverse of this underlining procedure is equally interesting. That is, we can define a process for deriving the Polish form of an expression from its tree representation. Making a clockwise traversal of the tree, we can tabulate from *right to left* the sequence of labels as they are encountered in a *descending motion*. A couple of examples will clarify this *Polish-tree-traversing* technique.

EXAMPLE 6.5   Suppose that we wish to convert the trees of Example 6.3 (Figure 6.3) to Polish form. Beginning from just above the root of the tree, we make a clockwise traversal as indicated in Figure 6.4. The arrows of Figure 6.4 indicate the encountering of labels in the downward motion. In Figure 6.4(a) we encounter (in this way) the labels $+, /, E, D, +, C, \cdot, B, A$ in sequence.

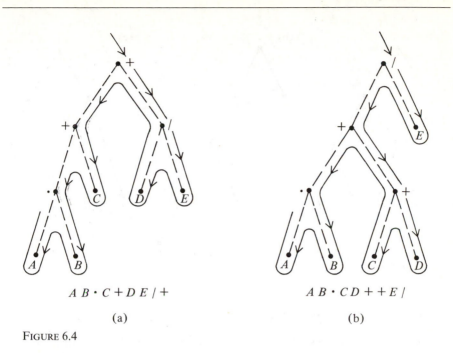

$$A\ B \cdot C + D\ E\ /\ +$$

(a)

$$A\ B \cdot C\ D + + E\ /$$

(b)

FIGURE 6.4

Stringing these symbols from right to left, we obtain the Polish expression $A\ B \cdot C + D\ E\ /\ +$. Similarly, for Figure 6.4(b) we obtain $A\ B \cdot C\ D + + E\ /$. △

EXERCISES

1 Obtain binary labeled trees for the following arithmetic expressions:
   (a) $(A/B) + C \cdot (D + E)$ (b) $(A/B) + ((C \cdot D) + E)$
   (c) $A/((B + C) \cdot (D + E))$ (d) $(A/(B + C)) \cdot (D + E)$

2 Obtain the Polish form of the expressions in Exercise 1.

3 Use the underlining procedure to transform the Polish expressions obtained in Exercise 2 into their corresponding binary labeled-tree representation.

4 Using the underlining procedure, obtain the binary labeled trees corresponding to the following Polish expressions:
   (a) $A\ B \cdot C + D\ /$ (b) $A\ B \cdot C\ D + /\ E +$
   (c) $A\ B \cdot C\ /\ D\ E + \cdot$ (d) $A\ B\ C + D\ E\ /\ \cdot\cdot$
   (e) $A\ B\ C\ D + \cdot E + /$ (f) $A\ B\ C \cdot + D\ E\ /\ +$

5 Use the Polish-tree-traversing technique to recover the corresponding Polish expressions from the trees obtained in Exercise 4.

6 From the trees obtained in Exercise 4 derive the corresponding parenthesized infix expressions.

7 Given the trees shown in Figure 6.5, obtain the corresponding infix and Polish forms.

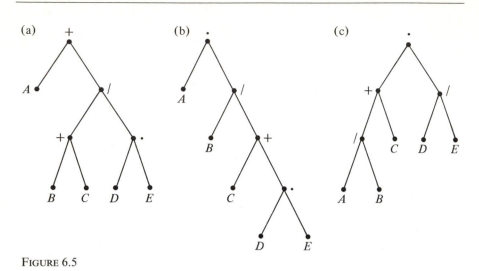

FIGURE 6.5

6.2 STACKS AND QUEUES

To say that we are interested in the Polish notation simply because the algebraic expressions can be written without parentheses does not tell the whole story. The full usefulness of the Polish form will come through only after one has seen the great advantage it provides in the compilation of expressions. We refer here to the internal generation of machine language instructions, once a translation into Polish form has been achieved. Thus the novel use of symbol-manipulating procedures stands out as one of the more significant technical advances in computer science. In order to give a reasonably detailed account here, we have to take a slight detour to describe the ingredients of the relevant symbol-manipulation schemes. In view of the fact that our expressions are nothing more than strings of symbols in an appropriate alphabet, we will find that the free monoids (Section 3.4) provide the necessary framework for the discussion.

Several of the procedures to be described in this chapter will make use of a "pushdown store" or *stack*, as it is commonly called. In effect, a stack is a growing memory which operates on a *last in – first out* basis. Anyone who has ever observed what happens to a stack of trays in a cafeteria should be able to have the right image in mind. Suppose that Σ is our alphabet of symbols. If $x \in \Sigma^*$ and we wish to add the symbol $\sigma \in \Sigma$ to the stack x, then the assignment

$$x \leftarrow x\sigma$$

may be viewed as our placing σ on "top" of the stack. This is the way that information is stored away in a pushdown stack. We say "pushdown" because a new symbol has taken the topmost position, just as a new tray would push the others down to a greater depth in a cafeteria stack. Also only the top of the stack is immediately accessible. If we remove the topmost symbol from a stack $x = \sigma_{i_1}\sigma_{i_2} \cdots \sigma_{i_k}$ to store it in the variable called σ, say, that is, $\sigma \leftarrow \sigma_{i_k} =$ top (x), then the stack is depleted accordingly and x becomes $\sigma_{i_1}\sigma_{i_2} \cdots \sigma_{i_{k-1}}$ as if the trays have popped up upon the removal of the symbol $\sigma_{i_k} =$ top (x). It follows that the assignment statement just given does not provide an accurate portrayal of the information transfer. Therefore, we will borrow a notation from Section 1.10 $[x \twoheadleftarrow \text{elt } (S)]$ and write

$$\sigma \twoheadleftarrow \text{top } (x)$$

to symbolize this "assignment and depletion."

A similar notation $\sigma \twoheadleftarrow \text{bot } (x)$ would refer to the transfer (with deletion) of the bottommost symbol of x to the variable σ. With this kind of access and the same mode of storage as before $(x \leftarrow x\sigma)$, we would have a *first in – first out* memory, sometimes called a *queue* because it exhibits the same kind of behavior as a line of customers waiting for service at a bakery counter.

EXAMPLE 6.6   To provide an elementary example of the use of stacks, let us show how we can write a word backward from the way it is given. We can do this by the following simple algorithm, with the reversal of the given word $x \in \Sigma^*$ accumulating in the variable y:

> **begin** $y \leftarrow \epsilon$;
>   **while** $x \neq \epsilon$ **do**
>   **begin** $\sigma \twoheadleftarrow \text{top } (x)$;
> $y \leftarrow y\sigma$
>   **end**
> **end**   △

EXAMPLE 6.7   If $x =$ DENVER, then a trace of the algorithm of Example 6.6 would be as in Table 6.1. △

TABLE 6.1

x	σ	y
DENVER		ϵ
DENVE	R	R
DENV	E	RE
DEN	V	REV
DE	N	REVN
D	E	REVNE
ϵ	D	REVNED

A less trivial example of the use of stacks is provided by an implementation (Algorithm XXI) of the Polish-tree-traversing scheme of Section 6.1. As in Section 3.5, T is a labeled binary tree. We assume that the labels are chosen from an alphabet Γ and may represent operation symbols of the given expression or variables occurring in it. The word $x \in B^*$ locates the positions of the tree during the traversal; it is the variable that is to be treated as a stack. Assuming that the mechanism of the traversal has been properly organized, we only have to accumulate the sequence of symbols in an output string z, as accomplished by the assignment statement $z \leftarrow T[x] \cdot z$. An example will help to clarify the details.

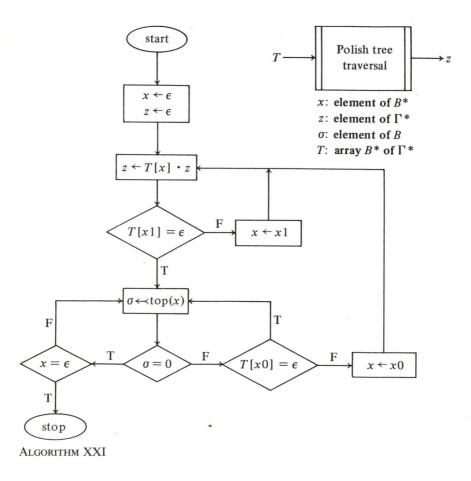

ALGORITHM XXI

EXAMPLE 6.8   As input to the Polish-tree traversal algorithm we take the tree of Figure 6.3(a). We should pay particular attention to the tree location variable x in studying a trace of the algorithm as shown in Table 6.2.

TABLE 6.2

x	σ	z
ϵ		ϵ
		$+$
1		$/\,+$
11		$E\,/\,+$
1	1	
10		$D\,E\,/\,+$
1	0	
ϵ	1	
0		$+\,D\,E\,/\,+$
01		$C\,+\,D\,E\,/\,+$
0	1	
00		$\cdot\,C\,+\,D\,E\,/\,+$
001		$B\cdot C\,+\,D\,E\,/\,+$
00	1	
000		$A\,B\cdot C\,+\,D\,E\,/\,+$
00	0	
0	0	
ϵ	0	

The algorithm produces the same result

$$z = A\,B\cdot C\,+\,D\,E\,/\,+$$

as was obtained in Example 6.5 [Figure 6.4(a)].   △

With this application under our belts, we are now ready to describe the use of stacks in compiling an algebraic expression given in Polish form. Once again the expression itself is nothing more than a string of symbols

$$x = \sigma_{i_1}\sigma_{i_2} \cdots \sigma_{i_k}$$

from the alphabet Γ. Assume that Γ contains the subalphabet

$$\Sigma = \{A, B, C, \ldots\}$$

which is the set of variable names. Then

$$\Gamma \sim \Sigma = \{+, \cdot, /, \ldots\}$$

is the collection of binary operation symbols that may be encountered in the various expressions to be processed or compiled. When we say "compile" here, we mean determine the required schedule of operations, say in the form

$$(L_j, O_j, R_j) \qquad j = 1, 2, \ldots$$

L_j and R_j being the left and right arguments, and O_j the jth operation symbol. When we are finished, it will be clear that we could just as well have generated a corresponding sequence of machine language instructions. In any case, we will find it necessary to introduce new variable names V_j as we proceed for the storage of intermediate results in an evaluation of the algebraic expression x.

In Algorithm XXII the pushdown stack y stores only variable names, waiting until an operation symbol is read off the input expression x. In fact, x is accessed as a queue, that is, from left to right, though we never store any additional symbols there. In studying the following example, the student should see that the algorithm merely simulates the underlining procedure described in the previous section, giving new variable names V_j to the groups of symbols previously "underlined."

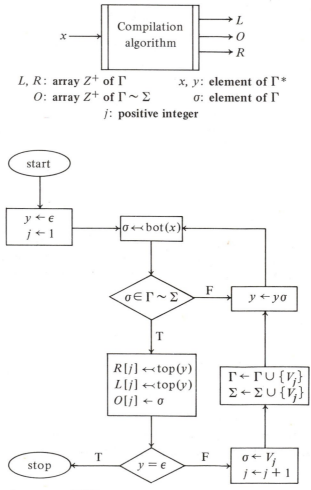

ALGORITHM XXII

EXAMPLE 6.9 Consider the assignment statement

$$F = A \cdot (B + C) + D/E$$

that makes use of the algebraic expression of Example 6.1. The equal sign is treated no differently from the other operation symbols in determining the Polish form: $F\,A\,B\,C + \cdot\,D\,E\,/\,+\,=$. A trace of the compilation is shown in Table 6.3.

TABLE 6.3

x	y	j	σ	L_j	O_j	R_j
$F\,A\,B\,C + \cdot\,D\,E\,/\,+\,=$	ϵ	1				
$A\,B\,C + \cdot\,D\,E\,/\,+\,=$	F		F			
$B\,C + \cdot\,D\,E\,/\,+\,=$	$F\,A$		A			
$C + \cdot\,D\,E\,/\,+\,=$	$F\,A\,B$		B			
$+ \cdot\,D\,E\,/\,+\,=$	$F\,A\,B\,C$		C			
$\cdot\,D\,E\,/\,+\,=$	$F\,A\,B$		$+$			C
	$F\,A$			B	$+$	
	$F\,A\,V_1$	2	V_1			
$D\,E\,/\,+\,=$	$F\,A$		$\cdot$			V_1
	F			A	$\cdot$	
	$F\,V_2$	3	V_2			
$E\,/\,+\,=$	$F\,V_2\,D$		D			
$/\,+\,=$	$F\,V_2\,D\,E$		E			
$+\,=$	$F\,V_2\,D$		$/$			E
	$F\,V_2$			D	$/$	
	$F\,V_2\,V_3$	4	V_3			
$=$	$F\,V_2$		$+$			V_3
	F			V_2	$+$	
	$F\,V_4$		V_4			
ϵ	F		$=$			V_4
	ϵ			F	$=$	

Looking now only at the sequence of values for (L_j, O_j, R_j), we find the schedule of operations:

$$V_1 = B + C$$
$$V_2 = A \cdot V_1$$
$$V_3 = D / E$$
$$V_4 = V_2 + V_3$$
$$F = V_4$$

which is suitable for evaluating the given expression. △

In spite of the obvious applications, the preceding discussion may have seemed to be a strange introduction to a discussion of the propositional calculus of mathematical logic. Our intention, however, is to develop the use of the Polish notation for the logical expressions, or *well-formed formulas*, as they are usually called. The reader will thus be able to discover immediate applications of these ideas to questions that he or she may have regarding Polish forms, trees, etc., as they relate to the algebraic expressions appearing in computer programming languages.

EXERCISES

*1 Devise a formal algorithm for deciding whether or not a given word $x \in \Sigma^*$ is a palindrome. [*Hint*: A *palindrome* is a word (for example, RADAR) that reads the same backward as forward.]

2 Devise a formal algorithm for transforming $x\sigma$ into σx for any $x \in \Sigma^*$ and $\sigma \in \Sigma$.

3 Use Algorithm XXI (and provide a trace) to obtain the Polish form corresponding to the trees of Exercise 7 of Section 6.1.

4 Repeat Exercise 3 for the trees obtained from the expressions of Exercise 1 of Section 6.1.

5 Use Algorithm XXII (and provide a trace) to compile the proper schedule of operations for the Polish forms of Exercise 4 of Section 6.1.

6 Repeat Exercise 5 for the Polish forms obtained in Exercise 2 of Section 6.1.

*7 Algorithm XXI transforms the tree of an algebraic expression into the Polish form. Devise a formal algorithm that will do just the opposite. [*Hint*: It may be more convenient to read the Polish expression from right to left.]

*8 Devise a formal algorithm for doing a sequential examination of a tree-sorted sequence of keys (Section 3.6).

6.3   WELL-FORMED FORMULAS

Propositional calculus, as originally intended by Boole, is quite literally a method for calculating with sentences or propositions. The sentences here are composed of elementary constituents called *propositional variables*, together with the elementary operations that represent the ordinary connectives of logical thought: and, or, not, etc. The sentences, moreover, are assumed to be declarative and not at all subjective. Thus it is reasonable to suppose (given the truth or falsity of the variables) that the sentences are either true or false; there should be no in between. All in all, we could say that mathematical logic is more concerned with the relationship between various propositions than with the abstract philosophical truth of isolated statements. Before we can properly examine these relationships, however, we must have a clear understanding of the correct way of formulating sentences out of the given elementary ingredients. An appreciation for the structure of these properly formed sentences or *well-formed formulas* is most easily gained by thinking in terms of the algebraic expressions of Section 1.3, which is also the best way of correlating the new material with the discussions of the two previous sections.

We are all familiar with sportscasters' penchant for prediction. As a sports season draws to a close, however, they will become more realistic. More and more they will begin to weigh the alternatives, trying to analyze for us all the possible outcomes, especially for the home team. Thus we hear, "Now if the Blue Blazers beat the Rough Riders and our Wildcats can at least tie in their game with the Hot Shots, then we have a chance for the pennant, provided that Biff Bruiser's knee holds up and the price of tea in China goes down, or . . . ." Indeed, one almost needs a degree in mathematical logic to be able to follow such trains of thought. But consider the following concrete situation. The current team standings of a four-team football conference, where each team has one game left to play are as follows.

	W	L	T
Oakland	6	4	1
Denver	5	4	2
Kansas City	6	5	0
San Diego	2	8	1

Suppose that in the final games of the season Oakland meets Kansas City, while Denver plays against San Diego. Let us see if we can anticipate the sportscasters' analyses by examining the circumstances that must prevail if Denver is to emerge as the conference champion (or at least cochampion). We assume that ties are counted as a half-game won and a half-game lost.

Now, Denver cannot afford to lose in their last game. A tie, however, would be equally fatal, because it would leave their record at $6\frac{1}{2}$–$5\frac{1}{2}$. Only an Oakland loss would bring Oakland's record that low, but in that case Kansas City would have won, putting them ahead of Denver. So in order to have a chance, Denver must beat San Diego. That is, *Denver must win and* (apparently) something else must happen, if Denver is to be a champion. Symbolically, we can write "$p \wedge \cdots$" if we take as propositional variables

p: Denver wins
q: Oakland wins
r: Kansas City wins

in reference to the final games. Further analysis will show that besides condition p, *Kansas City must win or tie* in its game with Oakland. Putting all the conditions together, we arrive at the proposition $p \wedge (r \vee (\neg q))$ as expressive of the conditions for a Denver championship. (Can you find an equivalent, yet simpler, formula?)

In such informal discussions as well as in the conditional statements of the flowchart language, indeed throughout the text, we have found it necessary to use the logical connectives

$\vee$ representing "or"

$\wedge$ representing "and"

$\neg$ representing "not"

These connectives are referred to by logicians as the operations of *disjunction*, *conjunction*, and *negation*, respectively. We use them to form complex sentences out of elementary ones, beginning with an alphabet of symbols

$$\Sigma = \{p, q, r, \ldots\}$$

called *propositional variables.* Since we are dealing with such sentences or propositions abstractly, we will rarely ask what it is that these symbols represent. As we have seen, they can be used to denote most anything, depending on the nature of the discussion. Ordinarily, however, the propositional variables are assumed to be elementary statements, not subject to further decomposition into simpler statements.

Again let us defer any questions as to the meaning of the more complex sentences and concentrate only on the syntax, that is, the proper formation of logical sentences out of the variables and connectives. If we do this, then we may just as well substitute the FORTRAN variables, say, and a convenient collection of arithmetic operations, for the propositional variables and logical connectives; the sense of our discussion will not change. Then instead of well-formed formulas, we will be speaking of the proper construction of arithmetic expressions. Formally, there will be no difference. So we now agree that a *well-formed formula* (abbreviated wff) of propositional calculus over the propositional variables $p, q, r, \ldots$ is either

 (i) a propositional variable (standing alone), or

(ii) $\alpha\,\beta\,\vee$, $\alpha\,\beta\,\wedge$, or $\alpha\,\neg$ for wff's α, β.

Then the formal similarity to the idea of an algebraic expression (particularly the Boolean expressions of Section 4.7) is immediately apparent. The main distinction is our use of Polish notation here; the connective is written after its two (or one in the case of negation) arguments. If we were to use the infix notation $(\alpha) \vee (\beta), (\alpha) \wedge (\beta)$, and $\neg(\alpha)$, then the analogy would be complete.

EXAMPLE 6.10   The expression

$$p\,q\,\neg\,\wedge\,r\,\vee\,p\,\wedge\,\neg\,p\,q\,\wedge\,\vee$$

is a well-formed formula, though we do not claim that this is immediately apparent. Until we have presented an algorithm for this purpose, we may have to convince the reader by stages:

p, q, r are wff's because of condition (i)

q ¬ is a wff because of condition (ii), *q* having already been found to be a wff

p q ¬ ∧ is a wff because of condition (ii), *p* and *q* ¬ having been found to be wff's

p q ¬ ∧ *r* ∨ is a *wff* because of condition (ii), etc.

Finally, *p q* ¬ ∧ *r* ∨ *p* ∧ ¬ and *p q* ∧ are wff's, so that the given expression is a wff, again because of (ii).   △

EXAMPLE 6.11   Let us try to construct a tree for the formula

$$p\ q\ \neg\ \wedge\ r\ \vee\ p\ \wedge\ \neg\ p\ q\ \wedge\ \vee$$

of Example 6.10. With one slight modification, we can use the underlining procedure of Section 6.1. Keeping in mind that negation is a unary operation, we obviously back up to locate only one operand in case we encounter a negation symbol. Otherwise, everything is the same as before. In the present example, we obtain the following underlined formula:

$$\underline{p\ q\ \neg}\ \wedge\ r\ \vee\ p\ \wedge\ \neg\ \underline{p\ q}\ \wedge\ \vee$$

which leads us immediately to the tree of Figure 6.6.

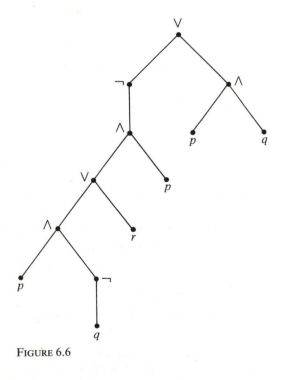

FIGURE 6.6

Using the tree, we may now express the formula in the fully parenthesized infix form $(\alpha) \vee (\beta)$, where

$$\beta = (p) \wedge (q)$$

$$\alpha = \neg((((p) \wedge (\neg(q))) \vee (r)) \wedge (p))$$

Altogether we obtain the infix formula:

$$(\neg((((p) \wedge (\neg(q))) \vee (r)) \wedge (p))) \vee ((p) \wedge (q))$$

No wonder Lukasiewicz became interested in the possibilities of a parenthesis-free notation.   △

In connection with these two examples, we must bear in mind that it is the reverse conversion, from infix to Polish, that is important in applications. That is because the arithmetic expressions in the computer programming languages are written in the infix notation, the latter being easier (supposedly) for human beings to understand. On the other hand, we saw in the preceding section that it is quite easy to compile the machine language code for an arithmetic expression if the expression is given in Polish form. The required infix to Polish conversion process will be the subject of the next section.

EXAMPLE 6.12   The following string of symbols

$$p \, q \, \neg \, p \, \wedge$$

is not a wff. Certainly it is not a propositional variable standing alone. So if it is to be a wff, it must take one of the forms $\alpha \beta \vee$, $\alpha \beta \wedge$, or $\alpha \neg$ for the wff's α, β. Now, the only possibility is $\alpha \beta \wedge$. But no matter how we parse the string $p \, q \, \neg \, p$:

$$p \, q \, \neg \, p = (p \, q \, \neg)(p) \rightarrow \alpha = p \, q \, \neg \text{ is not a wff}$$

$$= (p)(q \, \neg \, p) \rightarrow \beta = q \, \neg \, p \text{ is not a wff}$$

$$= (p \, q)(\neg p) \rightarrow \text{neither is a wff}$$

we do not obtain well-formed α and β.   △

We have somehow managed to work our way through Examples 6.10 and 6.12, finding that one of the expressions is a wff and the other is not. What is needed here, however, is an algorithm, if such determinations as we made are to be carried out more systematically. Note that if we write

$$\Gamma = \{ \vee, \wedge, \neg, p, q, r, \ldots \}$$

$$\Sigma = \{ p, q, r, \ldots \} \subseteq \Gamma$$

then in the sense of Section 3.8, we are speaking of a language $\lambda \subseteq \Gamma^*$, the *language of wff's*; that is

$$\lambda = \{ x \in \Gamma^* : x \text{ is a wff} \}$$

We are seeking an algorithm that will recognize or accept this language, much in the way that the finite state machines of Section 3.8 would function. Accordingly, the algorithm should provide us with a yes or no answer, depending on whether or not the given string x is a wff.

The required ingredients in such an algorithm are suggested by the following observations. For the symbols of our alphabet Γ, we associate *ranks* in accordance with an accumulated left to right "variable count" of a Polish formula x. That is, we take

$$1 = \text{rank } (p) = \text{rank } (q) = \text{rank } (r) = \cdots$$
$$-1 = \text{rank } (\vee) = \text{rank } (\wedge)$$
$$0 = \text{rank } (\neg)$$

Thus, keeping in mind the underlining procedure, we see that binary operations use up two variables to produce one, whereas a unary operator leaves the count unchanged.

EXAMPLE 6.13   We have rank $(p) = $ rank $(q) = 1$. We use up two arguments to create only one new one when we use a binary connective, as in $pq \vee$. Therefore, $\vee$ is assigned a rank of -1. Negation, on the other hand, produces a new argument from an old one, so it is assigned a rank of zero.   △

As these observations suggest, we ought to scan the word $x \in \Gamma^*$ (a candidate for being a wff) from left to right and accumulate the partial sums of the ranks of the various symbols. Thus, if we are given the word

$$x = \sigma_{i_1}\sigma_{i_2} \cdots \sigma_{i_n}$$

and we write

$$r_j = \text{rank } (\sigma_{i_j})$$

then the partial sums

$$S_1 = r_1$$
$$S_2 = r_1 + r_2$$
$$\vdots$$
$$S_n = r_1 + r_2 + \cdots + r_n$$

will play a role in determining whether x represents a well-formed formula. In particular, we expect to find $S_n = 1$ whenever x is a wff. (Why?) And so, it is no surprise to see this equation as one of the conditions in the following result.

Theorem 6.1   (Burks, Warren, and Wright*)

The string of symbols $x \in \Gamma^*$ is a wff if and only if
(1) each $S_i \geq 1$ (2) $S_n = 1$

---

* A. W. Burks, D. W. Warren, and J. B. Wright, "An Analysis of a Logical Machine Using Parenthesis-Free Notation," *Math. Tables and Other Aids to Computation*, **8**, 53–57 (April 1954).

Proof We prove only the necessity of the two conditions. This is done by induction on the length $|x| = n$. If $|x| = n = 1$, then $x = p$, a propositional variable (propositional variables being the only wff's of length one). Since rank $(p) = 1$ we have

$$r_1 = S_1 = S_n = 1$$

so that conditions (1) and (2) hold.

Now suppose that every wff of length $k \leq n$ has these two properties, and let $|x| = n + 1$. Evidently

$$x = yz \vee \quad (\text{or } yz \wedge) \qquad \text{or else} \qquad x = y \neg$$

for the wff's

$$y = \sigma_{i_1} \ \sigma_{i_2} \ \cdots \ \sigma_{i_k} \qquad z = \sigma_{j_1} \ \sigma_{j_2} \ \cdots \ \sigma_{j_l}$$
$$T_1 \ T_2 \ \cdots \ T_k \qquad\qquad U_1 \ U_2 \ \cdots \ U_l$$

(We let T's and U's denote the corresponding partial sums.) Taking the separate cases one at a time, suppose first that $x = yz \vee$ (the situation $x = yz \wedge$ would be the same). Then

$$x = \sigma_{i_1} \sigma_{i_2} \cdots \sigma_{i_k} \sigma_{j_1} \sigma_{j_2} \cdots \sigma_{j_l} \vee$$
$$\text{ranks: } t_1 \ t_2 \ \cdots \ t_k \ u_1 \ u_2 \ \cdots \ u_l - 1$$
$$\text{partial sums: } S_1 S_2 \cdots S_k S_{k+1} \cdots S_{k+l} S_{k+l+1} = S_{n+1}$$

Since $|y| \leq n$ and $|z| \leq n$, we have by virtue of the inductive hypothesis:

(1) each $T_i \geq 1$ each $U_j \geq 1$

(2) $T_k = 1$ $U_l = 1$

Now we see whether the same things hold for the partial sums of the word x.

(1) $S_i = T_i \geq 1$ $(1 \leq i \leq k)$

   $S_{k+j} = T_k + U_j = 1 + U_j \geq 2 \geq 1$ $(1 \leq j \leq l)$

(2) $S_{n+1} = S_{k+l+1} = -1 + S_{k+l} = -1 + (T_k + U_l) = -1 + 2 = 1$

Indeed, they do. We shall leave the easier case $(x = y \neg)$ as an exercise for the student. $\qquad\qquad\qquad\qquad\qquad\qquad\qquad\qquad\qquad\qquad\qquad\Box$

Obviously the theorem leads directly to the desired algorithm for recognizing well-formed formulas (Algorithm XXIII).

EXAMPLE 6.14   Certainly we expect the conditions of the theorem to be met for the formula of Example 6.10, and a simple calculation shows that, indeed, they are.

$$x = p \ q \ \neg \quad \wedge \ r \quad \vee \ p \quad \wedge \ \neg \ p \ q \quad \wedge \quad \vee$$
$$\text{ranks: } 1 \ 1 \quad 0 \ -1 \ 1 \quad -1 \ 1 \quad -1 \quad 0 \ 1 \ 1 \quad -1 \ -1$$
$$\text{partial sums: } 1 \ 2 \quad 2 \quad 1 \ 2 \quad 1 \ 2 \quad 1 \quad 1 \ 2 \ 3 \quad 2 \quad 1$$

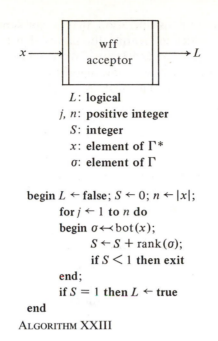

L: **logical**
j, n: **positive integer**
S: **integer**
x: **element of Γ***
σ: **element of Γ**

begin $L \leftarrow$ **false**; $S \leftarrow 0$; $n \leftarrow |x|$;
    for $j \leftarrow 1$ **to** n **do**
    begin $\sigma \leftarrow$ bot(x);
         $S \leftarrow S +$ rank(σ);
         if $S < 1$ **then exit**
    end;
    if $S = 1$ **then** $L \leftarrow$ **true**
end

ALGORITHM XXIII

These calculations are the same as those carried out by Algorithm XXIII. Note in particular that the partial sum S_j is most easily computed as $S_j = S_{j-1} + r_j$. In hand computation, as summarized above, we add "diagonally" to implement this rule:

$$\frac{1}{1} \Big/ \underset{+}{} \frac{}{2} \quad \frac{0}{2} \Big/ \underset{+}{} \frac{}{2} \quad \frac{-1}{2} \Big/ \underset{+}{} \frac{}{1}$$

In the formal statement of the algorithm, the logical variable L is set to $L \Rightarrow$ **true** only when each partial sum is greater than or equal to one, and the last partial sum is equal to one (consistent with the conditions of the theorem). We see that these conditions are met in the present example.   △

EXAMPLE 6.15   We should expect one or the other (or both) of these conditions to fail if x is the string from Example 6.12. Sure enough,

$$x = p\, q\, \neg\, p\, \wedge$$
$$\text{ranks: } 1\ 1\ 0\ \ 1\, {-}1$$
$$\text{partial sums: } 1\ 2\ 2\ \ 3\ \ \ 2$$

so we have $S_n = S_5 = 2 \neq 1$ so that x is not a wff.   △

EXAMPLE 6.16   Consider the Polish arithmetic expression

$$x = A\, B\, -\, \cdot\, A\, -\, -\, B\, /\, +$$

in which $(-)$ is a unary minus operation, that is, not subtraction. (There are certain technical difficulties in treating the minus sign as a binary or unary operation in one and the same expression. We would rather not get into the details, however.) With exactly the same algorithm we can test to see whether this arithmetic expression is well formed. As before, we compute

$$x = A\ B\ -\quad \cdot\ A\ -\ -\ B\quad /\ +$$
$$\text{ranks: } 1\ \ 1\ \ 0\ \ -1\ \ 1\ \ 0\ \ 0\ \ 1\ -1\ -1$$
$$\text{partial sums: } 1\ \ 2\ \ 2\quad 1\ \ 2\ \ 2\ \ 2\ \ 3\ \ 2\quad 1$$

It follows from Theorem 6.1 that x is a properly formed (Polish) arithmetic expression.

Using the underlining procedure, we have

$$\underline{\underline{A\ \underline{B\ -}\ \cdot\ \underline{A\ \underline{-\ \underline{-\ B}}}}\ /}\ +$$

so we obtain the tree of Figure 6.7. It follows that in infix notation we would write $A \cdot (-B) + (-(-A))/B$, with the assumption that the multiplication and division are performed before the addition. (Otherwise, more parentheses are required.) △

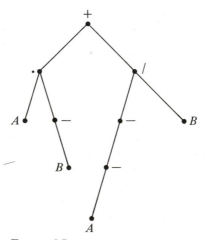

FIGURE 6.7

EXERCISES

1 Given the hypothetical football team standings discussed at the beginning of the section, obtain a logical proposition expressing the conditions required for
   (a) an Oakland championship   (b) a Kansas City championship

2 Show that the following strings are not wff's:
   (a) $q \vee p \neg$ (b) $q \neg p\, q \vee$ (c) $p\, q \vee \wedge$

3 By direct appeal to the definition, show that the following Polish forms are wff's of propositional calculus:

   (a) $p\,r \wedge q\,p \urcorner \urcorner \vee \wedge$ (b) $r \urcorner q \wedge p \vee \urcorner$

   (c) $r \urcorner q \vee p \urcorner q \vee \urcorner \wedge$ (d) $p\,q\,r\,p \vee \wedge q \wedge \vee \urcorner$

4 Use the underlining procedure to obtain binary labeled trees corresponding to the Polish expressions of Exercise 3.

5 From the trees obtained in Exercise 4 derive the corresponding parenthesized infix expressions.

6 Use Algorithm XXIII to show that the Polish forms of Exercise 3 are well formed.

7 Use Algorithm XXIII to decide whether or not the following are Polish wff's:

   (a) $r\,p\,r \vee p\,q \vee \wedge \vee$ (b) $p\,r \urcorner \wedge p\,q \urcorner \vee \urcorner \wedge \urcorner$

   (c) $r \urcorner p \urcorner q \vee p \urcorner \wedge$ (d) $p\,q\,r \vee \urcorner \wedge p \urcorner q\,r \vee$

8 Use Algorithm XXIII to decide whether or not the following Polish arithmetic expressions are well formed (if $-$ is the unary "minus"):

   (a) $A - C + B\,A\,/ - \cdot$ (b) $A\,C\,B - + B - A \cdot - /$

   (c) $A\,B\,C\,A + \cdot D + / -$ (d) $A\,B\,C + - \cdot B - A\,B\,/$

9 Complete the proof of the necessity of the conditions in Theorem 6.1.

6.4   CONVERSION TO POLISH FORM

In the best-known computer programming languages, arithmetic expressions as well as logical expressions are written in the infix notation. Yet we have seen (Section 6.2) a real advantage in the Polish postfix form in implementing a compilation procedure. If any use is to be made of this advantage, while keeping the infix form for communicating with the computer as it seems we must, we must have an algorithm or procedure for translating an expression from infix to Polish form. Of course, in an existing computer, such procedures are already built in. We will therefore be learning about the inner workings of the machine as regards the processing of arithmetic and logical expressions.

For purposes of continuity in our presentation, we shall phrase the translation algorithm in the language of propositional calculus. We have already noted that there is no real loss of generality in our so doing. Though our results will be most directly applicable to the processing of the logical expressions of a computer programming language, arithmetic expressions may be treated in an entirely analogous manner, as we saw earlier. But we shall make a slight concession in making sure that these ideas are easily applied to the arithmetic case. That is, we shall allow a priority or hierarchy of operations, parenthesis symbols, etc., just as in conventional arithmetic.

EXAMPLE 6.17   The infix expression of Example 6.16

$$A \cdot (-B) + (-(-A))/B$$

though not fully parenthesized, is not ambiguous because we ordinarily understand that · and / have a higher priority than addition.   △

EXAMPLE 6.18   In mathematical logic no complete hierarchy is universally accepted. In fact, that is one reason that the Polish form is so appropriate. Most people do agree, however, that conjunction has a higher priority than disjunction. Therefore, $p \wedge q \vee r$ would usually be interpreted as $(p \wedge q) \vee r$ and not as $p \wedge (q \vee r)$. The latter expression definitely needs the parentheses if it is to be properly understood. After all, that is the function of parentheses— to override the priorities that would otherwise take effect.   △

For purposes of introducing the conversion process, let us establish a hierarchy of logical symbols once and for all. We introduce two new binary connectives, ($\supset$) and ($\equiv$), for the sake of completeness. The reader does not have to place an interpretation on these just yet. We are still concerned only with the syntax; the meaning will come later. Our complete hierarchical system is as described in the following table:

$\Gamma \sim \Sigma$	(	)	$\equiv$	$\supset$	$\vee$	$\wedge$	$\neg$
priority	0	1	2	3	4	5	6

If $\sigma \in \Gamma \sim \Sigma$, we let $|\sigma|$ denote the *priority* of the symbol. Furthermore, we agree that in case of a tie, as could happen here only for two identical symbols, the leftmost occurrence is to take precedence.

The conversion or translation procedure is presented in Algorithm XXIV. A pushdown stack y is used to sort the operation symbols according to their priority. The given input formula (in infix notation) is denoted by x, and the output (Polish) expression is given by z. As we mentioned earlier, the proper interpretation of infix formulas requires considerable ingenuity, and this is reflected in the complexity of Algorithm XXIV. The reader, however, should pay particular attention to the handling of parentheses. A subexpression enclosed within them is to be evaluated in such a way that it takes the form of an argument for an operation symbol occurring outside the parentheses. So in entering the priority comparison $\langle |\tau| \geq |\sigma| \rangle$, a left parenthesis taken from the top of the stack y will have a lower priority (namely 0) than any symbol read off the input x; and unless that symbol is a right parenthesis, we see from the algorithm that both τ and σ are placed (replaced in the case of τ) on the stack. On the other hand, a left parenthesis read off the input is immediately placed on the stack (see the conditional statement in the upper left-hand corner) without regard to the priorities of any symbols that may already be there. Finally, note how the matching of left and right parentheses serves to dissolve the pair, as shown by the action of the conditional statement in the middle of the flowchart.

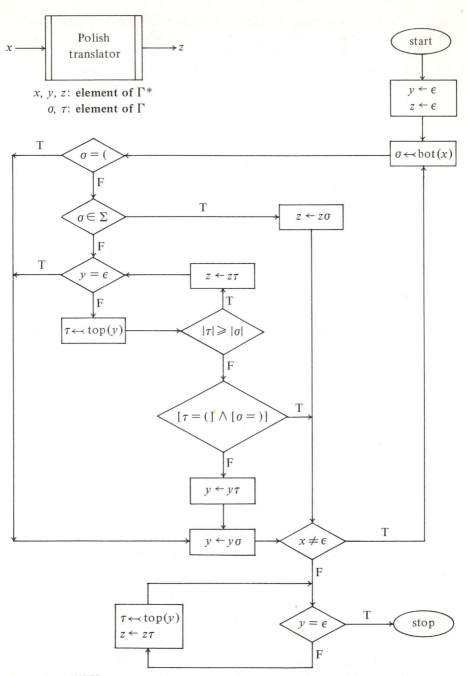

x, y, z: **element of** Γ^*
σ, τ: **element of** Γ

ALGORITHM XXIV

EXAMPLE 6.19   It is perhaps best that we begin with an elementary example, say the infix formula $x = p \wedge (q \vee r)$ discussed above. The reader should follow each detail of the trace given in Table 6.4. Then he or she will not have any problem with the intricacies found in the more complex examples. Note in particular that after the entire word x has been read, as indicated by $x = \epsilon$, the stack is simply emptied into the output z. The resulting Polish expression,

TABLE 6.4

x	y	σ	τ	z
$p \wedge (q \vee r)$	ϵ			ϵ
$\wedge (q \vee r)$		p		p
$(q \vee r)$	$\wedge$	$\wedge$		
$q \vee r)$	$\wedge ($	$($		
$\vee r)$		q		$p\,q$
$r)$	$\wedge$	$\vee$	$($	
	$\wedge (\vee$			
$)$		r		$p\,q\,r$
ϵ	$\wedge ($	$)$	$\vee$	$p\,q\,r \vee$
		$($		
		$\wedge$		$p\,q\,r \vee \wedge$

as given by the output z, is $p\,q\,r\ \vee\wedge$.   △

EXAMPLE 6.20   By way of contrast, the reader should run a trace of the infix formula $p \wedge q \vee r$ to see that it is translated into the Polish expression $p\,q \wedge r \vee$. (See the discussion of Example 6.18.)   △

EXAMPLE 6.21   Because of the similarity to Example 6.19, it is easy to see that the infix expression $p \supset (q \supset p)$ will be translated into the Polish form $p\,q\,p \supset\ \supset$.   △

EXAMPLE 6.22   All that we have to do to use Algorithm XXIV for arithmetic expressions is to establish a hierarchy of arithmetic operations. The following is typical:

$\Gamma \sim \Sigma$	$($	$)$	$=$	$-$	$+$	$/$	$\cdot$	$\uparrow$
priority	0	1	2	3	3	4	4	5

Here $-$ is the binary operation of subtraction, and $\uparrow$ is the binary operation of exponentiation (as in ALGOL). These priorities, along with a left to right precedence for breaking ties, establish conventions for processing expressions that are less than fully parenthesized.

Consider the infix expression $A + B \uparrow C / D \cdot E$ (with no parentheses whatsoever). Proceeding exactly as before, we obtain the Polish form $A B C \uparrow D / E \cdot +$ in the output z. See Table 6.5.

TABLE 6.5

x	y	σ	τ	z
$A + B \uparrow C / D \cdot E$	ϵ			ϵ
$+ B \uparrow C / D \cdot E$		A		A
$B \uparrow C / D \cdot E$	$+$	$+$		
$\uparrow C / D \cdot E$		B		$A\,B$
$C / D \cdot E$	ϵ	$\uparrow$	$+$	
	$+ \uparrow$			
$/ D \cdot E$		C		$A\,B\,C$
$D \cdot E$	$+$	$/$	$\uparrow$	$A\,B\,C \uparrow$
	ϵ		$+$	
	$+ /$			
$\cdot E$		D		$A\,B\,C \uparrow D$
E	$+$	$\cdot$	$/$	$A\,B\,C \uparrow D /$
			$+$	
	$+ \cdot$			
ϵ		E		$A\,B\,C \uparrow D / E$
	$+$		$\cdot$	$A\,B\,C \uparrow D / E \cdot$
	ϵ		$+$	$A\,B\,C \uparrow D / E \cdot +$ $\triangle$

EXERCISES

1 Use the hierarchical system for logical operation symbols to obtain the trees of the following (less than fully parenthesized) infix expressions:
(a) $(\neg\, p \supset r \wedge q) \vee \neg\, (p \supset \neg\, r \wedge q)$
(b) $\neg\, p \vee q \wedge (p \vee \neg\, q)$
(c) $p \vee q \wedge r \equiv \neg\, q$

2 Use the hierarchical system for arithmetic operation symbols to obtain the trees of the following (less than fully parenthesized) infix expressions (if $-$ is binary "subtraction"):
(a) $A + B \uparrow C \cdot D$ (b) $A \cdot C - A \cdot B / D$
(c) $A \cdot D \uparrow (A + B) - C$ (d) $(A + C) \cdot (B - E / D \uparrow A)$

3 Use Algorithm XXIV to translate the infix expressions of Exercise 1 to Polish.

4 Repeat Exercise 3 for the infix expressions of Exercise 2.

5 Repeat Exercise 3 for the following infix expressions:
(a) $p \wedge \neg\, (q \vee \neg\, r)$ (b) $(\neg\, r \vee q) \supset \neg\, (\neg\, p \vee q)$
(c) $\neg\, (p \wedge q \vee r) \equiv \neg\, q \vee r$ (d) $A + B \uparrow C / (D + E)$
(e) $A + B \cdot C + D \uparrow E$ (f) $A \cdot (B \uparrow C + D \uparrow E)$

6.5   SEMANTICS AND TAUTOLOGY

So far our brief excursion into logic has touched only on the syntax, the proper formation of formulas in propositional calculus. In order to associate meanings to the formulas, we have to introduce the idea of an interpretation, much as was done in Section 4.7 for the Boolean expressions. Despite the formal similarity of the well-formed formulas to the Boolean expressions, the interpretation given here will not coincide with the former. This time the reader may well keep in mind as a frame of reference the logical expressions for the conditional statements in a computer programming language (or in the algebraic flowchart language, for that matter). Given the truth or falsity of the appropriate propositional variables, one seeks only that yes or no answer which determines the proper branch of the control transfer.

Once again we attach no particular significance to the propositional variables. They merely constitute an alphabet of symbols $\Sigma = \{p, q, r, \ldots\}$, but now we admit our two new binary connectives into the collection

$$\Gamma \sim \Sigma = \{\neg, \vee, \wedge, \supset, \equiv\}$$

and the new operations are understood only to be abbreviations:

$$\alpha \beta \supset \quad \text{for} \quad \alpha \neg \beta \vee$$
$$\alpha \beta \equiv \quad \text{for} \quad \alpha \neg \beta \neg \wedge \alpha \beta \wedge \vee$$

(in Polish notation) for wff's α, β. At this stage a description of our interpretation becomes essential. Otherwise, the motivation for introducing these abbreviations would remain a mystery.

As before, we let $\lambda \subseteq \Gamma^*$ denote the language of wff's,

$$\lambda = \{\alpha \in \Gamma^*: \alpha \text{ is a wff}\}$$

Then we will say that an *interpretation* of λ is a function

$$I: \lambda \to B$$

into the Boolean algebra of Example 4.37, with the properties

$$I(\alpha \beta \vee) = I(\alpha) + I(\beta) \qquad I(\alpha \beta \wedge) = I(\alpha) \cdot I(\beta)$$
$$I(\alpha \neg) = I(\alpha)'$$

Note the similarity to the properties required of a homomorphism, as though λ (along with B) was a Boolean algebra.

Before giving examples, let us present one more important definition. Let α be in λ. Then we say that α is a *tautology* (written $\models \alpha$) if $I(\alpha) = 1$ for every interpretation I. We will soon see that the interpretations are "rows" of a gigantic truth table—that for all the wff's—over the propositional variables. Then a tautology is simply a wff that is true in every interpretation, that is, a wff whose truth table is a column of ones.

EXAMPLE 6.23   Over propositional variables p, q, r consider the wff's:

$$\alpha = p \neg p \vee$$
$$\beta = p \neg q \vee r \wedge$$

A particular interpretation $I: \lambda \rightarrow B$ might assign

$$I(p) = 0 \qquad I(q) = 1 \qquad I(r) = 0$$

in which case we must have

$$I(p \neg) = I(p)' = 0' = 1$$

The reader should begin to see that our definition of interpretations is consistent: "If p is false, then $p \neg$ is true."

Continuing, we compute

$$I(\alpha) = I(p \neg p \vee) = I(p \neg) + I(p) = 1 + 0 = 1$$

$$\begin{aligned} I(\beta) = I(p \neg q \vee r \wedge) &= I(p \neg q \vee) \cdot I(r) \\ &= (I(p \neg) + I(q)) \cdot I(r) \\ &= (1 + 1) \cdot 0 = 0 \end{aligned}$$

But this is only one interpretation! It should be clear that there are $2^3 = 8$ interpretations in all (for this example anyway), as shown in Table 6.6. The one just described appears in row three.

TABLE 6.6

$I(p)$	$I(q)$	$I(r)$	$I(p \neg)$	$I(p \neg q \vee)$	$I(\alpha)$	$I(\beta)$
0	0	0	1	1	1	0
0	0	1	1	1	1	1
0	1	0	1	1	1	0
0	1	1	1	1	1	1
1	0	0	0	0	1	0
1	0	1	0	0	1	0
1	1	0	0	1	1	0
1	1	1	0	1	1	1

This table of the eight interpretations proceeds horizontally ad infinitum, because there are infinitely many wff's, but the main thing to see is the consistency with the desired usage of negation, disjunction, and conjunction as "not," "or," and "and," respectively. Most important, we should note that $\alpha = p \neg p \vee$ is a tautology, as in "Denver wins or Denver does not win." The formula α is true in every interpretation. Accordingly, we may write

$$\models p \neg p \vee$$

or $\models \alpha$. The other formula (β) is more ordinary. Its "truth" depends on the interpretation, that is, on the truth or falsity of the propositional variables.

More than anything else, we see that the concept of interpretations here is simply the idea of truth tables all over again, but seen from a slightly different point of view.   △

EXAMPLE 6.24   We now show that if there are no interpretations I for which

$$I(\alpha) = 1 \quad \text{and} \quad I(\beta) = 0$$

then $\models \alpha \neg \beta \vee$. We compute

$$I(\alpha \neg \beta \vee) = I(\alpha \neg) + I(\beta)$$
$$= I(\alpha)' + I(\beta)$$

and ask the question, "How could this be zero?" Only if

$$I(\alpha)' = I(\beta) = 0$$

that is, if $I(\alpha) = 1$ and $I(\beta) = 0$. But this never happens, according to our assumption. Thus $I(\alpha \neg \beta \vee) = 1$ for every interpretation I, so that $\models \alpha \neg \beta \vee$ [if there are no interpretations I such that $I(\alpha) = 1$ and $I(\beta) = 0$].

The converse of our assertion is also true, as the student may verify. More important, however, is the fact that $\alpha \neg \beta \vee$ is the formula which we decided to abbreviate as $\alpha \beta \supset$. Also it will follow from the calculations of this example that $\alpha \beta \supset$ is the syntactic equivalent of *semantic implication* (the implication which we have informally denoted by $\Rightarrow$ throughout the text). All of this will be taken up in Theorem 6.3, and the resulting circle of ideas should provide a better understanding of the implications we have used ($\Rightarrow$ or "if . . . then . . .") constantly since Section 0.9.   △

EXAMPLE 6.25   Suppose that $I(\alpha) = I(\beta)$ for every interpretation I. Then, in analogy to Example 6.24, we will show that the formula $\alpha \neg \beta \neg \wedge \alpha \beta \wedge \vee$ is a tautology. Recall that this is the formula to be abbreviated as $\alpha \beta \equiv$. Again making use of the defining properties of interpretations, we compute

$$I(\alpha \neg \beta \neg \wedge \alpha \beta \wedge \vee) = I(\alpha \neg \beta \neg \wedge) + I(\alpha \beta \wedge)$$
$$= I(\alpha \neg) \cdot I(\beta \neg) + I(\alpha) \cdot I(\beta)$$
$$= I(\alpha)' \cdot I(\beta)' + I(\alpha) \cdot I(\beta)$$
$$= \begin{cases} 1 \cdot 1 + 0 \cdot 0 & \text{if} \quad I(\alpha) = I(\beta) = 0 \\ 0 \cdot 0 + 1 \cdot 1 & \text{if} \quad I(\alpha) = I(\beta) = 1 \end{cases}$$
$$= \begin{cases} 1 + 0 \\ 0 + 1 \end{cases}$$

which is 1 in either case. Thus with the given assumption, we have

$$\models \alpha \neg \beta \neg \wedge \alpha \beta \wedge \vee$$

as claimed.

Conversely, if

$$I(\alpha \urcorner \beta \urcorner \wedge \alpha \beta \wedge \vee) = 1$$

for every interpretation I, then we see from the above calculation that we can never have

$$I(\alpha)' \cdot I(\beta)' = I(\alpha) \cdot I(\beta) = 0$$

We must have either

$$I(\alpha)' \cdot I(\beta)' = 1 \quad \text{or} \quad I(\alpha) \cdot I(\beta) = 1$$

and the reader should see that this yields $I(\alpha) = I(\beta)$.

Again, in Theorem 6.2 we will find that $\alpha \beta \equiv$ is the syntactic embodiment of *semantic equivalence*. This is the notion informally expressed as $\Leftrightarrow$ or iff throughout the text. $\triangle$

In order to provide a better understanding of the ideas being approached through the last two examples, we must first introduce a few "semantic" definitions. Let α and β be wff's. We will say that β is a *semantic consequence* of α (written $\alpha \models \beta$) if for every interpretation I we have

$$I(\alpha) = 1 \Rightarrow I(\beta) = 1$$

Here for the first time we are confronted with a rather intricate situation, one which pervades the subject of mathematical logic. Our propositional calculus is a highly symbolic and formal language. At times we make statements within the language as well as statements about the language. Then we are using two different levels of discourse, the *language* and the *metalanguage*. The metalanguage by comparison is informal, as in our use of the implication symbol ($\Rightarrow$) above. Note that we do not write $\alpha \Rightarrow \beta$; instead, we use a new formal symbol and write $\alpha \models \beta$, even though we are primarily interested in clarifying the use of "logical implication," as discussed informally in Section 0.9. In order to preserve this distinction, we will call the relation defined here on wff's *semantic implication*. You will see, on paraphrasing the definition as "whenever α is true, then β is true," that we are on the right track.

Consistent with the preceding definition, we say that the wff's α and β are *semantically equivalent* (and we write $\alpha \dashv\vdash \beta$) if for every interpretation I, we have

$$I(\alpha) = 1 \Leftrightarrow I(\beta) = 1$$

Again we have formally injected the notion of logical equivalence into our language, using the metalanguage in order to help us phrase the definition.

With these intricacies out of the way, we can now proceed to develop theorems *about the* language or *theory* that is being developed: propositional calculus. Here again, however, we must make a careful distinction between these theorems and those of the next section, which are theorems *within the theory*.

EXAMPLE 6.26   Consider the wff's α and β, given as follows.

$$\alpha = p \qquad \beta = p \, q \, \vee$$

Then $\alpha \models \beta$, as we can see by looking at all the interpretations:

$I(p)$	$I(q)$	$I(\alpha)$	$I(\beta) = I(p \, q \, \vee) = I(p) + I(q)$
0	0	0	0
0	1	0	1
1	0	1	1
1	1	1	1

We have $I(\alpha) = 1 \Rightarrow I(\beta) = 1$, as required by the definition of $\alpha \models \beta$. $\triangle$

EXAMPLE 6.27   This time suppose that we are given

$$\alpha = p \qquad \beta = q \, p \, \supset$$

Again we can look at all the interpretations to see if $\alpha \models \beta$:

$I(p)$	$I(q)$	$I(\alpha)$	$I(\beta) = I(q \, p \, \supset) = I(q \, \neg \, p \, \vee) = I(q)' + I(p)$
0	0	0	1
0	1	0	0
1	0	1	1
1	1	1	1

We can write $p \models q \, p \, \supset$, since $I(\alpha) = 1 \Rightarrow I(\beta) = 1$. $\triangle$

EXAMPLE 6.28   An upcoming theorem (Theorem 6.3) claims that $\alpha \models \beta$ if and only if $\models \alpha \, \beta \, \supset$. Let us illustrate this principle by determining that $p \, q \, p \, \supset \, \supset$ is a tautology. (See Example 6.27.) In computations with the symbol ($\supset$) it is helpful to observe, in connection with Example 6.24, that $\alpha \, \beta \, \supset$ is true except when α is true and β is false. In this way we can avoid the substitution of $\alpha \, \neg \, \beta \, \vee$ for $\alpha \, \beta \, \supset$. We have

$I(p)$	$I(q)$	$I(q \, p \, \supset)$	$I(p \, q \, p \, \supset \, \supset)$
0	0	1	1
0	1	0	1
1	0	1	1
1	1	1	1

$\triangle$

Theorem 6.2

$$\alpha \dashv\vdash \beta \quad \text{iff} \quad \vDash \alpha \beta \equiv$$

Proof First, suppose that $\vDash \alpha \beta \equiv$. Then according to Example 6.25, we have

$$I(\alpha) = I(\beta) = 0 \quad \text{or} \quad I(\alpha) = I(\beta) = 1$$

for every interpretation I, that is, $\alpha \dashv\vdash \beta$. Conversely, if α and β are semantically equivalent, then $I(\alpha) = I(\beta)$ for every I. Again it follows from Example 6.25 that $\vDash \alpha \beta \equiv$. $\qquad\square$

Theorem 6.3

$$\alpha \vDash \beta \quad \text{iff} \quad \vDash \alpha \beta \supset$$

Proof If $\vDash \alpha \beta \supset$, then because of Example 6.24, we never have $I(\alpha) = 1$ and $I(\beta) = 0$. It follows that

$$I(\alpha) = 1 \Rightarrow I(\beta) = 1$$

that is, $\alpha \vDash \beta$. The converse is equally clear. $\qquad\square$

Corollary

$$\alpha_1, \alpha_2, \ldots, \alpha_n \vDash \beta \quad \text{iff} \quad \vDash \alpha_1 \alpha_2 \wedge \alpha_3 \wedge \cdots \alpha_n \wedge \beta \supset \qquad\square$$

The last result is not strictly speaking a corollary, but it does follow from the reasoning of the proof of Theorem 6.3. The reader can provide the details. Of course, we haven't defined what we mean by the left-hand side, that β is a *semantic consequence* of the wff's $\alpha_1, \alpha_2, \ldots, \alpha_n$. Naturally we would mean, by extending the definition of $\alpha \vDash \beta$, that for every interpretation I,

$$I(\alpha_1) = I(\alpha_2) = \cdots = I(\alpha_n) = 1 \Rightarrow I(\beta) = 1$$

With the corollary, we have established the existence of a well-formed formula $\alpha_1 \alpha_2 \wedge \alpha_3 \wedge \cdots \alpha_n \wedge \beta \supset$ that can be used to decide (depending on whether or not it is a tautology) whether such a consequence of meaning takes place. This result may then be used to analyze complicated arguments, insurance policies, legal documents, and the like. It is also used to establish the everyday "rule of inference" called *modus ponens* (Theorem 6.4).

EXAMPLE 6.29 (Lewis Carroll) Below you will find four statements involving:

$$p = \text{terriers}$$
$$q = \text{zodiac wanderers}$$
$$r = \text{comets}$$
$$s = \text{curly tailed objects}$$

(α) No terriers wander among the signs of the zodiac.
(β) Nothing that does not wander among the signs of the zodiac is a comet.
(γ) Nothing but a terrier has a curly tail.

(Consequence?)

(δ) No comet has a curly tail.

Translating into the formal language of propositional calculus, we have:

$$\alpha = q\,p\,\neg\,\supset$$
$$\beta = q\,\neg\,r\,\neg\,\supset$$
$$\gamma = s\,p\,\supset$$
$$\overline{}$$
$$\delta = s\,r\,\neg\,\supset$$

According to the corollary, δ is a semantic consequence of α, β, γ if and only if $\models \alpha\,\beta \wedge \gamma \wedge \delta \supset$. So the answer depends on whether

$$\tau = q\,p\,\neg\,\supset\,q\,\neg\,r\,\neg\,\supset\,\wedge\,s\,p\,\supset\,\wedge\,s\,r\,\neg\,\supset\,\supset$$

is a tautology, and this can be computed! We summarize the computation in Table 6.7 (omitting the *I*'s for clarity).

TABLE 6.7

p	q	r	s	α	β	γ	$\alpha\,\beta \wedge \gamma \wedge$	δ	τ
0	0	0	0	1	1	1	1	1	1
0	0	0	1	1	1	0	0	1	1
0	0	1	0	1	0	1	0	1	1
0	0	1	1	1	0	0	0	0	1
0	1	0	0	1	1	1	1	1	1
0	1	0	1	1	1	0	0	1	1
0	1	1	0	1	1	1	1	1	1
0	1	1	1	1	1	0	0	0	1
1	0	0	0	1	1	1	1	1	1
1	0	0	1	1	1	1	1	1	1
1	0	1	0	1	0	1	0	1	1
1	0	1	1	1	0	1	0	0	1
1	1	0	0	0	1	1	0	1	1
1	1	0	1	0	1	1	0	1	1
1	1	1	0	0	1	1	0	1	1
1	1	1	1	0	1	1	0	0	1

Since the consequence is indeed valid, the example shows the relevance of terriers to astronomy (or demonstrates once more the sense of humor of Lewis Carroll). △

Theorem 6.4 (Modus Ponens)

For wff's α and β,

$$\alpha, \alpha \beta \supset \models \beta$$

Proof We establish that $\alpha \alpha \beta \supset \wedge \beta \supset$ is a tautology. Then the result follows from the corollary. Over all interpretations I, we tabulate as follows:

$I(\alpha)$	$I(\beta)$	$I(\alpha \beta \supset)$	$I(\alpha \alpha \beta \supset \wedge)$	$I(\alpha \alpha \beta \supset \wedge \beta \supset)$
0	0	1	0	1
0	1	1	0	1
1	0	0	0	1
1	1	1	1	1

Since the formula $\alpha \alpha \beta \supset \wedge \beta \supset$ is a tautology, the proof is complete. $\qquad \square$

EXERCISES

1 Verify the converse of the assertion in Example 6.24.

2 For the interpretations in which $I(p) = I(q) = 0$ and $I(r) = 1$, compute $I(\alpha)$ given that:
  (a) $\alpha = p\,q \neg \vee r \wedge$
  (b) $\alpha = p\,q\,r \vee \wedge q \neg p \neg \vee \vee$
  (c) $\alpha = p \neg r \vee q \neg \wedge$
  (d) $\alpha = q \neg p \neg \wedge r \neg \vee$

3 Repeat Exercise 2 for the interpretation in which $I(p) = I(r) = 1$ and $I(q) = 0$.

4 Which of the following wff's are tautologies?
  (a) $p\,q \wedge p \neg r \vee \supset$
  (b) $p\,q \neg \wedge p \neg \neg q \equiv \supset$
  (c) $p\,q \neg \supset r\,q \neg p \vee \supset \equiv$
  (d) $p\,q \equiv r \neg \wedge p \neg q\,r \vee \wedge \neg \supset$
  (e) $p\,r \vee p \supset q\,q\,r \vee \supset \wedge$
  (f) $q\,r \supset p\,q \supset p\,r \supset \supset \supset$

5 For which of the following pairs α, β is it the case that $\alpha \models \beta$? Show directly and by appeal to Theorem 6.3.
  (a) $\alpha = p\,q \neg \supset r\,q \neg p \vee \supset \equiv$
  $\beta = p\,q \neg p \neg \neg q \equiv \supset$
  (b) $\alpha = p\,q \neg \supset \neg$
  $\beta = p\,q \neg \wedge p \neg \neg q \equiv \supset$
  (c) $\alpha = p\,q \wedge p \neg r \neg \wedge \vee$
  $\beta = r\,q \neg p \vee \supset$
  (d) $\alpha = p\,q \wedge$
  $\beta = p\,q \vee$

6 For which of the following pairs α, β is it the case that $\alpha \dashv\vdash \beta$? Show directly and by appeal to Theorem 6.2.
  (a) $\alpha = p\,q \neg \wedge p \neg \neg q \equiv \supset$
  $\beta = p \neg q \vee$
  (b) $\alpha = r\,q \neg p \vee \supset$
  $\beta = p\,r \neg \vee$
  (c) $\alpha = p\,q \neg \supset r\,q \neg p \vee \supset \equiv$
  $\beta = q \neg p \neg r \neg \wedge \vee$
  (d) $\alpha = p\,q \vee \neg$
  $\beta = p \neg q \neg \wedge$

7 Use Theorem 6.3 to show that the following are tautologies:

(a) $p\,q \land p\,q \lor \supset$　　(b) $p\,q\,r \land \land p\,q \land \supset$

(c) $p\,q\,q \neg \lor \supset$　　(d) $q\,r \supset p\,q \supset p\,r \supset \supset \supset$

8 Complete the proof of Theorem 6.3.

9 Prove the corollary.

10 (Lewis Carroll) Derive the symbolized conclusion from the symbolized premises:

(α) Nobody who really appreciates Beethoven fails to keep silence while the Moonlight Sonata is being played.

(β) Guinea pigs are hopelessly ignorant of music.

(γ) No one who is hopelessly ignorant of music ever keeps silence while the Moonlight Sonata is being played.

Conclusion?

(δ) Guinea pigs never really appreciate Beethoven!

6.6 AXIOMATICS

Up to this point our discussion of propositional calculus has centered around the two languages:

$$\lambda = \{\alpha : \alpha \text{ is a wff}\} \subseteq \Gamma^*$$

$$\lambda_T = \{\alpha : \alpha \text{ is a tautology}\}$$

the second being a sublanguage of the first. We have a procedure (Algorithm XXIII) for deciding whether or not a given string of symbols $\alpha \in \Gamma^*$ is a wff. There is an obvious algorithm for accepting the tautologies—compute the truth table of the formula α to see whether it is a column of ones!

For the time being, however, we shall ignore this concept of truth and approach the subject from a different angle. We will present a method for generating certain strings of symbols; that is, we will use the axiomatic method. We begin with a selection of wff's called *axioms* and prescribe certain *rules of inference* for generating new strings out of the axioms, much as in high school geometry. The collection of strings or words so generated will constitute the *axiomatic language* $\lambda_A \subseteq \Gamma^*$. Of course, our axioms will be true, that is, one can check that they are indeed tautologies. Therefore the plan we are describing is not unlike that discussed in Section 0.9 for developing meaningful mathematical theories.

As usual, we begin with an alphabet $\Sigma = \{p, q, r, \ldots\}$ of propositional variables. We find that we can get along with only three axioms. Thus we assume that the following are in the language λ_A for all variables p, q, r in Σ:

(A1) $p\,q\,p \supset \supset$

(A2) $p\,q\,r \supset \supset p\,q \supset p\,r \supset \supset \supset$

(A3) $p \urcorner q \urcorner \supset p \urcorner q \supset p \supset \supset$

Also we take only two rules of inference:

(R1) *Modus ponens*: α and $\alpha \beta \supset$ in $\lambda_A \Rightarrow \beta \in \lambda_A$
(R2) *Substitution*: $\alpha(p) \in \lambda_A$ and $\beta \in \lambda \Rightarrow \alpha(\beta) \in \lambda_A$

Quite obviously the first rule is motivated by Theorem 6.4. According to the second rule, we are permitted to substitute any wff β for a propositional variable p at *each* of its occurrences in a string α already generated.

The method for generating the formulas or strings of the language λ_A will now be described in a bit more detail. We say that the wff α is a *theorem* in propositional calculus (and we write either $\alpha \in \lambda_A$ or $\vdash \alpha$) if there is a finite sequence of wff's

$$\alpha_0 \rightarrow \alpha_1 \rightarrow \alpha_2 \rightarrow \cdots \rightarrow \alpha_n = \alpha$$

such that α_0 is an axiom, and each succeeding α_i either is an axiom or follows from the preceding α_j by one of the rules of inference. The entire sequence $\alpha_0, \alpha_1, \alpha_2, \ldots, \alpha_n$ is called a *proof* of (the theorem) α.

EXAMPLE 6.30   Let us give a proof for $\vdash p\, p \supset$. Since it is a tautology (as can be easily checked), we suspect that it might be a theorem.

0. $p\, q\, r \supset \supset p\, q \supset p\, r \supset \supset \supset$ by A2
1. $p\, p\, p \supset r \supset \supset p\, p\, p \supset \supset p\, r \supset \supset \supset$ by R2, (0)
   (We substituted the wff $p\, p \supset$ for q.)
2. $p\, p\, p \supset p \supset \supset p\, p\, p \supset \supset p\, p \supset \supset \supset$ by R2, (1)
   (p substituted for r)
3. $p\, q\, p \supset \supset$ by A1
4. $p\, p\, p \supset p \supset \supset$ by R2, (3)
5. $p\, p\, p \supset \supset p\, p \supset \supset$ by R1, (2), (4)
6. $p\, p\, p \supset \supset$ by R2, (3)
7. $p\, p \supset$ by R1, (5), (6)

Note the uses of modus ponens. In line 5 we compare the formulas of line 2 and line 4:

in order to derive $p\, p\, p \supset \supset p\, p \supset \supset$. Similarly in line 7.

Altogether, we have constructed a proof for $\vdash p\, p \supset$. Note that it follows by substitution that $\vdash \alpha\, \alpha \supset$ for every wff α. $\triangle$

EXAMPLE 6.31   Again the reader can check that $\models p \urcorner p \supset p \supset$, so let us see if we can construct a proof. It seems reasonable to begin with axiom A3, since it is the only one involving negation.

0. $p \neg q \neg \supset p \neg q \supset p \supset \supset$ by A3
1. $p \neg p \neg \supset p \neg p \supset p \supset \supset$ by R2, (0)
2. $p\, p \supset$ by Example 6.30
3. $p \neg p \neg \supset$ by R2, (2)
4. $p \neg p \supset p \supset$ by R1, (1), (3)

Here we have given only an abbreviation of a proof. One would have to substitute in line 2 the proof of Example 6.30 in order to obtain a proof in strict accordance with our definition. But in mathematics we are always quoting previous results in order to establish a proof of immediate concern. Otherwise, if we always had to argue from the first principles, proofs would grow to enormous lengths.   △

We have no intention, no hopes or desires, to make the student become adept at proving theorems in propositional calculus. The idea here is simply to impart an appreciation for the underlying framework and the style of the arguments involved, for several purposes. For one thing, the arguments here are similar to those found in the areas of automatic theorem-proving, computer chess-playing, etc., and in one way or another, these techniques play a role in many computer investigations. At the same time this axiomatic approach gives a useful introduction to the flavor of mathematical language theory, some elements of which we will discuss briefly in the remainder of this chapter. Finally, we should mention the importance of having some familiarity with the axiomatic framework in order to understand the various "completeness" (and incompleteness) results concerning logical mathematical theories. To give an example, one result on the positive side runs as follows:

Theorem 6.5   (Completeness of propositional calculus)

$$\lambda_A = \lambda_T$$

Proof   Before discussing the proof, let us look a little closer at the assertion. The claim is that

$$\vdash \alpha \quad \text{iff} \quad \models \alpha$$

or in words, "Every theorem is a tautology, and conversely, every tautology is a theorem (has a proof)." The converse statement in particular is a remarkable and, as we will soon see, an unusual result.

We are really prepared to consider only one-half of the argument: $\lambda_A \subseteq \lambda_T$ ($\vdash \alpha$ implies $\models \alpha$). First, every axiom is a tautology. We checked this statement in Example 6.28 for axiom A1. The truth of the other two axioms can be verified just as easily. Now, any application of modus ponens will preserve tautology because of Theorem 6.4. The same is true of substitution, as is easily seen. Since the axioms are tautologies and the rules of inference preserve tautologies, it follows that every line in a proof $\alpha_0, \alpha_1, \ldots, \alpha_n = \alpha$ is a tautology. So the theorem α itself is a tautology.

The converse argument, that every "true" statement has a proof, requires for its proof more background in logic than we have developed. So it will not be given here. The interested reader may consult one of the references given at the end of the chapter. And with that, we will be content if the student has begun to grasp the meaning and significance of it all. □

Such completeness results fit nicely into the philosophical framework contemplated by the great post-Renaissance mathematician, Gottfried Leibniz. What we now refer to as Leibniz' program, the dream of a universal combinatorial method of symbolic reasoning, was later interpreted by Whitehead and Russell* as a plan for providing a logical basis for all of mathematics. Since their work included the development of set theory, and even parts of geometry, the results were considered most encouraging, and at the late date of the early 1900s, Leibniz' dream was still very much alive. This can be seen in the remarks of David Hilbert, the foremost mathematician of that period. In his 1900 address,† especially, were considerable references to a program of developing a logical machinery for "generating" all of mathematics, and the somewhat surprising result of Theorem 6.5 would certainly be consistent with such a plan.

Even more astonishing, however, is the negative result obtained by Kurt Gödel‡ in 1931. He was able to show that any axiomatic theory that is sufficiently general as to include the elementary arithmetic of the integers (number theory), is incomplete! Presumably, such a theory would include axioms along the lines of the Peano postulates for the integers (see the exercises of Section 1.1). But regardless of the axiomatization, there must exist "true" statements that have no proof in the theory! Surely this is an amazing result, and coming at the time when Hilbert had in mind the development of a logical system in which all of mathematics could be generated, it was little short of devastating.

EXERCISES

1 Verify that each of the three axioms (A1), (A2), (A3) is a tautology.

*2 Give proofs for each of the following:

(a) $\vdash p \neg p \neg \neg \supset p \supset$ (b) $\vdash p \neg p q \supset \supset$

(c) $\vdash q \neg p \neg \supset p q \supset \supset$ (d) $\vdash p q \neg p q \supset \neg \supset \supset$

(e) $\vdash q r \supset p q \supset p r \supset \supset \supset$ (f) $\vdash p p \neg \neg \supset$

3 Show that the wff's in Exercise 2 are tautologies.

*4 Complete the proof of Theorem 6.5. [*Hint*: You may have to consult another source here, say Reference 6.2.]

* A. N. Whitehead and B. Russell, *Principia Mathematica*, 2nd ed. 3 vols., Cambridge University Press, Cambridge, 1925–1927.
† D. Hilbert, "Mathematical Problems," *Bulletin Am. Math. Soc.*, **8**, 437–479 (1901–1902).
‡ K. Gödel, "Über Formal Unentscheidbare Sätze der Principia Mathematica und Berwandter Systeme, I," *Monats. für Math. Phys.*, **38**, 173–198 (1931).

5 Give the complete proof for the theorem in Example 6.31 without appealing to Example 6.30.

6 Axiom (A3) is a formalization of the logical notion called the *contrapositive* (Section 0.9). Explain.

7 Prove that $\alpha \models \beta$ iff $\vdash \alpha\, \beta \supset$.

*8 Show that the rules of inference preserve tautologies.

6.7   TURING MACHINES

We have not yet acknowledged the full strength of Gödel's results, which will indeed sound more and more puzzling. In any axiomatic theory for the arithmetic of the integers, Gödel showed, there are statements which are neither provable nor disprovable! They are *undecidable*. Stated in another way, there are mathematical problems which are unsolvable within an axiomatic system, however that system of arithmetic may be formulated. Unsolvable problems will always remain. As so often happens, however, these discouraging results have led to an extremely fruitful and long-neglected area of inquiry, namely, the nature of effective methods of problem-solving. When do we have an effective method, an algorithm for solving a problem?

Throughout this text we claim to have presented algorithms for the solution of various problems, in the theory of graphs, groups, lattices, or wherever. In a sense, we have appealed to the good will of the reader in asking that he or she acknowledge the "effectiveness" of the methods employed. The question must naturally arise as to whether there are natural limits of "computability" or "decidability." In our preliminary discussions of effective decidability and computability in Section 1.4, we mentioned that these questions were first investigated by mathematicians of the 1930s, anticipating the development of the modern digital computers by a quarter of a century. Here we will be particularly interested in the work of Alan Turing,* and in the elementary mathematical machine model which now bears his name.

In his 1936 paper Turing presents and then goes on to defend the proposition, now called *Turing's thesis*: "Any process which could naturally be called an 'effective' procedure, can be carried out by some Turing machine." Considering the simplicity of his machine model, the thesis may seem overly optimistic. Nevertheless, the proposition has withstood the test of time. No one has come forth with any evidence to the contrary, and it seems safe to guess that there will never be such evidence.

There are any number of arguments in favor of Turing's position, but few are more compelling than that advanced by Turing himself. In order to give some idea of the reasoning, we shall quote freely from his 1936 paper in the

---

* A. Turing, "On Computable Numbers, with an Application to the Entscheidungsproblem," *Proc. London Math. Soc.*, **42**, 230–265 (1936).

following summary. The reader should bear in mind that when he uses the term "machine," he is speaking of a Turing machine. The "computer" he refers to is the human computer that he intends to replace by one of his machines.

Computing is normally done by writing certain symbols on paper. We may suppose this paper is divided into squares like a child's arithmetic book. In elementary arithmetic the two-dimensional character of the paper is sometimes used. But such a use is always avoidable, and I think that it will be agreed that the two-dimensional character of paper is no essential of computation. I assume then that the computation is carried out on one-dimensional paper, i.e., on a tape divided into squares. I shall also suppose that the number of symbols which may be printed is finite. If we were to allow an infinity of symbols, then there would be symbols differing by an arbitrarily small extent. The effect of this restriction of the number of symbols is not very serious. It is always possible to use sequences of symbols in the place of single symbols. Thus an Arabic numeral such as 17 or 999999999999999 is normally treated as a single symbol. Similarly in any European language, words are treated as single symbols (Chinese, however, attempts to have an enumerable infinity of symbols). The differences from our point of view between the single and compound symbols is that the compound symbols, if they are too lengthy, cannot be observed at a glance. This is in accordance with experience. We cannot tell at a glance whether 9999999999999999 and 999999999999999 are the same.

The behaviour of the computer at any moment is determined by the symbols which he is observing, and his "state of mind" at that moment. We may suppose that there is a bound B to the number of symbols or squares which the computer can observe at one moment. If he wishes to observe more, he must make successive observations. We will also suppose that the number of states of mind which need to be taken into account is finite. The reasons for this are of the same character as those which restrict the number of symbols. If we admitted an infinity of states of mind, some of them will be "arbitrarily close" and will be confused. Again, the restriction is not one which seriously affects computation, since the use of more complicated states of mind can be avoided by writing more symbols on the tape.

Let us imagine the operations performed by the computer to be split up into "simple operations" which are so elementary that it is not easy to imagine them further subdivided. Every such operation consists of some change of the physical system consisting of the computer and his tape. We know the state of the system if we know the sequence of symbols on the tape, which of these are observed by the computer (possibly with a special order), and the state of mind of the computer. We may suppose that in a simple operation not more than one symbol is altered. Any other changes can be split up into simple changes of this kind. The situation in regard to the squares whose symbols may be altered in this way is the same as in regard to the observed squares. We may, therefore, without loss of generality, assume that the squares whose symbols are changed are always "observed" squares.

Besides these changes of symbols, the simple operations must include changes of distribution of observed squares. The new observed squares must be immediately recognisable by the computer. I think it reasonable to suppose that they can only be squares whose distance from the closest of the immediately previously observed squares does not exceed a certain fixed amount. Let us say that each of the new observed squares is within L squares of an immediately previously observed square.

The simple operations must therefore include:

(a) Changes of the symbol on one of the observed squares.
(b) Changes of one of the squares observed to another square within L squares of one of the previously observed squares.

It may be that some of these changes necessarily involve a change of state of mind. The operation actually performed is determined, as has been suggested, by the state of mind of the computer and the observed symbols. In particular, they determine the state of mind of the computer after the operation.

We may now construct a machine to do the work of this computer.*

The machine model we are about to define differs in no essential way from that first described by Turing. It happens that there are several equivalent formulations, and each author chooses that particular variant which is best suited to his purposes. With this in mind, we agree that a *Turing machine* $Z = (S, \Sigma, M)$ is a system in which

(1) S is a finite set of *states*, with distinguished element $0 \in S$,
(2) Σ is an alphabet,
(3) M is the *transition* (partial) *function*

$$M: S \times \Sigma_\square \to S \times \Sigma_\square \times \triangle$$

S is the set of "states of mind" of which Turing speaks, Σ is his finite collection of symbols, and M describes the "simple operations" permitted. In this connection, $\Sigma_\square = \Sigma \cup \{\square\}$ where $\square \notin \Sigma$ is a blank symbol, and $\triangle = \{-1, 0, 1\}$. The latter allows for the motion of a "read-write head," left by one square, no motion, or right by one square, respectively, along the tape. Thus in terms of Turing's discussion, we are taking $L = 1$ and assuming that there is only one "observed square."

The transition function requires further explanation. First of all, M is permitted to be only a partial function; that is, there may be certain pairs (q, σ) at which M is undefined. And our understanding is that the machine Z will *halt* when it is in state q with its read-write head scanning the symbol σ, where $M(q, \sigma)$ is undefined. Otherwise, when we write $M(q, \sigma) = (q', \sigma', \partial)$ it is understood that the machine changes the observed tape symbol from σ to σ', and its state from q to q', and moves along the tape according to the above interpretation of $\partial \in \triangle$. All of this is consistent with Turing's idea of a "simple operation."

When Z halts, we will generally have the end of a computation. A computation will begin in the *initial state* 0 with the read-write head scanning the leftmost symbol of a word $x = \sigma_{i_1} \sigma_{i_2} \cdots \sigma_{i_k}$ written on the tape. Pictorially, we have the initial description shown in Figure 6.8 [later to be identified as $\alpha_0(x)$], and the computation proceeds as given by the transition function M, unless

---

* Turing, "On Computable Numbers," by permission of the Oxford University Press, Oxford.

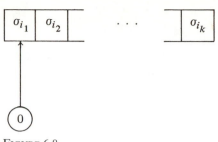

FIGURE 6.8

or until the machine halts. With respect to this partial function M, the important thing to see is that a Turing machine is *finitely specified*, in that a table with $|S|$ rows and $|\Sigma| + 1$ columns will suffice for a description of M.

EXAMPLE 6.32   Suppose that we let $Z = (S, \Sigma, M)$ with

$$S = \{0, 1, 2, 3\} \qquad \Sigma = \{0, 1\}$$

and the transition function tabulated below.

	0	1	□ ·
0	0, 0, 1	1, 1, 1	0, □, 0
1	1, 0, 1	2, 1, 1	1, □, 0
2	2, 0, 1	3, 1, 1	
3	3, 0, 1	0, 1, 1	3, □, 0

In the first two columns of the table we can see that Z is designed to perform exactly as a finite state machine (Section 3.8). As it reads a symbol, it will reprint it and move one square to the right (as seen from the second and third coordinates in the table entries). In fact, you will notice in looking at the first coordinates, that we have simply reproduced the capability of the finite state machine of Examples 3.38 and 3.39. If we take the tape $x = 01010$, as in Example 3.41, then we may diagram the transitions as shown in Figure 6.9.

In this representation, the state symbol is encircled and the pointer indicates the position of the read-write head on the tape. Alternatively, we may split the tape contents into left and right segments to illustrate the same thing as follows:

$$\alpha_0(x) = (\epsilon, 0, 01010) \rightarrow (0, 0, 1010) \rightarrow (01, 1, 010)$$

$$\rightarrow (010, 1, 10) \rightarrow (0101, 2, 0) \rightarrow (01010, 2, \square)!$$

In this alternative representation we favor the use of an exclamation point (!) to denote a halting configuration, and note that the state symbol appears as the middle entry of each of these "instantaneous descriptions."

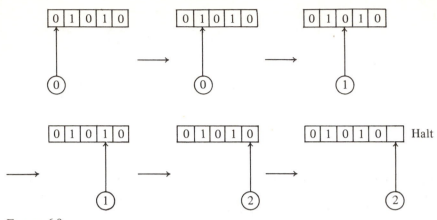

FIGURE 6.9

EXAMPLE 6.33   If we take the same machine but we present it with the tape $x = 0100$, then we will obtain the following sequence of instantaneous descriptions:

$$\alpha_0(x) = (\epsilon, 0, 0100) \rightarrow (0, 0, 100) \rightarrow (01, 1, 00)$$
$$\rightarrow (010, 1, 0) \rightarrow (0100, 1, \square) \rightarrow (0100, 1, \square) \rightarrow \cdots$$

With this tape x, the machine Z will never halt, even though it does appear to stand still.   △

We need this important distinction here if we are to say that a Turing machine Z *accepts* x (in which case, we write $x \in \lambda_Z$) provided that Z will halt (eventually) when it is started in the initial description

$$\alpha_0(x) = (\epsilon, 0, x).$$

This definition provides a sense for the *language λ_Z accepted by Z*, analogous to the notation λ_A in the case of finite state machines (Section 3.8). Can you describe λ_Z for the machine of Example 6.32?

Note that the machine of Example 6.32 can recognize the right-hand end of the input tape by virtue of the appearance of a blank symbol. It is simply a *Turing machine convention* that a blank symbol is to be "spliced" onto either end of the tape just as the head would tend otherwise to run off. In this way, the memory of a Turing machine is potentially infinite. Its finite set of internal states does not tell the whole story about its memory capacity. The tape itself can serve as an auxiliary storage. Therefore, we will see, as is not evident in the one example considered thus far, that the Turing machine computing and recognition capability far exceeds that of the finite state machines.

It is perhaps disquieting for the computer science student to learn that the capabilities of a Turing machine, with the simplicity of its description, go beyond those of any large-scale digital computer, past, present, or future! Yet that is what Turing's thesis is all about. In comparison with the finite state machines of Chapter 3, we have the two ends of the spectrum of automata, from A to Z. In order to fill in a few gaps in this spectrum, we will have to learn a bit more about mathematical language theory, and this will be done in Section 6.9.

Before presenting a variety of examples of the Turing machines, we need to introduce one more technicality. Again we let $Z = (S, \Sigma, M)$ denote a Turing machine. As we have already indicated, though not explicitly, the table M describing the moves or transitions of the machine will induce a relation ($\rightarrow$) on the *instantaneous descriptions* (abbreviated i.d.'s)

$$\alpha = (u, q, v)$$

Here $q \in S$ and $uv \in \Sigma_{\square}^*$, and you may recall that u and v are, respectively, the left and right segments of the word written on the tape. We are primarily interested in gaining a better understanding of this relation which leads from one instantaneous description to another, yet we hope that an example or two will suffice and that it will not be necessary to describe the relation in all of its details.

EXAMPLE 6.34   If Z has the entry $M(q, \sigma) = (q', \tau, -1)$, then for all u and v in $\Sigma_{\square}^*$ and all $\gamma \in \Sigma$, we have

$$(u\gamma, q, \sigma v) \rightarrow (u, q', \gamma\tau v)$$

because the entry in the table M has the following meaning: if the machine is in the state q scanning the symbol σ, it is to print τ (on the observed square) and move one square to the left, while making the transition to state q'. (See Figure 6.10.)

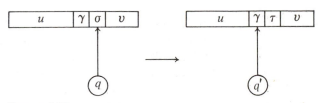

FIGURE 6.10

EXAMPLE 6.35   As an example of the example, suppose $\Sigma = \{a, b\}$ and the machine $Z = (S, \Sigma, M)$ has the entry $M(q, a) = (q', b, -1)$. Then with $u = ba$, $v = ab$, and $\gamma = b$, we would write

$$(bab, q, ab) \rightarrow (ba, q', bbb)$$

whereas

$$(\epsilon, q, ab) \rightarrow (\epsilon, q', \square bb)$$

in accordance with the convention on the splicing of blank squares. (See Figure 6.11.) $\triangle$

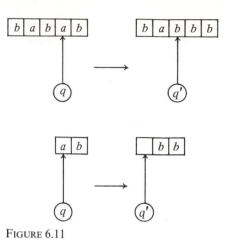

FIGURE 6.11

The student will need to study a number of examples in order to gain enough experience to design Turing machines for performing even the simplest tasks. It is a kind of "programming" skill which requires attention to be paid to the most minute detail. Though it is not essential to the formalism, it is certainly convenient in presenting examples to allow for the use of auxiliary symbols for marking purposes, etc. Thus we shall provide an enlarged alphabet $\Gamma \supseteq \Sigma$ in some of our examples. In studying these examples, the student should observe that at times we will use a Turing machine to accept or recognize a certain language, while at other times we will be designing a machine to compute a function. This dichotomy pervades the theory of Turing machines.

EXAMPLE 6.36   You may recall that in the sequential form of the flowchart language (Section 1.10), the compound concatenation statements are delimited with **begin** ... **end**. Some convention of this sort is found in most every computer programming language. The pairing of these delimiters is to be carried out as if they were left and right parentheses. The question naturally arises, here and with the direct use of parentheses in arithmetic expressions, as to whether there is an algorithmic procedure to check that a string of parentheses is well formed. Thus the string (()()) makes sense, but ())(() does not. So we see that it is more than a simple counting technique that is required.

A reasonable procedure here is to search to the right for a right parenthesis, then to back up and try to match it with a left parenthesis, checking both off as they are paired. (Note that we are not permitted to back up on the input string in a finite state machine.) If this is continued until there are no more pairs, we can decide whether the string is well formed according as to whether there are any unmatched parentheses left over.

Let us use the auxiliary symbol A for checking off pairs of parentheses. Then $\Gamma = \{(,), A\}$ and $\Sigma = \{(,)\}$ in the following Turing machine:

	(	)	A	□
0	$0, (, 1$	$1, A, -1$	$0, A, 1$	$2, □, -1$
1	$0, A, 1$		$1, A, -1$	$1, □, 0$
2	$2, (, 0$		$2, A, -1$	

If we have found a right parenthesis in state 0, then, transferring to the left-moving state 1, we may run off the left-hand side of the tape without finding a matching left parenthesis. Then the string is not well formed, and we "compute" forever at the $(1, □)$ entry of the table. Neither is it well formed if from state 0, we have run off the right-hand side, transferring to state 2, after which we find an extra left parenthesis. In this case, we are forever computing at the $(2, ()$ entry of the table. But if we are in state 2 and we run off the left-hand end of the tape, then the parenthesis string is well formed, and the machine will so indicate by halting at the $(2, □)$ entry.

Other recognition conventions are possible. We could arrange to transfer to states Y or N (for Yes or No), according as the string is well formed or not. This is done in the following alternative table

	(	)	A	□
0	$0, (, 1$	$1, A, -1$	$0, A, 1$	$2, □, -1$
1	$0, A, 1$		$1, A, -1$	$N, □, 0$
2	$N, (, 0$		$2, A, -1$	$Y, □, 0$

Since the rows Y and N do not have any entries at all, the machine would halt with the correct yes or no answer. △

EXAMPLE 6.37   To illustrate the operation of the machine in Example 6.36, let us try the well-formed string $(()())$. In the i.d.'s that follow, we omit the parentheses on the triples themselves for the sake of clarity.

$$\epsilon, 0, (()()) \rightarrow (, 0, ()()) \quad \rightarrow ((, 0,)()) \quad \rightarrow (, 1, (A())$$
$$\rightarrow (A, 0, A()) \quad \rightarrow (AA, 0, ()) \quad \rightarrow (AA(, 0,))$$
$$\rightarrow (AA, 1, (A) \quad \rightarrow (AAA, 0, A) \quad \rightarrow (AAAA, 0,)$$
$$\rightarrow (AAA, 1, AA \rightarrow (AA, 1, AAA \quad \rightarrow (A, 1, AAAA$$
$$\rightarrow (, 1, AAAAA \rightarrow \epsilon, 1, (AAAAA \quad \rightarrow A, 0, AAAAA$$
$$\rightarrow \quad \cdots \quad \rightarrow AAAAAA, 0, □ \rightarrow AAAAA, 2, A□$$
$$\rightarrow \quad \cdots \quad \rightarrow \epsilon, 2, AAAAAA□ \rightarrow \epsilon, 2, □AAAAAA□!$$

EXAMPLE 6.38   For any given symbol σ in the alphabet

$$\Sigma = \{\sigma_1, \sigma_2, \ldots, \sigma_n\}$$

the following Turing machine will perform the assignment statement $x \leftarrow x\sigma$, used so frequently in the earlier part of this chapter.

	σ_1	σ_2	$\cdots$	σ_n	$\square$
0	$0, \sigma_1, 1$	$0, \sigma_2, 1$	$\cdots$	$0, \sigma_n, 1$	$1, \sigma, 0$

The head just keeps moving to the right, looking for the end of the tape and finally placing the symbol σ there.   △

EXAMPLE 6.39   In Example 6.38 we have a Turing machine that computes a function. The value of the function replaces the argument on the tape. With the function presented here, the arguments are left on the tape, and the value of the function eventually appears at the right, separated from the arguments by a blank. These are only minor distinctions in the style of presentation, but we call them to the reader's attention because both conventions are found in the literature.

We recall (from Example 3.25) that when $\Sigma = \{1\}$, there is really no difference between Σ^* and the additive monoid of nonnegative integers. It is just that the integers are written in a "primitive" kind of notation ($4 = 1111$). Using this notation, we can design a Turing machine for performing addition. (As suggested in our next example, one could also design a machine for adding in the conventional notation.) We assume that the tape will initially contain the word $1^m \,\square\, 1^n$, and the idea is to transform the tape contents into $1^m \,\square\, 1^n \,\square\, 1^{m+n}$. The machine defined by Table 6.8 will serve this purpose.

TABLE 6.8

	1		$\square$	
0	1, $\square$,	1	7, $\square$,	1
1	1, 1,	1	2, $\square$,	1
2	2, 1,	1	3, $\square$,	1
3	3, 1,	1	4, 1,	-1
4	4, 1,	-1	5, $\square$,	-1
5	5, 1,	-1	6, $\square$,	-1
6	6, 1,	-1	0, 1,	1
7	8, $\square$,	1	12, $\square$,	1
8	8, 1,	1	9, $\square$,	1
9	9, 1,	1	10, 1,	-1
10	10, 1,	-1	11, $\square$,	-1
11	11, 1,	-1	7, 1,	1

In this machine, you will notice that we are able to use the blank as a marker. Beginning in state 0, the machine drops a marker ($\square$) to remember where the tally (1) came from. It then moves rightward down the tape, passing through states 1, 2, 3, and placing this tally at the right-hand end, as detected by the appearance of a blank symbol. Transferring to state 4, it moves leftward along the tape until it finds the marker left earlier. By this time the machine is in state 6, where it replaces the marker ($\square$) by a one. Then it returns to state 0, moves one square to the right, and begins the same series of moves all over again. The intermediate state transitions allow the machine to keep track of its position on the tape, whether it is in the first, second, or third string of ones. The transfer of tallies from the second to the third string of ones takes place in states 7 through 11, and is completely analogous to the preceding. $\triangle$

EXAMPLE 6.40 If we present the tape $11\square111$ to the machine of Example 6.39, then we will obtain the following sequence of i.d.'s:

$$(\epsilon, 0, 11\square111) \rightarrow (\square, 1, 1\square111) \rightarrow (\square1, 1, \square111) \rightarrow (\square1\square, 2, 111)$$
$$\rightarrow \cdots \rightarrow (\square1\square111, 2, \square) \rightarrow (\square1\square111\square, 3, \square)$$
$$\rightarrow (\square1\square111, 4, \square1) \rightarrow \cdots \rightarrow (\epsilon, 6, \square1\square111\square1)$$
$$\rightarrow (1, 0, 1\square111\square1) \rightarrow \cdots \rightarrow (11\square111\square, 12, 11111)!$$

This is a computation of $2 + 3 = 5$. $\triangle$

EXAMPLE 6.41 If we think of the symbols of an alphabet as ordered,

$$\Sigma = \{\sigma_1, \sigma_2, \ldots, \sigma_n\}$$

then an order is induced in the free monoid Σ^*, first according to length, then lexicographically within words of the same length. The following Turing machine computes the successor of a word x in this particular ordering. The style of presentation of the function is like that of Example 6.38, rather than that of Example 6.39.

	σ_1	$\cdots$	σ_{n-1}	σ_n	$\square$
0	$0, \sigma_1, 1$	$\cdots$	$0, \sigma_{n-1}, 1$	$0, \sigma_n, 1$	$1, \square, -1$
1	$2, \sigma_2, 0$	$\cdots$	$2, \sigma_n, 0$	$1, \sigma_1, -1$	$2, \sigma_1, 0$

The machine looks for the right-hand end of the word x by backing up one square after falling off the end. Then in state 1 it increments this rightmost symbol by one, unless it is the symbol σ_n. When the symbols are the digits of a number system, the computation is much like an addition of one. For this reason, we would want to change σ_n to σ_1 (the "zero" digit) and move left to "carry" one to the next position. This is precisely what is done by the machine. $\triangle$

EXAMPLE 6.42   Suppose that $\Sigma = \{0, 1, 2\}$. Then the student should run through a few i.d.'s in order to see that for the machine of Example 6.41, 102 is transformed into 110, 112 is transformed into 120, 222 is transformed into 0000, etc.   △

EXERCISES

1 Diagram (as in Figure 6.9) the transitions of the Turing machine of Example 6.32 for the input tape:
(a) $x = 01011$ (b) $x = 11011011$
(c) $x = 1101101$ (d) $x = 11110101$

2 Repeat Exercise 1 using, instead, the formal instantaneous descriptions.

3 In connection with Example 6.34, determine the transitions

$$(\epsilon, q \, \sigma \, v) \to ?$$

taking place for the entry $M(q, \sigma) = (q', \tau, -1)$. At the same time how do we determine

$$(u \, \gamma, q, \epsilon) \to ?$$
$$(\epsilon, q, \epsilon) \to ?$$

in case $\sigma = \square$?

4 Describe all the transitions of i.d.'s taking place for an entry
(a) $M(q, \sigma) = (q', \tau, 1)$ (b) $M(q, \sigma) = (q', \tau, 0)$
[*Hint*: See Exercise 3.]

5 Diagram the transitions of i.d.'s for the Turing machine of Example 6.36 with the input tape:
(a) $x = ())(()$ (b) $x = ((())()$
(c) $x = (())($ (d) $x = ((()))()$

6 Diagram the transitions of i.d.'s for the Turing machine of Example 6.38 with $\Sigma = \{a, b, c\}$, $\sigma = b$, and the input tape:
(a) $x = abbc$ (b) $x = bacb$
(c) $x = aabc$ (d) $x = bcaa$

7 Diagram the transitions of i.d.'s for the Turing machine of Example 6.39 in computing
(a) $1 + 2 = 3$ (b) $2 + 1 = 3$
(c) $2 + 3 = 5$ (d) $3 + 0 = 3$
(e) $0 + 2 = 2$ (f) $2 + 2 = 4$

8 Diagram the transitions of i.d.'s for the Turing machine of Example 6.41 with $\Sigma = \{0, 1, 2\}$ and the input tape
(a) $x = 112$ (b) $x = 121$
(c) $x = 222$ (d) $x = 202$
(e) $x = 1120$ (f) $x = 2122$

9 Design a Turing machine that will recognize (say, by always halting in one of the predesignated states Y = yes or N = no):
   (a) all palindromes (See Exercise 1 of Section 6.2)
   (b) all words beginning and ending in b
   (c) all words in which are embedded the sequence cab
   (d) all words of the form $a^n b^n c^n$
   (e) all words of the form $a^n b c^n$
   on the alphabet $\Sigma = \{a, b, c\}$.

10 Design a Turing machine that will recognize (say by always halting in one of the predesignated states Y = yes or N = no) all integers in the "primitive" notation of Example 6.39 ($\Sigma = \{1\}$) which are
   (a) even   (b) divisible by 3
   (c) greater than 5   (d) congruent to 1 (mod 3)

*11 Repeat Exercise 10 with the standard decimal notation.

12 Design a Turing machine for computing (in the "primitive" notation of Example 6.39):
   (a) $m \cdot n$ (b) $2n + 3$ (c) $n + 3$ (d) $3n$

*13 Design a Turing machine for performing addition, as in Example 6.39, but in the standard decimal notation.

*14 Design a Turing machine that will recognize (in the sense of Exercises 9 and 10) all well-formed formulas of propositional calculus, assuming a finite number of propositional variables.

6.8   UNDECIDABILITY

The theory of Turing machines has stimulated research and brought many interesting results outside of automata theory, touching on virtually every area of mathematics and computer science. Primarily, this development has come about because of the identification of the intuitive notions

(a) Effective decidability of languages, relations, and propositions
(b) Effective enumeration of the same
(c) Effective computability of (partial) functions

with the corresponding Turing machine concepts (assuming Turing's thesis). These corresponding Turing machine-theoretic definitions are generally referred to by the names *recursive* (languages), *recursively enumerable* (languages), and *Turing computable* (functions), respectively. The situation is not unlike that encountered in calculus of the intuitive notion of a function whose graph can be drawn "without lifting the pencil from the paper." It is generally agreed that the so-called continuous functions are the proper precise counterpart of the present idea. Of course, one cannot develop a theory on the basis of intuition alone, and so these correspondences are essential.

To limit the scope of our discussion, let us concentrate on the first two of the preceding notions, (a) and (b). We recall that a language $\lambda \subseteq \Sigma^*$ is really just a unary proposition on Σ^*. For this reason, the languages are easier to discuss than the relations (binary propositions) or the general propositions. As we said in Section 1.4, the effective decidability of a language λ should suppose the existence of a finite mechanical procedure for determining in every instance whether or not $x \in \lambda$. The formal counterpart here is the existence of a Turing machine Z that will halt (eventually) on every input tape x and provide a yes or no answer ($x \in \lambda$?), say by virtue of its arrival at distinguished predesignated states Y or N. (See Example 6.36.) Such a language is said to be *recursive*. It is natural to try to contrast these languages with the "regular" languages, those recognized or accepted by some finite state machine. For this purpose, we need the following result, usually attributed to Myhill*, though the ideas were extended somewhat by the work of Nerode†.

Theorem 6.6   (Myhill)

A language $\lambda \subseteq \Sigma^*$ is regular iff it is a union of equivalence classes resulting from a congruence of finite index on the monoid Σ^*.

Proof   By the index of a congruence relation, we simply mean the number of equivalence classes. So if we are given a congruence R of finite index on Σ^*, we will construct a machine having the quotient set Σ^*/R as its set of states, and having the initial state $[\epsilon]$. If we take the transition function

$$M([x], y) = [xy]$$

(well defined because R is a congruence), we see that our machine will accept λ if we let $\{[x] : [x] \subseteq \lambda\}$ be the set of final states. (Show this!) Since R is of finite index, this is indeed a finite state machine, and λ is regular.

Conversely, if A is a finite state machine accepting the language $\lambda \subseteq \Sigma^*$, then we can introduce the congruence relation R_A of Exercise 8 of Section 3.10, namely

$$x \, R_A \, y \quad \text{iff} \quad M_x = M_y$$

[that is, iff $M_A(s, x) = M_A(s, y)$ for all s in the state set S]. This congruence on Σ^* is of finite index because there are only finitely many distinct functions $M_x : S \to S$. It is easily shown that

$$\lambda = \bigcup_{M_A(0, x) \in I} [x]$$

which completes the proof.   □

* J. Myhill, "Finite Automata and the Representation of Events," *WADC Tech. Rept.* 57–624, Wright Patterson AFB, Ohio (1957).
† A. Nerode, "Linear Automaton Transformations," *Proc. Am. Math. Soc.*, **9**, 541–544 (1958).

We will find that this theorem is admirably suited for demonstrating that certain languages are nonregular, that is, accepted by no finite state machine (See Example 6.44). If we find that one of these is recursive, we will have established a definite "gap" between the regular and the recursive languages, and demonstrated that there are recognition problems that can be handled by Turing machines, but by no finite state machine.

EXAMPLE 6.43 Every regular language ($\lambda = \lambda_A$ for some finite state machine) is recursive. In order to see this, one has only to generalize the technique employed in Example 6.32 for simulating the behavior of a finite state machine by a Turing machine. (See Exercise 1 of Section 6.8.) △

EXAMPLE 6.44 According to Example 6.36, if $\Sigma = \{(,)\}$, then the language λ consisting of all well-formed parenthesis strings is recursive. If it were regular, then according to Myhill's theorem, there would be a congruence R of finite index on Σ^* with λ being a union of equivalence classes. Then the members of the infinite sequence of strings $\epsilon, (, ((, (((,\ldots$ could not be mutually inequivalent. We would necessarily have

$$(^n R (^m$$

for some $n \neq m$. Since R is a congruence, this implies that

$$(^n)^n R (^m)^n$$

On the one hand, $(^n)^n$ is well formed, whereas $(^m)^n$ is not. This contradicts the fact that λ is a union of equivalence classes, that is, we can't take $(^n)^n$ without $(^m)^n$. Consequently, λ is not regular. △

In summary, we have seen that regular $\Rightarrow$ recursive, but that the converse is definitely false.

Up to this point, we have not had much to say about effective enumeration. The curious thing here is that every language $\lambda \subseteq \Sigma^*$ is enumerable (countable), being a subset of a countable set. But can its members be "effectively" enumerated? By convention, such enumerability entails the existence of a mechanical procedure for actually listing the members, one after another, with the assurance that every $x \in \lambda$ will appear sooner or later on the list. Perhaps it will not be immediately apparent why it is that the formal counterpart here should be taken so as to include only those languages λ_Z that are accepted by some Turing machine Z. We recall that $x \in \lambda_Z$ means that Z will halt on x:

$$\alpha_0(x) = \alpha_0 \to \alpha_1 \to \alpha_2 \to \cdots \to \alpha_n!$$

In any case, we call such languages ($\lambda = \lambda_Z$ for some Z) *recursively enumerable*, analogous once again to the definition of "regular" in the case of the finite state machines (Section 3.8).

EXAMPLE 6.45   We just noted that the connection between the idea of being recursively enumerable and that of a mechanical listing, was not all that apparent. But one argument runs as follows. We imagine a list of all the words of Σ^* by order of their lengths, lexicographically within words of the same length. If nothing else, Example 6.41 shows that this can be done in a mechanical fashion. Now we take our Turing machine Z for which $\lambda_Z = \lambda$. We cannot run the machine indefinitely on successive words of Σ^*, since any one experiment might fail to terminate. But we can run Z as follows: one move on word x_1, two moves on x_1, one move on x_2, three moves on x_1, two moves on x_2, one move on x_3, etc. In this way, we can work our way through the infinite array as shown in Figure 6.12. At each position, we can test (look at M) to see whether we are in a halting i.d.(!) If so, we will read off the row heading as a word $x \in \lambda$ for our listing. In this way, purely mechanically, we can make every word of λ definitely to appear sooner or later on our list.   △

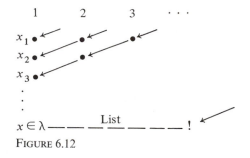

FIGURE 6.12

Our understanding of the two notions, recursive and recursively enumerable, is enhanced somewhat by the following comparison. Here and in much of the reasoning from this point on, we offer only informal arguments, since the details would sometimes tend to obscure the train of thought.

Lemma 1

Every recursive language is recursively enumerable.

Proof   We suppose that we have a Turing machine that will halt on every input $x \in \Sigma^*$ and provide a yes or no answer ($x \in \lambda$?) by virtue of its landing in state Y or N. We alter the machine just slightly, by causing it to "compute forever" from state N, which is easily accomplished. Calling the new machine Z, we have $\lambda = \lambda_Z$ so that the given recursive language λ is recursively enumerable as well.   □

The converse of this lemma is not true! In fact, it is one of the main results of the theory of automata that there exists a language L which is recursively enumerable but not recursive. Though its elements can be effectively listed, there is no finite decision procedure for answering the question of membership.

One might think that it would be possible to look at the listing in order to decide, but at any given time only a finite portion of the list has been generated. If the word x is not seen there, is it because the word is not on the list or because it has not yet been generated? We have no way of knowing. For the time being, we will take the existence of this language L for granted. Later we will try to give some indication as to its nature.

Lemma 2

The recursive languages form a Boolean algebra (of subsets of Σ^*).

Proof   We show that the complement (set difference) λ' of a recursive language is again recursive. Actually this fact is the only thing that we will use in the sequel. Given the Turing machine that decides about membership ($x \in \lambda$?), we simply interchange the roles of the states Y and N. Then the new machine will show that $\lambda' = \Sigma^* \sim \lambda$ is recursive along with λ. To get the full strength of the statement of the lemma, one would have to show the closure of the recursive languages with respect to set union (or intersection), in which case we could then use DeMorgan's laws (Theorem 4.3).   □

Theorem 6.7

λ is recursive $\Leftrightarrow$ λ and λ' are recursively enumerable.

Proof   First, suppose that λ is recursive. Then λ' is also recursive, according to Lemma 2. It follows from Lemma 1 that each of these in turn is recursively enumerable.

For the converse we give only an intuitive argument. Suppose that λ and λ' are recursively enumerable. Then each of them can be listed effectively:

$$
\begin{array}{c|cc}
 & \lambda & \lambda' \\
\hline
1 & x_1 & y_1 \\
2 & x_2 & y_2 \\
\vdots & \vdots & \vdots
\end{array}
$$

Our (effective) decision procedure for λ is as follows: Suppose that we are given any word x in Σ^*. Sooner or later it has to show up on one list or the other, because the lists are complementary sets. So we simply wait. If it turns up on the first list, we answer "yes"; if it turns up on the second list, we answer "no." The point is, we don't have to wait forever.   □

We recall that this discussion all began with the existence of a nonrecursive but recursively enumerable language L. We now use the same language to establish the following:

Corollary

There exists a nonrecursively enumerable language.

Proof Take the language L'. L is recursively enumerable. So if the same were true of L', we could use the theorem to conclude that L is recursive. But L is not recursive. So L' must be nonrecursively enumerable.   ☐

This corollary describes a very unusual situation. L' is a countable set, being a subset of the countable set Σ^*, but it is so "wildly" embedded in Σ^* that there is no way to list its members effectively! Naturally we would want to know how the original language L arises. Again we shall proceed on the intuitive level in order that the ideas be as transparent as possible. Given a Turing machine Z and an input tape x, it may be a long, long while before the computation is completed and the machine comes to a halt. In fact, as we know, for certain pairs Z and x this may never happen, and Z may go on computing forever, starting from $\alpha_0(x)$. It would certainly be nice to have a decision procedure for determining whether or not Z would halt on x. It turns out, however, that this question is (algorithmically) undecidable! This is one of the most important results of the Turing machine theory, the unsolvability of the halting problem.

Note that the most naive approach toward trying to devise a decision procedure for the halting problem meets with frustration. That is, given Z and x, we might simply try to run Z from $\alpha_0(x)$ and see what happens. Of course, if Z does indeed halt after a finite number of moves, then the answer to our question will be affirmative. But if we have run the machine through 17, 564, 323 moves and it still does not halt, then we cannot be sure whether it will never halt or we just haven't waited long enough. Now, the fact that the most obvious approach fails certainly does not prove anything, but it does lend credence to the proposition. An actual proof can be constructed along the following lines.

We first consider what seems to be an easier proposition. Suppose that we use a "coding" to represent each Turing machine Z by a tape d_Z called the *description* of Z. Then d_Z is to be a word in Σ^*, itself suitable as input to a machine. Obviously we are to imagine that such an association $Z \rightarrow d_Z$ be one to one and effective. The reader can undoubtedly construct such an encoding on his own, so we need not be more precise. Then the question is this: Is it algorithmically decidable whether an arbitrary Turing machine would halt if given the tape corresponding to its own description? Though the question may seem to be turned inward on itself, it is crucial to an understanding of the unsolvability results concerning Turing machines. We will begin to gain an appreciation for the unusual flavor of these results by considering the following negative answer to the question just posed.

Lemma

There does not exist a Turing machine H that always halts in one of two distinguished states Y, N giving the decision

$$H \text{ halts in Y if "}Z \text{ halts on } d_Z\text{"}$$
$$H \text{ halts in N if "}Z \text{ does not halt on } d_Z\text{"}$$

when presented with the input tape d_Z.

Proof It would be an easy matter to modify such a machine (as in the proof of Lemma 1 preceding Theorem 6.7) to allow

$$H \text{ computes forever if "}Z \text{ halts on } d_Z\text{"}$$
$$H \text{ halts (in } N) \text{ if "}Z \text{ does not halt on } d_Z\text{"}$$

given the input tape d_Z. Then comes the "punch line": We ask, "What would happen if we were to apply the tape d_H to H?" From the wording alone, we see that if H halts on d_H, then H doesn't halt on d_H; and if H doesn't halt on d_H, then it does. In either case, we have contradiction. It follows that the machine H cannot exist. $\square$

Theorem 6.8 **(Unsolvability of the halting problem)**

The problem of determining whether an arbitrary Turing machine Z halts on x is "effectively undecidable."

Proof Of course, in accepting Turing's thesis, we first identify the idea of effective decidability with the appropriate Turing machine concept. If there were a Turing machine I that, when provided with an appropriate tape representation (d_Z, x) would always halt in one of two distinguished states Y, N giving the decision:

$$I \text{ halts in Y if "}Z \text{ halts on } x\text{"}$$
$$I \text{ halts in N if "}Z \text{ does not halt on } x\text{"}$$

then we would be able to design an effective decision procedure for the earlier question. All we need to do is to modify I so that it first duplicates its tape, then proceeds as before. The modified machine I' then behaves as follows:

$$I' \text{ halts in Y if "}Z \text{ halts on } d_Z\text{"}$$
$$I' \text{ halts in N if "}Z \text{ does not halt on } d_Z\text{"}$$

when presented with the input tape d_Z. This is precisely the behavior of the machine H whose existence has already been denied. $\square$

Corollary

The language

$$L = \{d_Z : Z \text{ halts on } d_Z\}$$

is recursively enumerable but nonrecursive.

Proof   That L is not recursive follows immediately from the lemma. In fact, it is equivalent to the statement of the lemma. We leave the reader with the task of taking up some of the ideas of Example 6.45 to show that L is recursively enumerable.   □

The undecidability of the halting problem has spawned a search for undecidable questions in other areas. These investigations have proven to be most fruitful, even though the results are necessarily of a negative character. Some of the implications for computer science are especially worth noting. One can infer that it is impossible to design a uniform "debugging" procedure, that is, a computer program which can look at any computer program and give a yes or no answer to the question of whether that program will ever run to completion. Neither can there be a procedure for checking whether an algorithm will terminate for each set of initial conditions, for this question, too, can be shown to be undecidable. Of course, that does not mean that we cannot and should not seek to design such procedures as might prove to be entirely suitable for important subclasses of problems.

EXERCISES

*1 Show that every regular language is recursive. [*Hint*: Simulate the behavior of an arbitrary finite state machine with a Turing machine that halts in states Y or N.]

2 Show that the following languages are nonregular on the alphabet $B = \{0, 1\}$.
   (a) $\{0^n 1^n : n \geq 1\}$ (b) $\{0^n 10^n : n \geq 1\}$
   (c) All palindromes (see Exercise 1 of Section 6.2)   (d) $\{0^n 1^n 0^n : n \geq 1\}$

3 Show that the given language λ is accepted by the machine constructed in the proof of Myhill's theorem.

4 Complete the proof of Myhill's theorem by showing that

$$\lambda = \bigcup_{M_A(0,\, x) \in I} [x]$$

*5 Outline a procedure for obtaining an effective enumeration of the set of all Turing machines (or Turing machine descriptions) on the alphabet Σ.

*6 Show that the language L (in the corollary to Theorem 6.8) is recursively enumerable. [*Hint*: Use an effective listing of the Turing machines (Exercise 5) and proceed along the lines of Example 6.45.]

7 Give an informal argument to show that the union and the intersection of recursive languages are recursive.

8 Give an informal argument to show that the union and the intersection of recursively enumerable languages are recursively enumerable.

9 Show that the recursively enumerable languages do not form a Boolean algebra.

10 Informally describe a "mechanical procedure" for effectively listing the members of the languages described in Exercise 9 of Section 6.7. [*Note.* These effective listings tend to suggest that the given languages are recursively enumerable, whereas the results of Exercise 9 of Section 6.7 will show that they are in fact recursive.]

6.9   CHOMSKY GRAMMARS

With the appearance of large-scale computers, studies in mathematical linguistics have become intensified in several respects, and the attempts at the mechanical translation of the natural languages (English, French, etc.) have revealed the essential inadequacy of the classical linguistic theory, at least for this purpose. In the mid-1950s a project addressed to these and related questions was undertaken at the Massachusetts Institute of Technology under the leadership of Noam Chomsky.* The main result of this research, the development of the phrase structure grammars, has made an indelible imprint on computer science. Besides revealing the importance of phrase structure and context in the natural languages, these studies provided the much needed theoretical basis for the investigation, classification, and even compilation of the computer programming languages.

Instead of accepting the languages $\lambda \subseteq \Sigma^*$ by means of some mechanical device, we can now use the phrase structure grammars to generate languages through the application of a finite set of "productions." This generation scheme is reminiscent of the way that theorems were derived in our axiomatic presentation of propositional calculus (Section 6.6). We hope that just such a similarity was also made apparent in our treatment of Turing machine computations. Here, however, we have only one "axiom"; the productions play the role of the "rules of inference."

The grammars make use of auxiliary or *nonterminal* symbols, so we have to speak of an enlarged alphabet in the following definition. A *Chomsky grammar* $G = (\Gamma, P)$ is a mathematical system consisting of

(i) an alphabet Γ with a distinguished subset $\Sigma \subseteq \Gamma$ and a distinguished element S;
(ii) a finite set $P = \{u : v\}$ of *productions* (u and v in Γ^*).

The elements of Σ are called *terminal* symbols, whereas the "axiom" S may

---

* See Reference 6.6.

be thought of as the *sentence symbol.* It is used in the generation of "sentences" when the model is used in a natural linguistic setting. In this interpretation, the terminal symbols are the ordinary words (in English, French, etc.), and the nonterminal symbols are the names of various parts of speech or other grammatical classes, that is, NP = noun phrase, VP = verb phrase, V = verb, etc. In a computer programming language the latter become the program constituents such as decimal number, arithmetic expression, etc. The mechanics of the language generation is the same in either case.

If $G = (\Gamma, P)$ is a grammar, then P induces a relation $(\rightarrow)$ on the words of Γ^* according to the possibility of there being a substitution:

$$xuy \rightarrow xvy \qquad (\text{with } u : v \text{ in } P)$$

leading from one word to the next. Then the *language generated by* G (denoted by λ_G) is the set of all words (sentences) of Σ^* generable from the sentence symbol; that is,

$$z \in \lambda_G \Leftrightarrow S = \alpha_0 \rightarrow \alpha_1 \rightarrow \alpha_2 \rightarrow \cdots \rightarrow \alpha_n = z$$

it being understood that $z \in \Sigma^*$.

EXAMPLE 6.46   In the natural language setting the symbols of Σ are multiliteral. The same is true of the nonterminal symbols unless we choose to abbreviate the grammatical classes each by a single letter. Thus we may have

$$\Sigma = \{\text{a, an, big, cake, eats, football, has, imaginatively,}$$
$$\text{plays, Ron, skate, stodgy, sounds, sweet, swimmingly, the}\}$$

and as nonterminals ($\Gamma \sim \Sigma$):

NP = noun phrase	Adv = adverb	Pred = predicate
AP = adjective phrase	V = verb	N = noun
VP = verb phrase	Adj = adjective	Art = article

Consider now the following set P of productions, and observe that they are consistent with the rules of ordinary grammar.

S : NP VP	AP : Adj N
NP : N	VP : V
NP : Art AP	VP : V NP
NP : Adj AP	VP : V Pred
AP : Adj AP	Pred : Adv

N : cake	V : eats	Adj : big	Adv : swimmingly
N : sounds	V : has	Adj : stodgy	Art : a
N : football	V : skate	Adj : sweet	Art : an
N : Ron	V : plays	Adv : imaginatively	Art : the

In this grammar $G = (\Gamma, P)$ we may generate the following sentence:

$$S \to \text{NP VP} \to \text{Adj AP VP} \to \text{Adj AP V NP} \to \text{Adj Adj AP V NP}$$
$$\to \text{Adj Adj AP V N} \to \text{Adj Adj Adj N V N}$$
$$\to \text{Big Adj Adj N V N}$$
$$\to \text{Big sweet Adj N V N}$$
$$\vdots$$
$$\to \text{Big sweet stodgy Ron eats cake}$$

Inherent in the idea of a grammar is the nondeterministic or "multiple-choice" nature of the application of the productions. Thus one grammar can generate an infinite language, just as a Turing machine can accept infinitely many words, even though each of these is finitely specified. Thus we find that the same grammar considered here will also generate the sentences

> Ron plays football
> Big Ron eats the football

but unfortunately, it will also generate

> Sweet stodgy sounds skate swimmingly △

EXAMPLE 6.47 Just by implementing the definition of well-formed formulas with productions, we can give a Chomsky grammar for generating the language of wff's on a finite number of propositional variables. We only have to take

$$\Sigma = \{\neg, \vee, \wedge, p, q, r, \ldots\}$$
$$\Gamma \sim \Sigma = \{S\}$$

and the set of productions

$$
\begin{array}{ll}
S : S \neg & S : p \\
S : S S \vee & S : q \\
S : S S \wedge & S : r \\
& \vdots
\end{array}
$$

to generate all of the (Polish) wff's.

To illustrate, let us show the generation of the wff of Example 6.10:

$$S \to S S \vee \to S \neg S \vee \to S \neg S S \wedge \vee \to S \neg p S \wedge \vee$$
$$\to S \neg p q \wedge \vee \to S S \wedge \neg p q \wedge \vee \to S p \wedge \neg p q \wedge \vee$$
$$\to S S \vee p \wedge \neg p q \wedge \vee \to S r \vee p \wedge \neg p q \wedge \vee$$
$$\to S S \wedge r \vee p \wedge \neg p q \wedge \vee \to p S \wedge r \vee p \wedge \neg p q \wedge \vee$$
$$\to p S \neg \wedge r \vee p \wedge \neg p q \wedge \vee \to p q \neg \wedge r \vee p \wedge \neg p q \wedge \vee \quad △$$

EXAMPLE 6.48 If we have the terminal alphabet $\Sigma = \{0, 1\}$, then the grammar G with the productions

$$S : 0 \qquad S : 0 S \qquad S : 1 S$$

will generate that language $\lambda_G \subseteq \Sigma^*$ consisting of all words ending in a zero. △

EXAMPLE 6.49   With the same alphabet but the pair of productions

$$S : 0\,1 \qquad S : 0\,S\,1$$

we will generate the language

$$\lambda_G = \{0^n 1^n : n \geq 1\} \qquad \triangle$$

EXAMPLE 6.50   The grammar G which follows will generate the language

$$\lambda_G = \{0^n 1^n 2^n : n \geq 1\}$$

We take $\Sigma = \{0, 1, 2\}$ and $\Gamma \sim \Sigma = \{S, T, A, B\}$ with the set of productions:

$$
\begin{array}{ll}
S : 0\,T\,B & \\
S : 0\,B & B : 1\,2 \\
T : 0\,T\,A & A\,1 : 1\,A \\
T : 0\,A & A\,2 : 1\,2\,2
\end{array}
$$

The four productions on the left-hand side will transform

$$S \to \cdots \to 0^n\,A^{n-1}\,B$$

and the three on the right-hand side will complete the generation, transforming

$$0^n\,A^{n-1}\,B \to \cdots \to 0^n\,1^n\,2^n$$

As an illustration, let us compute:

$$
\begin{aligned}
S &\to 0\,T\,B \to 0\,0\,T\,A\,B \to 0\,0\,0\,A\,A\,B \to 0\,0\,0\,A\,A\,1\,2 \\
&\to 0\,0\,0\,A\,1\,A\,2 \to 0\,0\,0\,1\,A\,A\,2 \to 0\,0\,0\,1\,A\,1\,2\,2 \\
&\to 0\,0\,0\,1\,1\,A\,2\,2 \to 0\,0\,0\,1\,1\,1\,2\,2\,2 \qquad \triangle
\end{aligned}
$$

One can show that the languages λ_G generated by some Chomsky grammar are synonymous with the languages λ_Z accepted by some Turing machine, that is, they coincide with the recursively enumerable languages. In fact, it is much easier to see how an effective listing comes about from this point of view. The real contribution provided by the grammatical setting, however, concerns the classification of languages in a hierarchy of complexity. The idea, again due to Chomsky, involves the placement of increasingly stringent conditions or restrictions on the form of the productions. An unrestricted grammar is said to be of *type* 0. Otherwise, a grammar $G = (\Gamma, P)$ is said to be of

Type 1 (or *context sensitive*) if every production has the form

$$x\,A\,y : x\,w\,y$$

where w is a nonempty word in Γ^*.

Type 2 (or *context free*) if every production has the form

$$A : w$$

where again w is a nonempty word in Γ^*.

Type 3 (or *regular*) if every production has one of the forms

$$A : \sigma B \quad \text{or} \quad A : \sigma$$

where σ is in Σ and B is in $\Gamma \sim \Sigma$.

Throughout these definitions A is assumed to be a nonterminal symbol. We note that

$$\text{type } 3 \Rightarrow \text{type } 2 \Rightarrow \text{type } 1 \Rightarrow \text{type } 0$$

showing that the conditions are increasingly restrictive.

EXAMPLE 6.51   Examples 6.47 through 6.49 are type 2 grammars, but Example 6.50 is not. The production $A\,2 : 1\,2\,2$ is not context free but context sensitive. In a context-free production $A : w$, we are able to substitute w for A without regard for the context in which A is found. By contrast, we cannot make the substitution in the form $x\,A\,y : x\,w\,y$ unless A is found in the specific context or surroundings $x\,A\,y$. This is the justification for the terminology "context free" and "context sensitive."   △

EXAMPLE 6.52   In Example 6.48 we have a type 3 grammar. Now it should occur to the reader that this recognition problem—simply to decide whether a word composed of zeros and ones, ends in a zero—could be handled by a finite state machine (with just two states). This observation is consistent with the tabulation we are about to give.   △

Obviously the Chomsky type classification extends to the languages themselves. Suppose $\lambda \subseteq \Sigma^*$; then λ is said to be a *type i language* ($i = 0, 1, 2, 3$) when there exists a type i grammar $G = (\Gamma, P)$, with Σ as its set of terminal symbols, such that $\lambda = \lambda_G$. We noted earlier that the type 0 languages coincide with the recursively enumerable languages. At the other end of the spectrum, we can show that the type 3 languages coincide with the regular languages, those accepted by some finite state automaton (Section 3.8). Altogether, we may summarize with Table 6.9.

TABLE 6.9

Language type	Name	Accepted by
0	Recursively enumerable	Turing machine
1	Context sensitive	n. linear bounded automaton
2	Context free	n. pushdown automaton
3	Regular	finite state machine

As soon as the Chomsky grammatical hierarchy became widely known, untold numbers of researchers began to look for automata that would accept the intermediate context-free and context-sensitive languages. It was found that a machine operating on a one-way input tape but provided with an auxiliary pushdown storage stack (Section 6.2) was the proper automaton for recognizing the context-free languages. On the other hand, a Turing machine restricted to the use of only that portion of the tape which the input word occupies, was found to be appropriate for accepting the context-sensitive languages. These devices have become known as *pushdown automata* and *linear bounded automata*, respectively. The prefix (n = nondeterministic) in the table means that one must permit these machines to choose from several moves at each stage of the recognition process, and this choice is to be reflected in their transition function. Unfortunately, we cannot afford to present any more details concerning these so-called "intermediate machines," there being many other examples. Such a discussion would lead to the theory of "computational complexity."

What we have presented here is only a glimpse of what now amounts to an exceedingly vast complex hierarchy of machine and grammatical classes. Nevertheless, it is worth remembering that the existence of a grammar or a machine for generating or accepting a given language at one level of the hierarchy does not preclude its being done at a lower level.

EXAMPLE 6.53   From Example 6.36 we know that the language λ of well-formed parenthesis strings is recursively enumerable and in fact recursive. But the grammar with the productions

$$S : (\,)$$
$$S : (S\,S)$$
$$S : (S)$$
$$S : S(\,)$$

shows that λ is context free (type 2) and thus there is a pushdown automaton that will serve the same purpose as the Turing machine of Example 6.36.   △

EXAMPLE 6.54   Just because the grammar of Example 6.50 does not consist entirely of context-free productions is no guarantee that the language

$$\lambda = \{0^n 1^n 2^n : n \geq 1\}$$

is not of type 2 (though in fact it is not). Is λ a context-sensitive (type 1) language?   △

Among the Chomsky grammatical types the context-free languages have found the widest range of application in computer sciences. We will try to give some indication as to why this has happened. Suppose that $G = (\Gamma, P)$ is a context-free grammar. Then each production can be written $A : w$ where w is a nonempty word. In the *Backus normal form** of a context-free grammar we bring together all these productions, $A:w_1$, $A:w_2$, ..., $A:w_k$ for a fixed A, into a new format:

$$A :: = w_1 | w_2 | \cdots | w_k$$

which may be read, "The nonterminal symbol A may be replaced by w_1 or w_2 or ... or w_k." Besides the advantage of compactness, this notation, we should point out, is just the form of the official description of ALGOL.† We will present a few excerpts from this report as illustration.

EXAMPLE 6.55   In the Backus normal form, the productions of Example 6.46 are written as follows:

$$
\begin{aligned}
S &:: = \text{NP VP} \\
\text{NP} &:: = \text{N} \,|\, \text{Art AP} \,|\, \text{Adj AP} \\
\text{AP} &:: = \text{Adj AP} \,|\, \text{Adj N} \\
\text{VP} &:: = \text{V} \,|\, \text{V NP} \,|\, \text{V Pred} \\
\text{Pred} &:: = \text{Adv} \\
\text{N} &:: = \text{cake} \,|\, \text{football} \,|\, \text{Ron} \,|\, \text{sounds} \\
\text{V} &:: = \text{eats} \,|\, \text{has} \,|\, \text{plays} \,|\, \text{skate} \\
\text{Adj} &:: = \text{big} \,|\, \text{stodgy} \,|\, \text{sweet} \\
\text{Adv} &:: = \text{imaginatively} \,|\, \text{swimmingly} \\
\text{Art} &:: = \text{a} \,|\, \text{an} \,|\, \text{the}
\end{aligned}
$$
△

EXAMPLE 6.56   Consider the following excerpts from the ALGOL report:

$\langle\text{digit}\rangle :: = 0 \,|\, 1 \,|\, 2 \,|\, 3 \,|\, 4 \,|\, 5 \,|\, 6 \,|\, 7 \,|\, 8 \,|\, 9$

$\langle\text{letter}\rangle :: = a \,|\, b \,|\, c \,|\, d \,|\, e \,|\, f \,|\, g \,|\, h \,|\, i \,|\, j \,|\, k \,|\, l \,|\, m \,|\, n \,|\, o \,|$
$\quad p \,|\, q \,|\, r \,|\, s \,|\, t \,|\, u \,|\, v \,|\, w \,|\, x \,|\, y \,|\, z \,|\, A \,|\, B \,|\, C \,|\, D \,|$
$\quad E \,|\, F \,|\, G \,|\, H \,|\, I \,|\, J \,|\, K \,|\, L \,|\, M \,|\, N \,|\, O \,|\, P \,|\, Q \,|\, R \,|\, S \,|$
$\quad T \,|\, U \,|\, V \,|\, W \,|\, X \,|\, Y \,|\, Z$

$\langle\text{logical constant}\rangle :: = \textbf{true} \,|\, \textbf{false}$

$\langle\text{arithmetic operator}\rangle :: = + \,|\, - \,|\, \times \,|\, / \,|\, \div \,|\uparrow$

$\langle\text{identifier}\rangle :: = \langle\text{letter}\rangle \,|\, \langle\text{identifier}\rangle\langle\text{letter}\rangle \,|$
$\quad\quad \langle\text{identifier}\rangle\langle\text{digit}\rangle$

Here the angular brackets merely serve to remind us to regard the multiliteral expressions as single symbols of the alphabet.

The first four of these Backus normal forms are not particularly interesting since there is no recursion involved. The last one, however, makes recursive

---

* J. W. Backus, "The Syntax and Semantics of the Proposed International Algebraic Language of the Zurich ACM-GAMM Conference," *Proceedings of the International Conference on Information Processing*, Paris, 1959, pp. 125–132.

† P. Naur, ed., "Revised Report on the Algorithmic Language Algol 60," *Comm. A.C.M.*, **6**, 1–17 (1963).

use of the context-free production possibilities. It says that an identifier is any word composed of letters and digits, but beginning with a letter. Thus *A*, *A*2, *AB*43, *abba*14 are identifiers. The student should see that the grammar with these productions will generate these strings, starting from ⟨identifier⟩.   △

EXAMPLE 6.57   Another portion of the ALGOL report concerns the syntax for numbers. In the Backus normal form we have the following:

⟨unsigned integer⟩ : : = ⟨digit⟩|⟨unsigned integer⟩⟨digit⟩
⟨integer⟩ : : = ⟨unsigned integer⟩| + ⟨unsigned integer⟩|
              − ⟨unsigned integer⟩
⟨decimal fraction⟩ : : = . ⟨unsigned integer⟩
⟨exponent part⟩ : : = $_{10}$ ⟨integer⟩
⟨decimal number⟩ : : = ⟨unsigned integer⟩|⟨decimal fraction⟩|
              ⟨unsigned integer⟩⟨decimal fraction⟩
⟨unsigned number⟩ : : = ⟨decimal number⟩|⟨exponent part⟩|
              ⟨decimal number⟩⟨exponent part⟩
⟨number⟩ : : = ⟨unsigned number⟩| + ⟨unsigned number⟩|
              − ⟨unsigned number⟩

Of course, much of this is reminiscent of Example 3.40.

Let us show that $-16.2\,_{10}\,33$ is a number in ALGOL. We note that $_{10}\,n$ means, as in scientific notation, $\times\, 10^{n}$. In order to generate the given number, we employ productions as follows:

⟨number⟩ → − ⟨unsigned number⟩
      → − ⟨decimal number⟩⟨exponent part⟩
      → − ⟨decimal number⟩ $_{10}$ ⟨integer⟩
      → − ⟨decimal number⟩ $_{10}$ ⟨unsigned integer⟩
      → − ⟨decimal number⟩ $_{10}$ ⟨unsigned integer⟩⟨digit⟩
      → − ⟨decimal number⟩ $_{10}$ ⟨digit⟩⟨digit⟩
      → − ⟨unsigned integer⟩⟨decimal fraction⟩ $_{10}$ ⟨digit⟩⟨digit⟩
      → − ⟨unsigned integer⟩ . ⟨unsigned integer⟩ $_{10}$ ⟨digit⟩⟨digit⟩
      → − ⟨unsigned integer⟩ . ⟨digit⟩ $_{10}$ ⟨digit⟩⟨digit⟩
      → − ⟨unsigned integer⟩⟨digit⟩ . ⟨digit⟩ $_{10}$ ⟨digit⟩⟨digit⟩
      → − ⟨digit⟩⟨digit⟩ . ⟨digit⟩ $_{10}$ ⟨digit⟩⟨digit⟩
      → − 1⟨digit⟩ . ⟨digit⟩ $_{10}$ ⟨digit⟩⟨digit⟩
      → − 1 6 . ⟨digit⟩ $_{10}$ ⟨digit⟩⟨digit⟩
                ⋮
      → − 16.2 $_{10}$ 33

EXAMPLE 6.58   We shall present one more segment from the ALGOL report without comment, but the connection with the algebraic expressions of Section 1.3 is not difficult to see, because these productions concern the syntax of the arithmetic expressions of ALGOL, again in the Backus normal form.

⟨adding operator⟩ : : = + | −
⟨multiplying operator⟩ : : = × | / | ÷
⟨primary⟩: : = ⟨unsigned number⟩|⟨variable⟩|
                ⟨function designator⟩|(⟨arithmetic expression⟩)
⟨factor⟩ : : = ⟨primary⟩|⟨factor⟩↑⟨primary⟩
⟨term⟩ : : = ⟨factor⟩|⟨term⟩⟨mutliplying operator⟩⟨factor⟩
⟨simple arithmetic expression⟩ : : = ⟨term⟩|
                ⟨adding operator⟩⟨term⟩|
                ⟨simple arithmetic expression⟩⟨adding operator⟩⟨term⟩
⟨if clause⟩ : : = **if** ⟨Boolean expression⟩ **then**
                ⟨simple arithmetic expression⟩⟨adding operator⟩⟨term⟩
⟨arithmetic expression⟩ : : = ⟨simple arithmetic expression⟩|
                ⟨if clause⟩⟨simple arithmetic expression⟩ **else**
                ⟨arithmetic expression⟩   △

It should now be fairly apparent why it is that the context-free grammars have proven to be a useful formalism for guiding the design and documentation of computer programming languages. We can also begin to understand why they are so crucial in the design of syntax-directed compilers. The methods of checking for syntactic errors will employ pushdown stack techniques (because the grammars are of type 2) along the lines of some of the earlier examples in this chapter. Obviously we cannot give the details here, but it is clear that they would depend on the ideas that we have developed.

EXERCISES

1 In the grammar of Example 6.46 diagram the generation of the sentences
   (a) Ron plays football
   (b) Big Ron eats the football
   (c) Sweet stody sounds skate swimmingly

2 In the grammar of Example 6.47 diagram the generation of the wff's
   (a) $p \, r \wedge q \, p \wedge \neg \vee$ (b) $r \neg q \wedge p \vee \neg$
   (c) $r \neg q \vee p \neg q \vee \neg \wedge$ (d) $p \, q \, r \, p \vee \wedge q \wedge \vee \neg$

3 In the grammar of Example 6.50 diagram the generation of the words
   (a) 012   (b) 001122   (c) 000011112222

4 Design a finite state automaton that will accept the language of Example 6.48.

*5 Show that the language of Example 6.49 is nonregular.

6 Show that the language of Example 6.49 is recursive.

7 Design grammars that will generate the languages of Exercise 9 of Section 6.7.

8 Design grammars for generating the languages of Exercise 10 of Section 6.7.

9 In the grammar of Example 6.56 diagram the generation of the identifiers:
   (a) *AB*43   (b) *abba*14   (c) *a*12   (d) *X*min2

10 In the grammar of Example 6.57 diagram the generation of the numbers:
   (a) 3.14 $\widehat{_{10}}$ 0   (b) 3.14   (c) -22.8 (d) .5 $\widehat{_{10}}$ -12

SUGGESTED REFERENCES

6.1 Lyndon, R. *Notes on Logic.* Van Nostrand, Princeton, N.J., 1966.

6.2 Mendelson, E. *Introduction to Mathematical Logic.* Van Nostrand, Princeton, N.J., 1964.

6.3 Minsky, M. *Computation*: *Finite and Infinite Machines.* Prentice-Hall, Englewood Cliffs, N.J., 1967.

6.4 Hopcroft, J., and J. Ullman. *Formal Languages and their Application to Automata.* Addison-Wesley, Reading, Mass., 1969.

6.5 Nelson, R. J. *Introduction to Automata.* Wiley, New York, 1968.

6.6 Chomsky, N. "Formal Properties of Grammars," in *Handbook of Mathematical Psychology*, Vol. 2. R. D. Luce, R. R. Bush, and E. Galanter (eds.), Wiley, New York, 1965.

Our first reference is a beautiful little book, one which offers great insight into formal logic without smothering the reader in all of its details. The book by Mendelson is more of a textbook, but it is also quite well done. Both of these are oriented toward questions of computability and decidability and therefore should provide an appropriate point of view for computer science students.

There follow three references for automata theory are chosen because of their thorough treatment of Turing machines. Minsky's book is easily the most accessible, and he manages to inject a number of interesting comments on the theory. The book by Hopcroft and Ullman provides a broad coverage of languages and automata, with particular emphasis on the intermediate machines and the Chomsky hierarchy. These topics are not examined quite so thoroughly in the book by Nelson, which, however, is nevertheless an excellent book on automata theory, written from the point of view of mathematical logic. Finally, we could not fail to call attention to the classical article by Chomsky, outlining the whole grammatical theory.

Answers to Odd-Numbered Exercises

Section 0.1

1 (a) $A = \{$Truman, Eisenhower, Kennedy, Johnson, Nixon, Ford$\}$
  (f) $F = \{x: x = \sigma_1\sigma_2 \cdots \sigma_k, 1 \leq k \leq 6,$ each $\sigma_i \in \Gamma, \sigma_1 \in \Sigma\}$
   $(\Sigma = \{A, B, \ldots, Z\}$ and $\Gamma = \{A, B, \ldots, Z, 0, 1, \ldots, 9\})$
3 (c) $m \in F' \Rightarrow m$ is a perfect square, $m \leq 100$
   $\Rightarrow m$ is a perfect square $\Rightarrow m \in E$
  (f) $-1 \in D'$ but $-1 \notin C'$
5 There are 2^n subsets of an n-set, as can be proved by induction (see Section 1.1).
7 Define $|\varnothing| = 0$.

Section 0.2

1 $\varnothing, \{a\}, \{b\}, \{c\}, \{d\}, \{a, b\}, \{a, c\}, \{a, d\}, \{b, c\}, \{b, d\}, \{c, d\}, \{a, b, c\}$, etc.
3 $\{\{1, 2, 3, 4\}\}, \{\{1, 2, 3\}, \{4\}\}, \{\{1, 2, 4\}, \{3\}\}, \{\{1, 3, 4\}, \{2\}\}, \{\{2, 3, 4\}, \{1\}\}, \{\{1, 2\},$
  $\{3, 4\}\}, \{\{1, 3\}, \{2, 4\}\}, \{\{1, 4\}, \{2, 3\}\}, \{\{1, 2\}, \{3\}, \{4\}\}, \{\{1, 3\}, \{2\}, \{4\}\}, \{\{1, 4\},$
  $\{2\}, \{3\}\}, \{\{2, 3\}, \{1\}, \{4\}\}, \{\{2, 4\}, \{1\}, \{3\}\}, \{\{3, 4\}, \{1\}, \{2\}\}, \{\{1\}, \{2\}, \{3\}, \{4\}\}$
5 (a) $A \subseteq A \cup A$ by (a) in Example 0.10; $A \cup A \subseteq A$ as follows:

$$x \in A \cup A \Rightarrow x \in A \quad \text{or} \quad x \in A \Rightarrow x \in A$$

  (h) $A \cap (\sim A) = \{x: x \in A \text{ and } x \notin A\} = \varnothing$
7 Besides observing that the $A_i \cap B_j$ are the nonempty intersections, we must show that

   (I) $\bigcup(A_i \cap B_j) \supseteq S$ (II) $(A_i \cap B_j) \cap (A_k \cap B_l) = \varnothing$ $(i \neq k$ or $j \neq l)$

  I. $x \in S \Rightarrow x \in A_i$ and $x \in B_j$ for some i, j
    $\Rightarrow x \in A_i \cap B_j \Rightarrow x \in \bigcup(A_i \cap B_j)$
  II. $(A_i \cap B_j) \cap (A_k \cap B_l) = (A_i \cap A_k) \cap (B_j \cap B_l) = \varnothing$ (Why?)

9 (a) II $\Rightarrow$ 2 as follows: If (2) were false, then for some $i \neq j$ we would have $A_i \subseteq A_j$. With the A_i nonempty

$$A_i \cap A_j = A_i \neq \emptyset$$

contradicting II. Note that this is a contrapositive proof.

11 Write max $\{A_i \cap B_j\}$ where max deletes subsets which are contained in others and also eliminate duplications. In this way (2) will be satisfied.

Section 0.3

1 (a) $A \times C = \{(a, b), (a, e), (a, g), (b, b), (b, e), (b, g), (d, b), (d, e), (d, g), (f, b), (f, e), (f, g)\}$
  (e) $A \times \emptyset = \emptyset$

3 There are 2^{mn} relations from an n-set to an m-set. (See Exercise 5 of Section 0.1.)

5 (a) $R = \{(s_1, t_1), (s_1, t_2), (s_1, t_4), (s_2, t_2), (s_2, t_3), (s_3, t_2)\}$

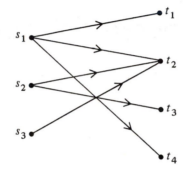

7 (d) $R = \{(x_1, x_1), (x_1, x_4), (x_2, x_1), (x_2, x_2), (x_3, x_1), (x_3, x_2), (x_3, x_3), (x_3, x_4), (x_4, x_2), (x_4, x_4)\}$

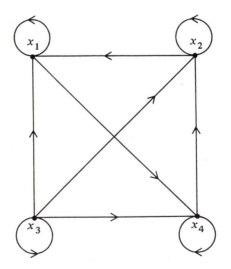

9 m^n (see Section 0.8)

Section 0.4

1 $\bigcup_{i=1}^{\infty} A_i = \{x \colon x \in A_i \text{ for some } i \geq 1\}$
Since $\Sigma^* \supseteq \{\epsilon\} \cup \Sigma \cup \Sigma^2 \cup \cdots$ is clear, let $x \in \Sigma^*$, say $x = \sigma_{i_1}\sigma_{i_2}\cdots\sigma_{i_k}$. Then $|x| = k$
and $x \in \Sigma^k$, showing $x \in \bigcup_{k=0}^{\infty} \Sigma^k$.

3 (a) 11, 12, 13, 14, 15, 21, 22, 23, 24, 25, 31, 32, 33, 34, 35, 41, 42, 43, 44, 45, 51, 52, 53, 54, 55
 (h) 123, 124, 125, 134, 135, 145, 234, 235, 245, 345

Section 0.5

1 (a) $[0] = \{\ldots, -5, 0, 5, 10, \ldots\}$
 $[1] = \{\ldots, -4, 1, 6, 11, \ldots\}$
 $[2] = \{\ldots, -3, 2, 7, 12, \ldots\}$
 $[3] = \{\ldots, -2, 3, 8, 13, \ldots\}$
 $[4] = \{\ldots, -1, 4, 9, 14, \ldots\}$

3 (i) $f \sim f$ since $D(f - f) = D(0) = 0$
 (ii) $f \sim g \Rightarrow D(f - g) = 0 \Rightarrow D(g - f) = -D(f - g) = 0 \Rightarrow g \sim f$
 (iii) $f \sim g, g \sim h \Rightarrow D(f - g) = D(g - h) = 0$
$$\Rightarrow D(f - h) = D((f - g) + (g - h))$$
$$= D(f - g) + D(g - h) = 0 + 0 = 0$$
$$\Rightarrow f \sim h$$
 $[f] = \{f + c \colon c \in R\}$

5 One must show that the collection $\pi = \{[a]\}$ is a partition of S, that is,

$$\text{(I)} \ \bigcup [a] = S \qquad \text{(II)} \ [a] \cap [b] = \varnothing \text{ for } [a] \neq [b]$$

But (II) is the statement of Lemma 2. For (I), note that for each $a \in S$ we have $a \in [a] \subseteq \bigcup [a]$. For the converse, see Exercise 10.

7 $x \bar{R} y, y \bar{R} z \Rightarrow x = x_1 R x_2 R \cdots R x_k = y$
$$y = x_k R x_{k+1} R \cdots R x_l = z$$
$$\Rightarrow x = x_1 R x_2 R \cdots R x_k R \cdots R x_l = z$$
$$\Rightarrow x \bar{R} z$$
Then show that if R is reflexive and symmetric, so is $\bar{R}$.

9 (a), (b), (d) are equivalence relations.

Section 0.6

1 (a)

	x_1	x_2	x_3	x_4
x_1	1	1	0	1
x_2	1	1	1	1
x_3	1	0	1	0
x_4	0	1	1	1

(g)

	x_1	x_2	x_3	x_4
x_1	1	0	1	0
x_2	0	0	0	0
x_3	0	1	0	1
x_4	1	0	0	1

3 Verifying the properties of Section 0.3, we have
 (i) $x R y \Rightarrow x R y$
 (ii') $R_1 \leq R_2, R_2 \leq R_1 \Rightarrow (x R_1 y \Leftrightarrow x R_2 y) \Rightarrow R_1 = R_2$
 (iii) $R_1 \leq R_2, R_2 \leq R_3 \Rightarrow (x R_1 y \Rightarrow x R_2 y \Rightarrow x R_3 y)$
$$(x R_1 y \Rightarrow x R_3 y) \Rightarrow R_1 \leq R_3$$

5 (a) $R_1 \leq R_1 \vee R_2, R_2 \leq R_1 \vee R_2$:
   $x R_1 y \Rightarrow (x R_1 y \text{ or } x R_2 y) \Rightarrow x (R_1 \vee R_2) y$
   $x R_2 y \Rightarrow (x R_1 y \text{ or } x R_2 y) \Rightarrow x (R_1 \vee R_2) y$
7 (a) All ones
9 (b) $p(R) = \{\overline{1, 3, 4}; \overline{2, 6}; \overline{5}\}$
11 (c) $\pi + \gamma = \{\overline{1, 2, 4, 5, 6}; \overline{3}\}$ $\pi \cdot \gamma = \{\overline{1}; \overline{2}; \overline{3}; \overline{4, 5}; \overline{6}\}$

Section 0.7

1 (a) Noninjective, surjective   (c) Bijective
3 There are m^n functions from an n-set to an m-set. Of these, $m!/(m - n)!$ are injective
   $(n \leq m)$.
5 (a) Noninjective, surjective
7 To see that the composition of injective functions is injective:
$$(f \circ g)(x) = (f \circ g)(y) \Rightarrow g(f(x)) = g(f(y))$$
$$\Rightarrow f(x) = f(y) \Rightarrow x = y$$
9 (a) $\{\overline{1}; \overline{2}; \overline{3, 4, 5}\}$ (g) $\{\overline{0}; \overline{-1, 1}; \overline{-2, 2}; \ldots\}$
11 There are only $26^3 = 17, 576$ distinct initials. Since the population of Denver is larger
   than this, the function
$$x \xrightarrow{f} f(x) = \text{initial of } x$$
   cannot be injective (use the pigeonhole principle).

Section 0.8

1 Nine students subscribe to none of the magazines.
3 $26^2 = 676$
5 14% order a hamburger "with everything"
7 $\binom{61}{50}$ by Rule 6
9 $\binom{8}{3}^3 = 175, 616$ according to Rules 3 and 6.
11 (b) 1 31 90 65 15 1
13 (a) Choose $a = b = 1$ in the binomial theorem:
$$(a + b)^n = \sum_{k=0}^{n} \binom{n}{k} a^{n-k} b^k$$

15 (a) 35   (d) 90
17 $\binom{n-1}{k-1} + \binom{n-1}{k} = \dfrac{(n-1)!}{(k-1)!(n-k)!} + \dfrac{(n-1)!}{k!(n-k-1)!}$
$$= \frac{k(n-1)! + (n-k)(n-1)!}{k!(n-k)!}$$
$$= \frac{n!}{k!(n-k)!} = \binom{n}{k}$$

Section 0.9

1 Obviously the sets $Q_k = \{n/k : n \in Z\}$ are countable, being in one-to-one correspondence
   with Z. In spite of the nonuniqueness of these representations (some expressions n/k
   may not be in the lowest terms), we may apply the third principle of countability to
   see that $Q = \bigcup_{k=1}^{\infty} Q_k$ is countable.

3 The intersection is a subset of a countable set, so the first principle of countability applies.

5 Example 0.71 shows that B^* is countable. So we may apply the first principle of countability.

7 Use induction on n (see Section 1.1). Otherwise, generalize the proof of the second principle of countability.

9 The set can be viewed as a subset of $Q \times Q$ for $Q = \{$rational numbers$\}$. But Exercise 1 and the second principle of countability show that $Q \times Q$ is countable.

11 The given set is a subset of Z^+, so the first principle of countability applies.

Section 0.10

1 First define addition and multiplication "inductively":

$$(n + 1) + m = (n + m) + 1 \qquad (n + 1) \cdot m = (n \cdot m) + m$$

identity: $1 \cdot 1 = 1$ and assuming that $n \cdot 1 = n$ for some $n \geq 1$, then

$$(n + 1) \cdot 1 = (n \cdot 1) + 1 = n + 1$$

distributivity: $1 \cdot (m + p) = m + p$ assuming commutativity already proved. Then assuming that $n \cdot (m + p) = n \cdot m + n \cdot p$ for some $n \geq 1$, we have

$$
\begin{aligned}
(n + 1) \cdot (m + p) &= (n \cdot (m + p)) + (m + p) \\
&= (n \cdot m + n \cdot p) + (m + p) \\
&= (n \cdot m + m) + (n \cdot p + p) \\
&= (n + 1) \cdot m + (n + 1) \cdot p
\end{aligned}
$$

again supposing that associativity has already been proved.

3 If $[a] \cap [b] \neq \varnothing$ and $[a] \neq [b]$, then by the latter there exists an element x which is in one class but not in the other, say

$$x \sim a \quad \text{but} \quad x \nsim b$$

By the first assumption, there is an element $y \in [a] \cap [b]$ so that

$$y \sim a \quad \text{and} \quad y \sim b$$

All together we have $x \sim a \sim y \sim b$, contradicting the assumption that $x \nsim b$.

5 It follows by the transitivity of implication.

Section 1.1

1 $\sum_{j=1}^{1} j^2 = 1 = 1(1 + 1)(2 \cdot 1 + 1)/6$ and if the proposition is true for some $n \geq 1$, then

$$
\begin{aligned}
\sum_{j=1}^{n+1} j^2 &= (n + 1)^2 + \sum_{j=1}^{n} j^2 = (n + 1)^2 + \frac{n(n + 1)(2n + 1)}{6} \\
&= \frac{(6n^2 + 12n + 6) + (2n^3 + 3n^2 + n)}{6} \\
&= \frac{2n^3 + 9n^2 + 13n + 6}{6} \\
&= \frac{(n + 1)((n + 1) + 1)(2(n + 1) + 1)}{6}
\end{aligned}
$$

3 If $|S| = 0$, then $S = \varnothing$ and

$$|\mathscr{P}(S)| = 1 = 2^0 = 2^{|S|}$$

Now, if the proposition is true for sets of size $n \geq 0$, and if

$$S = \{x_1, x_2, \ldots, x_n, x_{n+1}\}$$

we observe that S has two types of subsets, those which include x_{n+1} and those which do not. Since the two types are equinumerous, whereas the second type are subsets of $\{x_1, x_2, \ldots, x_n\}$, we have

$$|\mathscr{P}(S)| = 2 \cdot 2^n = 2^{n+1}$$

5 $\sum_{k=0}^{0} \binom{n}{k} = \binom{0}{0} = 1 = 2^0$ and if the proposition is true for some $n \geq 0$, then

$$\sum_{k=0}^{n+1} \binom{n+1}{k} = \binom{n+1}{0} + \binom{n+1}{n+1} + \sum_{k=1}^{n} \left[\binom{n}{k-1} + \binom{n}{k} \right]$$
$$= 1 + 1 + (2^n - 1) + (2^n - 1) = 2^{n+1}$$

7 $3 | 0 = 0^3 - 0$ and if the statement is true for some $n \geq 0$, then

$$n^3 - n = 3k \qquad (k \in Z)$$
$$(n+1)^3 - (n+1) = (n^3 + 3n^2 + 3n + 1) - (n+1)$$
$$= (n^3 + 3n^2 + 3n) - n$$
$$= 3k + 3(n^2 + n)$$
$$= 3(n^2 + n + k)$$

showing that $3 | (n+1)^3 - (n+1)$.

9 Assuming that associativity of addition has already been proved, we obtain the commutativity of addition:

$$1 + m = s(m) = m + 1$$
$$s(n) + m = s(n+m) = (n+m) + 1 = n + (m+1) = (m+1) + n$$
$$= m + (1+n) = m + s(n)$$

11 (a) 6   (c) 27

13 (b) The definition of gcd (a, b) is symmetric in a and b.
    (c) Since $n|n$ (and $n|n$), we have $n|$gcd (n, n) by (ii) in the definition of gcd. But gcd $(n, n)|n$ by (i), so gcd $(n, n) = n$ by antisymmetry of the divisibility relation.

Section 1.2

1 (a) $\langle (n, m): n, m \in Z^+, n < m \rangle \subseteq Z^+ \times Z^+$
    (e) $\langle (a, b, q, r): a, b \in Z^+, q, r \in N, a = bq + r, 0 \leq r < b \rangle \subseteq Z^+ \times Z^+ \times N \times N$
3 See the discussion of Section 3.1.

Section 1.3

1 (a) 3 is an algebraic expression ($Z^+ \subseteq \mathscr{E}$)
     n is an algebraic expression ($\mathscr{V} \subseteq \mathscr{E}$)
     $3 + n$ is an algebraic expression (ii)
     m is an algebraic expression ($\mathscr{V} \subseteq \mathscr{E}$)
     $(3 + n) \cdot m$ is an algebraic expression (ii)
     $((3 + n) \cdot m) \cdot n$ is an algebraic expression (ii)
3 See Exercise 1.
5 (b) $\varnothing$ is an algebraic expression ($\mathscr{P}(S) \subseteq \mathscr{E}$)
     B is an algebraic expression ($\mathscr{V} \subseteq \mathscr{E}$)
     $\varnothing \cup B$ is an algebraic expression (ii)
     $\{a\}$ is an algebraic expression ($\mathscr{P}(S) \subseteq \mathscr{E}$)
     $(\varnothing \cup B) \cap \{a\}$ is an algebraic expression (ii)

Section 1.4

1 (a) $| \,|: C \to R$ (d) $\#: \Pi(S) \to Z^+$
3 (a) type N (b) type R
5 (a) $|A \cup \{a, c\}| + 4 \cdot n$ corresponds to a function $f(A, n)$ such that

$$f: \mathscr{P}(S) \times N \to N$$

in which

$$(\varnothing, 1) \xrightarrow{f} |\varnothing \cup \{a, c\}| + 4 \cdot 1 = 6$$
$$(\{b\}, 7) \xrightarrow{f} |\{b\} \cup \{a, c\}| + 4 \cdot 7 = 31$$
$$(\{a, b\}, 9) \xrightarrow{f} |\{a, b\} \cup \{a, c\}| + 4 \cdot 9 = 39, \quad \text{etc.}$$

Section 1.5

1 (b)

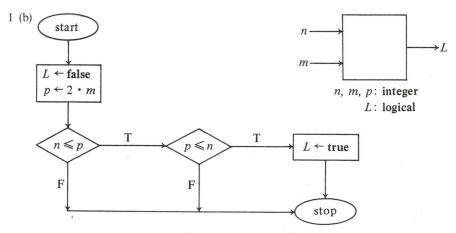

3 (a) $n = 7, m = 3$ (c) $n = 11, m = 121$
5 (b) The algorithm decides whether or not $\{A, B\}$ is a partition of S.

Section 1.6

1 (d)

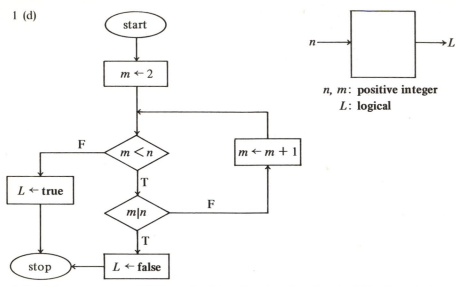

n, m: **positive integer**
L: **logical**

3 In each case we can exhibit an algorithm (that is, a flowchart) which always halts to yield the decision or functional value in question.

5 See Exercise 1(d).

7

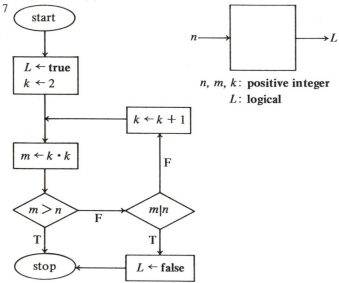

n, m, k: **positive integer**
L: **logical**

Section 1.7

1 (e) $n = 13, 40, 20, 10, 5, 16, 8, 4, 2, 1$

3 (c) n	A_1	A_2	A_3	A_4	i	T	L
4	$\{1, 3\}$	$\{2\}$	$\{6\}$	$\{3, 4, 5\}$	1	$\emptyset$	**false**
					2	$\{1, 3\}$	
					3	$\{1, 2, 3\}$	
					4	$\{1, 2, 3, 6\}$	

5 (a) a	b	r	temp	gcd
48	42	6		
42	6	0		6

7 (a) A	B	C	D	L
$\{1, 3, 7\}$	$\{2, 4, 5, 6\}$	$\{1, 2, 3, 4, 5, 6, 7\}$	$\emptyset$	**false**
				true

Section 1.8

1 a, b are consistent with Rule I.

3 All three are consistent with Rule III.

5 (a) $(p = n^2) \wedge (m = 3p + 1)$

   $p = n^2 + m = 4p + 1$

Section 1.9

3 (c) Take $F = r$

   (i) $r \neq 0 \Rightarrow F = r > 0$

   (ii) $F_{n+1} = r_{n+1} < r_n = F_n$ because the r_n are the successive remainders:

   $$r_{n-1} = r_n q_{n+1} + r_{n+1} \qquad (0 \le r_{n+1} < r_n)$$

5 This is an algorithm for computing the function $f(n) = n!$

Section 2.1

1 (b)

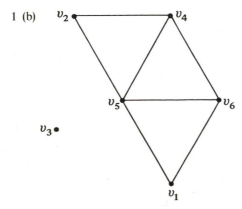

3 (a)

	v_1	v_2	v_3	v_4	v_5
v_2	1				
v_3	0	1			
v_4	1	1	1		
v_5	0	0	0	1	
v_6	0	1	1	0	1

5 (a)

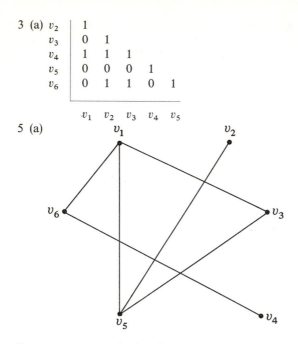

7 $n + n + \cdots + n$ (m times) $= m \cdot n$

Section 2.2

1 (a) $\langle v_1, v_2, v_3, v_4, v_5, v_6 \rangle$ is a simple path

$\langle v_1, v_2, v_3, v_6, v_2, v_3, v_4 \rangle$ is nonsimple

3 (a) $\langle v_1, v_2, v_3, v_6, v_2, v_4 \rangle$

5 (b) $x = v_1 \quad A = \{v_1, v_5\} \qquad B = \{v_1, v_5\} \quad p = 1$

$x = v_2 \quad A = \{v_2, v_3, v_4\} \quad B = V \qquad p = 2$

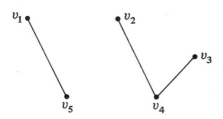

Section 2.3

1 (c) $c^4 \oplus c^5 \oplus c^6 = (1, 1, 0, 1, 1, 1, 0) \oplus (0, 0, 1, 1, 0, 1, 1) \oplus (1, 1, 1, 0, 1, 0, 1) = 0$

3 In addition to the circuits listed in Example 2.16, let

$$c^5 = (1, 0, 1, 1, 0, 0, 0, 1, 1, 0, 1, 0)$$

Then $\{c^1, c^2, c^3, c^5\}$ is independent, since each $c^i \neq 0$, each $c^i \oplus c^j \neq 0 \, (i \neq j)$, and also

$$c^1 \oplus c^2 \oplus c^3 = (1, 1, 1, 0, 1, 0, 0, 1, 0, 1, 1, 1) \neq 0$$
$$c^1 \oplus c^2 \oplus c^5 = (1, 0, 1, 1, 0, 1, 1, 0, 1, 1, 0, 1) \neq 0$$
$$c^1 \oplus c^3 \oplus c^5 = (0, 1, 0, 1, 1, 1, 0, 1, 0, 1, 1, 1) \neq 0$$

and finally

$$c^1 \oplus c^2 \oplus c^3 \oplus c^5 = (0, 1, 0, 1, 1, 0, 0, 0, 1, 1, 0, 1) \neq 0$$

5 (a) $p(G) = m - n + p = 8 - 5 + 1 = 4$

9 (a) $r(G_0) = 0 - 5 + 5 = 0$
$r(G_1) = 1 - 5 + 4 = 0$
$r(G_2) = 2 - 5 + 3 = 0$
$r(G_3) = 3 - 5 + 2 = 0$
$r(G_4) = 4 - 5 + 2 = 1 \quad \{e_2, e_3, e_4\}$
$r(G_5) = 5 - 5 + 1 = 1$
$r(G_6) = 6 - 5 + 1 = 2 \quad \{e_1, e_2, e_6\}$
$r(G_7) = 7 - 5 + 1 = 3 \quad \{e_5, e_6, e_7\}$
$r(G_8) = 8 - 5 + 1 = 4 \quad \{e_4, e_5, e_8\}$

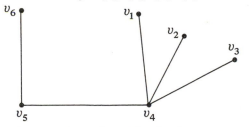

Section 2.4

1 (c) Remove in succession the edges v_1v_2, v_2v_3, v_3v_6, v_2v_6:

3 (c) Given the spanning tree shown, insert the chords v_1v_2, v_2v_3, v_3v_6, v_2v_6 to obtain circuits as follows:

$$v_1v_2: v_1v_2v_4v_1$$
$$v_2v_3: v_2v_3v_4v_2$$
$$v_3v_6: v_3v_4v_5v_6v_3$$
$$v_2v_6: v_2v_4v_5v_6v_2$$

5 Suppose that G is connected ($p = 1$). Since the tree has $n - 1$ edges (Theorem 2.2), we have removed $m - (n - 1) = m - n + 1 = p(G)$ edges in forming the spanning tree.

Section 2.5

1 A planar representation can be given.

3 (a) If K_n were planar, then we would have

$$r = m - n + 2$$
$$= \frac{n(n-1)}{2} - n + 2 = \frac{n^2 - 3n + 4}{2}$$

and according to Theorem 2.6, $3r \leq 2m$, that is,

$$\tfrac{3}{2}(n^2 - 3n + 4) \leq n(n - 1)$$
$$n^2 - 7n + 12 \leq 0$$

a contradiction, for all $n > 4$.

5 (c) $n - m + r = 6 - 12 + 8 = 2$

   (e) $n - m + r = 36 - 60 + 26 = 2$

Section 2.6

1 (b) There are 12 in all.

3 The following invariants

	n	m
a	5	6
b	5	7
c	5	8
d	6	

show that no two of these graphs are isomorphic.

5 The invariant m shows that no two of these graphs are isomorphic.

7 (a) $\beta = 2\ \{v_1, v_2\}$ (d) $\beta = 3\ \{v_1, v_2, v_3\}$

9 (c) $\alpha = 5\ \{v_1, v_2, v_3, v_4, v_5\}$

11 (e) $\beta' = 4\ \{v_2, v_4, v_5, v_6\}$ (f) $\beta' = 3\ \{v_1, v_2, v_4\}$

13 Use the fact that the V_i in a (smallest) proper coloring $V = \bigcup_{i=1}^{\kappa} V_i$ are independent sets. Also observe that we may always color the vertices of a maximal independent set with the same color.

15 (a) $\alpha_e = 3\ \{v_1 v_5, v_2 v_4, v_1 v_3\}$

   (d) Undefined, since the graph is not connected.

Section 2.7

1 (b)

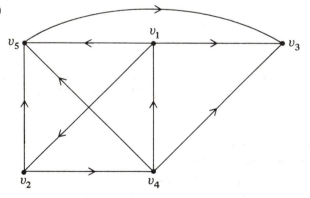

3 (a)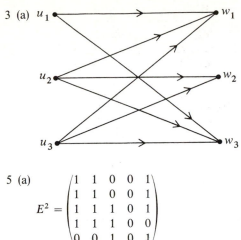

5 (a)
$$E^2 = \begin{pmatrix} 1 & 1 & 0 & 0 & 1 \\ 1 & 1 & 0 & 0 & 1 \\ 1 & 1 & 1 & 0 & 1 \\ 1 & 1 & 1 & 0 & 0 \\ 0 & 0 & 1 & 0 & 1 \end{pmatrix}$$

Section 2.8

1 (a)
$$E^* = \begin{pmatrix} 1 & 1 & 1 & 0 & 1 \\ 1 & 1 & 1 & 0 & 1 \\ 1 & 1 & 1 & 0 & 1 \\ 1 & 1 & 1 & 1 & 1 \\ 1 & 1 & 1 & 0 & 1 \end{pmatrix}$$

5 The algorithm can be devised as a simple extension of Warshall's algorithm, with a test at the conclusion of computing E^*.

7 See Exercise 5.

Section 2.9

1 (c) $\partial^-(v_1) = 2 \qquad \partial^+(v_1) = 0$ (sink)
$\qquad \partial^-(v_2) = 1 \qquad \partial^+(v_2) = 2$
$\qquad \partial^-(v_3) = 2 \qquad \partial^+(v_3) = 2$
$\qquad \partial^-(v_4) = 2 \qquad \partial^+(v_4) = 3$

3 Begin by choosing an arbitrary edge. Let $\langle v_1, v_2, \ldots, v_k \rangle$ be a path that has not duplicated any vertices. Then examine the two possibilities:

(a) Every vertex of the graph occurs in the path.

(b) There exists a vertex v not in the path.

5 *Hint*: Assume the existence of an array

$$D : \textbf{array} \ \{0, 1, \ldots, d[v]\} \ \textbf{of} \ \mathscr{P}(V)$$

where

$$w \in D(k) \Leftrightarrow d(w) = k$$

7 (b)

u	v	u^*	M	W	d
v_2	v_5	v_2	$\varnothing$	$\{v_2\}$	$d[v_2] = 0$
			$\{v_2\}$	$\{v_1, v_2, v_4, v_6\}$	$d[v_1] = d[v_4]$
				$\{v_1, v_4, v_6\}$	$= d[v_6] = 1$
		v_1	$\{v_1, v_2\}$	$\{v_1, v_3, v_4, v_6\}$	$d[v_3] = 2$
				$\{v_3, v_4, v_6\}$	$d[v_4] = 1$
		v_4	$\{v_1, v_2, v_4\}$	$\{v_3, v_4, v_6\}$	$d[v_3] = 2$
				$\{v_3, v_6\}$	$d[v_6] = 1$
		v_6	$\{v_1, v_2, v_4, v_6\}$	$\{v_3, v_6\}$	$d[v_3] = 2$
				$\{v_3\}$	
		v_3	$\{v_1, v_2, v_3, v_4, v_6\}$	$\{v_3, v_5\}$	$d[v_5] = 3$
				$\{v_5\}$	
		v_5			

Using the backtrack procedure, we find that $\langle v_2, v_1, v_3, v_5 \rangle$ is a shortest path (length 3) from v_2 to v_5.

Section 2.10

1 Use induction on "antecedents." Thus we choose any edge $v_1 E v_0$. (If there are no edges, then any vertex is a source.) Either v_1 is a source or there is an edge $v_2 E v_1$. Continuing in this way, we can never arrive at a vertex already chosen, because that would imply the existence of a circuit. Since there are only finitely many vertices, this process must terminate (in a source v_k).

3 (a) The vertices are consistently labeled in the following diagram:

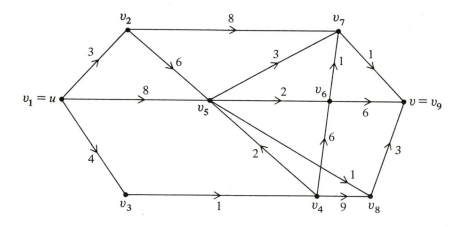

Then Algorithm VIII determines in succession

$$d(v_1) = 0$$
$$d(v_2) = \min \{\infty, 0 + 3\} = 3$$
$$d(v_3) = \min \{\infty, 0 + 4\} = 4$$
$$d(v_4) = \min \{\infty, 4 + 1\} = 5$$
$$d(v_5) = \min \{\infty, 0 + 8, 3 + 6, 5 + 2\} = 7$$

$$d(v_6) = \min\{\infty, 5 + 6, 7 + 2\} = 9$$
$$d(v_7) = \min\{\infty, 3 + 8, 7 + 3, 9 + 1\} = 10$$
$$d(v_8) = \min\{\infty, 5 + 9, 7 + 1\} = 8$$
$$d(v_9) = \min\{\infty, 10 + 1, 9 + 6, 8 + 3\} = 11$$

Using the backtrack procedure, we obtain $\langle v_1, v_3, v_4, v_5, v_6, v_7, v_9 \rangle$ as a shortest path (length 11) from $u = v_1$ to $v = v_9$.

5 (a) We determine

$$A(v_1) = \{v_3, v_4, v_5\}$$
$$A(v_2) = \{v_3, v_5\}$$
$$A(v_3) = \{v_1, v_2\}$$
$$A(v_4) = \varnothing \text{ (source)}$$
$$A(v_5) = \{v_2, v_3, v_4\}$$

Then a trace of the algorithm begins as follows:

V	n	k	u	U	L	
$\{v_1, v_2, v_3, v_4, v_5\}$	5	0		$\{v_1, v_2, v_3, v_4, v_5\}$	**true**	
		1	v_4	$\{v_1, v_2, v_3, v_5\}$		$\lambda[v_4] = 1$
					false	

At this point the proposition in the **while** statement is false. We have come up against a circuit.

Section 3.1

1 That $\mathscr{P}(S)$ is a monoid follows from Properties (P2) and (P6). Similarly with $\mathscr{R}(S)$. In $B = (B, \oplus, 0)$ we have

$$a \oplus (b \oplus c) = (a \oplus b) \oplus c$$
$$a \oplus 0 = 0 \oplus a = 0$$

for all a, b, c by exhaustive verification.

3 M, $\{0\}$, $\{0, 2, \infty\}$, and $\{0, 2, 3, \infty\}$ are submonoids.

7 All but (b) are monoids.

9 If

$$xe_1 = e_1 x = x \qquad xe_2 = e_2 x = x$$

for all $x \in M$, then $e_1 = e_1 e_2 = e_2$.

11 Define the mapping $\varphi: R^+ \to R$ by writing

$$\varphi(x) = \log x$$

Then φ is onto because $\varphi(e^y) = \log e^y = y$, and

$$\varphi(1) = \log 1 = 0$$
$$\varphi(xy) = \log(xy) = \log x + \log y = \varphi(x) + \varphi(y)$$

Since logarithm is a monotonic function, φ is one to one.

Section 3.2

3 (b)

$\circ$	T_0	T_1	T_2	T_3	T_∞
T_0	T_0	T_1	T_2	T_3	T_∞
T_1	T_1	T_2	T_3	T_∞	T_∞
T_2	T_2	T_3	T_∞	T_∞	T_∞
T_3	T_3	T_∞	T_∞	T_∞	T_∞
T_∞	T_∞	T_∞	T_∞	T_∞	T_∞

where $T_g(x) = x + g$

5 Denoting

we obtain the multiplication table

$\circ$	T_1	T_2	T_3	T_4
T_1	T_1	T_2	T_3	T_4
T_2	T_2	T_2	T_3	T_3
T_3	T_3	T_2	T_3	T_2
T_4	T_4	T_2	T_3	T_1

Section 3.3

3 $e \to 2, a \to 1$

5 (a) Because

$$T_2 \circ T_2 = T_0 : \begin{array}{l} 0 \to 0 \\ 1 \to 1 \\ 2 \to 2 \end{array}$$

which is the identity element of the transformation monoid, whereas we would have

$$\varphi(T_2 \circ T_2) = \varphi(T_2)\varphi(T_2) = b \cdot b = b \neq e$$

the given mapping does not extend to a monoid homomorphism.

9 The mapping $\varphi(x + iy) = x$ is a surjective homomorphism.

Section 3.4

1 Following the hint, we have

$$\begin{aligned}
\varphi_1(x) = \varphi_1(\sigma_{i_1}\sigma_{i_2}\cdots\sigma_{i_k}) &= \varphi_1(\sigma_{i_1})\varphi_1(\sigma_{i_2})\cdots\varphi_1(\sigma_{i_k}) \\
&= \varphi_2(\sigma_{i_1})\varphi_2(\sigma_{i_2})\cdots\varphi_2(\sigma_{i_k}) \\
&= \varphi_2(\sigma_{i_1}\sigma_{i_2}\cdots\sigma_{i_k}) \\
&= \varphi_2(x)
\end{aligned}$$

for all $x \in \Sigma^*$, that is, $\varphi_1 = \varphi_2$.

5 False

7 The symbols of the alphabet

$$\Sigma = \{A, B, \ldots, Z, 0, 1, \ldots, 9\}$$

are encoded by a mapping φ into words of an alphabet

$$\Gamma = \{\cdot, -, \square\}$$

that is, into a system of dots, dashes, and spaces. This is done in such a way that the extension $\varphi \colon \Sigma^* \to \Gamma^*$ to a monoid homomorphism is injective.

Section 3.5

1 (a) $\beta = 3$

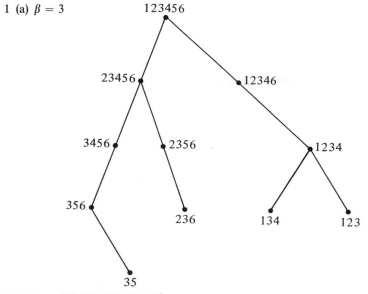

3 (c) $L = \{00, 01, 100, 101, 11\}$

5 *Hint*: Count the number of vertices in a "full" binary tree of length $|x|$.

Section 3.6

1 In the first pass through the exchange sort, the algorithm compares $y(j)$ with $y(j + 1)$ for $1 \le j \le n - 1$, exchanging them when they are out of order. In this way, the largest entry will be moved to the bottom in only one pass and will be ignored subsequently. Similarly, the next largest entry will be moved to the $(n - 1)$st position in the second pass. After $n - 1$ passes, the array will have been sorted. The setting of the logical variable $L \leftarrow$ **false** each time that an interchange is performed allows for an early termination of the algorithm in case the entries are already sorted.

5 (c) The successive rearrangements will be as follows:

$$
\begin{array}{ccccc}
19, & \underline{21}, & 6, & 20, & 25 \\
19, & 21, & \underline{6}, & 20, & 25 \\
6, & 19, & 21, & \underline{20}, & 25 \\
6, & 19, & 20, & 21, & \underline{25} \\
6, & 19, & 20, & 21, & 25
\end{array}
$$

7 (b) In interpreting the tree, we write

$$\text{tree} = \text{tree } \epsilon = (\text{tree } 0, \epsilon\text{-label, tree } 1)$$
$$= (\text{tree } 0, 103, \text{tree } 1)$$
$$= ((\text{tree } 00, 0\text{-label, tree } 01), 103, (\text{tree } 10, 1\text{-label, tree } 11))$$
$$= ((\text{tree } 00, 81, \text{tree } 01), 103, (\text{tree } 10, 115, \text{tree } 11))$$

etc., and finally, after disregarding the punctuation, we have the sorted sequence 32, 70, 81, 94, 103, 115.

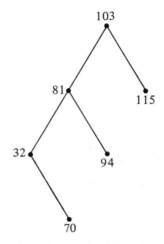

11 (a) After successive passes through the algorithm, the entries will be rearranged as follows (beginning with):

> our father who art in heaven hallowed be thy name
> father our art in heaven hallowed be thy name who
> father art in heaven hallowed be our name thy who
> art father heaven hallowed be in name our thy who
> art father hallowed be heaven in name our thy who
> art father be hallowed heaven in name our thy who
> art be father hallowed heaven in name our thy who

Section 3.7

3 *Hint*: First show that for each congruence R the congruence class $[0]$ consists of multiples of some fixed integer k. Certainly this is true if $[0] = \{0\}$. So suppose that $[0]$ contains nonzero (and hence also positive) elements, and choose k to be the smallest positive integer in $[0]$. Then use the division algorithm.

5 If $x\,R\,y \Leftrightarrow x = y$ then

$$x\,R\,u,\ y\,R\,v \Rightarrow x = u,\ y = v \Rightarrow xy = uv \Rightarrow xy\,R\,uv$$

On the other hand, if $x\,R\,y$ for all x, y, then

$$x\,R\,u,\ y\,R\,v \Rightarrow xy\,R\,uv$$

trivially.

7 The mapping

$$e \to E$$
$$a \to E$$
$$b \to E$$
$$c \to B$$
$$d \to B$$
$$f \to B$$

is a surjective monoid homomorphism $\varphi \colon M \to M'$. The kernel corresponding to the partition $\{\overline{e, a, b}; \overline{c, d, f}\}$ is easily seen to be a monoid congruence on M:

	e	a	b	c	d	f
e	e	a	b	c	d	f
a	a	b	e	d	f	c
b	b	e	a	f	c	d
c	c	d	f	c	d	f
d	d	f	c	d	f	c
f	f	c	d	f	c	d

Section 3.8

1 (c)

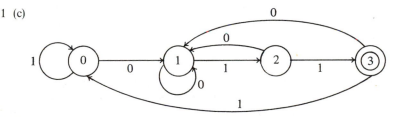

3 We have

$$M(s, x\epsilon) = M(s, x) = M(M(s, x), \epsilon)$$
$$M(s, x\sigma) = M(M(s, x), \sigma) \qquad (\sigma \in \Sigma)$$

for all x, owing to the definition of our extension of M to $M \colon S \times \Sigma^* \to S$. So assume that the result holds (for all x) and for all y of a certain length $|y| \geq 1$. Then

$$M(s, x(y\sigma)) = M(s, (xy)\sigma) = M(M(s, xy), \sigma)$$
$$= M(M(M(s, x), y), \sigma)$$
$$= M(M(s, x), y\sigma)$$

9 When $\varphi \colon A \to B$ is surjective, redesignate

$$I_A = \varphi^{-1}(I_B) = \{s \in S_A \colon \varphi(s) \in I_B\}$$

Similarly, choose

$$\Phi(\tau) \in \varphi^{-1}(\tau) = \{\sigma \in \Sigma_A \colon \varphi(\sigma) = \tau\}$$

for all $\tau \in \Sigma_B$. Then show that after Φ has been extended to a monoid homomorphism $\Phi \colon \Sigma_B^* \to \Sigma_A^*$, we have $x \in \lambda_B$ iff $\Phi(x) \in \lambda_A$.

11

	0	2	1
a	b	b	c
c	b	b	a
b	c	a	d
e	a	a	d
d	a	c	e

Section 3.9

1 Clearly $s \underset{i+1}{\sim} s'$ implies $s \underset{i}{\sim} s'$, but it also implies that $M_s^{(i)} = M_{s'}^{(i)}$. Otherwise, we would have

$$M(s, \sigma) \underset{i}{\sim} M(s', \sigma)$$

for some $\sigma \in \Sigma$. Then there would exist a word $x \in \Sigma^*$ of length $|x| \le i$ such that

$$M(M(s, \sigma), x) \in I \qquad M(M(s', \sigma), x) \notin I$$

that is

$$M(s, \sigma x) \in I \qquad M(s, \sigma x) \notin I$$

(or vice versa), and this contradicts the fact that $s \underset{i+1}{\sim} s'$. The converse is straightforward.

5 *Hint*: Find a surjective (noninjective) automatá homomorphism $\varphi\colon A \to B$ for automata with the same alphabet, in which

$$q \in I_A \Leftrightarrow \varphi q \in I_B$$

7 That $\tilde{A}$ is reduced is easy to show. If A_1 and A_2 are reduced and equivalent to A, then they are equivalent to each other. Using the connectivity of A_1, one sees that for each state $s \in S_1$, there is a (unique) state $s' \in S_2$ such that

$$M_1(s, x) \in I_1 \Leftrightarrow M_2(s', x) \in I_2$$

and conversely. The uniqueness follows from the fact that the A_i are reduced. It can then be shown that the mapping $s \overset{\varphi}{\to} s'$ is an automata isomorphism. That it is a homomorphism results from the facts that for $s \in S_1$ and $\sigma \in \Sigma$,

$$(M_1(s, \sigma)) \sim M_2(\varphi s, \sigma)$$

and A_2 is reduced.

Section 3.10

1 $M_A(s, \sigma) = ([M_A(s, \sigma)]_1, [M_A(s, \sigma)]_2)$

$= (M_{A/R_1}([s]_1, [\sigma]_1), M_{A/R_2}([s]_2, [\sigma]_2))$

$= M_{A/R_1 \times A/R_2}(([s]_1, [s]_2), ([\sigma]_1, [\sigma]_2))$

$= M_{A/R_1 \times A/R_2}(\varphi s, \varphi \sigma)$

3 Use the mapping

$$\varphi(s) = ([s]_{\pi_R}, [s]_\pi)$$

on states and the identity mapping on symbols. It will be injective because of the choice of π for which $\pi_R \cdot \pi = 0$.

5 (c) It is a monoid with 24 elements. Only the transformations that interchange two elements are missing from the full transformation monoid.

7 (b) The partition $\pi_R = \{\overline{0, 1, 4}; \overline{2, 3}\}$ represents a congruence R on the given machine A. The "head" machine is A/R:

	α	β	γ
[0]	[2]	[2]	[0]
[2]	[0]	[0]	[2]

In order to construct a "tail," we choose $\pi = \{\overline{0, 2}; \overline{1, 3}; \overline{4}\}$ and a corresponding machine A_π:

	([0], α)	([0], β)	([0], γ)	([2], α)	([2], β)	([2], γ)
[0]	[0]	[1]	[4]	[4]	[0]	[1]
[1]	[1]	[0]	[0]	[0]	[1]	[0]
[4]	[0]	[0]	[0]	$\longleftarrow$ arbitrary $\longrightarrow$		

Then the realization $A \rightarrowtail A/R \circ A_\pi$ is given by the mapping:

$$0 \to ([0], [0])$$
$$1 \to ([0], [1])$$
$$2 \to ([2], [0])$$
$$3 \to ([2], [1])$$
$$4 \to ([0], [4])$$

Section 4.1

3 (a) and (c) are posets.

5 By transitivity, we would have $x_1 \leq x_2$ and $x_2 \leq x_1$. By antisymmetry, $x_1 = x_2$, contradicting the fact that the x_i were distinct.

7 In considering $a < x < b$, we see that there can be only finitely many x in between (perhaps none, in which case we are done). Choose one (calling it x). Then there are even fewer elements between a and x. Continuing inductively, in a finite number of steps, we obtain x_1 covering $a = x_0$. Again proceed inductively on the intervals $x_i < x < b$.

9 For the antisymmetry, observe that in view of

$$(x_1, \ldots, x_n) \leq (y_1, \ldots, y_n)$$
$$(y_1, \ldots, y_n) \leq (x_1, \ldots, x_n)$$

we could not have $x_1 < y_1$ or $k < n$ and $x_{k+1} < y_{k+1}$, for they would contradict $(y_1, \ldots, y_n) \leq (x_1, \ldots, x_n)$. Consequently, $k = n$ and $x_i = y_i$ for all i, so that $(x_1, \ldots, x_n) = (y_1, \ldots, y_n)$. The order is total.

Section 4.2

1 (a), (b), (d) are lattices. Only (d) is distributive.

3 With the assumption $x \geq y$, we have

$$x \geq y, y + z \geq y \quad \text{so that} \quad x \cdot (y + z) \geq y$$
$$x \geq x \cdot z, y + z \geq z \geq x \cdot z \quad \text{so that} \quad x \cdot (y + z) \geq x \cdot z$$

using the definition of lub and glb. Finally

$$x \cdot (y + z) \geq y + x \cdot z$$

by the definition of lub.

5 Considering glb's, we have

(a′) $x \cdot y \leq x, x \cdot y \leq y$ since $x + x \cdot y = x, y + y \cdot x = y$

(b′) $w \leq x, w \leq y \Rightarrow w + x = x, w + y = y$
$$\Rightarrow x \cdot y = (w + x) \cdot (w + y) = w + x \cdot y$$
$$\Rightarrow w \leq x \cdot y$$

7 (b) Arrange to check just one of the distributive laws exhaustively (see Exercise 11).

9 (b)

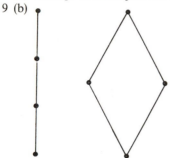

Section 4.3

1 (a) $\sigma_1\sigma_2 + \sigma_3 + \sigma_5$ (b) $\sigma_3 + \sigma_2\sigma_5$

5 (d) $f = v_1v_2v_3 + v_2v_3v_6 + v_1v_3v_4 + v_3v_4v_6 + v_3v_5$ and $\delta = 2$

Section 4.4

1 $\binom{n}{k}2^{n-k}$

5 (a) 1   (b) 3

7 Those in (b), (c), (d) are parallel.

9 *Hint*: It is best to assume that the cells will be given in the form $(c_1, c_2, \ldots, c_n)$; that is, introduce

$$c, d: \textbf{array } n^+ \textbf{ of } C$$

Section 4.5

1 For the zero element, consider $(\cdots ((x_1 \cdot x_2) \cdot x_3) \cdots) \cdot x_n$.

3 (b) $ab' + c(a' + b + d) = ab' + a'c + bc + cd$

$$= (ab' + ab'c) + a'c + (bc + abc) + cd$$
$$= (ab' + bc + a'c) + (ab'c + abc) + cd$$
$$= ab' + bc + (a'c + ac) + cd$$
$$= ab' + (bc + c + cd)$$
$$= ab' + c$$

5 (a) In the Boolean algebra $\mathscr{P}(S)$ for $S = \{a, b, c\}$, take

$$A = \{a, b\} \qquad B = \{a, c\} \qquad C = \{a\}$$

9 We can easily show that zero and one are unique in any Boolean algebra. Since they must exist, we may write our three elements as $0, x, 1$. Now $x' = 0$ would imply $x = 1$, a contradiction. Similarly, we cannot have $x' = 1$. So $x' = x$. But this gives

$$1 = x + x' = x + x = x$$

again a contradiction.

Section 4.6

1 (a), (c), (d)

x_1	x_2	x_3	$F + G$	F'	$G \cdot H$
0	0	0	1	1	1
0	0	1	1	0	0
0	1	0	1	0	0
0	1	1	0	1	0
1	0	0	1	0	0
1	0	1	1	0	1
1	1	0	1	0	0
1	1	1	0	1	0

7 Show that the truth table of each coatom L_i has exactly one zero entry.

9 To prove the analog of Lemma 1, that is,

(i) $a \cdot b = 0$ (for atoms $a \neq b$)

(ii) $\sum_{a \in A} a = 1$ (for the set A of atoms)

begin as follows. In (i), note that

$$a \cdot b \leq a \quad \text{and} \quad a \cdot b \leq b$$

But since a, b are atoms, we must have

$$a \cdot b = a = b \quad \text{or} \quad a \cdot b = 0$$

The former is a contradiction.

11 Use the result of Exercise 3 of Section 1.1.

Section 4.7

3 (b) Each has the truth table

x_1	x_2	x_3	
0	0	0	0
0	0	1	0
0	1	0	1
0	1	1	1
1	0	0	0
1	0	1	1
1	1	0	1
1	1	1	1

5 If f, g have the same minterm canonical form $\sum m_j$, then

$$f = \sum m_j = g$$

and certainly $I(f) = I(g)$. Conversely, if $I(f) = I(g) = F$, then they will have the same minterm canonical form by the uniqueness of the representation of F in the second lemma preceding Theorem 4.5.

Section 4.8

1 (a) 6   (b) 8   (f) 16
3 (b) and (c)
5 (b) $K(f) = \{(0, 0, 0), (0, 0, 1), (0, 1, 1), (1, 0, 0), (1, 1, 0), (1, 1, 1),$
$(I, 0, 0), (1, I, 0), (1, 1, I), (I, 1, 1), (0, I, 1), (0, 0, I)\}$
   basic cells: $(I, 0, 0), (1, I, 0), (1, 1, I), (I, 1, 1), (0, I, 1), (0, 0, I)$

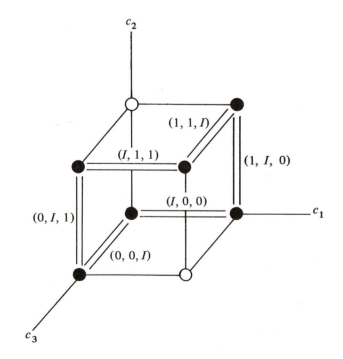

Section 4.9

1 (b)

K^0	K^1
$\sqrt{}000$	$0I0$
	$I00$
$\sqrt{}010$	—
$\sqrt{}100$	$01I$
	$10I$
$\sqrt{}011$	—
$\sqrt{}101$	$I11$
	$1I1$
$\sqrt{}111$	—

basic cells: $(0, I, 0), (I, 0, 0), (0, 1, I), (1, 0, I), (I, 1, 1), (1, I, 1)$

(c) basic cells: $(0, 0, I, 0), (0, I, 0, 0), (I, 0, 1, I), (I, 1, 0, I), (1, I, 1, I), (1, 1, I, I)$

3 (b) There are no essential cells. With

$$A = (0, I, 0) \qquad D = (1, 0, I)$$
$$B = (I, 0, 0) \qquad E = (I, 1, 1)$$
$$C = (0, 1, I) \qquad F = (1, I, 1)$$

Algorithm XIX yields

$$f = BCDE + ADE + ABEF + ACDF + BCF$$

We choose ADE or BCF to obtain the minimal coverings:

$$(0, I, 0), (1, 0, I), (I, 1, 1)$$
$$(I, 0, 0), (0, 1, I), (1, I, 1)$$

respectively, and the corresponding minimal sum-of-products expressions:

$$x_1'x_3' + x_1x_2' + x_2x_3$$
$$x_2'x_3' + x_1'x_2 + x_1x_3$$

(c) The cell $(I, 0, 1, I)$ is essential to vertex 3, and $(I, 1, 0, I)$ is essential to vertex 5. For the reduced matrix

	0	14	15
$A = (0, 0, I, 0)$	1		
$B = (0, I, 0, 0)$	1		
$C = (1, I, 1, I)$		1	1
$D = (1, 1, I, I)$		1	1

Algorithm XIX gives

$$f = (A + B)(C + D)(C + D) = AC + AD + BC + BD$$

Correspondingly, there will exist four minimal coverings with the associated minimal sum-of-products expressions:

$$x_2' x_3 + x_2 x_3' + \begin{cases} x_1' x_2' x_4' + x_1 x_3 \\ x_1' x_2' x_4' + x_1 x_2 \\ x_1' x_3' x_4' + x_1 x_3 \\ x_1' x_3' x_4' + x_1 x_2 \end{cases}$$

7 *Hint*: See Exercise 7 of Section 4.7.

Section 5.1

1 First show that an idempotent element, that is, an element a with $a^2 = a$, is an identity element in any group. Then note that $\varphi(e)\varphi(e) = \varphi(e \cdot e) = \varphi(e)$.

5 Consider a multiplicative group G generated by a single element x with distinct elements $e = x^0, x = x^1, x^2, \ldots, x^{n-1}$, and in defining

$$x^r \cdot x^s = x^{r+s}$$

take $x^n = e$ (see Exercise 2 of Section 5.3).

7 The group (call it G) has the subgroups:

$$G, \{e\}, \{e, a\}, \{e, b\}, \{e, f\}, \{e, c, d\}$$

To show that these are the only subgroups, use Lagrange's theorem.

9 (a) 1, 2, 3, 4, 6, 8, 12, 24   (e) 1, 17

11 (1) $Nx \cdot (Ny \cdot Nz) = Nx \cdot Nyz = Nx(yz) = N(xy)z$
$\qquad = Nxy \cdot Nz = (Nx \cdot Ny) \cdot Nz$

(2) $Nx \cdot Ne = Nxe = Nx$ and $Ne \cdot Nx = Nex = Nx$

(3) $Nx^{-1} \cdot Nx = Nx^{-1}x = Ne = Nxx^{-1} = Nx \cdot Nx^{-1}$

15 The group (call it G) has the subgroups G, $\{1\}$, $\{1, r^2\}$, $\{1, d\}$, $\{1, rd\}$, $\{1, r^2 d\}$, $\{1, r^3 d\}$ $\{1, r, r^2, r^3\}$, $\{1, r^2, d, r^2 d\}$, $\{1, r^2, rd, r^3 d\}$. Again use Lagrange's theorem to argue that these are the only subgroups.

Section 5.2

1 The isomorphism is given by the correspondence

$$
\begin{array}{ll}
a \rightarrow (12) & d \rightarrow (132) \\
b \rightarrow (13) & e \rightarrow 1 \\
c \rightarrow (123) & f \rightarrow (23)
\end{array}
$$

5 The realization resulting from the proof of Cayley's theorem is as follows:

$$a \rightarrow T_a = \begin{pmatrix} a & b & c & d & e & f \\ e & c & b & f & a & d \end{pmatrix} = (ae)(bc)(df)$$

$$b \rightarrow T_b = \begin{pmatrix} a & b & c & d & e & f \\ d & e & f & a & b & c \end{pmatrix} = (ad)(be)(cf)$$

$$c \rightarrow T_c = \begin{pmatrix} a & b & c & d & e & f \\ f & a & d & e & c & b \end{pmatrix} = (afb)(cde)$$

$$d \rightarrow T_d = \begin{pmatrix} a & b & c & d & e & f \\ b & f & e & c & d & a \end{pmatrix} = (abf)(ced)$$

$$e \rightarrow T_e = \begin{pmatrix} a & b & c & d & e & f \\ a & b & c & d & e & f \end{pmatrix} = 1$$

$$f \rightarrow T_f = \begin{pmatrix} a & b & c & d & e & f \\ c & d & a & b & f & e \end{pmatrix} = (ac)(bd)(ef)$$

9 According to Lagrange's theorem, a group of prime order can have no subgroups (normal or otherwise) except itself and $\{e\}$. So the result is immediate.

11 The alternating group A_4 has the elements

$$A_4 = \{1, (123), (132), (124), (142), (134), (143), (234),$$
$$(243), (12)(34), (13)(24), (14)(23)\}$$

The subgroup

$$N = \{1, (12)(34), (13)(24), (14)(23)\}$$

is normal.

Section 5.3

1 The isomorphism $\varphi: R \times R \rightarrow C$ is given by the mapping $\varphi(x, y) = x + iy$.

5 (a) In the isomorphism $\varphi: Z_{60} \rightarrow Z_4 \times Z_3 \times Z_5$ we have

$$14 \overset{\varphi}{\rightarrow} (2, 2, 4)$$
$$57 \overset{\varphi}{\rightarrow} (1, 0, 2)$$

$$\overline{\quad\quad\quad\quad}$$

$$11 \overset{\varphi}{\rightarrow} (3, 2, 1)$$

7 The hypothesis ensures that each group element x may be written in the form:

$$x = x_1^{r_1} x_2^{r_2} \cdots x_n^{r_n}$$

for certain generators x_i. The proof consists in showing that one can in fact find another set of generators y_i ($1 \le i \le m$) so that each x can be expressed *uniquely* as

$$x = y_1^{s_1} y_2^{s_2} \cdots y_m^{s_m}$$

(One does not claim that the s_i are unique, only that the factors $y_j^{s_j}$ are uniquely determined by x.)

Section 5.4

1 (b) The encoding $\varphi: Z_{18} \rightarrow Z_2 \times Z_9$ is given as follows:

$0 \rightarrow (0, 0)$	$6 \rightarrow (0, 6)$	$12 \rightarrow (0, 3)$
$1 \rightarrow (1, 1)$	$7 \rightarrow (1, 7)$	$13 \rightarrow (1, 4)$
$2 \rightarrow (0, 2)$	$8 \rightarrow (0, 8)$	$14 \rightarrow (0, 5)$
$3 \rightarrow (1, 3)$	$9 \rightarrow (1, 0)$	$15 \rightarrow (1, 6)$
$4 \rightarrow (0, 4)$	$10 \rightarrow (0, 1)$	$16 \rightarrow (0, 7)$
$5 \rightarrow (1, 5)$	$11 \rightarrow (1, 2)$	$17 \rightarrow (1, 8)$

3 (b) With the weights (Exercise 2)

$$x_1 = \varphi^{-1}(1, 0) = 9$$
$$x_2 = \varphi^{-1}(0, 1) = 10$$

we compute

$$\varphi^{-1}(1, 7) = x_1 \cdot 1 + x_2 \cdot 7 = 9 + 70 = 7 \,(\text{mod } 18)$$
$$\varphi^{-1}(1, 5) = x_1 \cdot 1 + x_2 \cdot 5 = 9 + 50 = 5 \,(\text{mod } 18)$$
$$\varphi^{-1}(0, 4) = x_1 \cdot 0 + x_2 \cdot 4 = 0 + 40 = 4 \,(\text{mod } 18)$$

5 One obvious difficulty is that monotonicity is not preserved by the encoding into a residue number system.

Section 5.5

1 (a) With $a = (0, 1, 1, 0)$ and $r = 2$

$$S(a, r) = \{(0, 1, 1, 0), (1, 1, 1, 0), (0, 0, 1, 0), (0, 1, 0, 0), (0, 1, 1, 1)$$
$$(1, 0, 1, 0), (1, 1, 0, 0), (1, 1, 1, 1), (0, 0, 0, 0), (0, 0, 1, 1), (0, 1, 0, 1)\}$$

3 The lemma shows that δ is surjective [each $\delta^{-1}(a) \neq \varnothing$]. And if $\epsilon(a') = \epsilon(a) \in \delta^{-1}(a)$ we have $a' = \delta(\epsilon(a')) = a$ showing that ϵ is injective.

5 Show that each sphere $S(\epsilon(a), 1)$ contains only one code word (at the center).

7 (b) $\Delta(\epsilon) = \min\limits_{i \neq j} \{\Delta(b^i, b^j)\} = 3$

as in the code words

$$b^i = (0, 0, 0, 0, 1) \qquad b^j = (0, 1, 0, 1, 0)$$

Then, according to Theorem 5.7,

$$\epsilon \text{ is } \begin{cases} 2\text{-detectable } (k + 1 = 3) \\ 1\text{-correctable } (2k + 1 = 3) \end{cases}$$

9 (a) $00 \overset{\epsilon}{\rightarrow} 00000\ 10000\ 01000\ 00100\ 00010\ 00001 \overset{\delta}{\rightarrow} 00$
$01 \overset{\epsilon}{\rightarrow} 01110\ 11110\ 00110\ 01010\ 01100\ 01111 \overset{\delta}{\rightarrow} 01$
$10 \overset{\epsilon}{\rightarrow} 10101\ 00101\ 11101\ 10001\ 10111\ 10100 \overset{\delta}{\rightarrow} 10$
$11 \overset{\epsilon}{\rightarrow} 11111\ 01111\ 10111\ 11011\ 11101\ 11110 \overset{\delta}{\rightarrow} 11$

Section 5.6

1 (a) Suppose that the table has n seats. Taking $X = A = n^+$, we may view a seating as a function $f: n^+ \rightarrow n^+$ (from seats to people, say). But the set of seatings is actually

$$F = \{f : P(f)\}$$

where P is the property of being a permutation (bijective function). Not wanting to distinguish two seatings when one is simply a rotation of the other, we take

$$f \sim g \Leftrightarrow g = \sigma \circ f$$

for a circular permutation $\sigma = (1\ 2 \ldots n)^k$, $0 \leq k \leq n - 1$.

3

	Mon	Tues	Wed
a	1	2	3
b	1	4	2
c	3	1	5
d	4	6	1
e	5	1	6
f	3	6	2
g	2	5	4
h	6	5	2
i	5	4	3
j	6	3	4

Section 5.7

1 If $\sigma\tau^{-1} \in G_x$, then (writing γ for an element of G_x):

$$\gamma\sigma \in G_x\sigma \Rightarrow \gamma\sigma = \gamma\sigma(\tau^{-1}\tau) = (\gamma(\sigma\tau^{-1}))\tau \in G_x\tau$$

Similarly, because we also have $(\sigma\tau^{-1})^{-1} = \tau\sigma^{-1} \in G_x$,

$$\gamma\tau \in G_x \Rightarrow \gamma\tau = \gamma\tau(\sigma^{-1}\sigma) = (\gamma(\tau\sigma^{-1}))\sigma \in G_x\sigma$$

Conversely, if $G_x\sigma = G_x\tau$, then

$$1\sigma = \gamma\tau$$

for some $\gamma \in G_x$, that is, $\sigma\tau^{-1} = \gamma \in G_x$.

5 (c)

x_1	x_2	x_3	f
0	0	0	0
0	0	1	0
0	1	0	1
0	1	1	1
1	0	0	1
1	0	1	1
1	1	0	0
1	1	1	1

(d)

x_1	x_2	x_3	f
0	0	0	0
0	0	1	0
0	1	0	1
0	1	1	1
1	0	0	1
1	0	1	1
1	1	0	0
1	1	1	1

7 (c) $G_{1110} = \{1, (12), (13), (23), (123), (132)\}$
   (d) $G_{0110} = \{1, (14), (23), (14)(23)\}$
9 (c) From the representation $g = x_1 x_2' x_3$ we recognize a symmetry (13) and predict
   that

$$|[g]| = [S_3 : G_g] = |S_3|/|G_g| = \tfrac{6}{2} = 3$$

Section 5.8

1 (c) $w(g) = 0^7 1^1$ (d) $w(g) = 0^5 1^3$
3 (a) The symmetric group S_4 consists of $4! = 24$ elements: the identity, six transpositions, eight 3-cycles, six 4-cycles, and three double transpositions. Thus

$$Z_{S_4}(z_1, z_2, z_3, z_4) = \tfrac{1}{24}(z_1{}^4 + 6z_1{}^2 z_2 + 8z_1 z_3 + 6z_4 + 3z_2{}^2)$$

5 (b) The group G is the cyclic group of order five, acting on 5^+. Besides the identity,
   we have just the four 5-cycles, so

$$Z_G(z_1, z_2, z_3, z_4, z_5) = Z_G(z_1, z_5) = \tfrac{1}{5}(z_1{}^5 + 4z_5)$$

According to Polya's theorem, the pattern inventory is

$$\begin{aligned}
\operatorname{inv}(F) &= Z_G(a + b + c, a^5 + b^5 + c^5) \\
&= \tfrac{1}{5}[(a + b + c)^5 + 4(a^5 + b^5 + c^5)] \\
&= a^5 + b^5 + c^5 + a^4b + ab^4 + a^4c + ac^4 + b^4c + bc^4 \\
&\quad + 2a^3b^2 + 2a^2b^3 + 2a^3c^2 + 2a^2c^3 + 2b^3c^2 + 2b^2c^3 \\
&\quad + 4a^3bc + 4ab^3c + 4abc^3 + 6ab^2c^2 + 6a^2bc^2 + 6a^2b^2c
\end{aligned}$$

where

$$a \leftrightarrow \text{student} \quad b \leftrightarrow \text{computer science faculty} \quad c \leftrightarrow \text{math faculty}$$

Section 6.1

1 (a)   (d)

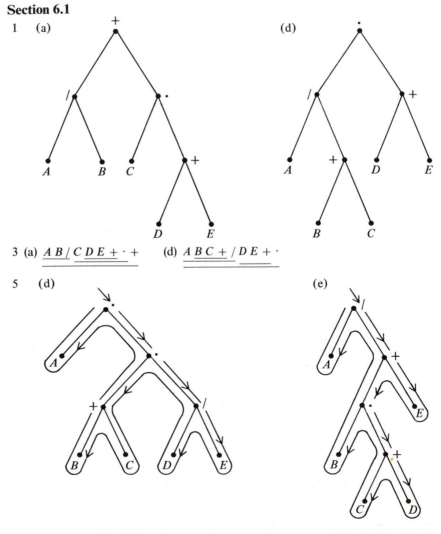

3 (a) $\underline{A\ B}\ /\ \underline{C\ D\ E} + \cdot +$ (d) $A\ \underline{B\ C} + /\ \underline{D\ E} + \cdot$

5 (d)   (e)

$A\ B\ C + D\ E\ /\ \cdot\ \cdot$ $A\ B\ C\ D + \cdot\ E + /$

7 (a) infix: $A + ((B + C)/(D \cdot E))$

Polish: $A\ B\ C + D\ E \cdot / +$

Section 6.2

1 Use the ideas of Example 6.6.

3 (a)

x	σ	z		x	σ	z
ϵ		ϵ		10		$+ D\ E \cdot / +$
		$+$		101		$C + D\ E \cdot / +$
1		$/ +$		10	1	
11		$\cdot / +$		100		$B\ C + D\ E \cdot / +$
111		$E \cdot / +$		10	0	
11	1			1	0	
110		$D\ E \cdot / +$		ϵ	1	
11	0			0		$A\ B\ C + D\ E \cdot / +$
1	1			ϵ	0	

7

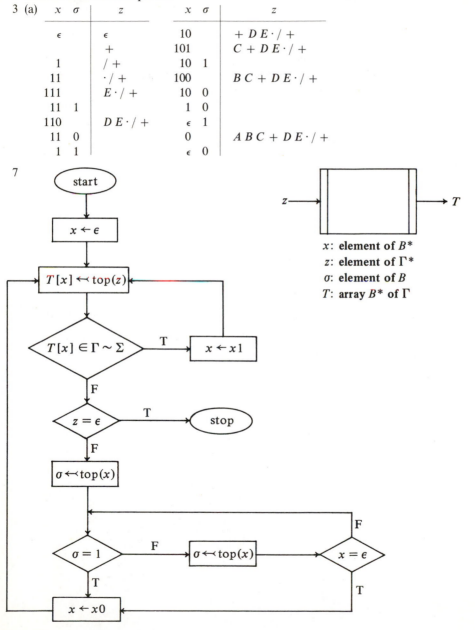

x: element of B^*

z: element of Γ^*

σ: element of B

T: array B^* of Γ

Section 6.3

3 (b) p, q, r are wff's by condition (i)

   $r \neg$ is a wff by (ii)

   $r \neg q \wedge$ is a wff by (ii)

   $r \neg q \wedge p \vee$ is a wff by (ii)

   $r \neg q \wedge p \vee \neg$ is a wff by (ii)

5 (b) $\neg((\neg r \wedge q) \vee p)$

   (c) $(\neg r \wedge q) \vee \neg(\neg p \vee q)$

7 (c) $x = r \neg p \neg q \vee \ p \neg \wedge$

   ranks: $1\ 0\ \ 1\ 0\ \ 1\ -1\ 1\ 0\ \ -1$ $\Rightarrow$ not a wff

   partial sums: $1\ 1\ \ 2\ 2\ \ 3\ \ \ 2\ 3\ 3\ \ \ 2$

9 Suppose that

$$x = \sigma_{i_1}\quad \sigma_{i_2}\quad \cdots\quad \sigma_{i_k}$$
$$\text{ranks:}\ t_1\quad t_2\quad \cdots\quad t_k\quad 0$$
$$\text{partial sums:}\ S_1\quad S_2\quad \cdots\quad S_k\quad S_{k+1}$$

with the hypothesis that

$$(1)\ \text{each}\ T_i \geq 1 \qquad (2)\ T_k = 1$$

Then we have

$$(1)\ S_i = T_i = 1\quad (1 \leq i \leq k)$$
$$(2)\ S_{k+1} = S_k + 0 = T_k + 0 = 1$$

showing that the required properties are true of $x = y \neg$ (when they are true of y).

Section 6.4

1 (b)

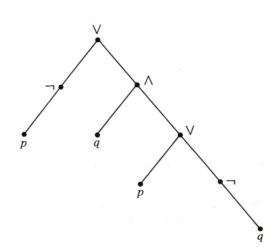

3 (b)

x	y	σ	τ	z
¬ p ∨ q ∧ (p ∨ ¬ q)	ε			ε
p ∨ q ∧ (p ∨ ¬ q)	¬		¬	
∨ q ∧ (p ∨ ¬ q)		p		p
q ∧ (p ∨ ¬ q)	ε	∨	¬	p ¬
∧ (p ∨ ¬ q)	∨	q		p ¬ q
(p ∨ ¬ q)	ε	∧	∨	
p ∨ ¬ q)	∨ ∧	(		
∨ ¬ q)	∨ ∧ (	p		p ¬ q p
¬ q)	∨ ∧	∨	(	
	∨ ∧ (∨			
q)	∨ ∧ (	¬	∨	
)	∨ ∧ (∨ ¬	q		p ¬ q p q
ε	∨ ∧ (∨	)	¬	p ¬ q p q ¬
	∨ ∧ (		∨	p ¬ q p q ¬ ∨
	∨ ∧		(	
	∨			p ¬ q p q ¬ ∨ ∧
	ε			p ¬ q p q ¬ ∨ ∧ ∨

Section 6.5

1 If $\models \alpha \,\neg\, \beta \,\vee$, then in view of the fact that

$$I(\alpha \,\neg\, \beta \,\vee\,) = I(\alpha)' + I(\beta) = 1$$

for all I, it is clear that we must have

$$I(\alpha)' = 1 \quad \text{or} \quad I(\beta) = 1$$
$$I(\alpha) = 0 \quad \text{or} \quad I(\beta) = 1$$

for all I. Otherwise stated, for no interpretation I can we have $I(\alpha) = 1$ and $I(\beta) = 0$.

3 (a) $I(\alpha) = I(p\,q\,\neg\,\vee\,r\,\wedge)$

$= I(p\,q\,\neg\,\vee\,) \cdot I(r)$

$= (I(p) + I(q\,\neg)) \cdot I(r)$

$= (I(p) + I(q)') \cdot I(r) = (1 + 0') \cdot 1 = 1$

5 (c)

$I(p)$	$I(q)$	$I(r)$	$I(p\,q\,\wedge)$	$I(p\,\neg\,r\,\neg\,\wedge)$	$I(\alpha)$	$I(q\,\neg\,p\,\wedge)$	$I(\beta)$
0	0	0	0	1	1	1	1
0	0	1	0	0	0	1	1
0	1	0	0	1	1	0	1
0	1	1	0	0	0	0	0
1	0	0	0	0	0	1	1
1	0	1	0	0	0	1	1
1	1	0	1	0	1	1	1
1	1	1	1	0	1	1	1

Since

$$I(\alpha) = 1 \Rightarrow I(\beta) = 1$$

we have $\alpha \models \beta$. Alternatively, we may note that $I(\alpha\,\beta\,\supset) = 1$ for all I and again conclude that $\alpha \models \beta$ on the basis of Theorem 6.3.

9 Suppose that

$$I(\alpha_1) = I(\alpha_2) = \cdots = I(\alpha_n) = 1 \Rightarrow I(\beta) = 1$$

Then we can never have

$$I(\alpha_1\alpha_2 \wedge \alpha_3 \wedge \cdots \alpha_n \wedge) = 1 \quad \text{and} \quad I(\beta) = 0$$

so that

$$I(\alpha_1\alpha_2 \wedge \alpha_3 \wedge \cdots \alpha_n \wedge \beta \supset) = 1$$

for all I; that is, $\models \alpha_1\,\alpha_2 \wedge \alpha_3 \wedge \cdots \alpha_n \wedge \beta \supset$. The converse is equally clear.

Section 6.6

1 For (A3), let $\alpha = p \neg q \neg \supset$ and $\beta = p \neg q \supset p \supset$.

$I(p)$	$I(q)$	$I(\alpha)$	$I(p \neg q \supset)$	$I(\beta)$
0	0	1	0	1
0	1	0	1	0
1	0	1	1	1
1	1	1	1	1

Since $I(\alpha) = 1 \Rightarrow I(\beta) = 1$, we have $\alpha \models \beta$. Using Theorem 6.3, we conclude that $\models \alpha\,\beta\,\supset$.

3 (a) Let $\alpha = p \neg p \neg \neg \supset$ and $\beta = p$.

$I(p)$	$I(p \neg)$	$I(p \neg \neg)$	$I(\alpha)$	$I(\beta)$
0	1	0	0	0
1	0	1	1	1

Since $\alpha \models \beta$, we obtain the desired tautology $\models \alpha\,\beta\,\supset$ from Theorem 6.3.

5 Simply insert the seven-line proof of Example 6.30 at the third line of Example 6.31.

7 Use Theorem 6.3 and Theorem 6.5.

Section 6.7

3 For the entry $M(q, \sigma) = (q', \tau, -1)$ we have the transitions

$$(\epsilon, q, \sigma v) \rightarrow (\epsilon, q', \square \tau v)$$
$$\left.\begin{array}{l} (u\gamma, q, \epsilon) \rightarrow (u, q', \gamma\tau) \\ (\epsilon, q, \epsilon) \rightarrow (\epsilon, q', \square\tau) \end{array}\right\} \text{ if } \sigma = \square$$

5 (a) $\epsilon, 0, ())(() \rightarrow (, 0,))(() \rightarrow \epsilon, 1, (A)(()$
$\qquad \rightarrow A, 0, A)(() \rightarrow AA, 0,)(()$
$\qquad \rightarrow A, 1, AA(() \rightarrow \epsilon, 1, AAA(()$
$\qquad \rightarrow \epsilon, 1, AAA(() \rightarrow \epsilon, 1, AAA(() \rightarrow \cdots$

7 (a) $(\epsilon, 0, 1\square 11) \rightarrow (\square, 1, \square 11) \rightarrow (\square\square, 2, 11)$
$\rightarrow \cdots \rightarrow (\square\square 11, 2, \square) \rightarrow (\square\square 11\square, 3, \square)$
$\rightarrow (\square\square 11, 4, \square 1) \rightarrow \cdots \rightarrow (\epsilon, 6, \square\square 11\square 1)$
$\rightarrow (1, 0, \square 11\square 1) \rightarrow \cdots \rightarrow (1\square 11\square, 12, 111)!$

9 (a)

	a	b	c	$\square$	A	B	C
0	$0\,a\,1$	$0\,b\,1$	$0\,c\,1$	$R\,\square -1$	$R\,A -1$	$R\,B -1$	$R\,C -1$
1	$1\,a -1$	$1\,b -1$	$1\,c -1$	$q_a\,\square\,1$	$q_a\,A\,1$	$q_a\,B\,1$	$q_a\,C\,1$
q_a	$0\,A\,1$	$N\,b -1$	$N\,c -1$		$Y\,A -1$		
2	$2\,a -1$	$2\,b -1$	$2\,c -1$	$q_b\,\square\,1$	$q_b\,A\,1$	$q_b\,B\,1$	$q_b\,C\,1$
q_b	$N\,a -1$	$0\,B\,1$	$N\,c -1$		$Y\,B -1$		
3	$3\,a -1$	$3\,b -1$	$3\,c -1$	$q_c\,\square\,1$	$q_c\,A\,1$	$q_c\,B\,1$	$q_c\,C\,1$
q_c	$N\,a -1$	$N\,b -1$	$0\,C\,1$				$Y\,C -1$
R	$1\,A -1$	$2\,B -1$	$3\,C -1$		$q_a\,A -1$	$q_b\,B -1$	$q_c\,C -1$
Y				$Y\,\square\,0$	$Y\,A -1$	$Y\,B -1$	$Y\,C -1$
N	$N\,a -1$	$N\,b -1$	$N\,c -1$	$N\,\square\,0$	$N\,A -1$	$N\,B -1$	$N\,C -1$

Section 6.8

3 Let the machine constructed be designated as A. Then

$$x \in \lambda_A \Leftrightarrow M_A([\epsilon], x) \in \{[x]: [x] \subseteq \lambda\}$$
$$\Leftrightarrow [x\epsilon] = [x] \in \{[x]: [x] \subseteq \lambda\}$$
$$\Leftrightarrow [x] \subseteq \lambda$$
$$\Leftrightarrow x \in \lambda$$

5 *Hint*: Let $|S|$ be the "size" of a Turing machine on the alphabet Σ. Use the fact that there are only a finite number of machines of a given size.

7 If λ, μ are recursive, then there are finite mechanical procedures (for instance, Turing machines) for deciding whether $x \in \lambda$ (respectively, $x \in \mu$). To decide whether $x \in \lambda \cup \mu$ we only have to run the two procedures to see if either of them gives an affirmative answer. Similarly, we check to see whether both answers are affirmative to decide whether $x \in \lambda \cap \mu$.

9 The complement of the language L is not recursively enumerable (otherwise, L would be recursive by Theorem 6.7). It follows that the recursively enumerable languages do not form a Boolean algebra.

Section 6.9

1 (a) $S \rightarrow$ NP VP $\rightarrow$ N VP $\rightarrow$ N V NP $\rightarrow$ N V N
$\rightarrow$ Ron V N $\rightarrow$ Ron plays N $\rightarrow$ Ron plays football

3 (b) $S \rightarrow 0\,T\,B \rightarrow 0\,0\,A\,B \rightarrow 0\,0\,A\,1\,2 \rightarrow 0\,0\,1\,A\,2$
$\rightarrow 0\,0\,1\,1\,2\,2$

5 If $\lambda = \{0^n 1^n: n \geq 1\}$ were regular, then according to Myhill's theorem, λ would be the union of equivalence classes resulting from some congruence R of finite index on Σ^*, where $\Sigma = \{0, 1\}$. In the infinite sequence of words 0^n we must then have $0^n R\ 0^m$ for some $n \neq m$. Since R is a congruence, we would then have $0^n 1^n R\ 0^m 1^n$. But $0^n 1^n \in \lambda$ whereas $0^m 1^n \notin \lambda$, contradicting the fact that λ is a union of the equivalence classes resulting from R.

7 (b) $T : aT$ $T : a$

$\qquad\quad T : bT$ $T : b$ $S : bTb$

$\qquad\quad T : cT$ $T : c$

9 (b) ⟨identifier⟩ → ⟨identifier⟩⟨digit⟩

$\qquad\qquad\qquad$ → ⟨identifier⟩⟨digit⟩⟨digit⟩

$\qquad\qquad\qquad$ → ⟨identifier⟩⟨letter⟩⟨digit⟩⟨digit⟩

$\qquad\qquad\qquad$ → ⟨identifier⟩⟨letter⟩⟨letter⟩⟨digit⟩⟨digit⟩

$\qquad\qquad\qquad$ → ⟨identifier⟩⟨letter⟩⟨letter⟩⟨letter⟩⟨digit⟩⟨digit⟩

$\qquad\qquad\qquad$ → ⟨letter⟩⟨letter⟩⟨letter⟩⟨letter⟩⟨digit⟩⟨digit⟩

$\qquad\qquad\qquad$ → $\cdots$ → *abba*14

Index